Fundamentals of Geographical Thought

Fundamentals of Geographical Thought

SUDEEPTA ADHIKARI

Orient BlackSwan

FUNDAMENTALS OF GEOGRAPHICAL THOUGHT

ORIENT BLACKSWAN PRIVATE LIMITED

Registered Office
3-6-752 Himayatnagar, Hyderabad 500 029 Telengana, India
e-mail: centraloffice@orientblackswan.com

Other Offices
Bengaluru, Chennai, Guwahati, Hyderabad, Kolkata,
Mumbai, New Delhi, Noida, Patna, Visakhapatnam

First published by Orient Blackswan Pvt. Ltd. 2015
Reprinted 2019, 2020, 2021, 2023

ISBN: 978-81-250-5949-3

Typeset in
Garamond 10/12
by PrePSol Enterprises Pvt. Ltd, Haryana

Printed in India at
B.B. Press, Noida

Published by
Orient Blackswan Private Limited
3-6-752 Himayatnagar, Hyderabad 500 029, Telangana, India
e-mail: info@orientblackswan.com

Contents

1

Introduction

HIGHLIGHTS

- Meaning and Scope of Geography
- Nature of Geography
- Approaches to the Study of Geography
- Interactions/Relations with Other Sciences
- Branches of Geography
- Major Schools of Geographical Thought

It was Eratosthenes, the great scholar of ancient Greece, who is identified to have coined the word geography, as *'the study of the Earth as the home of man'*. This definition still persists in a slightly different way–as *'the study of the Earth's surface as the space within which the human population lives'* (Haggett, 1981, 133). The word 'geography' comes from the Greek words *'geo' (the Earth)* and *'graphien' (to write)*. The discipline of geography is essentially a human creation. It begins with man's attempts to identify a particular extent of territory on the Earth's surface as his living space. The identification of the territory, or the discovery of living space, reveals a close manifestation of his own mind. With the spread of knowledge, man has attempted to discover and describe many different areas, depending upon his capacity to observe and to generalise what he has discovered and observed.

During the long record of man's efforts to gain more and more knowledge about the face of the Earth as the human habitat, there has been a continuing interplay between things and events observed through senses, and the mental images of things and events. *The direct observation through the senses is described as the percept, and the mental image of things and events is described as the concept.* Percepts are what people feel or sense, and describe as reality, e.g. the identification of the territory for habitation, which reveals a clear manifestation of their experiences about the nature of land to be settled/occupied. Concepts, on the other hand, are theoretical, implying that they are not real. People simply imagine of events, things and phenomena which they have not sensed and experienced or seen, and the derived deductions are based on pre-conceived opinions. Geography thus emerged as a result of an interplay between the percept and concept, and aimed at man's needs and requirements to gain more and more knowledge of the human habitat and of man's spread over the Earth.

MEANING AND SCOPE OF GEOGRAPHY

The word 'geography' has been in use since the time of Eratosthenes. At first it included all aspects dealing with description of Earth and its parts. As mentioned earlier, the word 'geography' consists of two Greek words, 'geo' (the Earth) and 'graphein' (to write). The former is more important and refers to 'the zone of contact of the solid, liquid and gaseous masses that make up the planet', while the latter refers to the description of these phenomena in relation to 'place, localisation and distribution'. *A description of these earth-bound phenomena is geography.* There are varieties of phenomena occuring on the surface of the Earth—some are mutually inclusive, some are mutually exclusive, others are interrelated and interacting, while some are purely physical phenomena and some others are purely human phenomena. But all these phenomena, whether physical or human, are relative, in as much as they have distinct 'place' of their 'localisation' and 'distribution'. A combination of both physical and human phenomena, occuring on the Earth's surface as mutually interrelated and interacting, appears to be the 'core' and the '*raison d'etre*' of geography.

The word 'geography' could be well explained and understood by the sentence *'where do the people live*'? The word 'where' refers to 'place and localisation' on the surface of the Earth that the people choose, and the word 'live' refers to 'distribution and concentration of the people'. Inherent in the sentence are the *'space and time factors*'. The Earth bound phenomena are ever-changing either by the natural calamities and disasters, or by the part that the people play in modifying the face of the Earth. Natural disasters intrinsically refer to 'space and time'—disaster will befall on a particular place at a particular time (place, localisation, distribution understood).

Similar connotation is inherent in 'the part that the people play in modifying the face of the Earth'. 'Play' and 'modifying' refer to 'particular time' because people here are identified with a particular set of time with a distinct way of living which is changeable in temporal perspective, while 'the face of the Earth' refers to that which forms the habitat or environment or space which is dependent entirely on man's selection, and the way changes and modifications of different sets of the earth-bound phenomena are done (place, localisation and distribution understood). All these earth-bound phenomena are properly enumerated in geography.

Geography attempts to bridge the widening gap between the changing physical and biological phenomena on the one hand, and the changing human phenomena on the other. There can be no geography without both, and without geography, knowledge about the Earth is fragmented and incomplete.

Geography has been rightly *defined as to provide accurate, orderly, and rational description and interpretation of the variable character of the Earth Surface* (Hartshorne, 1959). By 'variable character' is meant the spatial variation that occurs between the character of the Earth's surface at one location and another. This variation may occur at all map scales, from the globe itself (say, between continent and continent) down to a very local level (say, between one district and another within an area). By 'Earth surface' we mean that rather thin shell, only about one thousandth of the planet's circumference in thickness, that forms the habitat or environment within which lives the human population for survival.

Each different location on the surface of the Earth is an ensemble of celestial, terrestrial and human phenomena, and geography must represent an approach to the empirical knowledge of these phenomena which is necessary for philosophical research. *Geography essentially studies that man-environment relationship,* as developed in different locations on the surface of the Earth, in the form of an assemblage of interrelated phenomena.

This man-environment relationship with the interrelated phenomena, occuring on the surface of the Earth, has been approached in three different ways, as providing *three different types of geography*—the Geocratic type of geography; the Theocratic type of geography; and the We-ocratic type of geography (Taylor, 1967).

'Geocratic type of geography' suggests that the Earth (nature) itself plays a great part in determining the type of life which develops in a particular area. Alexander Von Humboldt attempted to demonstrate man's dependence upon his environment. The growth of Darwinism in the late 1850s placed greatest emphasis on environment and environmental changes. Friedrich Ratzel nourished the Geocratic geography in a

more scientific way, i.e. through a hypothetic-deductive method. As a result, the question 'why do people live as and where they do' is rationally and scientifically described/answered.

'Theocratic type of geography' seeks to understand various kinds of phenomena, observable on the surface of the Earth, in relation to their underlying purposes. Carl Ritter, a contemporary of Humboldt, attempted to combine a teleological (Greek teleos=purpose) approach with a simple deterministic model (geocratic approach) in which everything that occurs on the Earth's surface is directed to purpose whose controlling conditions are laid down by god. His ideas on the 'wholeness' of things were in accordance with the writings of the German 'idealist' philosopher Hegel whose attitudes amounted to comprehend the entire universe, to know the infinite and to see all things in god.

'We-ocratic type of geography' has developed in opposition to the geocratic type and the theocratic type. It seeks to suggest that the environment offers a number of possibilities, and their utilisation is dependent almost entirely on human selection. The we-ocratic type gave rise to the school of possibilism of which Paul Vidal de la Blache was the main architect. According to him, it is unreasonable to draw boundaries between natural and cultural phenomena; they should be regarded as united and inseparable. The areas over which such an intimate relationship between man and nature has developed through the centuries constitute a 'region', and each region is unique and particular.

The 'observable-interrelated phenomena' occuring in different locations, on account of non-environment relationships on the Earth's surface, have been accounted for in the above three different ways and approaches. All these approaches, therefore, attempt to provide rational description and interpretation of the variable character of man-environment relationships.

Geography studies variations in phenomena from place to place, and its value as an academic discipline depends on the extent to which it can clarify the connections between different features of the same area, i.e. 'the interrelated phenomena'. The central geographical question is 'why is it like this here'? All the phenomena which have a recognisable "spatial" distribution or can be shown on a map are basically of geographical interest.

The term *'spatial' describes a particular type of distribution*—one that is spread out over a surface, rather than categorises objects as in statistical distributions. Thus spatial distributions include objects (phenomena) that are spread out from each other in space on a surface that usually means the surface of the Earth. Each spatial distribution is made up of objects that are similar to each other. *The individual objects that make up a spatial distribution are arranged so that their pattern, density and dispersion can be analysed and mapped.* 'Pattern' refers to the geometric arrangements of the objects; 'density' means how many objects per unit area; and 'dispersion' means how spread out from each other the objects are. Geographers can compare one spatial distribution with another by comparing the three elements of pattern, density and dispersion.

Thus, the scope or field of geography is the ensemble of phenomena that occur in the zone of contact of the solid, liquid and gaseous masses that make up the planet, or the geosphere. These phenomena are analysed and rationally described in relation to place, localisation and distribution. Thus, the scope or field of geography is wide enough to include a double aspect—physical environment or nature and human life. To explain their experssion on the Earth's surface, one needs to borrow from both the natural sciences and humanities or social sciences. The field of geography contains eight sub-divisions, which in turn are linked with eight major disciplines. Thus geography links the four 'natural or environmental sciences' of geology, physics, astronomy and botany with the four 'human or social sciences' of history, anthropology, sociology and economics. There are vast uncharted areas on the borders of regional geography which merge into the eight subjects specified (ibid.).

Sciences do overlap and each one of the overlapping zones, which also represent specialised, systematic branches of geography, belong equally to some other science. The question is whether geography occupies anything more than these overlapping zones. Is anything left in the centre which is specifically geographical? What should be missing if the geography of an area were prepared and written by a group of people who were specialised in their disciplines and each of whom contributed his own chapter—the geologist writing about the rocks, the botanist about the plant life, the economist about the economic conditions, the sociologist about the social life, the physicist about climate, the pedalogist about the soil structure, the astronomer about the celestial phenomena and the motion of the heavenly bodies, and the demographer about population condition?

Such an account would surely fail to consider the overall interaction between recognisable phenomena occurring on the surface of the Earth. It is easy to see, for example, that the relationship between climate and soil type must have an important bearing on condition of agricultural production and that the development of industry in an area may not be due only to economic factors but also to the natural resources of the area, its population potential and its historical, political and social development. Fennman and Griffith Taylor argued that the interaction of these factors can primarily be studied within definite areas or regions and concluded that geography should attempt to cultivate the core, regional geography, as a safeguard against absorption by other sciences. (Halt – Jenson, 1981, 4–5).

Geography is 'a mother discipline' from which other specialised disciplines like geodesy, meteorology, soil science, plant ecology, and regional science have emerged. It is not like an inward-looking discipline as Fennman's geography suggests, but it has become an outward-looking discipline. This multi-disciplinary perspective may be regarded both as our raison d'etre and our life-raft sea of knowledge. If the periphery seems interesting, why not explore it; this would only widen the scope of geography. Geography does not border on the systematic sciences, overlapping them in common parts on a common plane, but is on a transverse plane cutting through them.

Some Specific Definitions

1. Geography is concerned to provide accurate, orderly, and rational description and interpretation of the variable character of the Earth's surface (Hartshorne, 1939/1959).
2. Geography has to be conceived as the science concerned with the formulation of the laws governing the spatial distribution of certain features on the surface of the Earth (Schaefer, 1953).
3. It is the study of spatial distributions and space relations on the Earth's surface (Ackerman, et al., 1965).
4. It seeks to understand the Earth as the world of man, with particular reference to the differentiation and integration of place (Broek, 1965).
5. Geography can be regarded as a science, concerned with the rational development, and testing, of theories that explain and predict the spatial distribution and location of various characteristics on the surface of the earth (Yeates, 1968).
6. Geography is the study of spatial organisation expressed as patterns and processes (Taaffe, 1970).
7. Geography offers a broad synoptic view of spatial relationships in human affairs (Smith, 1977).
8. Marxist geography is that part of a whole science which deals with the interrelationship between social processes on the one hand, and spatial processes on the other (Peet, 1977).
9. The focus of all geographical enquiry is place. This implies location on the Earth's surface, the relationship between it and other locations, and the processes affecting changes in those relationships (Jones, 1984).

Fundamental Concepts

It is clear from the above definitions that geography is based on certain fundamental concepts which intrinsically manifest its 'philosophy'. A brief idea of each of the major fundamental concepts is given below for a proper understanding of the philosophy of geography.

Earth's Surface. All those phenomena which occur on the surface of the Earth and which are recognisable (such as rock type and arrangement, configuration and drainage, climate, soil, flora and fauna, ocean currents, and human habitat) and which lead to the spatial variation that occurs between the character of the Earth's surface at one location and another, are treated in geography. This is the most fundamental concept. This concept makes geography worthy of its name, that it seeks to describe and interpret the differences among its different parts as seen at any given time, commonly the present.

Areal Differentiation. As inherent in the earlier concept that the phenomena occurring on the Earth's surface are variable in character, so there are 'differences in different areas on the surface of the Earth'. There are three basic reasonings to the concept: (a) the interrelation of different kinds of phenomena that are directly or indirectly tied to the earth; (b) the differential character of these phenomena and the complexes they form in different areas of the earth; and (c) the areal expression of the phenomena or complexes. These criteria lead to promote geography as 'chorology' whose highest expression is regional geography.

The Region. From the concept of areal differentiation (chorology) has evolved the concept of Region—that each individual unit of a given area is unique, and distinguishable from one another, by distinct interrelated and interwoven phenomena which show certain degree of homogeneity within boundaries that could be defined. A region manifests an intimate relationship between man and nature which has developed over an area through time. As a result, the physical phenomena and human features got interwoven and produced distinguishable, recognisable interrelated phenomena and features.

Inherent in the concept of Region is the 'ecological emphasis' on society-land relation. Here the stress is on the inter-relation of phenomena, the links between aspects of the natural environment of a particular area and the human population occupying and modifying. In this type of analysis, geographers shift their emphasis from spatial variations between areas (these may be thought of as horizontal bonds) to vertical bonds within a bounded geographic area. It is worth noting that the border may be two tier, e.g. the impact of people on land, as well as that of land on people. (possibilist and deterministic approaches); and that the bounded area may be anything from the globe itself to any part of it.

The regional concept, as a term, refers to the mental image of a portion of Earth's surface differentiated by an exceedingly complex fabric of interwoven strands and produced by diverse but interrelated processes.

Various kinds of regions have been identified. The *'uniform region'* is a discrete distribution that is defined in terms of specified criteria and is homogeneous throughout in terms of these criteria. In the 'Compage', which is a unifom region in the true sense of the term, all the features of the physical and societal environment are functionally associated with the human occupance of the Earth. The word compage implies a highly diversified through unitary complex of elements, and gives a greater precision to several facets of regional geography.

The *'nodal or functional region'* is defined by the contact relationships between a centre and its tributary surrounding the regions; or is an area that is tied functionally to a node or several nodes. The definition of such a region involves the measurement of movements, or spatial interaction, for example, the area functionally connected with a central place, the service area of a market, the area reached by a newspaper, or the territory that is tied to a centre of government by the lines of political authority. The nodal region is bounded by lines that mark the disappearance or weakening of the tie to its own focus in favour of some other focus. The lines that mark the extension of a nodal region are similar to isopleth in that they do not separate different kinds of things. Furthermore, the lines of connection to the node usually run at right angles to the boundary of the region.

Modern concepts of systems analysis and Marxist geography appear to be very much indebted to the idea of regional synthesis. Based on the approach (a) the systems analysis claims to be capable of addressing both horizontal (person-person) and vertical (society-nature) relations within a coherent and recognisably scientific framework; and (b) Marxist geography is concerned with the differential geographies of the division of labour, of successive layers of investment, and of uneven distribution, predicated on fundamentally materialist dialectic relations between society and nature.

Location. This concept is as old as the tradition of the discipline itself, and is logically linked up with the preceding concepts. When we say 'variable phenomena on the Earth's surface', we invariably talk of the 'location' of the specific phenomena (where?) because spatial variations in phenomena are recognisable and distinguishable from place to place. The word 'place' refers to location. Each place or each location on the Earth's surface is therefore different with regard to its site and situation. Each place is unique in its location; and phenomena there of are distinguishable and identifiable. So the question 'where is it' is significant when one studies the spatial section of the Earth's surface.

Geographical studies try to establish location accurately, to represent them effectively and economically and to disentangle the factors that lead to particular spatial pattern. Modern geography seeks to account for the location of economic activities: the determination of principles governing location of units of production. Fundamental to the concept of location is the philosophy of positivism, giving emphasis on Models and Laws of spatial arrangements.

Time. Geography provides scientific procedures for describing or explaining phenomena in relation to their development over time. Changes are more common on the surface of the Earth. An ensemble of phenomena on a section of the Earth's surface tends to manifest centuries of development (time factor understood). The development of a region is reflective of a change through time, that in the course of time man and nature adapt to each other like a snail and its shell, and the adaptation reflects centuries of development. Every situation on the Earth's surface denotes time, and every group of phenomena which we see 'today' becomes 'past' tomorrow, but that has to be explained in the light of its evolution (change through time). The stages of economic development or political development have to be viewed in relation to the situation in an industrial region or a political region to the relevant stages of development.

The concept of change over time is found in W. M. Davis's cyclic system for the development of a landscape through the stages of youth, maturity and old age. The French school of regional geography also took up the study of evolution within the cultural landscape.

The concept of time emphasises the importance of continuity and connectedness for sequences of events, which necessarily take place in situation bounded in time and space and whose outcomes are thereby modified by their common localisation.

Cause and Effect. It is one of the most fundamental concepts in geography and was developed in the nineteenth century. It seeks to explain that previous causes can explain observed phenomena. It can be explained in its simplest form that Cause 'A' leads to effect 'B', and that Cause 'B' cannot lead to result 'A'. This analogy can be further explained that 'the mountains have given man leg muscles of iron to climb the slope'. Here the 'mountains' refer to the Cause, and 'the leg muscles of iron to climb the slope' refer to the Effect. It further explains that inorganic elements of the Earth's surface, like mountains, etc. tend to control the organic inhabitants, their behaviours and activities which, as a result of the control, reflect their responses. The concept clearly manifests that if the cause is present, the effect will follow. It is a deterministic concept and seeks to emphasise the deterministic natural laws.

The concept, therefore, seeks to explain that 'a reasonable relation exists between some inorganic element of the Earth on which we live, acting as a control, and some elements of the existence or growth or, behaviour or distribution of the Earth's organic inhabitants, serving as a response' (Ratzel, 1882).

The assumption of cause and effect has given rise to the concept of ontography (organic half) in geography that 'the discipline is concerned with the combination of physiography and ontography, i.e. with the correlations between inorganic environment and organic response' (ibid.).

Sequent Occupance. It expresses the view of geography as a succession of stages of human occupance which establishes the genetics of each stage in terms of its predecessor. Each generation of human occupance leaves its imprints on the landscape. The life history of each manifests the inevitability of the transformation from stage to stage. This concept sharply contradicts the concept of cause and effect and tends to represent a form of cultural determinism. It also emphasises the significance of resources in any significant change in the culture of the community of the region, which is a possibilist concept.

Spatial Distribution. The term 'spatial' describes a particular type of distribution—one that is spread out over a surface, rather than categorises objects, such as population distribution. Thus, spatial distributions include objects (such as population, crops, etc.) that are spread out from each other in space, on a surface that usually means the surface of the Earth. Each spatial distribution is made up of objects that are similar to each other. A map of India showing nothing but the location of million-cities, would portray a spatial distribution. The objects are all of the same type, million-cities, and they are spread out over a surface—the surface of the map, which represents the surface of the Earth (India). Thus, the map is a visual representation of a spatial distribution.

Spatial distributions are collections (sets) of objects in which the objects are of a similar type, with each object having a particular location, a surface. The individual objects that make up a spatial distribution are arranged so that their pattern density and dispersion can be analysed. The geometric arrangement of the million-cities in. India would refer to their pattern; how many million-cities are located in per unit area would tell the density; and how the million-cities are spread out from each other would refer to the dispersion.

The concept of spatial distribution specifically deals with single phenomenon, and seeks to formulate empirical generalisation or law to treat the spatial distribution of the object or phenomenon.

Spatial Interaction. It refers to interdependence between geographic areas. This interdependence appears to be complementary to the society-environment interdependence within a single area, and it is a major focus of geographic enquiry. It includes the movement of goods, passengers, migrants, money, information, etc. between geographic areas. There are three bases for spatial interaction in terms of commodity flow between two regions: character of the regions; nature of commodities; and the existence of closer sources of supply, or markets. The strength of the original concept is that many forms of interaction are themselves interdependent, i.e. a flow of migrants will often stimulate subsequent flows or backflows of trade, passengers, money and information. The concept can be used to interpret the movement of people and ideas between two regions.

Geography as an Empirical Science. Both Alexander Von Humboldt and Carl Ritter insisted that geography should be empirical, in the sense that one should proceed from observation to observation in search for general laws, and not from preconceived opinions, to hypothesis, to observation. In geography, we deal with concrete data or phenomena which are recognisable also, and it studies what really exists (specially what exists at the present time). The observable phenomena, or for that matter concrete data, are unambiguously defined by natural science terminology and methods of measurement. Specifically, it is assumed that observational statements are the only ones which make direct reference to phenomena in the real world. The central part of geography, in the context of the present concept, is that it concerns itself with empirical questions, those with a factual content. Empirical questions are questions about how things are in reality. In this context, 'reality' is defined as the world which can be secured and experienced. This means that geography is concerned with objects in the world.

Synthesis. Synthesis has always remained the teleology of geography, the purpose that justified the raison d'etre of geography. Geography has strongest affiliations with the natural and human sciences. It integrates the materials, which other sciences study separately, in terms of the areal integrations which the heterogeneous phenomena form in different parts of the world.

In modern geography, the argument for the development of a geographical synthesis is practical in character. Geography has worked with practically all the relevant phenomena during its historical development and has devised methods which are central to the formation of synthesis. The planning of modern society requires a good deal of research on the development of synthesis. The concept of synthes is central to Ritter's concept of Ganzheit (wholeness). All interrelated phenomena, occurring on the Earth's surface, as a result of mutual interaction and interdependence of features, tend towards synthesising the whole. The concept of synthesis is therefore basic to geography's raison d'etre.

Modern quantitative or critical geography has not in fact given up the idea of synthesis, and research workers are specifically concerned in a search for synthesis through new forms of analyses.

Unity of Nature. The concept of unity of nature in geography refers to a causal inter-relation of all the individual features in nature. The phenomena of nature require to be studied in order to establish this coherence and unity. Humboldt and Ritter, the founders of modern geography, carried forward the heritage of the concept of unity of nature. For both it was axiomatic that the unity of nature included organic as well as inorganic, human and non-human, immaterial as well as material phenomena.

The concept of unity of all nature requires to be carried into the consideration of individual areas, that all the features of an area in their interconnection form a naturally unified complex. All the objects and forces of nature are interconnected to form a 'whole'. Every part of the world is a reflection of the unity of the whole. Nature in every corner of the Earth is a reflection of the whole. The 'wholeness' has to be established only by investigating the individual, single features in their relation to each other, and building these up in their actual relationships to form the whole. 'Unity in diversity of phenomena, a harmony blending together all created things, however, dissimilar in form and attributes, tends to create one great whole animated by breath of life' (Ritter, 1822–59). The concept of unity of nature is basic to the concept of synthesis and refers to holistic tradition.

Regional concept, the concept of synthesis and the concept of unity of nature, tend to reflect the interdependence and mutual interaction of phenomena, leading to the wholeness. These concepts are, therefore, complementary to each other, and refer to 'holism'. In fact, all the concepts which are fundamental to geography are logically tied up with one another, forming a greater circle of synthesis.

NATURE OF GEOGRAPHY

Geography occupies a logically defensible position among the sciences as one of the chorological studies, and attempts to consider not particular kinds of objects and phenomena in reality, but actual sections of reality. It further attempts to analyse and synthesise not processes of phenomena, but the associations of phenomena as related in sections of reality. The heterogeneous phenomena which other sciences analyse by classes are not merely mixed together in terms of physical juxtaposition on the Earth surface, but are causally interrelated in complex areal combination.

Geography attempts to integrate the materials that other sciences study separately, in terms of the actual integrations which the heterogeneous phenomena form in different parts of the world. In practice as well as in theory, any phenomenon studied in geography may at the same time be an object of study in some systematic field. But geography is not an agglomeration of pieces of the systematic sciences, it integrates these phenomena according to its distinctive chorographic point of view. Since geography cuts a section through all the systematic sciences, there is an intimate and mutual relationship between it and each of these fields (Hartshorne, 1976, 460).

Geography borrows from the systematic sciences all those generic concepts or type classifications that the sciences have developed, in order to effectively utilise them in making its description of phenomena and interpretation of their inter-relations as accurate as possible. However, geography attempts to discover its own generic concepts and system of classification when the borrowed generic concepts are found inadequate to account for geographic interpretations and purposes. Geography, on the other hand, contributes much to the systematic sciences. In its naive examination of the inter-relation of phenomena in the real world, it discovers phenomena which the sophisticated academic view of the systematic sciences may not have observed. It shows them to be worthy of study in themselves and thus adds to the field of the systematic studies.

Geography constantly emphasises one aspect of phenomena which is frequently lost sight of in the more theoretical approach of the systematic fields, namely, the geographic aspect. It serves, therefore, as a realistic critic whose function is to remind the systematic sciences that they cannot completely understand their phenomena by considering them only in terms of their common characteristics and processes. They must also note the differences in those phenomena that result from their actual location in different areas of the world. In order to interpret these differences, and to interpret the resultant world distribution of their phenomena; the systematic sciences take from geography something of the particular techniques which its point of view has required it to develop—notably the ·techniques of maps and map interpretation.

Geography adds the aptitude of comprehending the correspondence and correlation of facts, be they in the terrestrial milieu which includes them all, or in the regional milieu in which they are localised (Vidal, 1917). Geography is essential to the full understanding of reality while the systematic sciences necessarily destroy the essential character of reality.

It attempts to interpret the realities of areal differentiation of the world as they are found, not only in terms of the differences in certain things from place to place, but also in terms of the total combination of phenomena in each place, different from those at every other place.

Geography seeks to acquire a complete knowledge of the areal differentiation of the world, and therefore discriminates among the phenomena that vary in different parts of the world only in terms of their geographic significance, i.e. their relation to the total differentiation of areas. Phenomena significant to areal differentiation have areal expression—not necessarily in terms of physical extent over the ground, but as characteristic of an area of more or less definite extent. In order to make its knowledge of interrelated phenomena as accurate as possible, geography considers all kinds of facts involved in such relations and utilises all possible means of determining the facts, so that results obtained from one set of facts or by one method of observation, may be checked by those secured from other facts or from other observations.

On the basis of generic concepts, geography seeks to establish principles of relationships between the phenomena that are areally related in the same or different areas, in order that it may correctly interpret the inter-relations of such phenomena in any particular area.

Finally, geography attempts to organise its knowledge of the world into interconnected systems, in order that any particular fragment of knowledge may be related to all others that bear upon it.

APPROACHES TO THE STUDY OF GEOGRAPHY

It is clear from the preceding discussion that the subject matter of geography is logically indivisible, as it deals with 'phenomena' which occur in association on the Earth's surface, with distinct areal expressions and variations. The problem arises about how to deal with these variable phenomena and what approaches to be adopted for their study? There have been two distinct approaches for doing this: either one can lay emphasis on the description and elucidation of individual phenomena of each part of the Earth's surface in turn, region by region, because they are unique. This is *ideographical approach or 'regional geography'*. Or one can proceed with a significant theme of variable phenomena and trace it systematically over the whole of the Earth's surface, considering geography as a law-giving science. This is *'nomothetical approach' which is also called 'systematic geography'*. It requires the development and testing of theories and models in order to develop geographical laws.

Regional Geography

Geography is viewed as the study of areal differentiation, and it seeks to describe and interpret the variable character from place to place of the Earth as the world of man. As mentioned earlier, phenomena occur in association on the Earth's surface, and in that sense each particular location (region) on the Earth's surface is unique with regard to its individual phenomenon. Regional geography, therefore, attempts to organise knowledge of all interrelated forms of areal differentiation in localised individual units of area, which it must organise into a system of division and subdivision of the total Earth surface.

The description and elucidation of unique individual but interrelated forms of areal phenomena with distinct areal expression, proceeds with analysis and synthesis, the integration of all interrelated features and phenomena at individual units of places and must then express the integration of all such units of individual regions within a given area. Each unit must be considered homogeneous. Regional geography proceeds with more effective ways of measuring the interrelationships of phenomena in individual units of a given area; but at the same time, the explanatory description of features in the past must be kept subordinate to the primary purpose because that facilitates comprehension of the present.

Regional methods are characteristically broad-based and include consideration not only of the physical environment of the individual units of a given area but also of the population of the unit—its demographic characteristics, its density and distribution, its occupational structure, its social and political behaviour (e.g. migration and political choice). Stress is laid on the spatial dimension of each component, and such local associations of physical environment and human characteristics in particular areas give rise to distinctive sub-region within the overall area being considered. The Ganga plain is a region because of the uniqueness of the interrelated phenomena which have developed over centuries on account of intimate relationship and mutual interaction between man and nature, and which put a stamp on the landscape.

In order to measure the interrelationships of phenomena occurring in the Ganga plain, the regional geographer would attempt to identify the elements of physical environment (location, relief and geology, drainage, flora and fauna, climate, soils, minerals) and elements of human life (economic activity, political organisation, social organisation and settlement) of it. Then, with the help of analysis and synthesis of all recognisable interrelated features and phenomena occurring in the Ganga plain, he would try to find out what is unique and distinctive as well as particular about it. The description and elucidation of the unique phenomena of the Ganga plain must be comprehended by the explanatory description of the interrelated features of the historic past. By applying the regional method, the Ganga plain itself can be subdivided into the Upper Ganga plain, the Middle Ganga plain and the Lower Ganga plain.

Where the evidence of uniqueness of individual units is available, the stress is laid on the stability and instability of regional structures over time to underscore whether those being described are static or are undergoing changes and modifications. It is, therefore, characteristic of such regional methods in geography that they try to integrate a number of recognisable phenomena within a single area which have a complex array of links between themselves.

Formulation of scientific laws cannot be done in regional geography because it deals with the description and elucidation of complex integrations in unique units. Scientific laws can be best established in laboratory experiments which allow only a few independent variables to vary and suggest some kind of determinism, which is inappropriate to the human motivations which are in part the causes of landscape variations and changes. Instead of scientific laws, regional geography seeks to: (a) describe phenomena with the maximum degree of accuracy on the basis of objective empirical observations; (b) classify the phenomena, as far as reality permits, in terms of generic concepts; (c) interpret the phenomena through rational consideration of the facts thus secured and by logical processes of analysis and synthesis; and (d) arrange these findings in an orderly system (Hartshorne, 1959).

Systematic Geography

Systematic geography is organised in terms of particular phenomena of general geographic significance, each of which is studied in terms of the relations of its areal differentiation with the others. Its descriptive form is therefore similar to that of the systematic science. It seeks to formulate empirical generalisations or laws of the phenomena studied and their universal application, in terms of significance to areal differentiation. In systematic geography, each particular element or element-complex that is geographically significant, is studied in terms of its relations to the total differentiation of areas, as it varies from place to place over the world or any part of it.

Systematic method in geography attempts to take one or a few aspects of the human environment or the human population and studies their varying performance over a pre-defined geographical space. Such studies are usually labelled with reference either to the phenomenon concerned or to the subfield of the natural or social sciences with which it may be identified. Thus, a spatial study of voting pattern may be termed Electoral Geography (under the first) or Political Geography (under the second); or the spatial study of industrial pattern may be termed Industrial Geography (under the first) or Economic Geography (under the second).

Systematic geography treats the discipline as a science. A science is characterised by its explanations, and explanations require laws, and to explain the phenomena one has described means always to recognise them as instance of laws. As geography is concerned with the variation of phenomena over the surface of the Earth, the major phenomena or regularities or features which are described refer to spatial pattern. It is these spatial arrangements of phenomena or features, and not the phenomena themselves about which geographers should make law-like statements. Hence geography has to be conceived as the science concerned with the formulation of the laws governing the spatial distribution of certain features on the Earth's surface (Schaefer, 1953, 227).

Fig. 1.1 Integration of all branches of systematic geography, focused on a particular place on the Earth surface, is regional geography. (Hartshorne, 1939).

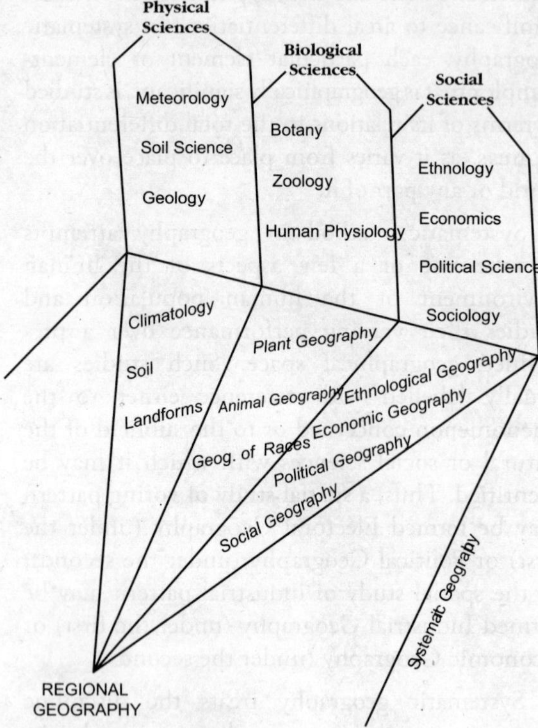

Source: Hartshorne, 1939.

Observation of phenomenon (e.g. the voting behaviour in Bihar) would lead to a hypothesis about the interrelationship between the spatial patterns concerning religious and caste cleavages on the one hand, and the electoral participation on the other hand, and this would be tested against large number of cases to provide the material for an empirical generalisation or law if it is thereby verified.

In order to develop geographical laws or generalisations, to account for the spatial arrangements of phenomena or features such as those of population mobility, migration, economic activity, drainage pattern and atmospheric circulation, etc., one is required to develop and test the theories and models through hypothetic-deductive methods. This requires the establishment of *a priori* models of the structure of reality. Systematic geography attempts to postulate a set of hypotheses, based on observations, which could be confirmed, corroborated or rejected by testing empirical data. A large number of confirmations lead to a verification of the hypothesis, which is then established as an empirical law whose validity is tested elsewhere. If the empirical validation of the model which is developed to account for the voting behaviour in Bihar is successfully tested to account for the voting behaviour in UP, Rajasthan, Tamil Nadu and other states, then its universal application is assured.

However, geography requires both the systematic and regional methods of study of phenomena and organisation of knowledge. Systematic geography is essential to an understanding of the areal differences in each kind of phenomena and the principles governing their relations to each other. This alone cannot provide a comprehension of the individual Earth units, but divests them of the fullness of their colour and life. To comprehend the full character of each area in comparison with others, it is necessary to examine the totality of related features found in different units of area, e.g. regional geography. These two methods are mutually and intimately related and essential to each other (Fig. 1.1). A distinction may be made between systematic geography which seeks to formulate generalisations or laws, and the study of uniqueness in regional geography whereby generalisations are tested so that subsequent theories may be improved.

INTERACTIONS/RELATIONS WITH OTHER SCIENCES

While defining the scope of geography we had seen that some parts of it have their strongest affiliations with mathematics and environmental (natural) sciences, others with history and social sciences.

While other sciences deal with distinctive types of phenomena, geography studies several kinds of phenomena, each already studied by another science. Thus, geography has firmly established itself as a discipline of synthesis, on the plane of empirical knowledge.

Relations with Natural Sciences

Astronomy and Geography. Astronomy basically deals with the celestial bodies (their motion in the universe) which include the different Suns and their planets, satellites, and their motion, constellations, as well as different kinds of phenomena occurring in the outer space. It is a pure natural science, and systematically deals with the concrete objects and phenomena which have a known situation in the Universe. Geography's link with astronomy has developed mainly because it also deals with concrete phenomena. Geography attempts to identify and measure the impact of the motion of the celestial bodies on the surface of the Earth. How have the phenomena on the spatial section of the Earth's surface been influenced by the objects of the celestial space? This has necessitated in geography to study the form, size and movement and motion of celestial bodies, especially of the Sun, the Earth and the Moon, and their corresponding position in the solar system. Geography also attempts to determine the position of the stars and the constellation in the horizon, the extent of the celestial sphere, and other celestial phenomena, such as to locate new planets, their satellites and stars.

It is the mathematical tradition in geography which has brought both astronomy and geography closer to each other. With the help of mathematics, the precise location, nature of motion and movement, extent of celestial sphere, form and size of celestial bodies, including those of the Sun, the Earth and the Moon have been successfully determined. Astronomy and geography both are equally concerned with these aspects, and in a sense, they are mutually interdependent.

The interaction of astronomy and geography within the mathematical framework has given rise to cartography in geography which treats specific phenomenon of location in relation to geodesy (form, shape, size of the Earth, latitude and longitude, etc.) and celestial objects.

Geology and Geography. Geology is the natural science which has traditionally been closest to geography, because both of them have strong links with the idiographic tradition. Geology describes and clarifies 'concrete' simple phenomena and puts them into a geological chronology and classification. Geography also attempts to deal with 'concrete' recognisable phenomena occurring on the surface of the Earth, classifying them into various categories. Both study such concrete phenomena which have known situation in time and space.

Geology is the study of the rocks, their layout and arrangements, types, minimal constituents, resistance and age, distribution, dip and alignment, etc. These are all concrete phenomena, and need not be quantified. Geography is essentially the study of the Earth's surface, especially the morphology of the surface. Morphology of the Earth's surface is the manifestation of the rock structure in different forms. So one has to study geology in order to explain the morphology. The analogy, *'landscape is the study of structure, process and stage'* clearly demonstrates the closest relation between geology and geography. The term 'structure' refers to geology and the words 'process' and 'stage' refer to geography. The sentence makes both complementary to each other. Geology and geography both study the distribution of various kinds of rocks and minerals, their mode of occurrence, chemical and physical components, hardness, specific gravity and lots of other things which are concrete and recognisable. Interaction between geology and geography leads to geomorphology or landform geography.

However, certain types of phenomena studied in geography are quantifiable in nature. They require the formulation, verification or rejection of hypotheses through experiments and the establishment of universal scientific laws. Because of this changing trend in geography, the discipline shows some deviation from its traditional idiographic link with geology.

Physics and Geography. Geography has long been associated with physics. As geography is the study of variable phenomena on the Earth's surface, the mechanism of the phenomena requires to be studied withi.n the framework of physics. The physics of atmosphere, called meteorology, uses the methods of physics to interpret and explain atmospheric processes. Similarly, the physics of hydrosphere is called oceanography, which too applies the principles of physics to explain and interpret hydrospheric processes. Even lithospheric processes and changes are studied within the framework of physics. The characteristics of phenomena studied in meteorology and oceanography show that objectives are quantifiable. Some of their phenomena are more abstract than those studied by geologists and biologists.

Geography, as a science of synthesis, seeks to draw within its circumference the basic principles of meteorology and oceanography, in an obvious attempt to explain and interpret 'the interrelated features' and 'terrestrial phenomena'. Interaction with meteorology has given rise to a systematic branch of climatology in geography. Climate is an important element of the physical environment of humankind. A geographer is interested in 'the processes' of exchange of energy and mass between the Earth and the atmosphere because that results in conditions for climate. It is the climate which is the prime factor for analysing and classifying the terrestrial phenomena occurring on the Earth's surface. In order to understand the variable phenomena, one has to take into consideration the basic principles of meteorology and hydrology, then to proceed with explanation, to account for the interdependence of phenomena.

Geography's inclination towards physics seems to have been further intensified on account of its increasing dependence on the hypothetic–deductive method. It has been mostly developed in physics. Geography's increasing tendency to adopt the hypothetic-deductive method has been mainly due to the type of questions regarding the variable phenomena to be answered and the nature of empirical data to be studied. It is characteristic of the phenomena now studied in geography and physics that objectives are quantifiable.

Some of the phenomena of theoretical geography are, however, much more abstract, and some of them cannot be directly observed, but the method of measurement has given a theoretical supposition of their existence. Theoretical geography, like theoretical physics, operates in an abstract milieu, seeking unity and association through mathematical and statistical hypotheses and postulates. It is from physics that theoretical geography has been able to develop into model-building, precisely because it works with abstract and quantifiable phenomena (see Chapter 13 for social physics which seeks to explain the isomorphic relations between social behaviour and the laws of physics).

Botany and Geography. It is the idiographic tradition in geography which seeks to suggest its strong link with biology and zoology. The systematic branches of botany and zoology have traditionally been confined to a classification and description of various kinds of species on the Earth's surface which tend to form unique features and phenomena. Geography, being the study of the spatial section of Earth's surface, also attempts to study the distributional aspects of floral phenomena and also provides their classification. Again, we come to the point that the distributional nature of floral species, right from the polar areas down to the equatorial lands, tends to form concrete phenomena. Both botany and geography seek to describe and classify the spatial pattern of concrete phenomena and put them into different classifications according to spatial conditions.

Geography is closely related to zoology, because both seek to interpret: (i) the distribution of the individual genera and species of animals; (ii) the distribution of the interrelated phenomena of genera and species of animals over the spatial section of the Earth' surface; and (iii) both are concerned with the difference in faunal equipment of the different lands.

It is the ecological tradition in botany and zoology which brought them much closer to geography. Ecology is the study of the relationships between animals, plants and their environments. Ecologists, therefore, study the natural relationships whereby particular species of plants and animals are dependent on each other 'and on the non-organic environment. This aspect of ecology exerts its influence on man's biological nature and his relationship to the natural environment.

Geography, especially population geography and synecology (which treats the development of plant and animal communities in an area), can be said to parallel traditional biological ecology. During recent decades, ecology has become a major interest of research in botany, zoology and geography. This is due to the development of a new direction within ecology—called systems ecology—which studies construction and function of the ecosystem. An ecosystem consists of the biological community at a specific place and the environmental physical circumstances which influence and are influenced by the biological community. It clearly manifests the interdependence and interrelationship between botany, zoology and geography.

The four elements of the ecosystem which interest the botanist, zoologist and geographer are: (i) It brings together the worlds of man, animals and plants within a single system wherein the interactions between the components can be analysed; (ii) it is structured in a more or less orderly, rational and understandable way; (iii) ecosystems include a permanent throughput of material and energy; (iv) it is a type of general system which uses and can make use of general developed themes of systems analysis.

Relations with Social Sciences

Economics and Geography. Economics and geography are long related to each other, and should be regarded united and inseparable because both study 'concrete phenomena which have a known situation and time', focussing on man.

Economics is basically concerned with man's economic activities, and the principles governing the location of units of production, forces of production, relations of production, division of labour, and a host of institutional aspects, including health, capital investment and assets, entrepreneurial skills, technology, investment funds, etc. In brief, economics is concerned with how human needs and wants are satisfied in a world of limited resources, where everyone cannot have as much as he or she wants of everything.

All the above phenomena concerning economics have a known situation and time. They must be studied in spatial and temporal context. Every human activity, dealing with satisfying needs and wants, must have a definite location (place), time and distribution. For example, mining activities refer to the place of occurrence of minerals, their distribution and the period of extraction. The purpose of the activity is production. Each unit of production on the Earth's surface is variable in the context of place, time and distribution.

Since geography is the study of ensemble of phenomena in relation to place, localisation, time and distribution, it is equally concerned with the variable character of the economic activities. Agricultural activities and industrial activities do have known places, localisation, time and distribution on the various spatial sections of the Earth's surface. Forces of production, relations of production, units of production, division of labour, the material condition and the entrepreneurial skill, all tend to have 'areal expression' with a known time and situation. This makes economics and geography interdependent.

Modern geography which focusses on testing, on the empirical validation of hypotheses concerning variable economic phenomena, has developed a much stronger affinity with economics. This has been productive of new ideas and techniques since the 1950s. The introduction of location theory into geography is based on concepts from

'Neoclassical Economics' (which forms the basis of the view of how economic activity functions) as conventionally adopted in capitalist society.

Geography's interaction with economics has given rise to economic geography, which is the study of the spatial variation on the Earth's surface of activities related to producing, exchanging and consuming goods and services.

Sociology and Geography. Sociology is mainly concerned with the institutional aspects of the society which broadly include social organisation of communities, family structure and system, rituals, culture, customs, social system, and overall the entire 'genre de vie' (way of living). Since all these aspects tend to constitute distinct phenomena which have known situation and time, they must be studied in spatial context.

Each spatial section of the Earth surface is identified by its distinct 'genre de vie, which results into 'social phenomena', and when one studies the variable character of its social phenomena on the Earth's surface, he seeks to integrate sociology with geography. For example, when a sociologist or a geographer studies the institutional aspects of the social organisation of the major communities of India, he takes into account the geographical aspects and bases of the social organisations of the major communities, each with a distinct way of life and having 'place', 'localisation' and 'distribution' in different sections of the country. This shows the interrelationship between sociology and geography.

Geography's traditional link with sociology appears to have developed mostly as a result of the idiographic tradition. However, the introduction of the location theory into sociology has further strengthened its ties with geography and vice versa. Geography has drawn some of its concepts from modern sociology which are associated with the formulation of empirical generalisations or laws through mathematical methods and statistical procedures. A number of studies of relations between social behaviour, of movement of people between urban centres, of spatial interactions between social groups, of the relations between innovation and tradition in rural and urban areas, have been made both in sociology and geography with the aid of models. Social geography is the logical expression of the interaction between sociology and geography as it studies social phenomena in spatial context.

Anthropology and Geography. Anthropology attempts to study human races, their physical characteristics, biological traits, cultural traits and organisations, and seeks to classify them accordingly. Physical and biological traits of human races and their organisation are products of nature. Races have been identified on the basis of various indices, such as head index, nasal index and cephalic index and skin colour, hair texture and stature, and are accordingly categorised as Negrito, Negro, Australoid, Mediterranean, Nordic, Alpine and Mongolic. Each of them has distinct biological traits and habitants. Both anthropology and geography seek to identify and classify them on the basis of their habitant and cultural traits and attempt to study the variable racial phenomena on the spatial section of the Earth's surface.

The reciprocal relationship between anthropology and geography has resulted in the development of 'anthropogeography' or ethnology which treats different human groups in terms of the natural conditions, and the geographical distribution of races. The study of 'apartheid' (the policy of spatial separation of races) as applied in South Africa forms an inseparable part of anthropogeography or ethnology.

History and Geography. History and geography fill up the entire circumference of our perception: history that of time, and geography that of space. History differs from geography only in the consideration of time and area. The former is a report of phenomena that follow one another and has reference to time. The latter is a report of phenomena beside each other in space. History is a narrative, geography is a description. History provides the framework into which the multiplicity of historical facts are ordered, the area provides the skeleton for geography, and both the fields are concerned with integrating different kinds of phenomena. Therefore, history and geography are traditionally idiographic in nature and are mutually interdependent.

There would be no history if human events never changed and were invariable from day to day. Similarly, there would be no geography if physical and human phenomena were distributed uniformly over the Earth's surface. It is because of the variation through time, and the variation on the Earth's surface that the disciplines of history and geography exist. There can be no history without geography. Any event, whether small or big, must represent a time, a place of its occurrence and localisation, and distribution.

It is firmly believed that the landscape changes are mainly due to communities' genre de vies (ways of living). Landscape changes are greatest where the level of genre de vie of the community is highest. Each major period in the past shall have to be seen in the light of its distinct phenomena occurring in sequence (temporal variations inherent) as a result of the community's action on the landscape, and the historian can identify such variation in phenomena accordingly. Each period would also manifest variation from place to place, in terms of natural and human condition. A geographer, or a historian, while attempting to study the Mughal period or the British period in India, seeks to integrate the temporal and the spatial phenomena of that period into one, in order to arrive at a conclusion with regard to the association of phenomena.

The interpretation of present geographical phenomena requires some knowledge of their historical development, whether physical or human features. For example, when we attempt to interpret the relief of India, we must take into account its geological development first, or if we attempt to analyse the present political phenomena of India, we must first try to acquire knowledge of the political development of India since independence. In this case, history is the means to a geographic end. Likewise the interpretation of historical events requires some knowledge of their geographic background; in this case *geography is the means to an historical end* (Hartshorne, 1976, 463).

Interpretation of political phenomena in their areal context or of historical phenomena in their spatial perspective may be logically called 'political geography' or 'historical geography', which is a direct product of the interaction between the disciplines.

BRANCHES OF GEOGRAPHY

Variable phenomena on the Earth's surface can be treated separately or in association. They are classified and categorised into physical phenomena and human phenomena, or are treated in association as interrelated phenomena. So geography has two main branches: Physical Geography and Human Geography.

Physical Geography

Physical geography is concerned with the study of the description of Nature or physical phenomena, encompassing the systematic sciences of geology, meteorology, astronomy, botany, zoology and chemistry. Physical geography has its origin in the antiquity when the ancient Greek and Roman scholars developed their interest in the study of Nature and its different attributes. It became a very popular subject during the last third of the nineteenth century. Physical geography has a number of sub-branches which deal with different kinds of physical phenomena.

1. **Astronomical Geography.** Astronomical geography studies the celestial phenomena which concern the Earth's surface. It is perhaps the oldest branch of geography developed by Thales, Anaximander, Eratosthenes, Aristotle and Ptolemy. Related to it is cartography which also developed along with astronomical geography in the antiquity, but its refinement took place during the period of Renaissance. However, some of the earlier erroneous

concepts concerning the motion of the celestial objects, including those of the Earth and other planets, were also rectified through observation during the period of Renaissance. Alexander Von Humboldt used the word uranography in place of astronomical geography.

2. **Mathematical Geography.** It studies the form, size and movement of the Earth, and of its position in the solar system. It has its origin in the works of Thales, Eratosthenes and Ptolemy. But their ideas were corrected by Copernicus, Kepler and Galileo.
3. **Geomorphology.** It is concerned with the study of the landforms on the Earth's surface. Though earlier credit of such studies goes to Herodotus and Polybius of ancient Greece, but the contributions of the medieval Arab scholars like Al-Biruni and Ibn Sina cannot be overlooked. By the end of the eighteenth century, studies of mountain structure and origin were done by the German and the French scholars.

In fact, it was Mary Somerville of Great Britain and Alexander Von Humboldt of Germany who established geomorphology. In Germany, Ferdinand Von Richthofen, Oscar Peschel, Alfred Hettner, Albrecht Penck, Walther Penck in the late nineteenth century and early twentieth century set forth a new direction in the field of geomorphology. W. M. Davis of the United States emerged as the leading personality in its development. In France, Emmanuel de Martonne carried forward the tradition of geomorphology into the country. Most of the German geomorphologists emphasised the importance of glaciology, a distinct part of geomorphology.

4. **Climatology.** It is the study of the atmospheric conditions and related climatic and weather phenomena. It was Eratosthenes, again, who for the first time provided a three-fold climatic division of the known world. However, Aristotle had earlier offered explanation with regard to climatic suitability of the human habitat. Moreover, substantial contributions to this branch of physical geography were made by the medieval Arab scholars, Al-Balkhi and Al-Masudi. Later, it was developed by Wladimir Koppen and Albrecht Penck of Germany, A. I. Voeikov of Russia and C. W. Thornthwaite of the United States.
5. **Oceanography.** It is concerned with the study of various types of phenomena and features, both on the surface of the ocean and on its floor, including relief of the floor; temperature of the surface water; distribution of salinity, depth, currents, tides, waves; coral reefs, etc. It was Posidonius (135–50 BC) who first investigated tides at Gades, measured the depth of sea off Sardinia, and sought to discover the origin of the Grau gravels. Oceanography was given a systematic treatment by an American, Matthew Fontaine Maury, in the mid-nineteenth century. However, he also offered a model of atmospheric circulation.
6. **Soil Geography.** It studies various soil-forming processes, their chemical, physical and biological constituents, their colour and types, distributions and carrying capacities, etc. It was V. V. Dokuchaiev of Russia who pioneered the study of soil geography, which was further developed by the American soil geographer, Marbut.
7. **Bio-Geography.** It is concerned with the biological phenomena in space, especially in terms of the distribution of various kinds of floral and faunal species. Bio-geography may be subdivided into plant or floral geography, and animal or faunal geography.

Theophratus, a pupil of Aristotle, is credited to have developed bio-geography which was later developed in a more meaningful way by John Ray, an English scholar. He was a major influence on the Swedish botanist Carolus Linnaeus. Lamarck and Buffon also made contributions to it. Similarly, the contributions of Humboldt and George Gerland cannot be overlooked.

Human Geography

Human geography is inseparable from physical geography because the variable phenomena on the spatial section of the Earth's surface have human elements in them. Broadly speaking, human geography is the synthetic study of the relationship between human societies and the Earth's surface. It is made up of three closely linked components: the spatial analysis of the human population (i.e. its numbers, its characteristics and activities, distribution over the Earth's surface); the ecological analysis of the relation between human population and its environment; and the regional synthesis which combines the first two themes in an areal differentiation of the Earth's surface.

Human geography developed along with the physical geography and ancient scholars like Hecataeus, Aristotle, Herodotus, Polybius, and Strabo pioneered the studies of the influence of the physical Earth on human affairs. Ibn Khaldun, a medieval Arab scholar, made a significant contribution to the field of human-environment relationship. Carl Ritter, one of the founders of modern geography, treated geography from the human point of view. George Perkins Marsh of the United States developed a novel approach to the study of the relation of humankind to the land and he also analysed its effect on nature. It was Friedrich Ratzel of Germany who is credited to have provided the guidelines for a comparative and systematic study of human geography. Paul Vidal de la Blache and his pupil Jeam Bruhnes of France treated human geography in the light of humankind's creativity in modifying the landscape. Human geography since then continued to prosper. Human geography has a number of sub-branches.

1. **Anthropogeography.** It studies racial phenomena in their areal context, or the geographical distribution of races. Agarthacides is identified to have classified the Ethiopian tribes on the basis of their diet and cultural traits. But it was Friedrich Ratzel who is credited to have coined the term 'anthropogeography' to study human geography.
2. **Cultural Geography.** It focusses on the impact of human culture, both material and non-material, on the natural environment and the human organisation of space. Herodotus pioneered the study of the cultural traits of the people strange to the Greeks. He is called the 'father of ethnography'. Friedrich Ratzel and his American pupil, Ellen Semple were influenced by ecological concepts of culture traceable to the biological sources. Otto Schluter of Germany coined the word 'Kultur landschaft', a landscape created by human culture. Carl O. Sauer of the USA also carried forward the same tradition as that of Schluter. Kant's 'moral geography' was nothing but 'cultural geography'.
3. **Population Geography.** It is the study of the ways in which spatial variations in the distribution, composition, migration, and growth of population are related to the nature of places. Achenwell (1748) and Sussmilch (1747) first pioneered the statistical studies of population in space. Since then its dimension has widened. However, it was Trewartha (1953) who is now credited for having offered a tentative scheme of the content and organisation of materials in population geography, and he (1969) stressed that 'population geography was concerned with understanding the regional differences in the Earth's covering of people.'
4. **Economic Geography.** It refers to the field of study focused on the location of economic activity at the local, national and world scale. It is also the study of economic phenomena in their areal context. Kant introduced commercial geography, which is the study of distribution of commodities. E. R. Johnson and J. R. Smith of the USA and N. N. Bavanskin of the Soviet Union were the pioneers of economic geography. Economic geography requires to be studied under the following heads:
   (a) Resource geography: the study of the resources, their distribution, production, utilisation and conservation;

(b) Agricultural geography: the study of spatial variations in agricultural activity;

(c) Industrial geography: the study of the spatial arrangement of industrial activity;

(d) Transport geography: the study of the role of transport in geography, including patterns and modes of transport, movement of goods and people and relationships between transport and other geographic factors.

5. **Political Geography.** It is the study of political phenomena in their areal or spatial context. The tradition of geographic studies of the state goes back to Aristotle, but Kant provided a classification of political geography. However, it is Friedrich Ratzel who is called the 'father of modern political geography'. It developed much in the last century. Somewhat different from political geography was 'geopolitics' which developed in Germany during the inter-war period.

6. **Geopolitics.** The origin of geopolitics can be traced back to the Swedish political scientist Rudolf Kjellen and the German natural scientist and geographer Friedrich Ratzel. While the former introduced the term, his interest was initially stimulated by the ideas of Ratzel, and it was through the latter that it became widely known. Geopolitica was earlier defined as 'the science of the state a realm in space'. Later, Kjellen redefined it as 'the theory of the state as a geographical organism or phenomenon in space'. In Germany, geopolitics was necessarily rationalised to justify the Nazi claims to a position of dominance in Eurasia. However, after the Second World War, Gennan geopolitics was lumped together with the rest of the ideological baggage of Nazism and the lot was banished. The return of geopolitics took place during the 1970s, the new geopolitics being pioneered in the USA and France. In the contemporary context, geopolitics is the spatial study of international relations.

7. **Historical Geography.** It is the geography of the past, or the study of the past events in their spatial context. It was Herodotus who provided some exciting examples of what we call historical geography, e.g. the re-creation of past geographies and the tracing of geographical change through time. Sir Halford J. Mackinder of Great Britain made substantial contribution to the field of historical geography. British geographers set forth the tradition of historical geography in the regional perspective.

8. **Social Geography.** It is the analysis of social phenomena in space. It is a recent sub-discipline: concern with social phenomena developed mainly after 1945. However, social geography was developed in Germany by Otto Schluter as a reaction against physical determinism. Interest in the studies of social well-being and social problems, and their spatial variations, and also the search for root cause of social problems has given rise to 'radical geography' and 'welfare geography'. Study of new spatial determinism of capitalist society led to the rise of 'Marxist geography'.

9. **Urban Geography.** It is the geographical study of urban areas, their spatial patterns and functions, origin and hierarchies. Its development took place in the 1940s but during the 1960s it was probably the most popular subfield within human geography.

The combination of both physical geography and human geography provides both the 'core' and the 'raison d'etre' of geography. There can be no geography without both, and without geography knowledge about the Earth is fragmented and incomplete. If both are separated and studied in isolation, the very synthesising character of the discipline would be lost.

MAJOR SCHOOLS OF GEOGRAPHICAL THOUGHT

The theoretical development of geography took place during 1750–1950. During the period, a number of schools, with their distinct concepts and ideas regarding the variable phenomena on the Earth's surface, came to be developed in

Germany, France, Great Britain, Soviet Union and the United States of America. The geographical thought is, therefore, made up of these schools. However, the year 1859 is a 'watershed' because it saw the death of the founders of modern classical geography. Alexander Von Humboldt and Carl Ritter, on the one hand, and the publication of Charles Darwin's 'Origin of Species' on the other, replaced the existing philosophy of geography by a materialistic scientific philosophy which emphasised natural laws and causality. However, the philosophy of modern geography is based on the following schools of thought.

German School of Ganzhiet, Chorology and *Landshaftskunde*

Concepts of *Ganzhiet (single whole)* and *chorology* were earlier developed by Immanuel Kant. He regarded the intricate interlacing of relationships between all the phenomena found in Earth's surface (organic or inorganic), as functional relation of the parts of a single whole. Within any particular area, the combination of all interrelated phenomena is not a mere aggregate, but an interrelated 'whole'. Later, this idea of Ganzhiet or whole was adopted by Carl Ritter in his geographical concept of synthesis, i.e. the unity in nature.

It is from the concept of Ganzhiet that has developed the concept of *'chorology'*. It is now widely believed that though Ferdinand Von Richthofen is credited to have applied the term 'chorology', the concept was earlier developed by Kant, and followed by Humboldt. Kant studied not only the features of the Earth produced by natural processes, but also the races of man and the changes on the face of the Earth resulting from human action and added that the phenomena be considered in their areal context. What Richthofen did was to specify the purpose, i.e. to throw light on the causal inter-relation among the diverse things in a particular area. Alfred Hettner defined geography as a chorological science, i.e. the study of the areal differentiations.

But in order to have a more balanced treatment of the inter-relation of things in particular areas, it was suggested that emphasis be given to the overall appearance of the landscape and that the concept of geography as *Landschaftskunde* came into being in the beginning of the present century mainly due to Otto Schluter's address at Munich in 1906. In fact, the concepts of 'chorology' and 'landschaftskunde' are concerned about the variations in the characters of the force of the Earth, which is the interpretation of 'the areal differentiation'. However, the German school of geography tends to identity itself with the traditional concept of geography as 'Landschaftskunde'.

French School of 'La Geographie Humaine'

French geographical thought is focused on Paul Vidal de la Blache, who is credited to have developed *we-ocratic tradition* in geography—an approach in human geography clearly distinguished by the central and active role it assigns to human awareness and human agency, human consciousness and human creativity (genre de vie) in modifying the landscape. According to him, the natural and human phenomena should be regarded as united and inseparable. In an area of human settlement, nature changes significantly because of the presence of man, but these changes are greatest where the level of human creativity is highest, where the people are more aware and conscious and proud of their culture. Vidal, emphasised the study of region which was a manifestation of an intimate relationship between man and nature developed through centuries over an area.

British School of Regional Studies

British school of regional studies has at least three different meanings in Great Britain. (1) There are regional studies which amount to descriptions of segments of the Earth's surface broadly synonymous with the German regional geography. (2) There are regional studies which attempt to divide the Earth's surface into either homogenous or functional areas of varying size which may be called regionalisation. (3) Regional studies denote regional specialisation to study different aspects of some parts of the world. One of the basic concepts derived from the regional studies was that of the regional survey of potential land quality and landuse as a basic input to plans for economic development. It was L. Dudley Stamp who pioneered landuse studies in Great Britain.

British regional studies were largely influenced by the French geographer Paul Vidal de la Blache, and to a greater extent by the French sociologist Frederic Le Play. In Great Britain itself, the tradition of regional studies was carried forward by Sir Padrick Geddes, Andrew J. Herbertson, Charles B. Fawcett and Herbert J. Fleure.

Russian School of Landscape Science (Landschaftovednie)

Russian geography is focused on applied geography as landscape science which appears to have developed as a result of V. V. Dokuchaier's 'law of zonality'. His concept of natural zones transformed by man is synonymous with Otto Schluter's concept of 'kulturlandschaft'. Earlier, A. I. Voeikov recognised the influence of man upon the environment.

Landscape in the Russian school is defined as a dynamic system in which matter and energy are circulating and in which there are rhythmic (seasonal) changes of heat and water balance and biological productivity. A landscape is a combination of interrelated environmental components (local climate, landforms, soils, plants and animals) occupying a discrete unit of territory. It exists objectively in the natural environment.

The Russian landscape science is synonymous with regional science, which attempts to link economics, geography and planning and is concerned with theoretical and quantitative analysis of regional economies and problems. The concept of landscape science in the Soviet Union was applied to practical purposes of building the socialist economy and contributed to the economic regionalisation and economic development planning.

American School of Spatial Science

The American school of geography has passed through several concepts, and not confined to a particular concept as those of the German, French, the Russian and the British schools. The American school of geography was *initially confined to 'ontography'*, a term coined by W. M. Davis, which dealt with the correlation between inorganic environment and organic response. The concept of ontography, with the logical inclination towards environmental determinism, continued for several decades of the late nineteenth and early twentieth centuries.

The concept of ontography gradually lost its academic and professional value in the American geography and was replaced by *human ecology* (application of ecological concepts to the study of the relations between people and their physical and social environment); *chorology* (the study of areal differentiation); *historical geography* (the attention was focused on the processes or sequences of events that provided an explanation of observed landscapes); *and the functional organisation of space* (which sought for functional explanation for the observed differences from place to place on the Earth's surface). These explanatory procedures have been applied in various topical fields.

American geography during the postwar period witnessed an intense rivalry between Hartshorne and Schaefer on whether geography should be idiographic or nomothetic. However, the regional paradigm in the American geography has been replaced by topical specialism with emphasis on spatial science and relations.

2

Ancient Classical Geography

HIGHLIGHTS

- Greek Geography
- Roman Geography
- Chinese Geography

The history of geographical ideas is the record of humankind's effort to gain more and more logical and useful knowledge of the human habitat and of humankind's spread over the Earth–logical in that the explanations of things observed could be so tested and verified that scholars could have confidence in them; useful in that the knowledge so gained could be used to facilitate humankind's adjustment to the varied natural conditions of the Earth, to make possible modification of adverse conditions, or even to gain a measure of control over them. During the long record of humankind's efforts to gain knowledge about the Earth as a human habitat, there has been a *continuing interplay between things and events observed through the senses (percept) and the mental images of them (concept).*

Concepts determine what the observer perceives, yet concepts are derived from the generalisation of previous percepts. One concept that is manifested through the history of ideas since the earliest records is that of an ordered, coherent, and harmonious universe. With the concept of an orderly world firmly established, the perception of evidence to give it support was a natural consequence. During the thousands of years since the earlier records of the history of ideas, learned people have accounted for the order they perceived in the universe in different ways. The accounts range along a continuum from arbitrary rule by men like deities, through rule by a deity subject to law, through various kinds of cause and effect relations, to abstract mathematical law. These do not represent successive stages of increasing sophistication, for all of them can be found in the thinking of the ancient Greek philosophers.

GREEK GEOGRAPHY

The Greek scholars provided a framework of concepts and a model[1], or paradigm[2] of scholarly method that guided Western thinking for many centuries. Some of the Greek concepts appear to have retarded the Western thinking to such an extent that European paradigms in geography could not emerge until the influence of Plato and Aristotle had been overcome. Some of the basic concepts and paradigms of the contemporary geographical philosophy seem to reflect a strong

[1] *Model:* An idealised and structured representation of the real. Model may be regarded as a selective approximation which, by elimination of incidental detail, allows some fundamental, relevant or interesting aspects of the real world to appear in some generalised form paradigm.

[2] *Paradigm:* The working assumption, procedures and findings routinely accepted by a group of scholars, which together define a stable pattern of scientific activity; this in turn defines the community which shares in it. A paradigm is 'an accepted problem solution which by its very solution poses the next problem.' Progress in science is achieved by problem solving.

inclination towards the tradition of the ancient Greek scholarship. It were the Greek scholars who attempted to distinguish between *kenos* (meaning void) and *cosmos* (which, according to them, refers to the universe conceived as a system of harmoniously related parts). The foundation of the modern scientific geography appears to have been laid down by the ancient Greek scholars.

Although the roots of the ancient Greek scholarship in the development of geographical ideas reach back to the 'observations', 'measurements', and 'generalisation' of the ancient Egyptians, the Phoenicians and the Mesopotamians, yet its organisation in the form of concepts or paradigm was essentially the achievement of Herodotus, Plato, Aristotle, Eratosthenes and Strabo.

Among the ancient Greek philosophers, almost all of whom made contributions to the study of geography, the two basic traditions of geographic study can be found: the *mathematical tradition* and the *literary tradition.* The Greeks developed the science of astronomy. Scholarly writers also produced topographical descriptions of places in the known world, discussing both natural conditions and the culture and the way of life of the people living there (Holt-Jenson, 1981, 9).

People believed that the Greek geographers credited Homer with being the father of geography, because his great epic poem, the *Odyssey*, initiated the literary tradition in the contemporary Greek geographical thinking. But he did not define any concepts or paradigms which could be universally acclaimed. In fact, he tried to combine the literary tradition with the topographic description of places of the known world.

However, with the increase in knowledge and information available in the contemporary Greece, the Greek sailors of the eighth century BC were able to distinguish four kinds of winds and their directions. *Boreas* was the north wind—strong, cool, with clear skies; *Eurus* was the east wind—warm and gentle; *Notus* was the south wind on the front of an advancing storm—wet and sometimes violent; and *Zephyrus* was the west wind—balmy, but with gab force. However, in the second century BC, the Athenians built a tower distinguishing eight wind directions with sculpture depicting the weather types associated with each of them.

Much of the ancient Greek geographical scholarship owed its origin in the town of Miletus in lonia on the eastern side of the Aegean Sea, near the mouth of the Meander River. Although it emerged as a major commercial centre and attracted sailors and merchants with different cultural backgrounds, it was soon flooded with a fund of information. Between 770 and 570 BC, there was not only a flow of geographical information, but there was also a group of people with intellectual and academic integrity who speculated about how all this miscellaneous information could be assembled in some kind of meaningful arguments and concepts. To Miletus also came reports on Egyptian geometry, Sumerian algebra and Assyrian astronomy, which laid the foundation of the mathematical tradition in the contemporary Greek scholarship.

The first of the Greek scholars to be concerned about the measurement and the location of things on the face of the Earth was **Thales,** who lived during the seventh and sixth century BC. He tried to combine his mathematical and geometrical precepts with a most accurate scientific precision. Thus, he offered a true paradigm, the relevance of which still persists in modern mathematical geography and astronomy.

There are six geometric propositions credited to him: (1) the circle is divided into ten equal parts by its diameter; (2) the angles at either end of the base of an isosceles triangle are equal; (3) whim two parallel lines are crossed diagonally by a straight line, the opposite angles are equal: (4) the angle in a semi-circle is a right angle; (5) the sides of the two similar triangles are proportionate; (5) two triangles are congruent if they have two angles and a side respectively equal (Sarton, 1964, 171).

The most important contribution of Thales was his recognition that the solution of practical problems of measurement was less of an intellectual accomplishment than the rational generalisation of the specific solutions. He held the view that *water in various forms constituted the prime substance/material* from which all observable features of the Earth were made. He conceived the Earth as a disc floating in water. With available information and data, Thales set forth an explanation that could be checked by new observation and measurement and which sharply contrasted with the unscientific contemporary traditional explanations.

Another great scholar who lived in Miletus and who seemed to be a younger contemporary of Thales was **Anaximander**. Though he was a contemporary of Thales, yet the tradition which Thales set forth in the form of a paradigm hardly had any effect on the scholarship of Anaximander. Instead, he brought into the contemporary Greek world a Babylonian instrument called the *gnomon,* which made it possible to make a variety of observations regarding the relative positions of the celestial bodies vis-a-vis varying season-to-season position of the noon-shadow. This made it possible to establish the time of the solstice and the equinox. Anaximander prepared a map of the world to scale, which was probably based on the pictorial maps, drawn by the Sumerians of some of their cities as early as 2700 BC. Anaximander's world map showed Greece at the centre and the known Eurasian parts lying around. There was an ocean that encircled the world.

Anaximander seemed to have offered an explanation with regard to the Thalesian tradition visualising the Earth as a disc floating in water as the contemporary scholars of Miletus were puzzled by the question: how could the Sun go under the water? He opined that somewhere to the north there existed a very high and huge mountain behind which the Sun made the trip back again to the east. The shadow, cast by these mountains, would account for the night.

He offered an ontological explanation with regard to the prime substance of the universe. This made the observable objects of the Earth, and this explanation is in sharp contrast to the Thalesian tradition which maintained that water is the prime substance/material. The ontological explanation of Anaximander appeared to have laid the foundation of idealism[3] and abstraction in the geographical paradigms of the twentieth century. He coined the word *'apeiron'* to symbolise the prime substance which could not be experienced through senses, but became a concept—a specific mental image that by the process of deduction could become a real substance. However, contrasting explanations put forward by Thales and Anaximander as to whether water or apeiron should be regarded as the prime substance or vice-versa, ultimately led to an apparent dichotomy in the contemporary Greek thinking.

Hecataeus, who was born at about the time of the deaths of Thales and Anaximander, and who died about 475 BC, inherited the conflicting explanatory legacies of his predecessors. He seemed to have initiated the literary tradition in the ancient Greek scholarship. He offered a *topographical explanation* of the information brought to Miletus not only from the known world of the Greeks, but also from the shadowy world beyond the Greek horizon. His major creative genius was the Gesperiodos, or Description of the Earth which blends his literary genius with the topographical-ecological tradition, and thus set in *'a new geography'* which has persisted for some 2500 years. Hecataeus divided his work into two parts, each dealing with one of the regional divisions of the Earth. One book deals with Europe, the other with the rest of the world—Asia and Libya. He set forth a different approach in the contemporary geographical scholarship as he preferred to organise all the available information and knowledge about the known world in usable form.

The contrast in the approaches of these scholars of Miletus more than 24 centuries ago, represents the apparent dualism between those who seek to formulate generalisations and those who seek to describe unique things. In the modern geographical

[3] *Idealism*: A philosophy which regards reality as residing in or constituted by the mind (metaphysical idealism) or which limits human understanding to perceptions or external objects.

philosophy, these two points of view are described as *nomothetic,* meaning 'law seeking', and *idiographic,* meaning descriptive of particular things. Down through the ages since Hecataeus there have been scholars who insist that geographical study must adopt one or the other approach.

Different approaches put forward by Thales, Anaximander and Hecataeus, in fact, dominated the ancient Greek scholarship for a century until a new approach in the geographical methodology was developed by **Herodotus** who ridiculed the mathematical tradition in the geographical studies. Instead, he *preferred a historical approach/ tradition.* He studied history from a true geographical standpoint. He was credited with the very old idea that all history must be treated geographically and all geography must be treated historically. Geography, to him, provided the physical background, the stage setting, in relation to which historical events took on meaning. In fact, the paradigm which Herodotus set forth in historical geography—'the recreation of past geographies and the teaching of geographical change through time'—seemed to have a deep effect on the scholarship of Sir Halford J. Mackinder, a noted British geographer of the twentieth century, whose thesis on the Eurasian land power vis-a-vis the struggle between the Asiatic tribes and the European natives, clearly manifested the historical tradition of Herodotus.

The contributions of Herodotus to geography were based on his own personal observations during many years of travel. Herodotus was of the view that the observational statements were the only ones which made direct reference to phenomena in the real world. During the course of his many years of travel across the Black Sea, the Russian steppes, and the Persian Empire, Herodotus probably came into contact with the people of varied cultural traits whom the Greeks had not met earlier. He observed the way of living of those people and portrayed their cultural traits and life style. He drew up the conclusions on the basis of direct observation, and in a sense he attempted to combine his concept with a most accurate percept regarding the cultural realm of the non-Greek people. Because of his vivid portrayal of the cultural traits of people strange to the Greeks, he is known as the *father of ethnography.*

Herodotus incorporated older sources of geographical information, including the existing maps, and he identified and criticised his sources. In this way, he handed on knowledge of earlier theories and description which would otherwise have been forgotten. He used the methods of historical geography to support the hypothesis that the Nile mud, deposited in the Mediterranean, had built the delta. He reconstructed the ancient shoreline: and showed that many former seaports were now far inland. He also pointed out the physical process of delta-building that occurred in many places. He not only described the regularity of the summer floods of the Nile, but also attempted to explain them.

Like all the Greek scholars, Herodotus accepted as a fundamental principle that the world must be arranged symmetrically. He accepted the Homeric view of the Earth as a flat disc over which the Sun travelled in an arc from east to west. Herodotus was the first scholar to have drawn a meridian on the world map. The meridian was drawn from Egypt to Cilicia (south coast of Turkey), peninsula of Sinope and the mouth of Ister (Danube). He had no interest in those mathematical and astronomical problems which later became associated with geography—measurement of the circumference of the Earth and the establishment of exact location for places.

With the concept of an orderly universe firmly established, the perception of evidence to give it support was a natural consequence, and it was during the time of Herodotus itself that an account of an ordered universe required the recognition of cause and effect sequences that occurred in accordance with some law. **Plato** (428–348 BC), who made important contributions to the development of geographical ideas, *offered a different interpretation of the cause and effect sequences.* He conceived of the world as having been created in perfection, but which is now in the process of decline from perfection. He developed the deductive procedures, and insisted that the observable things on the Earth were only poor copies of ideas, or perfect predicates from which observable things had degenerated or were in the process of degenerating. Plato's deductive procedures seemed to have been a major driving force in the growth of the natural sciences vis-a-vis geographical paradigms that developed in the mid-nineteenth century and continued upto the present century.

He pointed out with examples the decline of Attica in Greece, which once possessed a very productive soil, capable of supporting a huge population, and was equipped with immense forest and water resources to feed its occupants, but the relentless exploitation of these resources by its occupants caused destruction and degeneration of what existed earlier. Arguing from the general theory to the particular situation in Attica, Plato used this as an example of the degeneration of things from their original perfect state. If Plato had argued from the particular to the general, he might have realised that men made changes in the land they occupied and that soil erosion and land destruction were parts of cultural history and were repeated in many places.

He accepted the idea that *symmetry of form was one of the attributes of perfection,* and the most completely symmetrical form was a sphere. Plato seemed to be *the first philosopher to have given the concept of a round Earth* located in the centre of the universe with the celestial bodies in circular motions around it. Pythagoras, who lived in the sixth century BC, calculated some of the mathematical laws for the circular motions of the celestial bodies. Parmenides, who was a disciple of Pythagoras, applied these to observations made from the surface of a round Earth. Eudoxus of Chidus, a contemporary of Plato, developed the theory of zones of climate based on the increasing slope away from the Sun on a spherical surface. All these formulations were deductions from pure theory—the theory that all observable things were created in perfect form and that the most perfect form was a sphere.

The contribution of **Aristotle** (384–322 B.C.) had far-reaching impact on the development of geographical ideas. Like Plato, he also believed in an ordered universe and *offered a different interpretation of the cause and effect sequences.* He conceived of the world as striving towards perfection. He was the father of the teleological concept, which sees the universe planned by its creator. His *teleological explanation* seemed to have motivated the perception of the great German geographer, Karl Ritter of the classical period.

At the age of seventeen, Aristotle joined Plato's Academy near Athens and remained there until Plato's death, at which time he was thirty-eight years old. During the next 12 years, he spent his time travelling widely throughout Greece and around the shores of the Aegean Sea. In 335 BC he set up his own school, which he named the *Lyceum.* Aristotle, who *developed the inductive procedures,* preferred to formulate his concepts as generalisations of empirically observed 'facts'. He insisted on the importance of direct observation, and built the theory by reasoning from the particular to the general. Instead of making logical deductions from theory, he taught his pupils to observe things.

Aristotle formulated *four fundamental principles of scientific explanation*—that is, of: (a) answering the question; (b) finding what makes the thing the way it is, and that includes a description of its nature and essential characteristics, specifying the kind of matter, and the substance out of which it is made; (c) telling what caused the process through which the thing becomes what it is; and finally (d) the purpose the thing fulfils. *Aristotle felt that things were in the process of physical change leading to a final perfect state.* This model for scientific explanation constituted the world's first paradigm for the guidance of scholars for the succeeding generations.

He developed the *theory of natural places* and distinguished between the celestial space and the Earth space. In the former, ether was the prime substance, while the latter had four basic substances: Earth, water, fire and air. Aristotle insisted that mathematics could be used to explain the process of change that made things as they were, but it failed to answer the fourth question concerning the purpose. There is a tendency towards perfection that all things are not declining from an ideal state, but rather are developing towards an ideal stat. He accepted Plato's concept of a spherical Earth and began to seek explanation of it and to test the concept with observations. His explanation was derived from the theory of natural places.

Aristotle is credited with the concept of the *varying habitability of the Earth with differences of latitude.* The habitability, according to him, was the function of distance from the equator. He assumed that the parts of the Earth close to the equator (i.e. the torrid zone) were uninhabitable; that the parts of the Earth far away from the equator (i.e. the frigid zone) were constantly frozen and thus not habitable; and the temperate zone in between constituted the habitable part of the Earth. He visualised the existence of a south temperate zone which could not be approached from Greece because of the intense heat of the torrid zone (Glacken, 1956).

In *Politics*, Aristotle presented a model of the ideal or perfect state. He fully recognised the theoretical limitations involved in this, and thus spoke in terms of 'approximation' of the ideal state. *He introduced many of the ideas that later became important concepts in the field of political geography.* These include notions of ideal sizes of population and area for political viability and their relationship to changing technology, the distributional characteristics of the resident population, the locational and morphological problems of the capital city, including strategic and economic considerations, boundaries versus frontiers as limits of national space, and also notions of co-existence and interdependence within a larger system, both in the Hellenic sphere and, under the assumptions about the impact of temperature on political behaviour, in a global pattern (Kasperson and Minghi, 1970, 233).

Aristotle's paradigm for scientific explanation did not include anything about controlled experiments or the verification of premises, but only *the use of logic to formulate and give support to theory.* Some of his logical explanations appeared to be so unassailable in the fourth century BC, and have been so universally accepted for many succeeding generations since that time that his influence on the history of European ideas has been far-reaching and enormous. It has been rightly pointed out that modern science could not appear until Aristotle had been abandoned.

Aristotle' s inductive procedures which favoured the formulation of concepts as generalisations of empirically observed facts had a deep impact on Alexander the Great. In course of his world conquest, in the east beyond the Greek zone of influence, Alexander the Great observed the areal expression of different spatial phenomena of different lands he crossed. He was only a short distance from the eastern limit of the habitable part/*ekumene.* His conquests pushed back Greek knowledge of the Earth as far as the Indus. As a result, he sent back to the Greek world a wealth of new observations concerning what it was like beyond the Greek horizon and how far and in what direction it was necessary to travel to reach these strange places.

Hippocrates (460–376 B.C.) contended in his book *On Airs, Water, Places* that climate provided explanations for the physical and intellectual differences among people. Stressing the correspondence between the physical environment and national character, he paid particular attention to the intermediary role of human occupation. Hippocrates probably produced the world's first medical geography.

The voyages of **Pytheas** far to the north-west of the Greek world in western and northern Europe sometimes between 330 and 300 BC were significant in the sense that he went very close to the northern limit of the ekumene/habitable world. He brought with him a wealth of new observations concerning the customs and food habits of the people of Great Britain, the use of barns to thresh grain in wet weather and the change in the character of agriculture from south to north in Britain. He also pointed out the existence of an ice sea lying to the north of the land he visited. He was also the first Greek to inform about ocean tides and explained that *tides were related to the phases of the Moon.*

He probably sailed along the eastern shore of the North Sea, as far as modern Denmark and reported on the existence of a place called Thule. He is quoted as saying that at Thule the Sun remained above the horizon during the whole of the longest day, which would put this place well to the north in Norway or possibly Iceland. It is certain that Pytheas observed the angle of the Sun's shadow on a gnomon before he left for the voyage, and the latitude derived from this measurement was almost exactly correct. Although the Greek scholars of his time dismissed his observation as pure fantasy, yet modern scholars now believe that much of his observations and reports were correct. Now-a-days he is given a due place among the great explorers of all times.

Eratosthenes (276–194 B.C.) is rightly identified as the 'father of geography' because he was the first to coin the word 'Geography' and set forth a stamp on the study of the Earth as the home of humankind and that still persists. He succeeded in calculating the circumference of the Earth with remarkable precision. He may be regarded as the first scientific geographer who upheld the mathematical tradition in geography which was introduced by Thales much earlier in the seventh and sixth centuries BC.

With the help of an indigenous apparatus known as *gnomon (introduced to the Greek world by Anaximander),* Eratosthenes made two separate observations of the position of the Sun at the time of the summer solstice. One observation was done near Syene (Aswan). At this place there was a deep well, and at the bottom of the well, at the time of the summer solstice the image of the Sun was reflected in the water. This strange occurrence used to attract many tourists in ancient times. This meant that on that date the Sun was directly overhead. The second observation was done outside the museum in Alexandria, where there was a tall obelisk. Using the obelisk as a gnomon, he measured the length of the shadow at the time of the solstice. He was thus able to measure the angle between the vertical obelisk and the rays of the Sun.

With these two separate observations of the positions of the Sun at the time of the summer solstice in mind, Eratosthenes proceeded with the famous theorem of Thales that: 'When a diagonal line crosses two parallel lines, the opposite angles are equal.' The parallel lines were formed by the parallel rays of the Sun. The rays of the Sun at Syene, which were vertical, might be extended to the centre of the Earth. Also, the obelisk which was vertical at Alexandria could be extended to the centre of the Earth. Then the angle between the Sun's rays and the vertical obelisk at Alexandria must be the same as the opposite angle at the centre of the Earth.

The next question was: How much of the whole circumference of a circle is subtended by the angle at the centre of the earth? He measured this as one-fiftieth of the whole circumference. It was then only necessary to fill in the distance between Syene and Alexandria which was equivalent to about 500 miles (5000 stadia), and then multiply this distance by 50. Eratosthenes, therefore, found out that the whole Earth was about 25,000 miles in circumference (actually the circumference measured through the poles is 24,860 miles).

Eratosthenes pointed out that Alexandria was due north of Syene whereas, in fact, it is about 3°W of Syene. The length of the road between Syene and Alexandria, which was according to the Egyptian calculation 500 miles, is actually 453 miles, and Syene is actually at 24°5'N, a little north of the tropic of cancer. But all these errors cancelled out so that the resulting calculation was surprisingly close to the correct figure (Thomson, 1965, 159–61). Eratosthenes also attempted to determine the distance of the Sun and the Moon from the Earth. He concluded the distance of the Moon at 780,000 stadia (78,000 miles) and that of the Sun 40,00,000 stadia (4,00,000 miles).

He produced a book on the habitable part or the ekumene of the Earth and in the preparation of the book he very much relied on the assumption and accounts of Aristotle and Pytheas respectively.

He accepted the major division of Europe, Asia and Libya. He *provided mathematical boundaries to the five major climatic zones*—a torrid zone, two temperate zones and two frigid zones. The torrid zone was 48 degrees of the whole circumference (24 degrees north and south was calculated as the location of the tropics). The frigid zones extended 24 degrees from each pole. The temperate zone was between the tropic and the polar circles. His ekumene extended from Thule, near the Arctic circle in the north to Taprobane (Ceylon) in the Indian Ocean in the south and from the Atlantic Ocean to the Bay of Bengal which he thought to be the eastern limit of the habitable part.

Eratosthenes prepared a world map in which he made use of a frame of north-south and east-west lines, but these were not spaced regularly, In fact, he used the meridian of Alexandria extending it southward through Syene and northward through Rhodes and Byzantium, as the prime meridian. He also used the latitude of the Pillars of Hercules, which he assumed to have passed through Rhodes (Sarton, 1964, 106–08). He prepared his world map based on this frame of the north-south and east-west lines. Of equal importance was his development of systems of coordinates for the world, i.e. latitude and longitude, which he used in order to locate places and to measure distances. Geographical order now replaced the casual and unsystematic location measurements of earlier times. Eratosthenes' cartographical work was later developed by his students and successors at the museum in Alexandria.

Hipparchus succeeded Eratosthenes as the chief librarian when the latter died. The dates of Hipparchus' birth and death are not known, but if chronology is to be believed, then he was working at the library in 140 BC. He invented an instrument which was easier to handle than the gnomon. It was *astrolabe* which consisted of a circular dial marked off into 360 parts with a rotating arm fixed at its centre. Hanging on the rigging of a ship, the astrolabe made possible the measurement of latitude at sea by observing the angle of the polestar.

Hipparchus upheld the mathematical tradition of the Greek geographical idea which was introduced in the ancient Greek scholarship by Thales and carried on by Eratosthenes. Hipparchus was more of a mathematician and astronomer than a geographer. However, *he is identified as the first Greek scholar to have established the exact position of every point on the Earth's surface,* and divided the circle into 360 degrees, based on Assyrian arithmetic. He defined a grid of latitude and longitude lines. To him, the equator was a great circle that divided the globe into two equal parts, and the meridians which were drawn converging on the poles were also great circles. The parallels, on the other hand, became shorter and shorter as they approached the poles. Since the Earth makes one complete revolution in 24 hours and there are 360 meridians drawn from equator to poles, each hour the Earth turns through 15 degrees of longitude.

Hipparchus is *credited with showing the curved surface of Earth on a flat surface* as he devised the projection. He introduced to the Greek world two kinds of projection: *the stereographic projection* and the *orthographic projection.* He pointed out how to construct a stereographic projection by laying a flat parchment at a tangent to the Earth and extending the latitude and longitude lines from a point opposite the point of tangency; and an orthographic projection could be made by projecting the lines from a point of infinity. He also made it clear that on the stereographic map, the central portion would be too small in relation to the periphery, and on the orthographic map the central portion would be too large. However, both the projections, according to Hipparchus, would only show a hemisphere, not the whole world.

Upholding the mathematical tradition in geography, Hipparchus pointed out that it could be made more meaningful and precise through plotting of locations in the theoretical grid, and the instrument he invented to be more useful for its accurate results. In order to determine longitude, he pointed out that the local times of the start of an eclipse at different places could be compared and the time differences would provide a measure of longitude. But no such system of coordinated observations was even attempted for thousands of years after his time. Geography, during his time, became more mathematical and technical, and astronomy became the pivot of the discipline.

The history of ancient Greek scholarship would remain incomplete if the contributions of **Posidonius** are not mentioned. Posidonius lived shortly before the time of Christ. It was he who recalculated the circumference of the Earth and arrived at a much smaller figure than that of Eratosthenes. He observed the height above the horizon of Canopus (a star of the first magnitude) at Rhodes and Alexandria, which he assumed to be on the same meridian. He then estimated the distance between them based on average sailing time for ships. The figure he arrived at, for the circumference of the Earth, was 18,000 miles. He greatly overestimated the west to east distance from the western most part of Europe to the eastern extremity of the ekumene, then thought to be occupied by India. It is believed that Columbus used the smaller circumference estimated by Posidonius.

Posidonius assumed that the highest temperature and the driest deserts were located in the temperate zone near the tropics and the temperatures near the equator were much less extreme. He thus contradicted the view of Aristotle who had pointed out earlier that the equatorial part of the torrid zone was uninhabitable because of heat. However, the conclusion of Posidonius with regard to the habitability of the equatorial part of the torrid zone was purely a creation of his mental image, as he had no access to credible reports from anyone who had crossed the Sahara. The Sun pauses longest near the tropics and is overhead for a much shorter time at the equator. It is interesting to note that the incorrect estimate of the circumference of the Earth which Posidonius recalculated was widely accepted by those who upheld the mathematical tradition after him, but his correct assumption on the habitability of the equatorial land was not recognised by those who followed him.

ROMAN GEOGRAPHY

Much of the Greek traditions in geography were carried forward into ancient Roman scholarship. The Romans produced little that was new in the field of geography. Writing shortly before the time of Christ, one Marcus Terentius **Varro** prepared a compendium of geography that dealt with a theory of culture stages that remained almost unchallenged until the nineteenth century. Varro describes man's culture as moving through a regular sequence. Originally man derived his food from the things that the virgin Earth produced simultaneously. From this original state man advanced through a stage of pastoral nomadism, then through an agricultural stage and finally to the stage of contemporary culture.

Strabo (63 BC–24 AD), a Greek scholar and traveller writing at the height of the Roman Empire, was strongly influenced by the historical topographical tradition of Homer, Hecateaus, and possibly of Aristotle. It is often said that the historical tradition, introduced by the Greek scholars in geographical scholarship was finally summarised by Strabo. His assertion on the geographer is as follows: 'Accordingly, just as the man who measures the Earth gets his principles from the astronomer and the astronomer from the physicist, so, too, the geographer must in the same way take his point of departure from the man who has measured the Earth as a whole, having confidence in him and in those in whom he, in his turn, had confidence, and then explain, in the first instance, our inhabited world—its size, shape, and character, and its relation to the Earth as a whole; for this is the peculiar task of the geographer. Then, secondly, he must discuss in a fitting manner the several parts of the inhabited world, both land and sea, noting in passing wherein the subject has been treated inadequately by those of our predecessors whom we have believed to be the best authorities on these matters' (Strabo, trans. Jones, 1917, 429–31).

Strabo accepts Aristotle's zones of ekumene, as defined by Eratosthenes, and then goes on to assert that the limit of possible human life towards the equator is at latitude 12°–30' N, but does not explain the basis of his assertion. He also places the northern limit of the habitable Earth, where cold is the limiting factor, only 400 miles north of the Black Sea. Strabo accepts the calculation of the Earth's circumference done by Posidonius in about 100 BC. He gives a correct explanation of the floods of the Nile, attributing them to the heavy summer rains in Ethiopia (James and Martin, 1981, 36). Drawing from Hipparchus and Posidonius, Strabo concludes the geographer's need of a sound mathematical base, because that would give him an accurate description of the parts of the ekurnene.

Strabo wrote a work of 17 volumes called *Geographia*. This was largely an encyclopaedic description of the known world whose chief value was that it preserved for posterity many writings which he annotated and cited. *Geographia* also included attempts to explain cultural distinctiveness, types of governments and customs in particular places. The significance of natural condition for cultural development was discussed in relation to a number of places, especially in the description of Italy (Holt Jenson, 1981, 11).

The first two books of Strabo's work provide historical review of the progress of geography from the days of Homer whom be described as the founder of all geographical knowledge. He also discusses in detail the contributions of Eratosthenes, Hipparchus and Posidonius. He devotes eight books to Europe, six books to Asia, and one book to Africa, most of which deals with Egypt and Ethiopia.

Strabo's greatest contribution lies in the field of political geography, and in this field he carried forward the Aristotelian tradition. In his *Geographia*, he examines the pre-requisites for the successful functioning of larger political units and concludes that a strong central government and a single head of state are essential ingredients for success. Given the political environment of the times, it is not surprising that Italy, the core of the Roman Empire with its location in the then-known world in the Mediterranean, its 'ideal' climatic regime, and its balance of mixed resources, was best suited to fill this role. Again, ethnocentrism coupled with inherited deterministic notions concerning environmental political relationships account for these conclusions.

The period that followed Strabo's monumental work, *Geographia*, witnessed the compilation of extensive sailing materials for the guidance of ship captains and described the coastlines and ports with considerable detail and accuracy, such as the Periplus of Scylax for the shores of the Mediterranean and the *Periplus of Arrian* for the shores of the Euxine (Black Sea). The most complete work of this kind was an anonymous one that offered a guide for navigators and traders who covered the Red Sea, the east coast of Africa as far as Zanzibar (more than 6 degrees south of the equator), and the northern side of the Indian Ocean as far as the southern end of the Malabar coast in India. This was the famous *Periplus of the Erythraean sea,* which Bunbury dates as some 10 years after the death of Pliny in 79 AD. The merchants and sailors of the first century after Christ who had no access to the accounts of Aristotle, Posidonius, or Strabo, were happily not conscious of the horrible fate that would come to those who ventured within 120 of the equator or the impossibility of maintaining life in this central part of the torrid zone.

The mathematical tradition in ancient geographical scholarship, a heritage that the Greek scholars carried forward into the successive periods of academic innovations since Thales, really comes to an end with the monumental work of **Ptolemy** (Claudius Ptolemaus, AD 90–168). Much of his geography, which he claims to be *a science of the art of map-making,* seems to have been borrowed from the earlier works of Aristotle, Hipparchus, Posidonius and his teacher, Marinus of Tyre. Ptolemy tried to rationalise the earlier Greek geographical and astronomical ideas into organised categories preparing them in a scientific manner. His great work on classical astronomy—the Almagest—which long remained the standard reference work on the movements of celestial bodies, reveals his indebtedness to Aristotle, Hipparchus and Posidonius.

Ptolemy's concept of the universe is a mere reflection of Aristotle's tradition that the Earth was a sphere that remained stationary in the centre, while the celestial bodies moved around it in circular motions. This remained the accepted notion until the time of Copernicus in the seventeenth century. He adopted the grid of latitude and longitude lines developed by Hipparchus, based on the division of the circle into 360 parts. Every place, therefore, could be given a precise location in mathematical terms. His biggest mistake was to underestimate the size of the Earth, rejecting almost correct estimate of Eratosthenes in favour of Posidonius who recalculated the circumference of the Earth. The calculation of longitude in those days was only possible by estimating the length of journeys from one place to another and many of Ptolemy's locations of places were found to be incorrect. Therefore, each listing of latitude and longitude was, in fact, only a selection among estimates.

Ptolemy not only favoured a smaller estimate of the Earth's circumference, but also erroneously increased the eastward extension of the land area. Because calculations of longitude had to be based on rather unreliable travel distances, his otherwise accurate map of the then world included too many degrees of longitude. The map extended from a prime meridian in the Canary Islands to a 180 degree meridian which crossed inner China. (Holt-Jenson, 1981, 11). Ptolemy followed Marinus in taking his prime meridian a north-south line.

In the art of map making, the great contribution made by Ptolemy was his modification and improvement over the previously drawn maps and this he did by *adopting a projection for the world map showing the graticule of latitudes and longitudes.* He was, in fact, far ahead of his predecessors with regard to the mathematical construction of projections. He represented the equator and the latitudes by parallel curves, and the meridians as straight lines, cutting the equator at right angles that converge at a point (pole) situated beyond the limits of the map. Subsequently, he reduced the meridians also to a curved form, so as to make them correspond more nearly with reality. The map on which graticule of longitudes and latitudes were drawn was not a perfect hemisphere.

After the completion of the Almagest, Ptolemy produced a *Guide to Geography*, consisting of eight volumes. The first was a discussion of map projections together with a few corrections of the data from Marinus based on actual astronomical observation which he had himself done. Books II through VII contained tables of latitude and longitude. Book VIII contained maps of different parts of the world based on the gazetteer. Ptolemy rejected the hypothesis of Hecateaus, Herodotus and Strabo who believed that there was an ocean to the east of Asia. Strabo who pointed out that Sera and Sine (in China) were merely the most easterly points in Asia, just as Agisyamba was the most southerly. He also indicated on his map that the Indian Ocean was closed by land on the south, an idea he probably borrowed from Hipparchus. However, Ptolemy accepted Aristotle's concept that the parts of the Earth near the equator were uninhabitable due to heat. It is interesting to note that Ptolemy accepted the incorrect estimate of the circumference of the Earth done by Posidonius, but he failed to appreciate Posidonius' correct belief concerning the habitability of the equatorial region.

James and Martin have rightly observed: 'With the death of Ptolemy the geographical horizons that had been widened both physically and intellectually by the Greeks closed in again. It was many centuries before the effort to describe and explain the face of the Earth as the home of man again attracted the attention of scholars'.

Summing up

Undoubtedly, the ancient Greek scholarship laid the foundation of scientific geography, as important paradigms were set forth in the successive periods of academic innovation, based on the empirically observed facts. It was the Greeks who developed the procedures we describe as the scientific methods. If observed facts differ with general principles, the principles must be revised. The Greeks developed the science of astronomy. This was a tremendous step forward in the whole history of geographical ideas. The Romans, especially Ptolemy, made major improvement over the previously drawn maps with greater precision and accuracy. An apparent dichotomy seemed to have developed in the ancient Greek geographical scholarship between those who favoured mathematical tradition and those who sought the literary-historical tradition. Dualism in contemporary geographical ideas also came into being on account of contrasting approaches between those who tried to formulate generalisation and those who sought to describe unique things.

CHINESE GEOGRAPHY

When ideas pertaining to the geographical paradigms and models were developing in Greece and Rome, exactly at the same time, another major centre of geographic study in remote China was making rapid progress. Essentially, the Greek and the Chinese intellectual worlds remained isolated, each discovering the other step by step. Yet there are certain fascinating parallels in concepts and methods of study that seem to require the existence of contacts, however indirect and remote. *Chinese geographical work was based on the development of methods for making accurate observation and for using these in the construction of useful theories and paradigms.* The Chinese mathematicians had discovered the use of zero and had developed the decimal system, which was vastly superior to the sexagesimal system of Mesopotamia and Egypt. The decimal system was introduced in Baghdad about 800 AD from the Hindus, but the general belief is that the Hindus probably learned their decimal system from the Chinese.

The ancient Chinese scholars bad a basically different attitude towards the natural world than that held by their Greek counterparts. *An individual was conceived of apart of nature.* The ancient Chinese philosophers never held the teleological view and as such the Chinese believed that there was no law-giving deity who created the universe for man's use in accordance with a pre-conceived plan. The Chinese geography even in the ancient classical age was *concerned more with observable things and processes then with the formation of theory.* Here one can find a similarity with the deductive procedure of Aristotle. There are a number of weather reports, prepared on the basis of empirically observed facts, now preserved in the national archives, which probably date back to thirteen centuries before Christ.

The oldest piece of geographical writing is a survey of the resources and products of the nine provinces into which China was divided in the fifth century BC. For each province, the nature of the soil, the kinds of products, and the waterways that provide routes of transportation are described. By the fourth century BC, the Chinese had learnt about the nature of the hydrological cycle, and the Chinese philosopher Mengtzu (Mencius), a contemporary of Plato, who lived two centuries after Confucius, developed an identical view regarding the spatial consequences of deforestation and forest cutting, and maintained that forests once cleared from mountain slopes could not re-seed themselves as long as the slopes were grazed by cattle or goats. Chang Heng, in the second century after Christ, introduced the grid system in China.

In the second century BC, the Chinese engineers made accurate measurements of the silt carried by the rivers. In AD 2, the Chinese carried out the world's first census of population. Other technical inventions included the making of paper, the printing of books, the use of rain and snow gauges to measure precipitation and the use of the magnetic compass for navigation. Phei Hsiu *is identified as the father of Chinese cartography.* In AD 267, he produced a map of politically organised part of China on eighteen rolls of silk. To make the surveys on which the map was based, he measured several base lines and then located places distant from the baselines by the intersection of lines of sight, just as the Egyptians had done long before. Phei Hsiu made use of a grid of east and west and north-south lines crossing at right angles to provide the frame on which to plot the rivers, the coastlines, the mountain ranges, the cities and other features. The source of his idea of using triangulation to locate places and of using a grid of lines has not been established. It can be assumed that many of the ideas of triangulation might have been devised much earlier in China and diffused westward.

The Chinese travel books, consisting of the graphics of the geographical explorations, were probably compiled sometimes between the fifth century and third century BC. The books known as the *Travels of Emperor MU*, who ruled from 1001 to 945 BC, were found in the tomb of a man who ruled part of the Wei valley about 245 BC. There are evidences to show that because of the bad conditions of the books they were recopied as late as the third century BC. The books can be compared with the Odyssey of Homer, because they also reveal the story of high adventure.

There can be no doubt that at a very early date, the Chinese travellers had gone far beyond the original culture hEarth in the Wei Ho valley. The Chinese geographer Chang Ch'ien is believed to have discovered the Mediterranean civilisation in 128 BC. His book deals with the land route across inner Asia to Bukhara and then to Persia and the Mediterranean shore. It also describes that on this route traders travelled regularly and had probably made contacts with the West long before the West was discovered (Thomson, 1965, 177). It is interesting to note that neither Chang Ch'ien nor Hipparchus mentioned each other when the route to the Mediterranean across the Central Asiatic landmass was discovered about their time. This raises some doubts as to whether the Mediterranean civilisation was discovered by the Chinese about 128 BC.

Ancient Chinese geographers seem to have clear-cut views about nature and had a basically different attitude towards the natural world than that held by the Greeks. Ancient Chinese geography also represents two basic traditions of geographic study: the literary-historic topographic tradition, and the mathematical tradition. While the Greek philosophers stuck to the teleological concept, the Chinese basically relied on the mechanical explanation. They preferred to formulate their concepts as generalisations of empirically observed facts. The Chinese maintained perfection in the art of map making and they developed the idea of triangulation to locate places and of using a grid of lines, probably much earlier than their Greek counterparts in the West.

3

Medieval Geography and Renaissance

HIGHLIGHTS

- Geography in the Christian World
- Arab Geography
- Chinese Geography
- Impact of Voyages, Discoveries and the Renaissance

The medieval times, especially in Europe, were a dark period for the development of science and geographic ideas. At best, scholars made accurate, but sterile copies of the works of the ancients, rejecting anything which did not conform with the dogmas of the Church. Such an intellectual environment stifled any development of critical scientific analysis. Concepts of the world which had been developed in ancient times were reshaped to conform to the teachings of the Church. The Earth became a flat disc with Jerusalem at its centre.

However, it was only in the Arab world and the Chinese world that the tradition of the geographical scholarship on various aspects flourished. The journeys undertaken by the Arabs and Chinese scholars during the middle ages made significant contributions to the development of geographical knowledge. This was in contrast to the voyages and journeys undertaken by the Europeans during this period which made no significant contribution to the development of geographical concepts, models or paradigms. Around 1000 AD, the Norsemen sailed across the Atlantic to Greenland and North America, but the sagas of these voyages and journeys were only passed on by word of mouth and written down long afterwards in isolated Iceland. Some of these discoveries took a long time getting back to the rest of the world.

GEOGRAPHY IN THE CHRISTIAN WORLD

In the early medieval period, the European scholars could work only with Latin documents and only in the latter part of this period did a few of them master the Arabic language. The Greek materials remained entirely unknown, except in tradition. Geography in the Christian world was confined to the process of translation of the earlier geographical ideas of the classical age. Two medieval scholars—Martianus Capelle and Ambrosius Theodosius Macrobius—provided translation of Plato as early as the fifth century. Through their accounts, the medieval Christian scholars had come to know about the concept of a spherical Earth. In 1138, Plato of Tivoli translated that work of Ptolemy from Arabic into Latin which dealt with the effect of the positions of the celestial bodies on human affairs. In 1175, Gerard of Cremona translated Ptolemy's 'Almagest' into Latin. In fact, Ptolemy's geocentric model of the celestial universe remained the accepted model for many centuries, and most of the ideas still used by astrologers can be traced back to him (James and Martin, 198l, 42). Ptolemy's works were more popular in the medieval Christian world.

It was Albert Magnus who for the first time translated the geographic ideas of Aristotle from Arabic in the twelfth century, in which he gave emphasis on the Greek concept of equating habitability with latitude. Albert added few more explanations concerning the concept of ekumene that those who lived close to the limits of the habitable Earth turned black, but if dark-skinned people moved into the temperate latitudes they would gradually turn white. Isidore of Seville compiled a sort of geographic encyclopedia in the seventh century.

The *medieval Christian world witnessed the gradual deterioration of mapping* as the period lost the viability of showing alignments and boundaries with precision and accuracy. This had been the case also in the classical period when greater attention was given to accurate delineations of the coastlines. In fact, the medieval period witnessed the so-called *T-O* maps. The inhabited world was shown by a circular figure bounded by the ocean. The figure was 'oriented' towards the east. In the midst of the area, was a T-shaped arrangement of waterbodies. The stem of *T* represented the Mediterranean. The top of the *T* represented the Aegean and Black Sea on the one hand, and the Nile River and Red Sea on the other. Three divisions—Europe, Asia, Africa—were accepted as standard. The centre of the inhabited world, just above the centre of the *T*, was Jerusalem. At the far east, beyond the limit of the inhabited world was paradise.

It was during the crusades, that occurred between 1096 and 1270 when Christian Europe attempted to recapture the Holy Sepulchre at Jerusalem from the Arabs, that European men drawn from various parts of the continent brought with them exciting stories about strange people and strange landscapes beyond the geographic horizons.

However, the **Polo brothers** and the son of one of them, **Marco Polo**, did make valuable contributions to the Christian world about the Asian landmass. The Polo brothers visited China and stayed there from 1260 to 1269. They went across the great deserts and inhospitable surface of central Asia to reach China. In course of their sea voyages, they touched the shores of Java and Sumatra, and then continued to Ceylon and southern India. They moved along the west coast to the ancient port of Hormuz on the Persian Gulf. The Polos finally returned safely to Venice in 1295. They brought with them varieties of information, including the land and people of central Asia vis-a -vis China, the eastern limit of the habitable part, the nature of the coastline and the people living thereon. Their accounts discredited the Ptolemian view that beyond Sera and Sine (China), there lay the unknown land (Terra Incognita).

Marco Polo who also went with the expedition with his father and uncle to the eastern—most limit of the then assumed ekumene—also provided descriptions of life in China and of the perils encountered on the route to and from China. He also reported on Japan, and on the island of Madagascar. In Madagascar, he found ample evidence that the torrid zone was not torrid but was in fact inhabited. This empirically observed fact also discredited Aristotle who doubted the habitability of the torrid zone near the equator. It is a matter of dispute whether Marco Polo had known the concept of Aristotle, though in the twelfth century, Alberto Magnus had made the geographical ideas of Aristotle available to the Christian world. Marco Polo's book was thought to be a great record of geographic exploration, filled with much information and interesting stories.

Towards the latter part of the medieval period, a few scholars actually thought of the need to confront authority with reason and explanation. William of Corches, who died about 1150, portrayed a universe governed by law rather than by the unpredictable acts of the divine authority. He presented some remarkably modern ideas concerning the heating of the atmosphere from below and the formation of clouds by the cooling of air (Kimble, 1938, 79). Robert Grosseteste, the Bishop of Lincoln, contradicted the concept of a torrid zone that was uninhabitable with the availability of Arabic reports of an inhabited east coast of Africa extending at least as far as 20° S.

In the early fifteenth century, Cardinal Pierred Ailly published a book *Tractatus de imago mundi*, which represented a kind of summary and review of the period. In 1414, he made use of the Latin translation of Ptolemy's *Geography*, and accepted the smaller estimate of the Earth's circumference. He was among the first to insist that India could be approached by sailing west, which influenced Columbus to do just this. However, he contradicted Ptolemy's contention of an enclosed Indian Ocean and offered different opinions concerning the habitability of the equatorial torrid zone. Pope Pius II (Aeneas Silvius) wrote a book sometimes between 1458–64 on Europe and Asia in which he tried to show that ekumene was also extended across the torrid zone, and the Indian Ocean was not enclosed on the south.

The late fourteenth century witnessed a gradual recovery from the state of deterioration in the field of mapping as was the situation in the early period of the medieval Christian world. There had been a notable improvement in the art of map making. The *Portolano charts* became standard equipment for the sea captain. These charts were covered by a network of overlapping lines radiating from several centres in different parts of the chart. The radiating lines conform to the eight or sixteen principal directions of the compass, each corresponding to a wind direction. Sailors laid out compass courses along these lines. With the lines to indicate direction from key points, the coastlines, especially around the Mediterranean, could be drawn with considerable accuracy. The famous *Catalan map of the world,* made in 1375, incorporated the material for numerous Portolano charts. It also included the west coast of Africa, south of cape Bojador, which had not then been reached by European sailors. It also showed east and south-east Asia based on reports of Marco Polo. This was the first map ever to give a proper outline of Ceylon and the Indian peninsula (Kimble, 1938, 193).

The late fourteenth century also witnessed a notable improvement in the art of navigation. The instrument astrolabe, which was invented by Hipparchus, had been improved and came into common use as an aid to navigation by making it possible to fix the altitude of the polestar more accurately. The map-makers were a vital necessity in the field of navigation, and thus most of the map-makers were working for the merchants and sailors though there were differences of opinion with regard to the land beyond the equator.

Geography in the Christian world of the medieval period appears to have very much relied upon Ptolemy's works and ideas, but very little attempts were made to revive the Greek tradition in geography. Christian scholars of the period had different interpretations of Ptolemy's works, and their opinions differed on various aspects of Ptolemy's geographic ideas. The Christian world cannot boast of having developed any concepts, models or paradigms for the succeeding generations.

ARAB GEOGRAPHY

Arab geographers in the medieval period preferred to *formulate their concepts as generalisations of empirically observed facts* and insisted on the importance of direct observation. Unlike their counterparts in the Christian world, they made valuable contributions to geographical knowledge, though some Arab scholars made logical deductions from existing theories and their conclusions conformed to near realities. They would be long remembered for some of their concepts, models and paradigms which still govern the method of geographical teachings.

It would be appropriate to have a brief outline of the driving force which seemed to have enriched and widened the Arab geographic horizon between 800 and 1400. The geographical writings of the period were based on a much greater variety of sources than those of Christian scholars. The spread of the Arabs in the medieval period was one major event that enlarged and widened the geographic horizon of the religion of Islam. Islam became the 'raison d'etre' of the Arab world because it commonly held together the Arabic-speaking people of Arabia who were previously grouped in small isolated wandering tribes and had no feeling of unity.

The followers of Islam embarked on a conquest is of the world outside Arabia. In 641, they conquered Persia and, in 642, they took control of Egypt.

They moved across the Sahara and by 732 took possession of the Great Desert. They marched through the Iberian Peninsula into France. For some nine centuries, the Arabs had ruled most of the Iberian peninsula. Arab rule was also extended eastward into India and eventually to some of the islands of south-east Asia. The Arabs organised military expeditions across the Black Sea into the Russian Steppes. Successive military expeditions in Europe, North Africa and Asia helped them to acquire stimulating knowledge about the land and people they conquered, and to unfold the ancient Greek tradition in geographical scholarship preserved at the museum in Alexandria.

Arab geography flourished in the city of Baghdad which was founded by the Arabs near the ruins of Babylon in 762, and for more than a century it was the centre of the intellectual world. Baghdad may be compared with the Greek intellectual centre, Miletus, which remained a centre of academic innovations between 770 and 570 BC.

With the patronage of the caliph Harun al-Rashid, a project was started for the translation of the works of the Greek philosophers and scholars into Arabic. Materials pertaining to geographical ideas were collected from all available sources, and the translators were paid the weight of their books in gold (Ahmed, 1947, 5). From Baghdad, therefore, a flood of new ideas from varied sources began to spread throughout the Arab world. Eventually, the innovations were carried forward into Christian Europe as a result of Latin translations from Arabic.

Among other innovations was the use of the decimal system in arithmetic, which was brought into Baghdad from the Hindus, who learnt if from the Chinese. It is equally important to note that the Arab geographers/scholars showed a genuine fascination for Greek traditions and concepts rather than for the Roman heritage. They preferred to adopt the Greek concepts about the shape and size of the Earth and conceived of the Earth as being located at the centre of the universe with the celestial bodies in circular motions around it.

At the direction of Al-Mamum, the scholars of the Baitul-Hukma (Academy) attempted to recalculate the circumference of the Earth. They made use of the same method as devised by Eratosthenes some ten centuries before. On the level plain of the Euphrates they set up a north—south line and fixed the latitude at either end by observations of the stars. They then measured the distance between the fixed points and decided that the length of a degree was 56-2/3 Arabic miles. The scholars attempted several other measurements, but arrived at almost the same results. These values were much too small, owing to errors in the linear measurements. The circumference of the Earth was, thus, found to be 20,160 miles, much less than what Eratosthenes had calculated. The Arabs assumed that the Mediterranean Sea defined the western limit of the known ekumene while the eastern limit was defined by Sila (Japan).

The Arabs in medieval Arabia had access not only to translations from Greek, but also to the reports and accounts of their own travellers. As a result, they had a much more accurate knowledge about the world than their counterparts. One of the earliest Arab traveller was Ibn-Haukal who travelled through some of the most remote regions of Africa and Asia. He sailed along the east African coast to a point some 20° South of the equator. To his surprise, he found people in large numbers living in those latitudes. This observation made him discard the Greek view that the equatorial torrid parts were uninhabited. In spite of this observation of Ibn-Haukal, the Greek concept of the habitable zone persisted for a longer period and appeared in different forms.

Arab scholars in the early medieval period made *substantial contributions to the field of climatology and physical geography* and some of their concepts still have relevance in modern geographical thinking; They made some important observations regarding climate, and processes shaping landforms. In 921, **Al-Balkhi** collected information regarding climatic data from the accounts of different Arab travellers who made direct observation in course of their expeditions and journeys. On the basis of the collection of climatic data, AI-Balkhi prepared the world's first climatic atlas Kitab al- Ashkal.

Al -Masudi, who was born in Baghdad towards the end of the ninth century and died in 956, had gone south as far as Mozambique, south of the equator, and prepared a very good description of the monsoon. His concept regarding the monsoon seemed to have been formulated as a generalisation of empirically observed facts of the climate he witnessed during his voyage along the east African coast. He described the evaporation of moisture from the water surfaces and the condensation of the moisture in the form of clouds. He had a conception of the sphericity of the Earth and believed that the surface of the Earth must be curved. He provided description about the effect of environment on the mode of life and attitudes of people, and held an opinion on environmental determinism (Husain, 1988, 91).

In 985, **Al-Maqdisi** prepared a new climatic map of the world with 14 climatic regions which was an improvement over the map of Al-Balkhi. Al-Maqdisi was of the opinion that climate varied not only as regards latitude, but also according to east and west position. He also presented the idea that the southern hemisphere was mostly open and that most of the world's land area was in the northern hemisphere.

If Al-Balkhi, Al-Masudi and Al-Maqdisi could be known for their contribution to climatology, then **Al-Biruni** and Ibn-Sina would be known for their *contributions to geomorphology.* AI-Biruni, who had been to India during the time of Mahmud Ghaznavi, prepared his great geography of India 'Kitab-al-Hind' in 1030. The book deals with the processes shaping land forms under normal conditions. He identified the significance of the rounded stones he observed in the alluvial deposits south of the Himalayas. The stones became rounded, as they were rolled along in the torrential mountain streams. He also observed coarse alluvial materials near the mountain and finer alluvial materials farther away from the mountain. One of the interesting observations credited to him is that '*the south pole night ceases to exist*'—an information which he probably collected from the accounts of those explorers who might have voyaged to the far south before the eleventh century.

Avicenna or **Ibn Sina** is credited with the idea of *landscape erosion.* He had the opportunity to observe mountain streams in the act of cutting down their valleys in the mountains of central Asia. He deduced a hypothesis on the basis of this observation that mountains were being constantly worn down by streams, and the highest peaks occurred where the rocks were especially resistant to erosion. Mountains are raised up and are immediately exposed to this process of wearing down, a process that goes on slowly but steadily. Ibn Sina also observed the presence of fossils in the rocks in high mountains, which he interpreted as examples of nature's effort to create living plants or animals that had ended in failure. It was eight centuries later that James Hutton presented similar ideas concerning the process of erosion. He had probably never heard of Ibn Sina, and could not read Arabic.

It was **Al-ldrisi** or **Edrisi** who for the first time made extensive corrections of the erroneous ideas of the Roman geographer Ptolemy. AI-Idrisi was of the opinion that there were uncertainties concerning the actual arrangement of mountains, rivers, or coastlines. On the basis of much new information, he was able to prepare a new geography. In 1154, he produced a book with the title *Amusement For Him Who Desires to Travel Around the World*, in which he corrects the idea of Ptolemy regarding an enclosed Indian Ocean, and the idea of the Caspian Sea as a gulf of the world ocean. He also makes correction of the courses of rivers, including the Danube and the Niger, and the alignment of several major mountain ranges. He had shown that the Greek division of the world into five climatic zones did not correspond to reality and suggested a more sophisticated world climatic system (Holt-Jenson, 1981, 11).

One of the great travellers of the medieval Arab world was **Abdullah Muhammad** surnamed

Ibn-Batuta. He was born at Tangier in 1304, and at the age of 21 in 1325, went out to make the usual pilgrimage to Mecca where he proposed to complete his studies of the contemporary Arab law. But his interests focused on the land and people of North Africa and Egypt which he passed through to reach Mecca. He was so fascinated by nature that he decided to give up his study of law, and switched over to travelling across the then Arab territory. He carefully avoided following the same route twice. He went to many parts of Africa never before visited by any person. He sailed along the Red Sea, visited Ethiopia, and then moved southward along the coast of East Africa as far as Kilwa, nearly 10°S of the equator, where he learnt of an Arab trading post at Sofala in Mozambique, south of the modern port of Beira and more than 20° south of the equator. His visit to the south of the equator confirmed what Ibn-Haukal had implied in the tenth century that the torrid zone in East Africa was not torrid, but it was occupied by a large native population that justified the raison d'etre of Sofala.

From Mozambique, Ibn-Batuta again sailed for Mecca, and after a brief stop at Mecca, he again left for journeys to Baghdad and Persia, and the land around the Black Sea. He travelled across the steppes of Russia and then went to Bukhara and Samarkand. He crossed the Asiatic mountain through Afghanistan and moved into India. He was in the court of the Mongol emperor in Delhi for several years and had the opportunity to travel widely in India. The emperor appointed him as ambassador to China, but certain reasons delayed his reaching China on time. However, before going over to China, Ibn-Batuta visited the Maldive Islands, Ceylon, Sumatra, and eventually went to China where he stayed for a shorter period. He finally left India and in 1350, returned to Fez the capital of Morocco. In 135l, he made a trip to Spain, and again came back to Fez. During 1351–53, Ibn-Batuta travelled across the Sahara and reached Timbuktu on the Niger river, gathering important information about the culture traits of Arab Negro tribes living in that part of Africa. In 1353, he finally settled at Fez, where at the Sultan's command, Ibn-Batuta devoted himself to writing a lengthy account of his travels (Ibn-Batuta, 1358). During some thirty years, he visited a linear distance of about 75,000 miles, which in the fourteenth century was world record. Unfortunately his book, written in Arabic, made little impact on the Christian world as no attempt was made to translate it into Latin.

The last great Islamic geographer of the medieval period was **Ibn-Khaldun** (writing between 1382–1405). He established the foundation of political geography in those of his writings which analysed the rise and fall of the empires. Like Ibn-Batuta, he was born on the Mediterranean coast of north-west Africa. He spent most of his life in the cities of what is today Algeria and Tunisia and also for a time in the Arab part of Spain. In 1377, when he was 45, he completed a voluminous introduction to his world history, known as Muqaddimah (or, Introduction to History). It is the most detailed autobiography available in the medieval Arab history and reflects among other things the politico-geographical thinking of the late fourteenth century Arab world. The book begins with a discussion of humankind's physical environment and its influence, and points out the characteristics of humankind's that are related to her/his culture or way of living rather than to the environment.

Ibn-Khaldun concentrated on the most powerful units in the political hierarchy of his own times—the tribe and the city which were epitomised in the struggle between nomadic and sedentary states. Both Bedouins (nomadic or savage) and urban (sedentary) peoples were considered 'natural' and interrelated groups. He explained their differences deterministically in terms of choice of livelihood: agriculture and animal husbandry necessitate desert life, while commerce and industry demand an urban location. Both groups, however, were held to be related on the evolutionary scale, with the Bedouins of the desert preceding the urban folk. The latter, descendants of immigrant nomads, had reached the last stage of civilisation in their cities and hence were at the point where decay sets.

Ibn-Khaldun also discussed the group's functioning in space, and it is here that the *nature of environmental-political relation and of territoriality* come into play. He suggested that the qualities of courage, alertness and loyalty prevailed among the Bedouins, but the sedentary people by contrast represented immobile concentrations of wealth, lack of courage, loyalty, common lineage and tribal cohesion. Hence, the sedentary states had to develop new and complex systems of political organisation for security against attack and for their own self government. Where contiguous, the two types of socio-political organisations were interdependent, not only in an evolutionary sense, but also functionally. However, the sedentary state dominated in this functional relationship both economically and politically.

Ibn-Khaldun was of the opinion that war-like nomads often founded large states, but after a while the nomads were absorbed by their permanently settled subjects as peasants and townmen, the rulers lost their warlike spirit and eventually their kingdoms collapsed. He both predicted and lived to see the collapse of the Islamic state he lived in.

At the fall of Damascus in 1400, he actually met Tamerlane, the conqueror and devastator (Holt-Jenson, 1981, 11–12). Ibn-Khaldun is credited with having *formulated the first concept of the life cycles of the State* which still holds some relevance in tribal Africa and Asia.

Though his generalisations with regard to the Bedouin and sedentary states/political organisation were based on the empirically observed facts, his views on the ekumene and other aspects of the climatic determinism seemed to have been deduced from the earlier Greek notions.

He accepted the traditional *seven climatic zones* running parallel to the equator. He repeated the idea of Aristotle that 'there existed an uninhabitable zone along the equator because of intense heat, and an inhabitable polar zone because of cold' (a view he borrowed from the Norsemen). Ibn-Khaldun shared the view of Albertus Magnus and pointed out that the people turned black when they lived too close to the Sun, and when the black people moved to the temperate zone they gradually turned white or produced white children. 'Ibn-Khaldun may be considered to have discovered the true scope and nature of geographic inquiry, but the fact remains that his knowledge of the physical Earth is based largely on Greek theory; and his ideas about environment influence are not highly sophisticated' (Kimble, 1938, 180).

CHINESE GEOGRAPHY

The Chinese had a rich tradition in geography, and the heritage that developed in the ancient Chinese world of geographic horizon, seemed to have been passed on to succeeding scholars. The Chinese paradigm in the medieval period was also concerned more with observable things and processes and with the formulation of theory. This is evident from the view of the Chinese geographer **Shen Kua** who presented an idea in 1074 about the erosion of mountain: 'considering the reasons for these shapes, I think that (for centuries) the mountain torrents have rushed down, carrying away all sand and Earth, thus leaving the hard rocks standing alone ... standing at the bottom of the ravines and looking upwards, the cliff face seems perpendicular, but when you are on the top, the other tops seem on a level with where you are standing. Similar formations are found right upto the highest summits.'

The Chinese had a *rich tradition in the art of map making* which seems to have been inherited by the scholars of medieval China. Two beautiful examples of Chinese maps were carved in stone in 1137 AD based on data that probably had been surveyed before 1100 (Needham and Ling, 1959, 547–48). The first map entitled 'A Map of China and the Barbarian Countries', shows a wider area extending from the Great Wall of China north of Peking, southward to the island of Hainanand westward to the mountains of inner Asia. The second map, known as 'The Map of the Tracks of YU the Great' covers essentially the same area, but is even more accurate in showing the courses of the great ·rivers and the coastline from the Gulf of Chihli, north of the Shantung Peninsula, to the island of Hainan. Both the maps were drawn with north at the top.

In medieval China, the scholars *maintained the tradition of expeditions, explorations, and voyages.* Two of the most distinguished medieval Chinese travellers in the seventh century were the Buddhist monks, Hsuan Tsang who reached India across the wind-swept plateau of Tibet and the world's highest mountain; and I. Ching who in the same century reached India by sea, stopping first for eight months in Sumatra. However, Hsuan Tsang was known as the Chinese discoverer of India. He brought with him to China a large collection of Buddhist relics and manuscripts. Chiang carried with him more than 10,000 rolls of Sanskrit Buddhist texts which he undertook to translate into Chinese.

In 1220, a Chinese crossed the deserts of inner Asia and met with the Mongol leader, Genghis Khan, in Samarkand. He described in his accounts about horse and camel mobility, of the tribesmen whom he found ruthless and cruel. Rabban Bar Sauma, a Nestorian Christian monk made a pilgrimage to Rome in 1287–88. However, he visited Genoa, Paris and Bordeaux. He made contacts with the kings of France and England. This was several decades before similar journeys of the Polo brothers in China. In 1296, another Chinese traveller, Chou Ta-Kuan made a trip to Cambodia and prepared an account of the customs of the Cambodians.

The Chinese voyages were basically confined to South-East Asia. Seldom did they set out far into the Pacific. However, Macro Polo found Chinese sailors in the Persian Gulf port of Honnuz. One of the Chinese admirals Cheng Ho organised seven separate expeditions between 1405 and 1433. His expeditions went to South-East Asia, Ceylon and the west coast of India in south Asia, the Persian Gulf, the Red Sea, and the east coast of Africa, south of the equator. He also noticed the presence of numerous native people in the equatorial torrid zone, which he described in his accounts. Cheng Ho also went as far as Taiwan in the east. It is even possible that he sent one of his ships to the north coast of Australia and prepared an account of his voyages.

Summing up

The medieval period witnessed remarkable progress in geographical concepts, models and paradigms. New data were collected by direct observation, but the lack of close contacts among different people, in large part resulting from the barrier of language, meant that geographic knowledge gained by one group was diffused only slowly to the other. Geography in the Christian world was basically confined to the translation of the earlier Greek and Roman works, and deductions were made purely on theoretical grounds. But the geographical explorations and voyages of the Europeans definitely widened and broadened their geographical outlook. Inspite of that, the same old Greek concept equating habitability with latitude persisted in the Christian perception. The Christians made cartographic innovations that helped them to obtain maximum benefits from expeditions and voyages.

The Arab geographical writings, on the other hand, were based on much wider variety of sources than were those of Christians. But the earlier Arab scholars had a fascination towards the works of Ptolemy and his wrong notion, about an enclosed Indian Ocean which persisted until the mid-twelfth century, when that erroneous idea was discarded. However, the Arab geographers made valuable contributions to climatology and geomorphology and some of the notions still hold relevance. Arab scholars also travelled wide in the Arab world and collected facts from observations which also discarded the Greek concept of habitability. Arab geographers are identified as having drawn attention specifically to human-environment relations vis-a-vis environmental determinism.

The Chinese made outstanding contributions to geomorphology and cartography, but their geography was more concerned with observable things and processes. The medieval Chinese geography basically emphasised on studies of people, descriptions of the regions of China, descriptions of foreign countries, accounts of travels, drainage pattern of China, coastal geography, relief features, cities and palaces, and geographical encyclopedias. Geography in contemporary China seemed to be multi-faceted, but it always showed an orientation towards the topographical tradition.

IMPACT OF VOYAGES, DISCOVERIES AND THE RENAISSANCE

It is surprising to note that the exploring activities among the Arabs and the Chinese gradually came to an end. There was no great Arab traveller after Ibn-Batuta in the fourteenth century, and in China, the end came after the seventh expedition of Cheng Ho in 1433. But the exploring activities in Europe all of a sudden increased in the fifteenth century and this time they were well-planned and supported by governments or by merchant companies, and *expedition to the open oceans was preferred.* It was a major turning point in the world geographical scholarship as new discoveries largely removed the doubts created by the earlier notions and concepts.

These exploring activities were motivated by religious and economic factors: the zeal to spread the Christian faith and the need to replenish the European supplies of precious metals and spices, and the desire for personal wealth. The desire to liberate the greater part of Spain and Portugal from the Arab rule, as a part of a unified national consciousness, accelerated the zeal to spread the Christian faith. Increasing trading activities with a view to provide financial support for trade, necessitated the need for a supply of gold and silver or valuable gems.

Prince Henry of Portugal is credited with organising voyages and expeditions to the open seas. The fall of Arab stronghold on the southern side of the strait of Gibraltar under his command in 1415, widened the horizon of the Portuguese world. Prince Henry is also credited with the establishment of the world's first geographic research institute in 1418. At Sagres, he built a palace, a chapel, an astronomical observatory, buildings to store collections of maps and manuscripts, and houses for the institute staff. He may be compared with Roger II of Sicily who founded Palermo for the academic activities. Like Roger II, Prince Henry brought scholars of all faiths to Sagres from all around the Mediterranean. Master Jacome of Majorca was the chief geographer.

The purpose behind the setting up of Sagres was to improve and teach the methods of navigation to Portuguese sea captains, to teach the new decimal mathematics, and to shift the evidence from documents and maps concerning the possibility of sailing southward along the African coast and thence to the Spice Islands. Prince Henry's captains successfully sailed to the three groups of islands of the Atlantic off North Africa and Europe that included the Canary Islands (Ptolemy's Fortunate Islands), the Madetra Islands, and the Azores. The discovery of these islands by the Portuguese sailors brought about changes in the perception about the high seas, and the bearings which were wrongly shown on the Portolano Chart of 1351 were made accurate.

In 1434, one of the trusted captains of Prince Henry, Gil Eannes sailed south of the latitude of Bojador and turned eastward. He made his first attempt to sail to Cape Bojador in 1433, but was forced to return to Sagres. The contemporary belief was that the strong current of water along the coast formed eddies with much foam off the end of the cape and the water was boiling, and the people would become black if they moved beyond the cape. It is reported that when Gil Eannes reached the shore, the water was not boiling and no one turned black. The '*cape Boja.dor puzzle*' thus came to an end. The very next year, Portuguese ships sailed 350 miles south of Bojador.

Under Prince Henry's constant patronage, the Portuguese sailors had sailed far enough south to reach the southern zone of transition between the desert and the humid country beyond. In 1473, a Portuguese ship crossed the equator without burning. This expedition successfully encountered what the Greek geographers had said earlier. Between 1486 and 87, Bartholomew Dias sailed southward from the equator and to avoid stormy weather, he sailed far out westward away from the land and reached the coast of southern Africa at Algoa Bay (Port Elizabeth). While coming back, he passed Cape Agulhas, the southern most point of Africa, and then the cape of Good Hope.

Between 1497 and 1499, **Vasco da Gama** made his famous voyage across the Arabian Sea to reach Calicut on the west coast of India. He made a wide circle out into the Atlantic before turning eastward along the latitude of Cape Agulhas and voyaged along the coast of Mozambique and then sailed eastward. By the time Vasco da Gama returned to Lisbon, he had sailed 24,000 miles in more than two years. In 1510, the Portuguese took Goa on the west coast of India. In 1511, they set up a base at Malacca on the strait between Malay Peninsula and Sumatra. Japan was approached in 1542. In 1557, the Portuguese leased Macao from the Chinese and in 1590 they reached Taiwan and gave it the Portuguese name, Formosa. Undoubtedly, the geographical explorations, conducted by the Portuguese in the fifteenth and sixteenth centuries, seemed to have collected a variety of information about the places, never visited by any European, and removed the doubts about certain areas in the Atlantic Ocean, created by the wrong notions of the earlier Greek geographers.

The voyages of **Christopher Columbus** during 1492–1504, across the North Atlantic Ocean also marked a turning point in the world's history of expeditions. He was to have discovered Asia as he thought earlier about the possibility of sailing westward to Asia. He read Ptolemy's 'Geography', Cardinal Pierre d' Ailly's 'Imago mundi' and the books of Aeneas Silvius (Pope Pius II). He was familiar with the accounts of Marco Polo. But ever since he started his voyage, he carried forward with him certain wrong notions about the circumference of the Earth, and the eastward extension of the known lands of the Earth.

Columbus spent a few years in Portugal to study what was left of the institute at Sagres founded by Prince Henry in 1418. He reckoned that the east coast of Asia was located just about where the east coast of Mexico is actually located. However, his belief was strengthened when he found the southern coast of Cuba and the coast of central America trending towards south-west, just as the coast of Asia was shown on Ptolemy's map. When he heard from the Indians of central America that there were sources of gold only a short distance west and that there was another great ocean beyond, he was sure it must be the Indian Ocean. He also thought that there must be a strait connecting the Caribbean with the Indian Ocean as he witnessed the great flow of water along the northern coast of South America. He was the first explorer to discover and make use of the wind system of the Atlantic. Columbus had known about the presence of easterly winds in the low latitudes and of westerlies in the higher latitudes.

Columbus is credited with the Treaty of Tordessilles which was agreed upon by Spain and Portugal in 1494. Under the treaty, the world was divided between these two countries, and the dividing line was to be drawn 270 leagues west of the Azores or 370 leagues west of the Cape Verde islands. Portugal got exclusive rights over the land east of that line, Spain to the west of it. The treaty gave Portugal a free hand in the Indian Ocean and Columbus got a free hand in the lands he discovered west of the Atlantic. The treaty of Tordessilles marked the *beginning of the concept of the geometric boundary* which in the nineteenth century became accepted principle for the European powers for the division of their colonial territories in Africa and elsewhere.

Magellan, or Fernao de Magalhaes, was the first European explorer to have reached eastern Asia by sailing west. His expedition began in 1518 and he sailed along the Brazilian coast, and in 1520 he found the entrance to the strait that bears Magellan's name. It took him 30 days to pass through the 360 miles strait. He moved in a north-westerly course for 98 days before he reached the

Island of Guam. He reached the Philippines in 1521. Magellan, on a previous voyage for Portugal, had reached the Spice Islands (the Moluccas) which are east of the Philippines. Therefore, at this point he had become the first man to sail all the way around the Earth.

Certain inventions in the sixteenth, seventeenth and eighteenth centuries further enhanced the contemporary knowledge about the Earth, as some of the major problems faced by the explorers had to be solved. To achieve perfection and accuracy, navigation was the major problem that the sailors had to be confronted with. The preparation of the isothermic map by Edmund Halley who in 1699–1700 noted the magnetic variation he plotted on the map during a voyage, suggested that it was not the same for each longitude. In the sixteenth century, the cross-staff was invented. The measurement of the Sun's altitude by its shadow was made possible by the invention of the back-staff by John Davies. John Hadley invented the octant in 173l. The sextant, still in use, was a yet later improvement based on the principle first adopted by Davies. The British sailors are credited with the invention of the log which solved the problem concerning the measurement of speed at the sea (Taylor, 1957, 201–02). This invention solved the problem of measuring distance and direction and fixing the position on the surface of the Earth, and the determination of latitude, as it involved observations of the height of certain stars of the Sun above the horizon. These inventions, in fact, revolutionised the navigation as a result of which newer and correct ideas flourished with regard to the high seas and oceans.

Inspite of some achievements in tackling the problem of perfecting the accuracy of navigation, there was more to be done because, until the longitude could be precisely determined, navigation remained dangerously inaccurate. The measurement of longitude needed some way of keeping accurate time at sea. It was felt as early as 1522 that a dependable time piece would solve the longitude problem. It was in 1657 that Christian Huygens invented the pendulum clock, but it was no good at sea. The need for an accurate clock was felt in 1707 when an English fleet was wrecked on the Scilly Island because the navigator had a wrong estimation of longitude. In 1714, the British Parliament offered a reward to any person or persons who could devise a pendulum clock to measure time at sea with sufficient accuracy. John Harrison is credited with the invention of a pendulum clock which proved to be so accurate that for the first time in humankind's long effort to gain useful knowledge about the Earth, an explorer could tell about its exact position. Harrison was given the reward in 1775. By this time French and Swiss clock-makers were almost ready with their models. After this date, navigators could measure longitude with the same precision with which they could measure latitude.

One of the *major consequences of the age of discoveries and invention was that of an organised cartographic revolution* which occurred simultaneously both at Venica and Genoa. The need was felt to correct the works of the ancient cartographers. It was as early as 1.459 that Fra Mauro for the first time prepared a map showing the Indian Ocean open to the south, thereby breaking the Ptolemaic tradition of enclosing that ocean. It is not known how he could come to know about an open Indian Ocean, but his map like the other maps of his time was oriented towards the south. Martin Behaim of Nuremberg was the first cartographer to have devised the world's first globe in 1490. He was helped by a painter named Jorge Glockendon in the preparation of the globe which was based on the small estimate of the Earth's circumference, but it was strange enough that Columbus could never have known of Behaim's globe.

In 1500, Juan de la Cosa prepared a map using the observations obtained from the first three voyages of Christopher Columbus and also from John Cabot's voyage to North America. Martin Woldseemuller in 1507 produced the first world map showing America as a separate continent and not the eastern continuation of Asia. The word America was used by him to denote the new continent. This was largely because he thought that Amerigo Vespucci had reached the new continent before Columbus or because Amerigo was definitely the first explorer to identify the newly discovered land as a separate continent. However, for navigation, this map was not found to be more useful than other maps of this period that made use of the Portolano principle of design. Explorers had already found that when they followed any of the straight lines on these maps for long voyages, they did not arrive at expected destination. The contemporary need was for a new kind of projection which could make it possible to show the curved surface of the Earth on a flat paper or parchment (Bagrow and Skelton, 1964).

In 1530, Peter Apian prepared a heart-shaped map of the Earth with the curved lines of latitude and longitude. But Apian's projection was found to be defective because distance and direction on the projection appeared to be distorted. **Gerard Krema**, or **Gerardus Mercator**, produced a graticule for the world by joining two heart-shaped projections, one for each hemisphere. Mercator is credited for his projection which he produced in 1569 for the world. This projection was the only contemporary projection/map being used for navigation in the low and middle latitudes. Though the projection offered solution to the problems of navigation and enabled the navigators/sailors to reach the desired destination by sailing the great circle (shortest) route, it was not widely adopted probably because of it being too mathematical. The sailors who were not trained in mathematics found difficulties in believing that a short line was not the shortest distance between two points on the surface of a sphere. It was not until 1599 when Edward Wright produced the trigonometric tables that made it possible for other people to produce Marcator's projection. It is also believed that in 1511, Erhard Etzlaub of Nuremberg prepared a map of Europe and Africa on the same principle as that of Mercator. It is not known whether Mercator knew about this or not.

Abraham Ortelius of Antwerp produced the Ortelius atlas—'Theatrum orbis terrarum' in 1570. Amsterdam gradually grew up as the major centre for the publication of atlases and wall maps, especially in the seventeenth century, breaking the tradition of Venice and Genoa in the art of map making. Nicholas Sausond' Abbeville was the first producer of world atlases in France, and founded a 'dynasty of cartographers' in the seventeenth century which produced maps and atlases for over a century. During the seventeenth and eighteenth centuries the map-makers only slowly revised their maps to include the newest information and to acknowledge the existence of gaps in their information.

The period between 1768 and 1779 was a landmark in the history of geographical thought. Erroneous idea which the Christian Europe inherited from Ptolemy during the Middle Ages after his geography was translated into Latin, stipulated the presence of a great southland which he called terra australis incognita which enclosed the Indian Ocean on the south. Though the Arab traders had known that the Indian Ocean was open to the south and the map of Fra Mauro showed it, there were many who either did not know about the Arabic writings or disputed the travel accounts of the Arab scholars. It was during his first voyage (1768–71) that Captain Cook found no land at about 40°S latitude. His second voyage (1772–74) finally removed the wrong notion about Ptolemy's great southland, as he sailed as far as 71°10' south latitude. However, through Captain Cook it became known that there was an ice-covered land farther south, which he had been unable to reach. In 1778, Captain Cook again sailed into the Arctic Ocean through the Berring Strait, but was unable to move further because of ice-floes at 70°44' N.

Voyages, inventions and cartographic precision and innovations seemed to have solved certain major problems regarding navigation, determination of longitude, projection, and those arising out of many erroneons concepts of Ptolemy. But inspite of all these major achievements in an era of discoveries, the problem concerning the shape of the Earth remained to be solved. The problem was whether to accept what Isaac Newton and Christian Huygens said in 1687 that the Earth must be flattened at the poles and it must bulge at the equator. In 1720, Jacques Cassini contrasted what Newton and Huygens said about the shape of the Earth. The French Academy decided to settle the controversy by carrying out measurements of the arc of the meridian at different latitudes. Two such surveys were done between 1735 and 1748, and these surveys proved that the Earth was flattened at the poles and did bulge at the equator. These surveys removed the doubts created by Cassini who had challenged the mathematical concept of Newton and Huygens.

Summing up

These discoveries, inventions and revolution in the art of map making may be collectively called the Renaissance which brought about renewed interest in the geographical knowledge and removed some earlier misconceptions. As a result, the world picture visualised by ancient geographers (including Ptolemy) had to be changed. At first the explorers tried to describe everything they found in the strange lands. But, having no concepts regarding the new phenomena they found, an attempt was made to describe things by analogy. Geographers began to ask questions such as how things observed could be related to some generalisations of the empirical perception.

The Renaissance in the geographical thinking also brought about a revival in another branch of ancient geography, that of topographic description. The jumble of information related to place names, included in such topographies, gives notes on natural and physical condition juxtaposed with miscellaneous information on folk lore, etymology and history. It seems to place them outside the field of science. The popularity of these works might well puzzle us today until we realise that we are often prisoners of contemporary logic and are unable to see those qualities of old works which cannot be integrated into our system.

4

Foundation of Scientific Geography

HIGHLIGHTS

- Rise of Dualism
- Scientific Discoveries
- Classical Roman Tradition Revived
- Beginning of Scientific Geography
- Special Geography vs. General Geography
- Geography in the Eighteenth Century

The Renaissance caused the unprecedented flood of new information coming from the voyages of discovery, inventions and cartographic precision which seem to have greatly revolutionised the contemporary geographical scholarship, resulting in a number of concepts, models and paradigms that appear to have laid the foundation of modern geography of the classical period.

As the inheritors of the geography of classical antiquity and that of the Renaissance, the geographers of the seventeenth and eighteenth centuries considered all kinds of phenomena whose differences in different parts of the world helped to acquire knowledge of the world. The pre-classical period examined here was cine of groping, when humankind struggled to shake loose from the weight of old paradigms and to find new answers to old questions about order, harmony and meaning. There were some who asked questions and looked for answers, and others who were fully occupied and satisfied with the task of establishing the validity of the facts that were reported by the explorers and of putting together a general picture of what things were like.

RISE OF DUALISM

The geography of pre-classical period included both 'general' studies of particular kinds of phenomena on the Earth's surface and description of many kinds of phenomena found in particular areas. The inclusion of these two forms of study within the single field has no doubt been the cause of more controversy than any other single problem in the methodology of geography. It is significant, therefore, that it was not introduced into modern geography as a result of the chance combination of Humboldt and Ritter. On the contrary, the same difference is found in the works of the geographers of antiquity (Hartshorne, 1976, 41).

This period witnessed apparent dichotomy or dualism which is still a characteristic of geography. Some scholars have regarded it as the essential justification for the role of geography while others have argued for a division of the subject into physical and human geography on the ground that their respective methodologies must be different. In studies of natural phenomena, including climate, geology and landforms, it is possible 'to use the methods of natural sciences and to draw conclusions with a large measure of scientific precision. The methods of natural science, however, do not lend themselves very well to the study of social and cultural phenomena. Our generalisations about human groups must be limited in time and space, and must relate to statements of probability rather than certainty' (Holt Jenson, 1981, 13).

SCIENTIFIC DISCOVERIES

Conceptual developments took place in different kinds of problems and tended towards the development of special branches of knowledge. The German scholars made significant contributions in geography and offered valuable concepts, models and paradigms, some of which are still relevant. However, some more developments took place in Great Britain, France, Russia, Poland and Italy. The idea of a heliocentric, universe proved by Polish scholar Nicolaus Copernicus (where he calculated the movement of the celestial bodies, with the Sun as a fixed centre) discarded the Ptolemic tradition of a geocentric universe. However, he still followed the Ptolemic tradition of a circular orbit of the planets around the Sun regarding their movements.

In 1618, Johannes Kepler, a German astronomer, developed the laws of motion, in which he recognised the elliptical planetary motion of the celestial bodies around the Sun. In 1623, Galileo recognised the idea of a heliocentric universe of Copernicus. He, for the first time, formulated the concept of a universal mathematical order, a universe that could be described in terms of a mathematical law rather than in the verbal and logical terms of Aristotle. In 1686, Isaac Newton presented his law of gravitation.

Copernicus, Kepler, Galileo and Newton sowed the seeds of scientific revolution that marked the beginnings of specialisation. So deeply had the mathematical work of these scholars impressed the German geographer Varenius that he defined geography as a branch of mixed mathematics and rebuked those who tended to limit it to a description of various countries (Tatham, 1967, 32). Let us examine the contribution of the individual scholars of the period who identified themselves with different traditions, and laid the foundation of Scientific Geography.

CLASSICAL ROMAN TRADITION REVIVED

Though the Christian world was much influenced by the scientific discoveries, which tended to ridicule the Ptolemic tradition, yet attempts were made to revive the classical Ptolemic tradition and also the classical Roman tradition of Strabo.

Peter Apian and Sebastian Munster are credited with having produced the great geographical compendia at the beginning of the Age of Renaissance, which for a century after their publication in the sixteenth century, remained the chief standard theoretical works: the one for its popular exposition of astronomy and mathematical geography and the other for its descriptive geography modelled on Strabo.

Peter Aplan was an astronomer and cartographer, and was credited with two published works: one was an astronomical treatise, and the other dealt with those aspects of geometry and astronomy which are relevant to geography. His second work was published in 1524. Apian seemed to have conceived of a geocentric universe and also recognised the Aristotelian concept of the ekumene or the zone of habitability.

Sebastian Munster was born at Ingelheim near Mainz in 1489. He studied at Heildelberg and Vienna, and in 1536 was appointed to a chair of Hebrew at Basle where he remained till his death in 1552. He translated several important geographical works from Latin into German, but Munster is identified by his 'Cosmographia Universalis' which was published in 1544. It is a compilation from many contemporary authorities rather than a carefully arranged treatise. History and genealogical tables take up a large part of the treatise. The work very much follows the geographic tradition of Strabo as it almost excludes mathematical and physical geography.

Munster recognises the geocentric universe and maintains the Ptolemic traditions. His book also deals with the dispersal of humankind over the Earth after the great flood. It provides us with a human and political geography of Germany on a chronological basis. The last part of the book deals with Asia, Europe and Africa. However, he was amazingly uninformed about Asia and it is now held that he must have had access to only such information as was available in Germany. Cosmographia Universalis was regarded as the authoritative work on world geography for more than a century.

Cluverius (Philipp Cluver) (1580–1622) was the first German geographer to have produced universal geography, because he had better access to the new materials available in Germany. He carried forward the historical tradition of Sebastian Munster. In 1616, he published a book on the historical geography of Germany. He approached geography through the classics and history. His general work 'Introduction in Universam Geographiam' was published posthumously in 1624, in which he also maintained the Ptolemic tradition of a geocentric universe. The compendium starts with a brief and inadequate account of mathematical geography with a distinction between geography and chorography and then passes on to a regional description of the countries of the world which occupy four-fifth of the whole. He set forth a paradigm in regional geography that for long was unsurpassed (Tatham, 1967, 32).

BEGINNING OF SCIENTIFIC GEOGRAPHY

However, the classical Roman tradition of Strabo and Ptolemy, which was carried forward by Apian, Munster and Cluverius, despite the scientific discoveries about the Universe, gradually disappeared with the beginning of the scientific geography in the early seventeenth century.

More influential, however, in the development of the scientific geographical thought was the work of **Bernhard Varenius** (1622–1650) who profoundly influenced the content and scope of geography for more than a century. He was born in 1622 at Hitzacker near Hamburg. In 1640, he entered the University of Hamburg and studied philosophy, mathematics and physics and after three years went to the University at Konigsberg to study medicine and then moved to Leiden to pursue the same work. In 1649, he published a book entitled Descripto regnilaponiac et Siam. The book contains five parts: a description of Japan; a translation into Latin of a description of Siam by J. Schonter; an essay on religion of Japan; some excerpts from the writings of Leo Africanus on religion of Africa; and a short essay on governments. This was immediately followed by a companion volume on the religion of Japan. At this time, Varenius probably thought that descriptions of particular places could have no standing if they are not related to a coherent structure of general concepts.

Varenius started work on his famous work *Geographia generalis* in the fall of 1649 and completed it in the spring of 1650. The work should have been followed by a second volume, but it remained unfinished owing to his premature death in 1650. It is sometimes said that his *Geographia generalis* was a legitimate successor of Munster's 'Cosmographia Universalis'. It was the standard text for more than a century.

Geographia generalis is divided into three parts: (1) the absolute or terrestrial section which describes the shape and size of the Earth and the physical geography of continents, seas and the atmosphere; (2) the relative or cosmic section which treats of the relation between the Earth and other heavenly bodies, especially the Sun and its influence on world climate; and (3) the comparative section which deals with the location of different places in relation to each other and the principles of navigation.

Varenius asserted that the *Geographia generalis* must be followed by what is called special geography in which descriptions of particular places should be based upon (a) celestial conditions, including climates and climatic zones; (b) terrestrial conditions with descriptions of relief, vegetation and animal life; and also (c) human conditions, including trade, settlement and forms of government in each country. In fact, he showed little enthusiasm to embark on the study of human geography as it was not possible to treat it in an exact way. He explained that he had included this last group of human conditions as a concession to traditional approaches to the subject (Holt-Jenson, 1981, 14). Varenius for the first time brought together contemporary knowledge of astronomy and cartography and subjected the different theories of his time to a sound critical analysis.

Varenius contended that the methods of the natural sciences could be successfully used to draw conclusions with a large measure of scientific precision for the treatment of natural phenomena, including climate, geology and landforms. But these methods could not successfully be applied to treat the social and cultural phenomena. The generalisation about human groups must be limited in time and space, and must relate to statement of probability rather than certainty. He was thus one of the first scholars to point out these essential differences in the character of physical and human geography.

To him, geography focuses attention on the surface of the Earth, where it examines such things as climate, surface features, water, forests and deserts, minerals, animals and the human habitats. The human properties of a place include a description of the inhabitants, their appearances, arts, commerce, culture, language, government, religion, cities, famous places and famous men.

Varenius recognised the idea of the heliocentric universe of Copernicus, Kepler and Galileo. He was the first to point out the difference in the amount of heat received from the Sun in the equatorial regions as compared with that in the higher latitudes. The Sun's heat, he suggested, thins the air close to the equator and therefore the cold, heavy air of the polar regions must flow towards the equator. This was the first step towards the explanation of the world's wind systems.

SPECIAL GEOGRAPHY vs. GENERAL GEOGRAPHY

These two terms were used by Bartholomew Keckermann in his lectures at Danzig in 1603, and later published in a book in 1617. It is commonly believed that Varenius made use of Keckermann's work in an organised manner and provided a clean demonstration of the relation between these two points of view (Baker, 1963, 113).

Varenius distinguished the relationship between geographical writings that dealt with the characteristics of particular places and those that described the general and universal laws or principles that could be applied to all places. Thus, Varenius partitioned geography into 'Special Geography' and 'General Geography'. In *special geography,* what Varenius intended was that with the exception of celestial features (climate), things must be proved by experience (by direct observation through the senses), but in *general geography* most things could be proved by mathematical or astronomical laws.

This intellectual problem of the relation, between the specific and general became of major importance in the early pre-classical period of modern geography, as a result of the redundance of new information about specific places and the effort to generalise this information. However, Varenius observed that there was a need to demonstrate the close inter-connection between special geography and general geography so as to avoid contradiction. To him, special geography has great practical importance for government and commerce, but it leaves out the fundamentals of this field of study; and general geography provides these fundamentals, but to be of maximum utility they must be applied. Therefore, special geography and general geography are not just a dichotomy, but rather two mutually interdependent parts of a whole as he conceived of the whole world as a unit in his *Geographia generalis*. Special geography was primarily intended as a description of individual countries and world regions. It was difficult to establish laws in the special part, for explanation must be descriptive where people are involved (Holt Jenson, 1981, 13–14).

The terms 'general' and 'special' geography later became the standard terms in Europe. The frequent use by German writers of the adjective 'systematisch' in describing 'general geography' supports the common use of the term : 'systematic geography'. The term special geography was largely replaced in German literature by 'Landerkunde' which, in spite of obvious disadvantages, is commonly favoured over non-Germanic terms like 'special' or regional geography. Hartshorne observed that the distinction/*division between general or systematic geography and Landerkunde or regional geography*, as inducted into geography by Varenius, represented a form of dualism that was characteristic of geography throughout its initial period of development as a modern science. The distinction between the two aspects of geography was more explicitly stated by Gattener (1773–75), Krug (1800) and particularly by Bucher (1812).

GEOGRAPHY IN THE EIGHTEENTH CENTURY

The century after Varenius witnessed no advance in the scientific geography. *Interest was now focused on the natural sciences.* This was the period when the prodigious mass of empirical knowledge, accumulated to an extent under the guidance of the Scientific Academies, was organised on systematic lines. Such unprecedented expansion of knowledge paved the way for a new advance in geography. The light thrown on the nature of physical and biological phenomena by natural sciences made it possible by the middle of the eighteenth century to give a more scientific description of the Earth's surface than ever before. The problem of man's place in nature, as to whether or not the Earth was a stage purposefully created for the development of man and his culture, made such a description necessary. Equally important was the growing desire for some synthesis of the data produced by the systematic sciences (Tatham, 1967, 33–34).

Systematic studies were conducted to account for the origin of the landforms and that *replaced catastrophism by uniformitarianism.* This made the theories more viable and scientific which were based on the notion that all the processes of change observable today also operated in the past and could account for all the world's surface features. One of the earlier European scholars was Leonardo da Vinci who argued that running water could level off the heights of land until the Earth was a perfect sphere. John Ray (1627–1705), who is credited with the development of systematic study of botany and provided an empirical classification of plants, pointed out that water running down the slopes could slowly wash away the mountain. John Strachey in 1719 offered an explanation to show how landforms reflect the underlying structures.

Domenico Guglielmini studied the laws of river flow. Louis Gabriel Comte de Bant of France developed the concept of 'graded profile' in 1786, and mathematically proved to show how the flowing water of a river could establish an equilibrium between velocity and the load of alluvium being carried forward. It was James Hutton (1726–97), the Scottish geologist who pioneered the concept of the geomorphic processes shaping the surface of the Earth resulting into the evolution of landforms. His ideas were properly organised by John Playfair in 1802. These developments further tightened the bond between geology and geography and at the same time gave it a higher status in the contemporary scholarly world.

Johann Rheinhold Forster and his son Johann George Forster pioneered empirical explanation and approached geography from a practical standpoint. They accompanied Captain Cook on his second voyage (1772–75) in which he sailed far to the south in the Indian and Pacific Oceans.

The Forsters carried out botanical observations and collected data which helped them to offer generalised explanations, George Forster was the first to identify the pattern of temperatures on the eastern and western sides of continents at the same latitude and to point out the climatic similarity of western Europe and western North America. The Forsters seemed to have .derived their ideas of physical geography from Buffon, and their ideas about the classification of plants from Linneaus. They had an advantage over their predecessors in that they could make their own observations and were no longer dependent on descriptions of what other observers had seen and made. They were critical of ideas about the influence of climate on man, supporting their criticism with careful observation of the Dutch in South Africa and of the Polynesians.

Endowed with keen powers of observation and a scientific attitude of mind, George Forster collected facts, compared and classified them, drew from this classification the generalisations for which he then sought a causal explanation. His systematic treatment of material is well shown by the classification of his observations in the South Seas (Taylor, 1967, 37). Because of his empirical explanations of the observed facts and the skill of classification and generalisation, *Forster is known as 'the first great German methodological geographer in the modern sense'.*

The problem of developing scientific methods for the study of the life forms also received attention in the late eighteenth century. Leibnitz and Buffon attempted to arrange all life forms in a scale from the simplest organisms to the most complex. Linnaeus, on the other hand, arranged all life forms into taxonomic groupings. The passion for the scientific collection of specimens on voyages of exploration resulted in remarkable assemblage of accounts and new methods of classification that offered new hypotheses.

The great work of George Louis Leclerc **Comte de Button** *offered explanation regarding the laws governing processes of change.* He was director of the Jardin du Roi, the botanical garden in Paris, between 1739 and 1788. There he had access to a vast number of specimens of plants and animals and description written by travellers and explorers from all over the world. In his study, Buffon followed the inductive procedure which aimed at finding some kind of order in the flood of new information. His approach was non-mathematical.

Buffon's 'Historie naturelle, generale et particuliere' was a monumental work, published between 1749 and 1804. It was completed after Buffon's death by La Cepede. He *formulated the concept of man as agent of change* as he had been changing the face of the Earth in the process of developing a civilisation. He believed in the creative power of man. Buffon wrote much about forest clearance and the need for conservation. He opined that the removal of forests and the drainage of marshes might lead to temperature increase. As evidence of the beneficial results of forest cleaning, he pointed out that although Quebec and Paris are both at about the same latitude, Paris which is a cleared area is much warmer than forested Quebec. Buffon was one of the early writers to offer an explanation on soils. These he grouped into clays, calcareous Earth and vegetable earths, the last falling into two forms—the terrean (leaf mould) and the limon which is the residue in the decomposition of the terrean.

Glacken (1960) has evaluated Buffon's work as follows: 'The Histoire naturelle satisfied a hunger for concrete and detailed knowledge that was not dependent on mathematics or Cartesian deductive reasoning, but on study, travel, observation, and description.' This offers a background for understanding the philosophical views of scientific developments in this remarkable era, that was marked by a prodigious enthusiasm for exploration, travel, and the collection of data about the Earth, its flora and fauna, and its primitive societies.

During the early part of the eighteenth century, it was the *utilitarian attitude* which prevailed in geography. Geography was only of value for the light it could cast on historical events or for its aid in the science of government. But the status of geography which it held as subordinate to history had to be abandoned in the light of the rapidly increasing mass of subject material and empirical knowledge. Geography took its rightful place as an interdependent science and *there emerged an interest in reine geographie or pure geography.* This meant the use of natural physical land units instead of political units as a basis for description of both land and people. The two kinds of study were known as *Landerkunde* and *Statenkunde,* the seeds of which were sown long ago by Varenius.

There was a growing frustration among the contemporary geographers because of the predominance of the utilitarian attitude, which seemed to have stood in the way of rational basis for description and orderly presentation of real phenomena and the related spatial expression. It was in 1756 that a major breakthrough was achieved in the direction that caused the frustration to disappear.

Reine (Pure) Geography

The French geographer Philippe Buache and the German philosopher and writer of geography Anton Friedrich Busching, attempted to describe the areal similarities and differences in an orderly manner, and provided rational explanation. In his book published in 1756, **Buache** developed the concept of an Earth marked off into major basins bordered by continuous ranges of mountains. On land there are drainage basins, and the mountains form the drainage divides between different river systems. The basins continue under the oceans, and here the mountains form strings of islands or submerged sand banks. The persistence of his idea that drainage basins are bordered by mountains is amazing in view of the easily accessible examples in Europe of rivers that rise in one basin and flow through mountains to other basins. It was Johann Christoph Gatter of Germany who carried forward the view of Buache and identified the drainage basins as natural regions and used them as the frame of organisation for geographical test. The river basin was widely used as the framework for the identification of what would be now called systems of interrelated elements.

Anton Friedrich Busching produced his famous Neue Erdbeschreibung in six volumes in 1792, in which he sought to provide a description of the known surface of the Earth. In many ways, the work reflects the Munster tradition, as *political units were chosen for regional description.* It could be said that Busching made extensive use of the statistical materials provided by Anchenwall and Sussmilch. The work was, therefore, no different from that of the old masters such as Strabo. His contribution lay in his insistence on the critical handling of all source material and the setting up of a high standard of accuracy. Busching was very much influenced by the 'political-statistical school' of his time which led him to make use of population density as a geographic element (Peschel, 1865). He is credited with having formulated the theory of economic interdependence among countries and favoured the water transportation as it involved low transfer cost. *Undoubtedly, Busching's work started the trend towards pure Geography (reine geographie).*

The idea of the natural division of the Earth's surface in the late eighteenth century stirred the contemporary geographical scholarship to such an extent that the geographers became more and more involved in the development of the idea pertaining to reine geographic. H. G. Hommeyer carried the concept to its logical conclusion. He, however, abandoned political boundaries altogether, and divided his areas into 'terrains', natural regions which were in most cases river basins. German writers of the twentieth century also believed that geographical study should have its basis in natural rather than in political units. Johann August Zeune sought to discover the interrelationships of plants, animals, and men and to define his sub-divisions of the Earth's surface by reference to several factors, including climate and vegetation, not just relief. He offered an explanation which sought for an orderly presentation of natural units in which the areal similarities and differences could be properly organised, and further strengthen the base of the regional concept.

Universal Geography

Conrad Malte-Brun (1775–1826) is credited with the introduction of the first universal geography. He was born in Denmark and his name was Malthe Conrad Brunn. He was banished from Denmark in 1800 for liberal activities, and finally settled in France where he changed his name. His 'Precis de la Geographic Universelle' was published in eight volumes between 1810 and 1829 (though the work was completed by colleagues after his death).

He provides a vivid description of the history of geography, its ideas, concepts and paradigms that developed in the past three centuries. The work also provides the 'general theory of geography' which consists of its 'mathematical, physical and political principles. It then passes to 'the leading features of nature' and then to animals, plants, 'all the beings that are nourished in the exhaustless bosom of the Earth in their native region'. This introduction concludes with man in his natural and his political condition, meaning races, languages, beliefs and the laws which mark the progress of civilisation. In the second volume, he gives an outline of geographic concepts, including the shape of the Earth, the type of projections and the astronomical relation. Malte-Brun in his universal geography also reviews the various theories regarding the origin of the Earth and the controversy between the catastrophists and the uniformitarianists and then observes that the only way to treat physical geography in a useful way is to remain purely descriptive.

The second half and the next five volumes contain general sections on the world's major divisions. These divisions are selected on a basis that is essentially the same as the procedure of Strabo. Description is on the basis of countries, variously defined. For example, Asia is divided as follows: Caucasian countries, Turkey-in-Asia, Arabia, Persia, Caspian Sea, Afghanistan, Tartary, Siberia, Central Asia, Manchuria and Korea, Japanese Islands, China, Tibet, Industan, Indo-China and Oceanica. Brun is of the opinion that the description of an extensive country must be chronographic and ethnographic. He also discusses the land hemisphere concept and places the centre of the land hemisphere in France, to the west of Paris. However, be makes no reference to Buache.

The work of Malte Brun, in spite of its inadequacies and shortcomings, provides a landmark in the history of geographical description, for it offers a description of the lands and peoples of the Earth. However, it shows little, if any, advancement on the conceptual framework and procedure of Strabo who preceded him by almost a thousand years.

Philosophical Foundation of Scientific Geography

Immanuel Kant (1724–1804). A somewhat different stimulus to the development of scientific geography throughout this period came from the lecture course on 'Physical Geography' that the philosopher Immanuel Kant presented at the University of Konigsberg during 1756 to 1796. To him, the study of geography represented only an approach to empirical knowledge necessary for his philosophical consideration. But finding the subject inadequately developed and organised, he devoted a great deal of attention to the assembly and organisation of materials from a variety of sources, and also to the consideration of a number of specific problems, for example, the detection of wind direction resulting from the Earth's rotation (Hartshorne, 1976, 38).

His interest in physical geography was not stimulated by actual experience of the variety of nature in different parts of the Earth. It arose through his philosophical investigation of the whole field of empirical knowledge. His great contribution to geography, however, is arguably that he *laid the philosophical foundation* for the belief that the subject has a significant scientific contribution to make.

In contrast to Johann R. Forster, Kant was an armchair geographer. For this reason Kant's contribution was more philosophical than Forster's, since it consisted of his definition of the nature of geography and its relationship to the natural sciences. The definition given in the introduction of his lectures describes so completely the scope of geography that it has affected directly or indirectly all succeeding methodological discussion (Taylor, 1976, 38).

In his famous Critique of Pure Reason (1781), Kant rejected the teleological idea of final causes and insisted that explanations must be sought in what is chronologically antecedent. Physical geography, according to Kant, included not only the features of the Earth produced by natural processes, but also the races of man, and the changes on the face of the Earth resulting from human action. The later part of the philosophy conceiving of *'man as an agent of change',* seems to have been borrowed from Comte de' Buffon's 'Historic naturelle'.

Kant found that knowledge about the Earth as the home of humankind was a necessary support for his philosophical studies. 'Knowledge, according to him, is obtained: (a) either by the existence of pure reason, or (b) through the senses. Sense perception is of two kinds—those perceived by the inner and those by the outer senses—and together they furnish the whole of humankind's empirical knowledge of the world. The world, as perceived by the inner senses, is Soul (Seele) or Man (Mensch), that is the self; as perceived by the outer senses, is Nature. Anthropology (he uses authropology in the modern sense of Psychology) studies the Soul or Humankind's, physical geography studies Nature. Physical Geography is thus the first part of knowledge of the world; indeed it is the essential preliminary for understanding our perception of the world.

Kant's conception of the nature of geography and of its location within the sciences as a whole has provided the basis for a series of major disagreements. He considered that knowledge could be classified in two ways: either in accordance with their nature, or in accordance with their position in time and space. The former is a logical classification, the latter is a physical one. The *logical classification* collects all individual items in separate classes according to similarities of morphological features; it could be called something like an 'archieve', and will lead, if pursued, to a 'natural system' (Buttner and Hoheisal, 1980). The logical classification lays foundation for the systematic sciences—the study of animals is zoology, that of the rocks is geology, that of social groups is sociology (Holt-Jenson, 1981, 15).

The *physical classification,* in contrast, collects individual items which 'belong to the same time or the same space'. In this connection, Kant asserted: 'History differs from geography only in the consideration of time and (space). The former is a report of *phenomena that follow one another (nachienander)* and has reference to time. The latter is a report of *phenomena besides each other (nebeneinander) in space.* History is narrative, geography is descriptive. Geography and history fill up the entire circumference of our perceptions: geography that of space, history that of time' (Hartshorne, 1939).

Kant's views on geography appear to be broadly similar to those of Humboldt and Hettner. They appear to have had 'no direct influence' other than 'as a form of confirmation'. Indeed, they were not explicitly endorsed in any major programmatic statement of the scope of geography until Hartshorne's account of the nature of geography which accepted that geography's basic task was essentially Kantian:

'Geography and history are alike in that they are integrating sciences concerned with studying the world. There is, therefore, a universal mutual relation between them, even though their bases of integration are in a sense opposite: geography in terms of each space, history in terms of periods of time' (Hartshorne, 1976).

To Kant, 'knowledge about the spatial location of objects is quite distinct from knowledge about their time, nature and the natural laws governing them. The latter sort of knowledge are eternal and universal, are truly scientific ... (whereas) spatial and temporal coordinates are separate and rather secondary attributes of objects, and spatial and temporal arrangements of objects are not a matter of science.'

Kant is regarded as the *'father of exceptionalism'*[1] and was thus *opposed to the explanation and generalisation (rather than mere descriptions)* required if geography were to be reconstituted as a spatial science. It was Schaefer (1953) who traced out the exceptionalist view in the Kantian geography. But when Kant was working, Schaefer claims, history and geography were cosmologies, not sciences, and cosmology is not rational science, but at best thoughtful contemplation of the universe.

Hagerstrand attempts to revitalise Kant's basic distinction with regard to the classification of empirical knowledge. The contrast Hagerstrand draws is between compositional theory and contextual theory which clearly corresponds (in its essentials) to that between 'logical' and 'physical' classification.

Kant appears to have emphasised the structuring activity of the thinking subject in his 'Critique of Pure Reason': 'Space is not something objective and real, neither is it a substance or an accident nor a relation, but is subjective or ideal and proceeds from the ·nature of mind by an unchanging law, as a scheme for coordinating with each other absolutely all things externally sensed' (Hartshorne 1959, 1972).

This stress upon 'the epistemic structuring of the world by the human actor (is) the essence of the Kantian heritage, so it is claimed, and constitutes the common theme which has, in practice, been distilled from the variety of humanistic philosophies to which geographers of subjectivist orientation have turned in their endeavour to transcend the dichotomy inherent in subject-object relations' (ibid).

In the nineteenth-century Germany, there emerged what may be called *Neo Kantianism,* and those who maintained the tradition of Neo-Kantianism substituted a distinction between the cultural and historical sciences on the one hand, and the natural sciences on the other. The cultural and historical sciences deal with an intelligible world of 'non-sensuous objects of experience' which have to be understood and which are thus concerned with the Idiographic. The natural sciences deal with the 'sensible world of science' which could be explained and which are thus concerned with the Nomothetic.

The *Idiographic or empirical approach* places primary emphasis on the description of particular groupings of nations and people in terms of lands, seas, countries and places. It does not seek to develop laws, but to find out how phenomena account for the genus loci, the character of a place and its relation with other places. The *Nomothetic (deductive) approach* seeks to establish theories relevant to the location and interrelation of places and to establish laws and make deductions on the basis of laws. These are the basic approach and the traditional approach in all geographical inquiry and their contrast and conflict have become more marked and difficult to bridge as knowledge of the surface of the Earth has increased. In fact, the apparent dichotomy between Idiographic and Nomothetic approaches appears to have evolved at Miletus more than twenty-four centuries ago, at the time of Hecatacus but, in geography, the dualism developed only in the late eighteenth century in an organised manner.

Immanuel Kant is said to have pioneered other branches of geography which he defines as:

1. Mathematical Geography—'which treats of the form, size and movement of the Earth, and of the position in the solar system.'

[1] *Exceptionalism:* The belief that geography and history are methodologically distinct from other sciences because they are peculiarly concerned with the study of the unique and the particular. The idea is closely associated with Kantianism, but in geography the term is usually identified with Schaefer's (1953) posthumous challenge to what he regarded as the *Idiographic orthodoxy.*

2. Moral Geography—which discusses the different customs and characters of men, i.e. examines the contrast of Oriental civilisations where patricide is a most fearful crime, with customs in Lappland where a father, if wounded while hunting, expects his son to kill him.
3. Political Geography—'the study of the relationships between political units and their physical background'. For example, in ancient Persia two states existed whose mutual independence arose from, and was assumed, by the Kerman desert which divided them.
4. Commercial Geography—'which examines the reasons why certain countries have a superfluity of one commodity while others have a deficiency, a condition that gives rise to international trade'.
5. Theological Geography—'which studies the changes theological principles undergo in different environments (Boden). For example, one would make a comparison of the form of Christianity in the Far East with that in Europe, and of the variation of Christian beliefs in different parts of Europe.'

Evaluation. Kant gave geography a central place amongst the sciences, and geographers have often reiterated his views in justification of the existence of the subject and its special position amongst sciences (Holt-Jenson, 1981, 15). Hettner admitted that for a long time he had not paid sufficient attention to Kant's exposition of geography, but had rejoiced to discover a close correspondence between the conclusion of the great philosophers and his own. Hartshorne pointed out that Kant's approach led to satisfactory understanding of the nature of geography and answered all its basic questions. Hettner believed that the unity of geography can only be maintained through Kant's concept of a chronological science, i.e. a subject which studies things which are mutually coordinated, not subordinated, in space. Within human geography, Kantianism may be attributed to, among other things, the *Possibilism* of French School of Human Geography, and to *modern humanistic geography.*

Today, it is widely considered to be impossible, and to some extent philosophically untenable, to draw such sharp divisions between the 'sciences' as Kant did. The systematic sciences study phehomena with reference to time and space, and it is difficult to separate time and space in studies of human geography. Understanding of geographical situations is always improved when we consider their development over time. Just as there are geographers who wish to study the cultural landscapes of former times without necessarily using their knowledge to illuminate contemporary condition, so historians cannot neglect the study of the distribution pattern of phenomena. Here the individual sciences overlap each other.

During the period between the death of Kant in 1804 and Hettner's book in 1927, this methodological rule had not only been broken, but almost systematically opposed, since Kant's philosophical guidelines for the subject had been virtually forgotten until Hettner and Hartshorne reviewed them. Kant not only included man as one of the features 'encompassed in the Earth surface', but also considered man as one of the principal agents affecting changes on the Earth. Indeed, Kant's 'Physical Geography', both in purpose and in content, might be considered as 'anthropocentric', a point of view which Ritter appeared to have inherited from Kant.

Although the new science of geography sought its foundation in the more permanent physical features of the Earth, it was characteristic of its time in considering these primarily in relation to humankind. Throughout the period most writers, consciously or unconsciously, followed Kant in regarding physical features in terms of their ultimate importance to man. Muller in 1785, and Kayser in 1810 specifically emphasised the study of the Earth as the dwelling place of humankind. 'In large part, however, the new science of "reine geographie" represented the expression of purposes and hopes rather than of accomplishment. In spite of the insistence on establishing facts, the geographers of this period, with few exceptions, continued to set-up *a priori* systems of facts without putting them to the test, and many of these concepts we still inherit—or have constructed anew' (Hartshorne, 1939).

Russian Geography

The Russian geography of the pre-classical period also appears to have been rooted in the new sciences of reine geographie. As the new geography found its basis in physical rather than political features, it did not exclude the specific problems of social and political geography. M. V. Lomonosov is credited with the systematic collection of information about the physical character of the land, the population, social condition and the economic situations. He was made the head of the Department of Geography of the Russian Academy of Sciences in 1758 (Gerasimov, 1968).

The contemporary Russian geography was very much influenced by the Germans. The great German philosopher and writer in geography, Anton Friedrich Busching who was the pastor of a German Lutheran Church in St. Petersberg from 1761 to 1776, provided the Russian 'reine geographie' with some kind of dynamism because his *Neve Erdbeschreiburg* dealing with Russia was translated into Russian. The Russian geographers of the time laid greater emphasis on region as the basis of organisation for geographical work and insisted that regions were the real entities which could be objectively defined. However, the political units were used for regional base, and description as extensive materials, based on empirically observed facts, were made available by Ivan Kirilov, the first Russian named head of the cartographic office. Expeditions were organised to establish the location of rivers, coasts, mountains, places, political units and to identify places where furs or precious metals could be found.

Viewed as a whole, the work of the eighteenth century geographers is very notable. Academic controversy between political, statistical and pure geographers levelled the barriers of traditional thought and opened the way for fresh unimpeded advance. The foundations were thus laid on which during the next 50 years the edifice of scientific geography arose.

5

Classical Period of Modern Geography

HIGHLIGHTS
- Alexander Von Humboldt (1769–1859)
- Carl Ritter (1779–1859)

If the geographers of the late eighteenth century developed the greater part of the theoretical concepts of the new science of geography, the transformation of their 'ideas, demands and wishes into facts' was largely the work of **Alexander Von Humboldt** (1769–1859) and **Carl Ritter** (1779–1859). It is an extraordinary fact that modern geography in all lands should owe so much to these two men living at the same time in the same country for over 30 years in the same city. (Many writers refer to Humboldt and Ritter as the founders of modern geography. But there are also good reasons for thinking of them as bringing the period of classical geography to an end).

Using the large volume of new information resulting from the voyages of exploration, Humboldt and Ritter, each in his own way, produced massive syntheses. Although these syntheses made use of the new concepts, and methods of study developed during the preceding two centuries, they, nevertheless, sought to present universal knowledge, just as Strabo had done and as had been attempted during the Age of Exploration by Munster, Varenius, Busching and others.

Both the founders of the classical modern geography depended in large part on their predecessors whom they had in common. Kant gave geography a philosophical foundation and the Forsters demonstrated method of research and a literary style, but Humboldt and Ritter developed the subject as an independent branch of knowledge. Both of them had many views in common and were united in their criticism of the causal and unsystematic treatment of geographical data by other predecessors. The similarity of their opinion was not accidental—Ritter regarded himself as a student of Humboldt and Humboldt described Ritter as 'his old friend'. They differed markedly, however, in temperament and character.

The geography which Humboldt and Ritter demonstrated was defined as 'classical' by Hartshorne (1976), because it dominated the foundation period of the subject and because its methods were uniform and simple. Both believed that science must be founded on the objective descriptions of observed facts rather than on logical propositions like those proposed by Kant and others; geographers should collect empirical data and must shape order and association from this material in order to arrive at an inductive explanation. They both died in 1859, the year of the publication of Charles Darwin's *Origin of Species*. Their ideas and purposes, although formulated before the great impact of evolutionary thought, and before the enormous extension of exploration and mapping in the latter half of the century, have continued to serve as guide posts in the field of geography.

ALEXANDER VON HUMBOLDT (1769-1859)

Humboldt was the last of the great polymaths. He mastered a number of disciplines and put all his energy into travel and research in order to understand the whole complex system of the universe. He contributed to so many branches of science that his work is almost as difficult to summarise as Carl Ritter's. At various times, he did research in botany, geology, physics, chemistry, anatomy, physiology, history and all aspects of geography. On several occasions Humboldt had to defend himself against the charge of being too versatile.

Life and Works. He was born in Berlin in 1769 and belonged to the family of a Prussian land-owning aristocracy. His father was an officer in the Prussian army. The unity of human races, an idea which Humboldt pioneered in the latter part of his academic career, seems to have been developed on account of his frequent meetings (arranged by his mathematics tutor) with the Jewish philosopher, Moses Mendelssohn. With him and others, Humboldt discussed matters relating to the contemporary social inequities of an aristocratic society and drew up plans to do something about these things. He also attended the lectures on scientific subjects, delivered by the noted physician, Marcus Herz, who was a disciple of Immanuel Kant. These lectures had a profound impact on the thinking of Humboldt. He also witnessed the demonstration of scientific experiments which Marcus performed.

At the age of 18, Humboldt entered the University of Frankfurt, but there he stayed for six months only. In 1789, he entered the University of Gottingen, where he had the opportunity to meet George Forster who had lately returned from his voyage around the world with Captain Cook. Before Humboldt came into contact with George Forster, he was very fascinated by geology, and this fascination led him to organise a short scientific tour, in the autumn of 1789, to the Rhineland Basaltic region. He published a short monograph and several articles on the Rhineland Basalt and attempted to uphold the current theory of the aqueous origin of basaltic rock.

However, in the company of George Forster, Humboldt made his first foreign tour in 1790 through Holland, Belgium, central and southern England and northern France. Forster's interest in geography, his method of careful observation and critical treatment of facts, and above all his talent for artistic though scientific description of landscape made an indelible impression on his young companion. Humboldt said later that his interest in geography started with his acquaintance with George Forster.

After he came back from the foreign tour, Humboldt joined the School of Mines at Freiberg in Saxony in 1791 where he met the famous geologist and scholar A. G. Werner. There, Humboldt attended lectures in physics, chemistry, geology and mining. In 1792, he was appointed as the director of mines in the Prussian state of Franconia. His official position provided him opportunities to travel across southern Germany and carry out extensive investigations in botany, geology and meteorology. His active mind was always formulating new questions about almost everything that caught his imagination and attention. He is said to have studied the effect of different rocks on magnetic declination. He also carried out scientific experiments on the subterranean plants which he found growing in the underground mines and in 1793 published his first scientific paper, based on the results of these findings. There was no limit to the range of his curiosity and he wanted to travel and see for himself what the different parts of the world were like.

Humboldt visited Bavaria, Austria, Switzerland and Italy, on which trip he observed the rock structure of the Alps and tested some of the ideas of the Swiss scholar, Horace Benedict de Saussure, who thought that the deep Alpine valleys had been cut by the rush of water in the receding flood. Visits to his brother Wilhelm, who was then living in Jena, brought Humboldt into contact with Goethe and Schiller, which strengthened his aesthetic appreciation of and philosophical approach to nature.

Humboldt gave up his government job in 1797 after the death of his mother, and began to plan for travelling. In Paris, he had access to a variety of instruments and learned how to make use of them. There he met Pierre Simon Laplace and learnt from him how to use the aneroid barometer to determine elevations above sea level. Humboldt had to call off his expedition to Egypt and also to the Pacific due to some unavoidable circumstances. In 1798, he sailed with the French botanist Aime Bonpland across the Mediterranean Sea. The purpose of the voyage was to go to Marseilles and then to sail for Algiers, from where they intended to travel overland to Egypt. However, the plan could not succeed as the ship was wrecked off the coast of Portugal before it ever reached the shores. They then decided to set out for the city of Madrid. On the way to Madrid, Humboldt made daily observations of temperature and altitude, and he was the first to make an accurate measurement of the elevation of the Spanish Meseta. In Madrid, Humboldt's position in the Prussian aristocracy gave him access to the ruling aristocracy and he succeeded in getting permission to visit the Spanish colonies in America.

Humboldt and Bonpland sailed for America in June 1799. They returned to France in 1804 and Humboldt lived in Paris until 1827. In Paris itself, Humboldt was much influenced by one of the five methodological precepts of August Comte's 'le reel' which meant that the scientific status of knowledge had to be guaranteed by the direct experience of an immediate reality, and this required a particular conception of causality in which causal relations amounted to regular associations between phenomena (Gregory, 1987, 26). It can be said that Humboldt was impressed by Comte's positivist approach.

Exhausted financial position forced him to go back to Berlin in 1827. He was offered a post of Chamberlain to the King of Prussia which he accepted. In 1829, Humboldt was invited by the Russian Tsar to explore the virgin lands of Siberia across the Ural mountains.

Expeditions to America

Humboldt and Bonpland landed at Cumana, in what is now Venezuela, and then went to Caracus. One of the first places they visited was the Basin of Volencia in the midst of which is the Lake Valencia. Humboldt found Lake Valencia being shrunk and crops were being grown on the flat lake-bed soil from which the lake water had receded. He presumed the existence of a continuous cover of tropical forest around the Basin of Valencia which had now been cut down and the lands were being used for agriculture. Humboldt, who had earlier learnt about Buffon's theory that the removal of forests caused the drying up of rivers and climatic change, was the first to test this theory by confronting it with the empirically observed facts which he noticed around the Basin of Valencia.

In 1800, they explored the Orinico and established the truth of its connection with the Amazon. They mapped some 1,725 miles of the Orinio River, mostly through uninhabited forests. Humboldt established the exact latitudes of places they visited and came close to correct longitudes. They collected some 6,000 plant specimens and rock specimens, all of which were transported to Caracas and then to Cuba. In 1801, they arrived in the Colombian port of Cartagena and from there began the exploration of the Andes of Columbia, Ecuador and Peru.

Humboldt drew sections along the Andes based on 1500 of his own measurements and started the practice of using these profiles to show the altitudinal belts of vegetation, an idea conceived during his visit to Teneriffe (1799). He also used them to show geological structure and, though not the first to do so, he demonstrated their value and geological cross-section so successfully that their invention is frequently attributed to Humboldt. He provided the first scientific description of the relation of altitude, air temperature, vegetation and agriculture in tropical mountains. He is said to have disputed the idea of A. G. Werner regarding the origin of rocks when he closely observed the rocks of the Andes.

Humboldt collected, classified and interpreted plants, animals and rocks with respect to their origin and their graphic distribution, and he was the first to draw cross-sections of relief in order to show the altitudinal limits and zones of correspondence of the phenomena which he everywhere systematically observed. He also collected measurements of heights, areas, length and so forth to discover the areal distribution and interrelation of phenomena.

He climbed most of the volcanoes of Ecuador and in 1802, along with Bonpland, he scaled the Mt. Chimborazo, reaching an altitude of 19,286 feet which was the highest known peak of that time. Humboldt offered explanation regarding the effect of altitude on human beings and deduced that the feeling of dizziness resulted from low air pressure. He pointed out the symptoms of mountain sickness.

Humboldt and Bonpland finally reached Lima where Humboldt made some astronomical observations. On a sea voyage from Callao in Peru to Guayaquil in Ecuador, he measured the temperature of the ocean water and for the first time described the movement of ocean water, including the upwelling of the cold water from below. He named this the Peruvian current. In 1803, both sailed from Guayaquil to the Mexican port of Acapulco.

In 1804, they reached Havana, Cuba, but there Humboldt was pained to see the deplorable condition of the slaves and the deteriorating socio-economic conditions of Cuba. He deplored the racial gradations among men, prevalent in Cuba, which he described as unkind and inhuman. Earlier, he was much upset by the social inequities which he observed in Prussia, and now he was pained to see the deteriorating condition of the human race, vis-a-vis the slave trade in Cuba. These observations about the quality and condition of the human race led Humboldt to vigorously maintain the concept of the 'unity of the human race.'

They reached Philadelphia in 1804. There they visited the American Philosophical Society and then started for Washington by way of Baltimore. At Washington, Humboldt had the opportunity to meet Thomas Jafferson, the American geographer whose interest in the geographical research had already been known. Both became close friends. Humboldt was very influenced by the liberal ideas of Jefferson. On June 1804, he and Bonpland sailed for Bordeaux. On returning to France in 1804, Humboldt lived in Paris until 1827. It was in Paris that Humboldt wrote the 30 volumes of the results of his expeditions and explorations to the Spanish colonies.

Publications of Results of American Expeditions. The publications of the results of his expeditions in 30 volumes are included under the general title *Voyage aux regions equinoxiales du Nouveau continent*. Humboldt's books provided the reports of careful scientific investigation, the seeking of answers to question about the inter-connections among the phenomena grouped together in rich diversity on the face of the Earth. He was always concerned with the areal associations of natural and organic phenomena. In handling of selected natural phenomena, he invariably turned to their distribution and their impact on the spatial arrangement of other phenomena.

Humboldt's *Relation historique* (vols. 28–30, of which the fourth volume was never published) had profound impact on the contemporary world. It usually dealt with his own experiences and hardships, provided report on scientific problems which had been investigated, and on the results achieved. In vol. 27, he distinguished between the distribution of individual plant species and the numerical association of plants in particular areas. He examined the numerical distribution of plants in different isothermal zones. In this volume, he presented a synthesis of his detailed findings as a basis for the study of plant geography. He was also the first to divide the regions he explored in the Spanish colonies in Latin America, into botanical provinces, maps of which were published in his *Atlas, Geographique du Nouveau Continent*, which came out in 1814–19.

Humboldt's *Essai Politique Sur Ie Royaume de la Nouvelle Espange* (vols. 25–26) appears to have introduced a new period of regional geography because he attempted to recognise the interdependence of areal phenomena and the need for explaining any set of spatially distributed phenomena in relation to their spatial content. The book dealt with the resources and products of New Spain (or Mexico) in relation to population and political conditions.

Humboldt was moved by the far greater prosperity of New Spain (or Mexico) than the countries of northern South America. His interpretation of this difference was based on the theory that the only proper way to increase the general prosperity of a country depended on the rational utilisation of the natural resources, and New Spain (or Mexico) seemed to have done it effectively. His conclusion was based on the availability of the well organised statistical data he found in Mexico. Humboldt is said to have suggested that a canal should be dug across the Isthmus of Central America and that the best place to do this would be in Panama.

The *Essai Politique Sur l'le de Cuba* (published after 1826) is another regional monograph in which Humboldt attempted to explain the deplorable conditions of Cuba, and pointed out that elimination of slavery and an effective use of natural resources would help Cuba recover from the socio-economic hardships. He outlined the fact that the phenomena arising out of the slavery and lack of effective utilisation of the available resources in Cuba were inter-connected spatially to create such a state of socio-economic uncertainty there.

Russian Expedition and Generalisations

Humboldt first went to St. Petersburg, and from St. Petersburg travelled on horseback through Kazan, Bogosloski, Tabolsk, Tara, Bersk, Omsk and reached upto the borders of China. While returning he passed through Omsk, Orenburg and Astrakhan and made a survey of the coastal lowlands of the Caspian Sea. Throughout his Russian expedition, he kept on taking observations of temperature and air pressure at different places, and ultimately deduced that temperatures varied at the same latitude in accordance with distance from the ocean. On return to St. Petersburg, he urged upon the Tsar to set up a network of weather stations at which weather data could be recorded regularly and in accordance with standard procedures to make the results comparable. During his Siberian expedition, he wrote to a member of the Russian Government: 'The Ural mountains are a true El Dorado, and I am confident from the analogy they present to the geological conformation of Brazil ... that diamonds will be discovered in the gold and platinum washings of the Ural mountains'. A few days later diamonds were found in the gold and platinum washing.

On the basis of the temperature data he collected from different weather stations set up by the Tsar, Humboldt constructed the first world map of average temperature in which he used lines to connect points of equal temperature, called isotherms. From a compilation of isothermal lines, he was able to demonstrate that near the equator they are parallel to it, but nearer the poles they spread north over the sea and south over the land and that they reach further north on the west coast of continents. He also saw the need for seasonal isotherms. On the basis of the isothermal maps, Humboldt developed the concept of continentality that continental climates are warmer in summer and colder in winter than places near the oceans at the same latitude. Climates of islands and coasts differ from the centre of continents in that the islands have milder winters and colder summers than the coasts.

It was largely due to his Siberian expedition, which helped him to formulate generalisations about the climatic phenomena, that Humboldt probably introduced the term 'climatology' for the first time. To him, climatology deals with all variations of atmosphere—temperature, humidity, barometric pressure, winds, purity and degree of visibility. He is also credited to have coined the term 'perma-frost' to describe the permanently frozen soil of Siberia. He is said to have expressed doubts over the idea of a universal Ice Age which was developed by the Swiss scholar, Louis Aggasiz.

Geographical Concepts

Geography, according to Humboldt, was the means of comprehending *'the harmonious unity of the cosmos'* as a living whole, 'a unity of multiplicity'. Insight into the cosmic organism creates a spiritual enjoyment and an inner freedom which even under fate's hard blows cannot be destroyed by external power (Hartshorne, 1976, 65). Humboldt's concept of geography was basically the same as of Ritter's. In an early work (*Flora Fribergenis*), he briefly touched upon the limits of the various sciences and distinguished between 'Physiographia' (the systematic natural sciences), 'Naturgeschichte (natural history) where the emphasis was on the development of things in time, and 'Geognosie' or 'Weltbeschreibung', which discussed spatial distribution. The science of spatial distribution thus established was not limited to the Earth's surface, it was not a description of the Earth but of the world, i.e. a science of the cosmos (Tatham, 1967, 51–52).

Humboldt points out: 'the uncommon but definite expression of the science of the Cosmos recalls to the mind of the inhabitant of the Earth that we are treating of a more widely extended horizon; of the assemblage of all things with which space is filled. From the remotest nebulae to the climatic distribution of those delicate tissues of vegetable matter which spreads a variegated covering over the surface of our rocks … if scientific terms had not long been diverted from their signification, the present work ought rather to have borne the title *Cosmography* divided into Uranography and Geography' (Humboldt, 1845, 53).

Uranography was descriptive astronomy dealing with the celestial part of cosmos. Geography, or physical geography, dealt with the terrestrial part. Its ultimate aim was 'to recognise unity in the vast diversity of phenomena, and by the exercise of thought and the combination of observations, to discern the constancy of phenomena in the midst of apparent changes'.

The Humboldtian geography seems to reflect the Kantian concept of geography, but there is no evidence of Kant being quoted by Humboldt. However, the question normally arises: to what extent is his thought derived from Kant? Humboldt's Flora Fribergenis was published nine years before Kant's lectures on geography, but Humboldt was so acquainted with Kant's philosophy and scientific opinions from the age of sixteen onwards that it is highly probable that he had some notion of Kant's concept of geography before he wrote the Flora. Whatever the actual facts, it was the Kantian concept of geography that Humboldt held and which he expounded in *Cosmos* (Tatham, 1967, 52–53). He was also influenced by the philosophical approach of Rousseau which dealt with the descriptions of 'the harmonies of nature', and which had profound effect on Humboldt's philosophical foundation.

It is assumed that Kant' s logical classification of knowledge made room for the specialist in the study of particular processes without reference to time or space. That this was a generally accepted division of the world of scholarship and not an invention of Kant is made clear by Humboldt's earlier studies (1793) of the subterranean plants in the mines at Freiberg. In the introduction to this monograph, he pointed out that he was not studying plants as such, but rather the plants in relation to their surroundings.

Humboldt strongly *held the idea of the unity of human race,* 'a concept which he further developed and strengthened, and manifested his concern about the deteriorating human condition both at home and abroad, especially in Cuba' (Humboldt, 1811 and 1826). The degeneration of Christianity by great wealth was a great socio-economic phenomenon that seemed to have rocked the continent much earlier and which persisted even during his time. The opposite 'hypothesis of racial gradations among men' he characterized as 'not only unkind but also false'.

Humboldt believed that all the races of humankind had a common origin and that *no race was necessarily inferior* to others; all races, he insisted, were equally destined for freedom, individually or in groups. In his master work, the *Cosmos*, he remarked: 'We wish to note our idea which is visible in ever-increasing validity through the whole of history .... the idea of humanity to treat the whole of humanity, without considerations of religion, nationality and color as one great closely related race, as one whole existing for the attainment of one purpose, the free development of inner powers' (Humboldt, 1858). Although 'Humboldt's thesis would find more support among the anthropologists than that on which Frobel as well as Peschel based their criticism, the criticism itself was essentially justified, his views on the question were strongly influenced by non-scientific consideration' (Hartshorne, 1976, 64).

To Humboldt, the *concept of unity of nature presumed a causal interrelation of all the individual features in nature.* The phenomena of nature were studied in order to establish this coherence and unity. For him, it was axiomatic that the unity of nature included the organic as well as inorganic, human as well as non-human, material as well as immaterial. The exclusion of any part would be not only arbitrary, but would destroy the coherence and unity of the whole. Humboldt remarked: 'My attention will ever be directed to observing the harmony among the forces of nature, to remarking the influence exerted by inanimate creation upon the animal and vegetable kingdom.' He further continued: 'Nature, considered rationally, that is to say submitted to the process of thought, is a unity in diversity of phenomena, a harmony blending together all created things, however dissimilar in form and attributes, one great whole animated by the breath of life. The most important result of a rational inquiry into nature is therefore to establish the unity and harmony of this stupendous mass of force, and matter, to determine with impartial justice what is due to the discoveries of the past and to those of the present, and to analyse the individual parts of the natural phenomena without succumbing beneath the weight of the whole' (*Cosmos*, vol. I, 2–3).

To establish this unity, the relationships of organic life (including man) to the inorganic surface of the Earth must be investigated. For Humboldt, this *unity of nature was teleological, anthropocentric unity*—he wrote of the comprehension of 'the inner, secret play of natural forces', where Ritter spoke of discovering the 'divine secrets'—but was a balanced unity of the whole of nature, of which man was a part.

An important factor in the evolution of Humboldt's thought was close personal relationship with Goethe. With him, and with the entire generation of the 'romanticists' (in the widest sense), Humboldt shared the idea of an organic coherence of all phenomena. It was through Goethe that he received the concept of the 'Landschaft' as conceived by the 'romantic' movement (Hartshorne, 1976, 65).

He shared the concept with Hegel that envisaged the Earth as an inseparable organic whole, all parts of which were mutually interdependent. This was the concept of the physical world or the field of moral philosophy, the Earth and the universe, in so far as accessible to the senses, as opposed to the view of humankind's inner world or the field of moral philosophy. This synthesis Humboldt regarded as a harmonious unity, a living whole of all parts. He derived aesthetic satisfaction through the scientific analysis of the ways in which things and phenomena on the Earth's surface depended upon each other, an idea which runs right through the *Cosmos* and was clearly in his mind as early as 1797 as the physique du mode. His field of inquiry was the observable world of phenomena, 'the total impression' (Dickinson, 1969, 23).

Humboldt did not pursue idealism[1] as far as Ritter, for his concept of the unity of nature was more aesthetic than religious. In the writings of

[1] *Idealism:* A philosophy which regards reality as residing in or constituted by the mind (metaphysical idealism) or which limits human understanding to perceptions of external objects (epistemological idealism).

Rousseau and St. Pierre, it was the descriptions of the harmonies of nature that appealed to him very much and through Goethe he learnt of the concept of the 'Landschaft' as conceived by the romantic movement. To Humboldt, every part of the world was a reflection of the unity of the whole.

In the handling of his multitudinous observations, Humboldt demonstrated far more clearly than Ritter *the value of comparative geography.* The essay on 'Steppes and Deserts' in *Views of Nature*, is full of comparisons of steppes and ocean, of all the steppes· of the world, the heaths of central Europe, Llanos, Pampas, North American prairies, African desert, central Asiatic steppes—all are compared so as to bring out the peculiar character or physiognomy of each as determined by diversity of soil, climate and elevation above the sea. He attempted to compare his every new observation with previous new observation of similar kind and thus recorded similarities and differences. His object was not simply to measure one kind of phenomenon, but to bring out the ways in which the variety of observable phenomena of the landscape are associated and interconnected with each other at different places. This was the essence of his concept. Humboldt combined this comparative method with exceptionally sharp observation and thus deduced by objective descriptions of observations of nature.

'The Cosmos'

Humboldt would be known for his book, the *Cosmos*, which be produced during the last years of his life. The first two volumes were published in German in 1845 and 1847, the third and fourth in 1850 and 1858, and the fifth was published posthumously in 1862. Written in superb literary style, the *Cosmos* became the most prestigious scientific work ever produced upto that time. It put together in one unified work all the various interests and discoveries of Humboldt's life time.

The first volume of the *Cosmos* provides a general presentation of the whole picture of the universe. The second volume begins with a discussion of the portrayal of nature through the ages by landscape painters and by poets and then continues with a history of humankind's effort to discover and describe the Earth since the time of the ancient Egyptians. The third volume deals with the laws of celestial space, i.e. astronomy. The fourth volume deals with the Earth, not only with geophysics, but also with man.

In the introduction of the *Cosmos* one reads: 'The most important aim of all physical science is this: *to recognise unity in diversity,* to comprehend all the single aspects as revealed by the discoveries of the last epochs, to judge single phenomena separately without surrounding to their bulk, and to grasp nature's essence under the cover of outward appearances.... The purpose of this introductory chapter is to indicate the manner in which natural science can be endowed with a higher purpose through which all phenomena and energies are revealed as one entity pulsating with inner life. Nature is not dead matter, but She is the sacred and primary force' (*Cosmos*, vol. I).

The philosophy of the *Cosmos* comprises four main themes:

1. the definition and limitation of a physical description of the world as a special and separate discipline;
2. the objective content, which is the actual and empirical aspects of nature's entity in the scientific form of a portrait of nature;
3. the action of nature studies through media such as travel descriptions, poetry, landscape painting, and the display of contrasting groups of exotic plants; and
4. the history of natural philosophy and the gradual emergence of concepts pertaining to the cosmos as an organic unit.

Humboldt seemed to have been impressed by the natural philosophy founded by Schelling, as he felt the need for something better and higher to which everything could be referred. He was convinced that 'a true natural philosophy could not harm empirical research'; on the contrary, 'such a philosophy leads findings back to principles and is likewise the foundation for new findings'. Humboldt's special treatment to Schelling's natural philosophy finds expression in his *Cosmos.*

Humboldt *described man as a part of nature.* He remarked: 'The general picture of nature which I have endeavoured to delineate, would be incomplete, if I did not venture to trace a few of the most marked features of the human race, considered with reference to physical gradations to the geographical distribution of contemporaneous types, to the influence exercised upon man by the forces of nature, and the reciprocal, although weaker, action which he in his turn exercises on these natural forces. Dependent, although in a· lesser degree than plants and animals; on the soil, and on the meteorological processes of the atmosphere with which he is surrounded—escaping more readily from the control of natural forces by activity of mind, and the advance of intellectual cultivation, no less than by his wonderful capacity of adapting himself to all climates—man everywhere becomes most essentially associated with terrestrial life' (Humboldt, 1858). Humankind was thus considered as an element in the balance of the physical world and his capacities allowed him to contemplate with aesthetic satisfaction.

Humboldt's *Cosmos* is a monument of compilation and in it are embodied the conclusions drawn from a lifetime of travel and diligent research. The method of presentation consists of the art of collecting and arranging a mass of isolated facts and rising thence by a process of induction to general ideas (Dickinson, 1966, 29). Humboldt's *Cosmos* was one of the earliest modern attempts to provide a theoretical conception capable of systematising scientific knowledge about humankind and nature in geographical terms. For all its Kantian trappings, it spoke directly to Comte's Ie reel and Ie relative (Gregory, 1978, 45).

Systematic Geography

Humboldt categorised the knowledge of the natural or physical phenomena into three categories:

1. First were the phenomena examined taxonomically as form and content, that is phenomena which were categorised according to analogous characteristic such as botany, zoology and geology. These were the *systematic sciences.*
2. Second, there were the *historical sciences* dealing with existing groups of phenomena—the history of the development of animals, plants and rocks.
3. Third, there was *geography or Earth science,* which Humboldt defined as *Erdkunde,* or physical geography. This is concerned with the distribution or arrangement of phenomena on the Earth's surface. This is the study of phenomena in their spatial distribution and spatial relationship and interdependence.

Humboldt pointed out that in order to establish the unity of the total cosmos, it seemed more important to make systematic studies of particular kinds of phenomena in their interrelations in areas, than to prepare complete studies of individual areas. His effort to cover the whole of 'physical geography' in the *Cosmos,* demonstrated this view. In the first volume of the *Cosmos,* Humboldt described physical geography as the study of 'that which exists already (im Raume) together; it studies phenomena as arranged in areas in their mutual relation to all other phenomena with which they form a natural whole' (*Cosmos,* vol. I). Thus it included all Earth phenomena, organic and inorganic. It was the inter-connection of different categories of phenomena which made the systematic studies much more intrinsic to understand the natural whole.

Humboldt *distinguished between natural sciences and geography.* Natural sciences which include physical and biological sciences, according to him, were concerned with Earth phenomena—study the forms, construction and processes of individual animals, plants, solid objects or fossils and seek to arrange these in classes and families according to their internal analogies. Geography, on the other hand, was concerned with these objects as they exist together, related to each other causally in an area. In contrast with physics or botany, in which the objects of nature were divided according to kinds of objects, Humboldt pointed out that geography regarded all the objects as a natural whole as they stood in areal connection, in part with the Earth body, in part with the universe (Hartshorne, 1976, 76–77).

He recognised that to comprehend a whole consisting of a multitude of things and their interrelation, it was necessary to understand first the relation of some of the things to all others. Humboldt concentrated on particular classes of phenomena—particular vegetation—in relation to all other phenomena in areas. The development of plant geography required the combination of means of measuring elevation and temperature with a knowledge of plants, and had been delayed until the methods of the physicists could be combined with those of the botanists. Humboldt pointed out that though a systematic study in geography must be focused on one category of objects, it was in no sense to be restricted to the objects of that category, but was to consider those objects in their relationship to other geographic phenomena. Though at various places in his writings, he presents brief systematic considerations in anthropological geography, his studies in systematic geography were primarily limited to non-human geography.

Humboldt's contribution to the development of systematic geography is therefore to be found not only in the value of the works which he produced, but also in the fact that he first clearly portrayed the distinction between systematic and regional, but chronological studies in geography and systematic studies in spatial sciences. The fact that, rather than attempting systematic studies in the spatial fields, he maintained the chronological point of view, enabled him to establish several branches of systematic geography, notably climatology and plant geography.

Regional Geography

Humboldt can also be described as a regionalist, for his description of areas seemed to have initiated a new period in regional geography. He *recognised the interdependence of areal phenomena* and the need for explaining any one set of spatially distributed phenomena in relation to their spatial content. His 'Essai politique sur Ie Royaume de la Nouvelle Espagne' and 'Essai politique sur I' ile de Cuba', are the two major works which raised geography to a scientific status in the regional description. These regional descriptions and studies reflect Humboldt's interest in natural condition (which he considered subordinate to socio-economic condition) and much of the material cannot be included in geography as they dealt with political economies.

However, the manner in which Humboldt analysed the differences in economic, social and political condition of different areas in relation to the differences in natural conditions was followed neither by economists or political scientists, but by geographers who made this form of regional study one of the functions of geography. Humboldt made significant contribution to regional economic geography which was based on the empirically observed facts. He described in detail about Mexico and Cuba which he visited and so was able to provide a relatively complete picture. He not only considered each of the significant features of an area, but combined their individual characteristics for each distinctive part of the area, then showing the relation of the different parts of the area in the total. However, *Humboldt's regional geography was not anthropocentric because the human factors appeared to be subordinated to the natural factors.*

Methods and Approach

Humboldt emphasised the need for careful observation of nature in the field and for the careful and precise measurement of observations. He outlined the importance of *the empirical method of research.* He pointed out: 'I limit myself to the domain of empirical ideas. Facts remain; even the hastily erected edifice of the theory has long fallen into ruins. I have always kept my facts distinct from my conjectures. This method of dealing with the phenomena of nature appears to be the one best grounded and most likely to succeed' (*Cosmos*, vol. I, 349).

'One of the first (and most precocious) attempts to qualify geography's commitment to the classical models of Baconian empiricism was Humboldt's *Cosmos*, but despite its incorporation of elements of Kant's transcendental idealism Humboldt remained, like so many of his predecessors and his heirs, *a strong advocate of inductive approach*' (Gregory, 1978, 37). Over one hundred years later, the quantitative revolution continued to strengthen the hold of inductivism, and still more recently it has received a new impetus from the development of spatial techniques whose auto-projective methods can be subseemed within the inductive-statistical model.

Contribution Undoubtedly, Humboldt was the last of the great polymaths. He mastered a number of disciplines and put all his energy into travel and research in order to understand the areal associations of diverse categories of physical and human phenomena of the whole complex system of the universe. He is credited with having *made geography an original and distinctive science*, rather than a collection of facts from the physical and biological sciences. He appeared to have combined his aesthetic standpoint with a most critical scientific explanation to explain the natural whole.

He *set forth a new paradigm* which laid the scientific foundation to the contemporary geographical scholarship not only in Germany, but also in the whole of Europe. Humboldt may be described as an academician of versatile genius who had skilfully widened the knowledge and scope of the geographical horizon in the nineteenth century. Humboldt may be identified with the introduction of reductionism in geography as he offered explanation in the *Cosmos* which sought to catalogue separately every facet of the physical environment and all the known facts of human distribution and activity about the places he visited in course of his expeditions. His studies in physical geography were specifically carried out using the methods of contemporary physical and biological sciences.

CARL RITTER (1779–1859)

Carl Ritter exercised a much more direct influence on the development and growth of geography in Germany than did Humboldt. Ritter was born in Quedlinburg in 1779. His father was a physician and when he died his widow lacked any means of support for her family of five. Inspite of acute financial trouble, Carl was fortunate enough to have the guidance of a geographer named J. C. F. Guts Muths whose major interests were in nature study and in geography and who had earned a name for his contributions to the development of geography.

Carl Ritter received his early training in a school in Schnepfenthal near Gotha, where the teaching was based on the principles of Rousseau and Pestalozzi. Rousseau in his novel Emile (1762) devised a new educational procedure which contradicted the traditional method, as the purpose of the new procedure was to encourage a child to develop inborn potentialities. The Swiss scholar and educator, Johann Pestalozzi, further developed the ideas and pointed out that concepts must be based on the observation of phenomena and the young children must be exposed to nature so that they could observe the close involvement of man with his immediate surroundings. Salzamann was enthusiastic about these new suggestions and founded his school at Schnepfenthal in the Thuringerwald to experiment with them. Space relations were basically emphasised. Students were taught to observe the relationships of things in their immediate vicinity.

At an early age, Ritter was able to observe the close relationship between humankind and his environment. From the richly varied landscapes of the region of hills and low mountains, to which Ritter was exposed by his teacher, he derived the *idea of unity of nature which became a basic theme of his geographical scholarship.* At the age of seventeen he entered the university of Helle, where he worked with Professor Niemeyer. He learnt Latin and Greek and read widely in geography and history. It was here at Frankfurt-on-the Oder that Carl picked up the habit of frequent field trips around the city. He greatly increased his competence as an observer that helped him to develop his own methods. He went to Switzerland and Italy, etc. making careful and detailed observation. Ritter came into close contact with many eminent scholars of the day, of whom the most significant were the anatomist Sommering, the geologist Ebel, and the leader of educational reform Pestalozzi.

Ritter's preparation for research covered an extremely wide field, in which on the one hand natural sciences (particularly the observational methods of nature study) predominated, but in which, on the other hand, his interest was increasingly human problem, i.e. history. Geography maintained the connection between his earlier field of studies and his ultimate interests. Ritter's meeting with Von Humboldt in 1807 was significant in many ways because Humboldt clearly demonstrated to him the importance of Earth conditions to humankind which had a profound impact on the scholarship of Ritter.

The Works of Ritter

Ritter's initial geographical works included two volumes on Europe which appeared in 1804 and 1807. These were supplemented in 1806 by a series of six maps of the continent. One of these books provided a brief study of the systematic geography of Europe which was reviewed in France. In 1810, Ritter prepared a more complete systematic geography of the world, a '*Handbuch der Physischen Geographie*', but it was never published probably because of the criticism offered by Von Buch. Parts of this manuscript were used by various colleagues with whom he freely exchanged views and the work thus appeared in a number of tents of the time. It is in this unpublished manuscript that Ritter expressed his strong orientation towards the inductive method. He *considered geography to be an empirical and descriptive science.*

In 1814, he went to the University of Gottingen after the death of one of his pupils of the Hollweg's family. At that University between 1814 and 1816, Ritter studied geography, history, pedagogy, physics, chemistry, mineralogy and botany.

The *Erdkunde***.** It was at the University of Gottingen that Ritter published the first volume of the *Erdkunde* in 1817. This dealt with Africa and volume II on Asia appeared in 1818. The publication of these two volumes in 1817 and 1818 made a tremendous impression in the contemporary academic world and he was recognised as the reformer of geography, as the master who first brought the field of geography into science. His approach in the *Erdkunde*, Ritter said, had entirely changed, and he claimed that its distinctive theme was a comparative one which aimed at showing the connection between history and nature, both organic and inorganic. To him geography had a right to be considered a sharply defined science of kindred dignity with the others.

Humboldt is believed to have described Ritter's *Erdkunde* as the most inspired work of this kind (*comparative geography*) and further commented: 'our country has yielded: it is the first work in which is presented the influence which the surface-view has had on the peoples and their fates.' Humboldt also admired Ritter's work on Asia and added that such an important work had not appeared in thirty years. In 1822, Ritter published a second edition of volume I and in 1832 a second edition of volume II. He realised the magnitude of the work he had begun. He appeared to have given up many of his positions which he had held, such as his engagement with the Prussian Military School and his association with the scientific commission on geography and history. Since 1831, he had devoted himself fully to the completion of the *Erdkunde*. Between 1832 and 1838 Ritter completed six more volumes, and between 1838 and 1859 11 more volumes were completed. Yet the 19 volumes of the *Erdkunde*, he actually completed covered Africa and part of Asia.

In 1845, Humboldt described in the first volume of his *Cosmos*, about Ritter's the *Erdkunde* which, to him, 'has shown that the comparative geography can attain thoroughness only when the whole mass of the facts that have been gathered from various zones, is comprehended in one view, is placed at the disposal of the integrating intelligence'. In the same volume, Humboldt referred to Ritter's work as the reflection of a part of the plan of Varenius, which in the earlier period of scientific development had not been accomplished. He further remarked: 'It was reserve for our time, to see the comparative geography cultivated in masterly fashion, in its widest compass, indeed in its reflex on human history, on the relation of the form of the Earth to the direction of the characteristics of peoples and the progress of civilization' (Hartshome, 1939, 54).

Geography, according to Ritter, deals with local condition (Lokalverhaltnisse) and *embraces the attributes of a place with respect to topical, formal and material characteristics.* (1) The first attribute was topographical, i.e. it dealt with natural divisions of the Earth's surface and was intended for the searching presentation according to the principles of Pestalozzi. (2) The second included the distribution and movements of water, sea and atmosphere: the bases of human life. (3) The material conditions were described as the geographical aspect of natural history; this covered the distribution of minerals, plants and animals. The central principle of geography is the relation of all the phenomena and forms of nature to the human race, examined and organised within the framework of the unique geographical association of land and man on the Earth's surface (Dickinson, 1969, 35–36). It is in the *Erdkunde* that Ritter provides the above explanatory definition of geography and *makes his geography humankind-oriented or anthropocentric.*

Ritter's humankind-oriented geography clearly states: 'The Earth and its inhabitants stand in the closest reciprocal relation and one cannot be truly presented in all its relationships without the other. History and Geography must always remain inseparable. Land effects the inhabitants and the inhabitants the land' (Tatham, 1967, 44). Because of this remark, it would be wrong to describe Ritter as a determinist as some commentators have done in the past.

Erdkunde, Ritter pointed out, was a more fitting name for scientific geography. This he used in the title of his major work, and in defining it observed: '*Erdkunde* should strive to embrace the most complete and the most cosmical view of the Earth, to sum up and organise into a beautiful unit, all that we know of the globe', and to show the 'connection of this unified whole with man and his creator'. The concept of regional individuality, a development from the ideas of Zeune and Forster, combined with the concept of a ganzheit or whole derived from Kant, became one of the motifs of the *Erdkunde* (ibid.: 47).

Geographical Ideas and Methods

Carl Ritter is believed to have stated that he was teaching *a new scientific geography* 'in contrast to the traditional lifeless summary of facts about countries and cities, mingled with all sorts of scientific incongruities'. His scientific geography was based on the concept of terrestrial or spatial unity (Raumbegriff). To him, the concept of spatial unity presumed a causal interrelation of all the individual features in nature. *The phenomena of nature had to be studied in order to establish the coherence and unity.*

It may be noted that for Humboldt and Ritter, it was axiomatic that the terrestrial unity included organic and inorganic, human as well as non-human, material and immaterial. The exclusion of any part would be not only arbitrary, but would destroy the coherence and unity of the whole; all the features of an area in their interconnection form a naturally unified complex, whether or not that is to be considered as a unit whole or merely a part of the one natural whole of the world. Ritter emphasised that the unity of nature could be recognised in theory and in practice and this could be established scientifically only by investigating the individual, single features in their relation to each other and building these up in their actual relationships to form the whole. The function of the scientific geographer was to perceive these features, not separately, but in their interrelation so that he could thereby reproduce intellectually the unified whole that was Nature.

Ritter emphasised that the enormous multiplicity of observations must be organised according to the *chorological principle* which should be considered both in a horizontal and a vertical sense. However, he attempted to show the 'coherent relation' of the different features in terms of cause and effect, and the formation of the multiplicity of features into the essential character of an area. By proceeding from the simple to the compound, Ritter tried to establish the totality of interrelated features as the distinctive character of each area.

Ritter seemed to have rejected Gattener's and Hommeyer's idea of pure Geography (reine geographie). However, his concept was somewhat akin to that of Zeune though much broader and logically more consistent. He objected to the procedure of dividing only by mountain ridges or the equally simple by drainage systems. According to his theoretical statements, his divisions were made in terms of relief (in particular he divided the upper, middle and lower portions of stream basins), but in practice he also recognised other factors, as in his separation of the Sahara and in its sub-divisions in terms of climate and vegetation, etc. Thus, he was of the opinion that *one must look to nature for its principles rather than to evolve a theory of natural divisions of the lands.*

Ritter also strengthened the *horizontal concept of unity of particular areas,* as individual wholes, and emphasised on the individuality of the continents which at times he referred to as 'organs'. However, he conceived of each continent as a natural whole in terms of all its characteristics. Similarly, in some cases he wrote of the division of the continents as individuals ('members') rather than mere parts of the 'organisms' of the continent. The individual area had its own particular determined value. Nature in every corner of the Earth is a reflection of the whole. The forms of the organism repeat themselves in every different combination.

As in the Idealism of Fichte, Schelling and Hegel, *the concept of 'wholes' was linked with a Teleological View.*[2] Ritter's concept of the unity of nature was shaped by his deeply religious outlook and by the accepted natural philosophy of his time. His ideas on the 'wholeness' of things were in accordance with the writings of the German idealist philosopher Hegel whose attitudes amounted to an attempt to comprehend the entire universe to know the infinite and to see all things in God.

The scientific stance of Ritter was teleological. He studied the working of nature in order to understand the purpose behind its orders. His views of science sprang from his firm belief in God as the planner of the universe. He regarded the Earth as an educational model for humankind, where nature had a God-given purpose which was to show the way for humankind's development. To him, the shape of the continents was determined by God, so that their form and location enabled them to play the role designed by God for the development of humankind.

The *teleology in Ritter's geography seemed to be an attempt to interpret philosophically that which science could not explain.* Ritter observed that these were those fundamental facts of geography for which science could not offer any explanation. These were the uniqueness of the Earth in the universe; the Earth as the home of that unique creature, man; and finally, the explanation of a host of geographic facts—the differentiation in the character among

[2] *Teleology.* A theory that events can only be explained as stages in a movement towards a pre-ordained end; the end may be defined by those involved in the event or it may be externally defined as in many religions. It seeks to understand events in relation to their underlying principles.

the major land units of the world. The idea of a geographical whole, as in the integral nature of place and people, reflects Ritter's attempt to develop a more scientific holism. His philosophy seems to have an all-embracing explicatory holism, particularly in the notion of teleology—the theory that the evaluation of anything can only be validated by consideration of the purpose to which it is in the end directed. His philosophy seeks to understand the whole rather than the parts and suggests that the whole is more than the sum of its parts, i.e., it has emergent properties not predictable from knowledge of the constituent parts.

Ritter's *purpose* always led from the individual facts towards a 'whole' of all phenomena, rather than to trace the inner relation of phenomena to a last 'cause'. His purpose was to find the coherence of forces in a whole, and these ultimately to indicate the purpose of the whole. This purpose must be found most particularly in humankind (highest of Earth creatures), the only one who could conceive of an organisation of the whole. *This concept makes geography much more human kind-oriented*.

In his lectures before the Berlin Academy of Sciences (one in 1826 and the others in 1850), Ritter attempted to describe the size, form, construction and climatic conditions of the different continents. He sought to *construct 'a law of the arrangement' of the parts of each continent to the whole;* and particularly endeavoured to show how these conditions have determined the development of the peoples of the different continents at different periods of history. Each continent was designed to a particular function which each had played in the course of world history. Ritter in his research did not proceed from pre-conceived opinions, but his scientific procedure was directed throughout on temperate, purely factual, comprehension of the facts and their relations. In the knowledge of the Earth, he attempted a comprehension of the divine world plan in no other way than the natural scientists pursue the thought of evolution.

Ritter was among the first to show that the theory of a continuous network of mountains was opposed by the records of observed facts and that, likewise, there was no general correspondence (as often assumed) between the crestlines of the mountains and the divisions of the drainage basin. *He was convinced that there were laws governing the relation of human and non-human phenomena on the Earth.* He opined that if he could bring together all the facts and relationships observed in areas, these would make it possible to state such laws.

In order to achieve the highest degree of accuracy, Ritter's plan was to bring together on each question the greatest number and variety of reliable evidence of all times and from all peoples, so that their testimony might be seen together, whether in agreement or for comparison. The material was by no means confined to the works of travellers, whether scientific or otherwise, but was drawn also from the works of large number of specialised scientists. Thus, he combined a basic teleological standpoint with a most critical scientific precision, and remarked that *'my system builds on facts, not on philosophical arguments'.* 'Accumulation of facts was not to be an end in itself.... The plans of God, which give purpose and meaning, could only be discovered by taking into account all facts and relationships in the world as objectively as possible'.

Ritter *defined geography as an empirical science,* rather than one deduced either from rational principles of philosophy or from *apriori* theories of general geography. He felt that the geographic studies must proceed from observation to observation, not from opinion to hypothesis, or pre-conceived notion to observation. He was convinced that the phenomena of the Earth's surface were governed by laws, but these would only become apparent when all facts and relationships had been observed in all parts of the Earth.

Causal relationships had to be sought by the comparative method that had proved so successful in other sciences, particularly anatomy. Hence the *Erdkunde* carried the sub-title: All gemeine vergleichende geographie *(general comparative geography).* All geography that went beyond mere description was comparative and the scheme of a general comparative geography would give in broad outline the system of an *Erdkunde* (Tatham, 1967, 46).

Ritter sought for relations or connections, between sets or phenomena in the same area and between one place and another. The task of the geographer is to get away from mere description to the law of the thing described; to reach not a mere enumeration of facts and figures, but the connection of place with place, and the laws which bind together local and general phenomena of the Earth's surface. The task imposed upon geography is to reach the connection which exists between parts; or, in other words, 'to get at the relation between places and what fills and occupies them'. *The geographer traces 'causation' and 'interdependence' of the spatially distributed phenomena,* and the relations of everyone to the country which supplies its condition of being. These words are the keynote to Ritter's approach.

Ritter emphasised on the relative values of wet and dry boundaries, such as rivers or mountains, which he conceived of as a means towards the real purpose of geography the understanding of the content of areas. The areas of the Earth surface are to be studied in terms of the particular character resulting from the phenomena, interrelated to each other and to the Earth, which fill the areas. The great function of geography, according to him, is to show the unison of human social phenomena with the complex of natural forces in the locality.

Most of Ritter's earlier works point towards *his zeal in the systematic studies.* One of the earlier publications on Europe (which appeared between 1804 and 1807) was a brief study in the systematic geography, based on individual categories of phenomena. It is presumed that the interest in the systematic studies that developed in young Ritter just have stemmed from the influence of Humboldt's writings. In 1810, Ritter produced a more complete systematic geography of the world, which he considered to be the completion of geographic work and a bridge in history. But this work was not published.

Ritter insisted repeatedly that the proper method for the study of a science of the Earth was 'to ask the Earth itself for its laws'. It must use the objective (empirical) method, calling attention to the main types of formation of nature, by investigating the relationship grounded in nature itself; this method leads to a 'natural system'. This clearly reflects Ritter's zeal in the systematic studies of regional geography which also implies the importance of systematic studies of individual features and phenomena.

In reaction to those who had specialised in systematic studies and never reached the actual formulations of interrelated factors in areas, Ritter urged them first to study the areas filled with interrelated phenomena. When this had been achieved and done for the whole world, the material would then be available for a more successful development of the general principles of systematic geography. His earlier studies on Europe are often exemplified as the model for systematic studies and as the handbook of general (systematic) geography.

He was *one of the founders of regional geography* and stressed that geography should first concentrate on the study of all the interrelated phenomena to be found in each of all the areas of the world, and then on their basis systematic studies could be made on the relations of individual type of phenomena. 'Areal phenomena are so interrelated as to give rise to the uniqueness of areas as individual units.' Ritter, therefore, insisted that a real synthesis and description must precede the world-wide analysis of particular sets of features and phenomena. 'The collection of facts was not an end in itself, the systematisation and comparison of data, region by region, would lead to a recognition of unity in apparent diversity.'

Ritter's search for the unity in diversity led him to make use of the *regional approach to geography* rather than the systematic study of individual features. Yet, he realised the importance of systematic studies and acknowledged his indebtedness to Von Humboldt whose general studies made Ritter's special studies of regions possible. His ambition to produce a regional geography led him to consider regions of relatively large size. His major geographic units were the continents or Erdteile, and he applied the traditional definition for the continents as the major regions. Each continent was subdivided by its orography into highland core, surrounded by terrace lands, drained by major rivers and peripheral coastal low lands. Ritter believed that each continent had a similar built. Therefore, these are the divisions of a second order arrived at deductively.

Ritter appears to have *envisaged each continent as a complex of individual units* whose whole structure and extent could be arrived at by inductive procedures from the smallest areas of associated phenomena within the framework of his regional geography. He attempted to formulate generalisations concerning the continents or Erdteile; and their population. The continued use of continents as major regional entities not only for the teaching of geography, but also for the formulation of concepts has retarded the progress of geographical scholarship (James and Martin, 1981, 129).

Ritter's regional geography was centred on humankind; the aim was to study the Earth's surface from an anthropocentric standpoint; to seek to relate humankind and nature and to see the relationship between humankind and his history and the ground on which he lived. Thus, a dynamic element entered his teaching. 'He saw he must not just describe the Earth's surface and subdivide it into natural regions, but he must understand, it as the fundamental cause of events' (Penck, 1928).

Evaluation

As mentioned earlier, Ritter's concept of the unity of nature simply reflected his teleological standpoint and he saw the unity as an evidence of God's divine plan. Even in the arrangement of the regions (i.e. the continents), he saw an evidence of God's plan. Asia, said Ritter, represents the sunrise where the early civilisation of man flourished. Africa represents the noon—because of the smoothness of outline and the uniformity of climate; the inhabitants are induced to slumber and to shun outside contacts. Europe represents the sunset, or the end of the day, the culmination of humankind's greatest accomplishments. The polar regions represent mid-night, where land and people are locked in eternal sleep. He enlarged the concept of the land hemisphere and remarked that this was also a part of God's plan. Only in this central location among the Earth's land areas that a world-conquering civilisation could arise.

Ritter's teleological standpoint with regard to the major regions of the world was subject to criticism as his contemporaries felt that religion could not provide explanations of natural phenomena, especially in the arrangement of the major regions (i.e. continents). His view that 'among all the creatures on the Earth only man could comprehend the existence of a divine plan and so could adjust his life to it and make maximum use of god's gifts', was also criticised as valueless and unscientific because it did not take into account the concept of struggle and survival.

Ritter's significance as a scientist lies in his thorough and critical study of sources and his ability to systematise extensive material. He seems to have laid the foundation of the holistic tradition in the nineteenth-century geography.

Ritter's influence in the development of the methodology of geography during the middle of the nineteenth century was of much greater importance. He exerted great influence beyond the range of his immediate students as he was a geographer in the true sense of the term, whereas Humboldt's influence was on the non-geographers. Ritter repeatedly expounded his views on the nature and probiems of geography in methodological papers, Humboldt's numerous discussions of such questions were scattered through his general writings. For this reason, his concept of relation of geography to other sciences was lost sight of for nearly a century. Undoubtedly, Ritter gave to geography its systematic frame which remained valid for the successive periods of modern geography. The classical period in the development of modern geography came to an end with the death of both Humboldt and Ritter in 1859.

6

Impact of Darwinism and the New Scientific Geography

HIGHLIGHTS

- Inductive-Teleological Philosophy
- Impact of Darwinism on Geographical Thinking
- Ritterian Tradition in Geography

The 'classical period' in the development of geography may conveniently be considered as terminating with the death of beth Humboldt and Ritter in 1859 (Hartshorne 1939, 1976, 84). The following decade experienced a whole range of new directions in geography. The consensus which had characterised the subject before 1859 gradually disappeared, although the viewpoints of the school which Ritter had founded lived on. The year 1859, which saw the deaths of both Humboldt and Ritter, was also the year of publication of Charles Darwin's *Origin of Species.* The foundation of Ritter's approach was gradually replaced by 'a materialistic scientific philosophy' which emphasised 'natural laws' and 'causality'; 'mechanical' rather than teleological explanation.

Many scholars appeared to have criticised Ritter's work as being valueless and non-scientific because it could not be accommodated in Darwin's concept of struggle and survival, and to his belief that evolutionary changes resulted from random variation. The new scientific method that came to dominate the philosophical foundation and research in the post-Ritterian period, appeared to have sharply contrasted the inductive approach which sought to explain the existence of God's plan and design with regard to the natural phenomena and their areal associations. The new scientific method gradually replaced the inductive teleological philosophy of Ritter, sought to offer a mechanical explanation, or more precisely the *'hypothetic-deductive method'* which sought 'prime causes by deduction'. It set up hypotheses to see whether they could explain the reality of the observable phenomena. The testing of hypotheses led to the formulation of natural laws.

INDUCTIVE-TELEOLOGICAL PHILOSOPHY

Ritter used the inductive method as framework for his presentation of data and as a means to arrive at some simple empirical generalisation. The empirical generalisation of the classical period in modern geography of the mid-nineteenth century Germany appeared to have been valid for a specific time and space. However, Ritter's teleological philosophy cannot be tested empirically and therefore does not qualify as scientific explanation because religion cannot provide explanation of natural phenomena.

Ritter's most active pupil and a close disciple, Elisee Reclus (1830–1905) does not seem to have been very much influenced by the teleological philosophy of his teacher. He achieved recognition for his systematic physical geography and is best known for his 19-volume regional world geography

(1875–1894). Reclus' standpoint showed a major departure from the traditional empirical generalisation towards more objectivity. His method of regional description may be compared with the method of Malte-Brun 50 years earlier and of Vidal de la Blache 50 years later (Dickinson, 1969, 224–25).

Whatever may be the inherent methodological and philosophical weaknesses in the Ritterian tradition of geography, it does have the characteristics of a paradigm. *Geography was in the pre-paradigm phase until the time of Darwin.* It is a fact that the teleological framework of explanation is possible without a metaphysical assertion. A teleological explanation is generally taken to mean that a phenomenon is explained in relation to the purpose it is believed to serve. A mechanical or causal explanation, on the other hand, depends on pre-existing causes to explain the observed phenomena. Holistic syntheses and concepts of organic relationships are strongly related to teleological explanatory models.

Because of Ritter's teleological outlook, it is difficult to identify the professionalism[1] phase in geography. The first apparently active school of geography did not lead the subject into the first phase of paradigm. The development of Darwinism led to the rejection of ideas which might have developed into a paradigm and a discussion actually began as to whether geography as a whole could be regarded as a science or not. It should be emphasised that Darwin did not represent a complete break with the major ideas upon which geography was founded.

After Darwin, scientists started looking for the controlling laws of nature and the materially conditioned social laws, and to a *considerable extent adopted a nomothetic or law-making approach.* This development was especially characteristic of natural sciences which were making major advances. Inductive arguments had to be increasingly replaced by hypothetical-deductive method. The post-Ritterian geographers, starting from inductive arrangements of their observation or from intuitive insights, tried to devise for themselves *a priori* models of the structure of reality.

IMPACT OF DARWINISM ON GEOGRAPHICAL THINKING

In geography, Darwinism was interpreted primarily as evolution, in the sense of a 'continuous process of change in a temporal perspective long enough to produce a series of transformations'. **Darwin** was primarily concerned with the mechanism of the change, or as *The Origin* was subtitled, 'the preservation of favoured races in the struggle for life'. This 'element of struggle' was applied in a deterministic way, particularly in human geography, at about the same period. In both physical and human geography, supposedly Darwinian ideas were applied in an eighteenth rather than a nineteenth century fashion, and geographers were still applying essentially Newtonian views of causation well into the twentieth century. The Darwinian evolution seemed to have provided fresh incentive to concepts of biological origin which date back to Ritter and before, and the subsequent development of ecology led to new insights in some branches of geographical scholarship.

Stoddart (1965) suggests that the following *four main themes* from Darwin's work are to be taken as significant contribution to the contemporary geographical thought from biology:

1. the idea of change through time;
2. the idea of organisation and ecology;
3. the ideas of selection and struggle;
4. the randomness or chance character of variations in nature.

1. The strongest and most explicit impact of the **idea of change through time** or **evolution** was in the study of landforms, a field in which Darwin had worked during the 'Beagle Years' when he developed

[1] *Professionalism* :it refers to the stage of scientific maturity from the pre-paradigm period.

his famous 'subsidence theory' regarding the coral reef formation. The initial deduction and subsequent development of this theory closely resembles the later development of Darwin's biological ideas and it could serve as the archetype for the 'cyclic' ideas later developed in geomorphology. Though Huxley wrote on the new subject of 'physiography' in the 1870s, but it was W. M. Davis who considered evolution to be the source of his inspiration in the idea of the geographical cycle.

In his first article on the development of landforms, Davis referred to a cycle of life and used such terms as birth, youth, maturity and old age, and which he channelised into the restricted field of denundation chromology (Sauer, 1925, 32). Davisian geomorphology was deductive, time-oriented, and imbued with mechanistic notions of causation. Closely similar views were being developed at about the same time in plant geography and particularly in ecology. Hooker seemed to have worked on this aspect. Clements occupied in plant ecology a position similar to that of Davis in geomorphology. Plant ecologists and pedologists followed Davis' biological analogy for change through time. They have also adopted terms such as 'infancy', 'youth', 'maturity' and 'old age' to describe development through time.[2]

Change through time has been a dominant theme in much of geography, particularly in the work of the Berkeley School on the settlement of the American southwest and other areas. Here Sauer's influence has been a dominant factor and it is interesting that he himself worked at Chicago under Salisbury and Cowles. Another pupil of Cowles, who also studied under Davis and published in both geomorphology and plant geography, was the historical geographer Ogilvie who carried their emphasis on time into regional studies. The influence of plant ecology and the historical viewpoint was also clear, both in concept and language, as also in Whittlesey's ideas of sequent occupance in the development of landscape (Stoddart, 1965).

2. Darwin's second major theme which made significant contribution to geography was **the idea of organisation and ecology** which dealt with the inter relationships and connection between all living things and their environment. This theme developed in Haeckel's new science of ecology (Haeckel, 1869). Darwin seemed to have been impressed by the exquisite adaptation and interrelationship of organic forms in nature and the theme of ecology is implicit in many of his writings. Perhaps Darwin's most significant contribution to ecological thinking was to include humankind in the living world of nature. Huxley in his 'Man's Place in Nature' (1863) attempted to show how humankind had emphatically become a subject for scientific speculation, and Darwin treated modern man on the same level as other living things.

The term 'ecology' was used by Haeckel in 1869 and from about 1910 *'human ecology'* was used for the study of humankind and environment not in a deterministic sense but for humankind's place on the web of life or the economy of nature. Park defined the scope of human ecology, as dealing with the web of the balance of nature, concepts of life, the balance of nature, concepts of competition, dominance and succession, biological economics and symbiosis—all concepts taken from plant and animal ecology (Stoddart, 1965, 58). Human ecology seemed to have investigated the processes involved in biotic balance in which 'man interacts with nature through culture and technology'.

---

[2] Cowles developed what may be called 'physiographical ecology' in which he brought together Davisian geomorphology and Clementsian ecology, following field work in the Chicago area on the coincidence of plant formations and physiographic units. A few workers, some of them influential in geography, maintained a developmental framework, ranging from the 'unilinear' school of White to the 'multilinear' evolution of Steward. Childe studied historical interpretation of technological development, particularly among the English geographers. The botanist Geddes's work on cities influenced early ideas on 'urban geography'. Taylor applied developmental principles to the study of race and cultural history, and Beaver and others attempted to introduce cyclic ideas into the interpretation of economic landscapes though with little success.

For Barrows, 'geography is the science of human ecology.... Geography will aim to make clear the relationships existing between natural environments and the distribution and activities of man.... The centre of geography is the study of human ecology in specific areas. This notion holds out to regional geography a distinctive field, an organizing concept throughout, and the opportunity to develop a unique group of underlying principles'.

Ecology has, however, become increasingly empirical in method. In doing so, it has run counter to, and has superseded, the synthetic geographical tradition of explanation by analogy, which attempted to understand the complexity and interrelationships of phenomena by reference to the even greater complexity of living organisms. *In geography as a whole, the organism analogy operated on three distinct levels—those of the Earth, its regions, and its states*—and on each level it uses long predates of the Darwinian evolutionary theory. The idea of the organic unity of the Earth basically manifested the Ritterian tradition in geography, because to him the Earth was an organism, with a divine intent, to fit the needs of man to perfection. Humboldt also conceived of the idea of the 'harmonious unity of the cosmos' as a 'living whole', a unity in multiplicity.

For both Ritter and Humboldt, *unity, harmony and interdependence of parts constituted the organic analogy.* Half-a-century later, Vidal de la Blache arrived at a similar conclusion and recognised his debt to Ritter, both at the Earth and at the regional level (Stoddart, 1965).

The idea of organism analogy also appeared to have profoundly influenced the contemporary regional geography as the idea served as a unifying theme, making it an increasingly particularistic discipline. Herbertson, in 1905, applied the term 'macro-organism' for the complex unity of physical and organic elements of the Earth's surface. To him, 'the natural regions are definite associations of inorganic and living matter with definite structures and functions, with as real a form and possessing as regular and orderly changes as those of a plant or an animal' (Herbertson, 1913–14, 158–59).

The region—that particular field for geographers—has been regarded as a unique functional complex mutually interdependent, which despite a steady stream of material and energy, 'is in apparent equilibrium and constitutes a whole (or Ganzheit) which is more than the sum of its parts. The understanding of the region is found in the French School of regional geography.

Ratzel made use of the organism analogy in his *Politische Geographie*, published in 1897, in which he compared the state to an organism—the state as an organism attached to the land. The organic quality of states depends upon organisation and interdependence of parts; it then assumes properties of growth and competition, and in doing so goes beyond the organic analogies of the Earth and the geographical region.

The major objection to the organic approach in geography is methodological, for it is a synthetic notion which gives no assistance in actual investigation, and is an essentially idiographic concept in an increasingly nomothetic science. The concept is thus reduced to a metaphor of dubious value, hinging on gross formal and functional comparisons between living matter and complexly interrelated facts in areas, and as such has dropped out of geographic work since 1939, except in occasional mention of Herbertson and Vidal de la Blache.

3. Darwin's third major theme, **selection and struggle,** was much more influential than his other themes as *it did lead towards both 'scientific determinism' and 'possibilism'* in the contemporary geographical concepts. The philosophy of selection and struggle seemed to have been based on cause-effect relationships, as it talked about the environmental influence, selection and adaptation, depending upon the way of life. Most Darwinian writers on the effects of environment were content to look for cause-effect relationships without enquiring too closely into the process. The theme of the effects of environment on the course of human

activities was taken up by Ratzel in the first volume of *Anthropogeographie*, and later developed by his pupil Semple and Demolins. The theme of 'selection and struggle' also found mention, of course, in a different manner, in the possibilist philosophy of Vidal Blache who treated it in the light of the 'genre de vie' or the way of life of the people.

Fleure stressed the need for the physiological study of environmental effects on humankind, and in his typology of human regions (regions of difficulty, of effort, of increment) came close to applying Darwinian ideas of natural selection through environmental influence on human groups. Huntington analysed the problem of natural selection, effects of environment, and human population on a world scale. Taylor explored the same theme in a series of studies of race, people, states and towns, emphasising their development through time under the influence of environmental factors.

It was *in political geography that the theme of 'selection and struggle' found its legitimate expression* because it led to the 'increasing space consciousness and conception' on a national level. In 1896, Ratzel developed his seven laws of the growth of states, from which he derived the powerful concept of 'lebensraum'. 'Just as the struggle for existence in the plant and animal world always centres about a matter of space, so the conflicts of nations are in great part only struggles for territory.' It is clear that the organic analogy for Ratzel not only provided a simple and powerful model in analytical political geography, but also an apparently scientific justification, in Darwinian selection, for political behaviour.

Mackinder argued that 'European civilisation is the outcome of the secular struggle against Asiatic invasion.' He further remarked: 'through the steppe there came from the unknown recesses of Asia in all the centuries from the fifth to the sixteenth, a remarkable succession of Turanian nomadic peoples ... against the settled peoples of Europe. A large part of modern history might be written as a commentary upon the changes directly or indirectly ensuing from these raids.' This argument of Mackinder also reveal the ideas of struggle and selection on a super-continental level, that the Asiatic nomads attempted to expand their space through a series of raids upon the fringes of the pivot to which they belonged.

Similarly, Mackinder's 'Heartland' concept also deals with the struggle for the territorial aggrandisement of that part of the World Island which was inaccessible to the maritime powers, but had immense strategic value for its vast resources. In this thesis, Mackinder seemed to have incorporated the idea of selection and struggle in an organised manner, though his philosophy was highly deterministic in approach.

Ratzel's concept of 'Lebensraum' and Mackinder's 'Heartland' thesis appeared to have formed a symbiosis which led to the development of the Geopolitik in Europe between 1919 and 1945. For Kjellan, states were biological manifestation, endowed not only with the mortality but also with 'organic lusts'. The political usage made of the organic view of the state, and the ideas of struggle and Lebensraum, brought the subject into intellectual disgrace in the 1930s, and modern political geography is at pains to dissociate itself from any kind of organic analogy.

4. The theme of **'randomness'** *or chance character of variation in nature* appears to have made no significant contribution to the geographical thought. No agreement was reached as to whether variation and development occurred by chance or were determined. Any discussion of the biological impact on geographical thinking must hinge on the central question: Why was Darwinism, a theory for the selection of randomly occurring variants, interpreted in a deterministic and not a probabilistic sense? Why was chance omitted in geography? The problem is more remarkable in the sense that the study of random processes in the nineteenth century was by no means limited to Darwinian biology. 'The study of this blind chance in theory and practice is one of the greatest scientific performances of the nineteenth century' (Stoddart, 1965).

In both the natural and social sciences, the 'probabilistic sense' found its legitimate expression. Why, then, in such an intellectual atmosphere, was the geographical interpretation so deterministic? Darwin's theory made a .clear distinction between the way in which evolution was effected and the course of evolution itself; geography emphasised on the latter and ignored the former. Darwin began with the idea of the selection of chance variations, which are undoubtedly governed by 'laws' which he failed to point out or discover. Nowhere does he use the word 'random' and in the fourth chapter of *The Origin* he points out that the rise of word 'chance' is wholly in concept. He explained adaptation in nature by variation and natural selection, but he could not offer any explanation of the basic variation.

The theme of random variation remained neglected in the contemporary geographical tradition because of religious and scientific reasons. It is interesting to note that the method which includes and incorporates randomness is now being increasingly used by geographers.

Appraisal. Darwinism caused a methodological shift in the contemporary geographical thinking, especially in the field of human geography. Darwin set up a sphere of scientific enquiry free from *a priori* theological ideas in which he made humankind a fit object for scientific interpretation. The geographical thinking since the mid-nineteenth century appears to have relied on the Darwinian tradition as it laid the foundation of 'Scientific Geography'.

However, Darwinism had no impact on the contemporary Russian geography. The four basic themes of *The Origin* were more influential in Britain, Germany and France, but were perhaps less intoxicating in Russia because of the earlier studies of evolution by the Russian biologist, K. F. Rul'ye. In any case, the Russian rejected the more extreme forms of environmental determinism stemming from Herbert Spencer and also the use of the biological analogy to describe sequences of landforms as suggested by the American geographer Davis.

RITTERIAN TRADITION IN GEOGRAPHY

Before going over to 'New Geography' that came to be developed in the latter part of the nineteenth century in different parts of Europe, and across the Atlantic Ocean, it is necessary to provide an analysis of the heritage of geography of the period of Humboldt and Ritter. Contribution of Ritter to the development of geography would be incomplete without a discussion of the contributions of his students and colleagues. His most famous disciples were Elisee Reclus and Arnold Guyot.

Elisee Reclus (1830–1905). Elisee Reclus was a French geographer and anarchist who attended Ritter's lectures in Berlin in 1849–50, but his concepts and writings advanced far beyond what he learnt from his master. Departing from France following revolutionary activities in 1852, he spent the next five years travelling in Great Britain, the United States and Columbia. He returned to France in 1857, was imprisoned in 1871 and was banished for 10 years in 1872. In 1879, the ban was lifted.

Elisee Reclus was a professor of comparative geography at the University of Brussels in Belgium in 1892. However, his appointment was cancelled on account of his revolutionary activities. From 1894 until his death in 1905, he was director of the Institute of Geographique, which was set up at the New University of Brussels.

Though Reclus was one of the most active pupils of Carl Ritter, but leaned less towards history and teleological thinking than most of the others. He is credited with his *work on systematic physical geography* called *La Terra* (1866–67). The two volumes of *La Terra* usually revealed the strong influence of Ritter because it was prepared in the Ritterian tradition. However, Reclus is best remembered for his 19-volume regional world geography *Nouvelle Geographie Universelle* (1875–1894). With the completion of this kind of universal geography that Ritter had started, Reclus fulfilled the dream of Ritter. This 19-volume *'new universal geography'* was in a sense the last echo of the classical period when one scholar could present all available knowledge about the Earth as the home of humankind (James and Martin, 1981, 147). This work became a model for a range of encyclopaedic studies of geography of the world and of particular countries.

Reclus' descriptive material is lucid, detailed, systematic and balanced, while his philosophical approach reflects his time. Thus, all the facts of primitive history are explained by the disposition of the geographical theatre in which they have taken place.

Like Ritter and Humboldt, he also *believed in the 'harmony of nature'*, but pointed out that humankind has destroyed the natural plants and wild animals and replaced them with his own cultivated crops and domesticated animals, and thus upset the balance of nature. *The theme of man's role in changing the face of the Earth is the dominant conception of Reclus.* He was much concerned with the idea of 'destructive exploitation' and advised to maintain the harmony of nature. However, his geography put major emphasis on the study of the physical features of the Earth's surface.

In *La Terra*, Reclus recognised that the inequalities of man are due to the inequalities in the external environment: 'Each period in the life of peoples corresponds to a change in their environment. It is the inequalities of the planetary traits that create the diversity in human history. Life corresponds to the environment. The Earth, climate, habit of work, type of food, race, kinship and mode of social grouping—these are the fundamental facts which play their part in influencing the history of man." Reclus recognised clearly that climate is a major environmental influence. He gave more attention than the earlier writers to aridity versus excessive humidity.

Arnold Guyot (1807–1884). Arnold Guyot was born in Switzerland. In 1839, he was a colleague of Louis Agassiz at the University of Neuchatet. He was fascinated by the effect of the glacial action in producing distinctive kinds of mountain landforms. Guyot went to Germany and studied with Carl Ritter. Later, he became one of his devoted disciples. He was also a friend of Von Humboldt. Guyot left his native Switzerland in 1848 and at the suggestion of Louis Agassiz had settled in America.

Guyot is credited with his *'new geography'* which he outlined in a series of lectures delivered at different centres of the United States. In 1849, he published his lectures in a book form which was designed to make Ritter's ideas known to the Americans. He was against the traditional descriptive geography which, he felt, simply made encyclopaedic collection of facts and students were required to remember them. In this sense, he was more close to Carl Ritter.

For him, the new geography should not only describe, but also compare and interpret, and 'should rise to the how and wherefore of the phenomena that it describes'. He taught his pupils to observe their surroundings and to match their perceptions with the word symbols they used to describe them. Like Ritter, he also favoured that generalisation should be made on the empirically observed facts. He urged upon his pupils to get acquainted with the topographic maps because this would serve the study of distant regions.

Guyot was the first professor of physical geography arid geology at the College of New Jersey (Princeton) from 1854 to 1880, and remained a staunch supporter of the ideas of Ritter. It is interesting to note that Guyot was not influenced by the idea of Darwinism. Even as the widespread acceptance of the concepts of evolution and scientific methods swept away the philosophical ideas of teleologist, he remained undisturbed and unshaken.

Guyot seemed to have leaned more towards the philosophy of teleology than Reclus and said, 'The Earth, as an individual organization, with definite structure, character and purpose, is the subject of geographical science. A careful study of physical geography tends to lead the mind to the conclusion that the great geographical constituents of our planet—the solid land, the ocean, and this atmosphere—are mutually dependent and connected by incessant action and reaction upon one another; and hence, that the Earth is really a wonderful mechanism all parts of which work together harmoniously the purpose assigned to it by an all-wise Creator' (Guyot, 1849, 57–58).

It is clear from the above statement that Guyot also held the Kantian concept of chorology that the constituents of the Earth were mutually coordinated, not subordinated in space. His 'new geography' *probably attempted to combine the Kantian concept of chorology with the Ritterian philosophy of teleology.* After his retirement in 1880, the 'new geography' lost much of the credibility because its philosophical basis was discredited in the light of the new paradigms, largely on account of the Darwinism vis-a vis the hypothetical deductive method.

Julius Frobel (1805–1893). He taught geography and mineralogy at Zurich from 1832 to 1842. He was a contemporary of Carl Ritter but was a *bitter critic of the Ritterian tradition in geography.* The correct interpretation of 'comparative' produced much methodological controversy and he criticised during Ritter's lifetime the use of this term and asked for its definition. Ritter's reply indicated that the term lacked clear meaning, and he felt that it was just a convenient term for distinguishing the geography of the nineteenth century from that of the eighteenth. But Frobel appeared never to understand the term fully and hence he was critical of it.

Frobel was trained in topographic mapping and map drawing and seemed to have been involved in the study of natural sciences other than geography. Before joining the assignments at Zurich, he prepared a geographic handbook on Peru, Bolivia and the Plata countries which was completed in the 20-volume book named *Handbuch der Erdbeschreibung'* and edited by Gaspari, Gutsmuth and others. In the handbook, he tried to give short treatments of the natural regions of each country he dealt with and also attempted to write on minor political units. However, this handbook was not in accordance with 'reine geographie'—a movement in the contemporary geographical world initiated by the German.

Frobel attempted to construct a science of geography based not on a study of geography itself, but rather on a comparison with the natural sciences. He *defined geography as 'the science of the phenomena of the Earth (Erdwelt),* insofar as these constitute this Earth by their association in area'. Earlier, he divided geography into two categories: Geography as a natural science and 'applied geography' or 'historical-philosophical geography'. Geography as a natural science consisted of separate studies of each of the features of the Earth, classified by kind. Following Humboldt's point of view as well as his terminology, he included the study of humankind as a feature of the Earth in his 'natural science', in terms of all objective facts of man and the Earth that represent relations between them (Hartshorne, 1976, 103).

The category he recognized as a form of 'applied geography' or 'historical-philosophical geography' included those aspects of geography which could not be fitted into his frame of a 'natural science'. The first category, according to Frobel, is analytic and the second synthetic. Each can be built into a separate system, but the two apparently cannot be combined. Geography is, therefore, not merely 'dualistic', but is completely separated into two kinds.

While at Zurich, Frobel again divided geography into four different kinds: (1) 'Pure geography' in which the phenomena are studied 'in so far as this association has purely scientific interest'; (2) 'political geography' or, possibly 'statistics' in which the interest is 'ethical-practical'; (3) 'historical-philosophical geography' in which the interest is aesthetic; (4) pure scientific geography which included objective studies of human races and peoples and the geography of states as Earth phenomena.

Frobel was more close to Humboldt because of the latter's aesthetic explanation of the unity of nature. But, at the same time, he rejected Ritter's teleology, and saw no reason to explain unity and order in nature as a God-given system to further humankind's development. 'Since geography was concerned with the Earth's surface in a systematic way, hence there was no place in geography for Ritter's Landerkunde, nor for the method of synthesis such studies required' (Tatham, 1967, 59–60). As early as 1832, Frobel pointed out the similarity between the ideas of Kant and Humboldt with respect to the concept of chorology, but his writings were overlooked and only rediscovered many decades later.

He was banished from Germany for his revolutionary ideas. In 1848, he went in exile to America where he spent 10 years as a roving journalist. He produced in two volumes the story of his travels across North American and Central American countries. Though a large part of the work dealt with the contemporary socio-political problems in the United States, there was one short chapter that provided a 'systematic study' of the mountain system *of* western North America. That study could fit into what he had defined as scientific geography. More valuable are the five descriptions of areas which could fit into his division of applied geography.

Though Frobel wrote a number of articles on the contemporary political situation in Europe, but his short study on the relation of America to the European politico-geographical situation greatly influenced Humboldt. Indeed one may observe that the same Frobel, 'who in his first geographic phase appears as a herald of the late nineteenth century purely scientific geography, in his later, most important study stands as the forerunner of the twentieth century post-War school of geo-politik' (Hartshorne, 1976, 103–106).

Ludde. Less critical disciples of Ritter were equally confused about the exact meaning of comparative geography and offered many definitions. The most logical development of Ritter's work was probably that of geographers who studied the interrelation of the various phenomena (relief, climate, vegetation, animal and humankind) in a particular area. Ludde was one of them who pointed out (1841–50) that there were *three types of comparison in geography:*

1. The comparison of one element with another in a given region, i.e. of vegetation with climate and soils;
2. The comparison of the present conditions *of* a region with those of the same region at an earlier stage of development;
3. 'The comparison of one region with another, each being considered as a whole.'

All these were essential, no single one by itself could claim the title 'comparative geography'. Indeed, it was only when one studied the physical and biological features of a region along with humankind, his history and culture (including the ethical, religious, and aesthetic forms), that one could really claim to be doing comparative geography in the Ritterian tradition.

However, Ritter's approach was continued by several scholars of no mean distinction.[3] They all

[3] *Henrich Kiepart* (1818–99) was a follower of Ritter and was also a link in the University of Berlin between Ritter and Richthofen. He prepared several maps and atlases of the classical period and wrote a textbook on classical geography in 1879. He upheld the tradition of Ritter, believing in the geographical philosophy which Ritter incorporated in his geographical concepts. J. G. Kohl was a geographer, and a staunch follower of Ritter. He is credited with his early deductive work on the relation of the settlements to the Earth surface. The work was prepared in 1841. He also prepared a work on the geographical location of capital cities in 1874. *Ernst Kapp* in 1845 discussed the distinction between general and special geography and prepared a broad portrayal of nature and humankind in major world areas. G. B. Mendelssohn described the German lands in terms of their physical build and their people. *Johann Edward Wappaeus* attempted to incorporate the statistical tradition into geography. In 1847, he started working on the completion of a standard handook of geography and statistics. The book was prepared in 10 volumes. He gave much the same definition in his physical geography, where he pointed: 'Vergleichende Erdkunde is geography when in considering the Earth surface it does not separate the physical and historical viewpoint, but presents in every unit the physical and ethical relation in their mutual dependence and interaction' (Tatham, 1967, 60–61). Throughout his life, he clung to the ideas of Carl Ritter, and stood strongly against the ideas of the critic Oscar Peschel who chose to restrict comparative morphological study of the Earth's surface to its physical features.

had the same objectives, namely, to clearly portray the Earth in its relation to humankind, to establish the influences exercised by the Earth as the stage and determinant of human activity on the destiny of individuals and peoples and on the course of cultural development.

Summing up

It is interesting to note that, although Darwinism made inroads into the contemporary geographical thinking and revolutionised the geographical paradigm not only in Germany, but also outside it, yet the idea of Darwinism miserably failed to obliterate the impact of the Ritterian tradition. Quite a good number of geographers, unaffected by Darwinism, preferred to remain close to the Ritterian tradition; they nourished it and maintained its relevance. It is equally significant to observe that the growth of Darwinism creates a sort of dualism in geography in the sense that those who clung to the idea began to see in the differentiation of humankind the operation of natural laws, while those who upheld the tradition of Ritter in geography favoured an anthropocentric approach.

'Many geographers in the last decades of the nineteenth century clung to a dualistic concept of geography—on the one hand, the study of the Earth as a natural body, and on the other, as the home of man' (Mackinder, 1895). Despite the view that a science cannot have two distinct fields but must have single objective gained increasing momentum, this duality of geography between the two distinct disciplines, physical geography and human geography, is a fundamental problem that will always be with us (Dickinson, 1969, 52). However, the foundations which Humboldt and Ritter had established for geography did not provide a clearly unified field to the extent to which their followers exaggerated certain aspects of the views of each of the founders or attempted to introduce new concepts of the nature of the field; geography was for a time split in several directions and its position as a branch of knowledge thereby brought it into serious question. (Hartshorne, 1976, 86).

7

Modern Geography of Germany

HIGHLIGHTS

- New Dimension
- Systematic Studies
- Chorology and Landscape Morphology
- Geopolitics
- Contemporary Trends

The post-Ritterian period in Germany felt the need of a professionally accepted paradigm to serve as a guide to the study of geography. There was no clear-cut definition to account for the purpose of geography, which could be universally accepted. Those who taught geography were trained in history, geology, botany, zoology, mathematics, engineering or journalism. None of them were professional geographers, as a result of which each of them defined geography in different ways. There was no unanimity among the scholars with regard to the definition, purpose and methodology of geography, and the first half of the nineteenth century in Germany witnessed this major problem. The mid-nineteenth century witnessed the rise of the materialistic scientific philosophy in other branches of knowledge, which emphasised natural laws and causality, as well as mechanical explanation. *This made 'professionalism' an academic necessity.* In the light of this development, geography could not remain isolated. The need was professionalism in geography which required a new dimension for the purpose of geography.

NEW DIMENSION

In the absence of any guidelines regarding the field of geography, each new professor felt the need to set forth his own ideas concerning the scope of the field. Each tried to provide a definition of geography that would give it unity and would establish its position among other academic disciplines. All around the world in the late nineteenth century, the question raised was: what is geography? The growth of professionalism in geography provided the answer. It was no longer confined to the plotting of information on maps and to the publication of clearer maps of finer design. Geography acquired an academic status and became a permanent independent discipline.

The post-Ritterian period, however, proved to be a brief interlude before the last quarter or third of the century brought a rapid development in academic geography in Germany. In many respects, this period may be regarded as the critical period in the development of the field. The foundation which Humboldt and Ritter had established for geography did not provide, in appearance certainly, a clearly unified field. To the extent to which their followers exaggerated certain aspects of the views of each of the founders, or attempted to introduce new concepts of the nature of the field, geography was for a time split in several directions and its position as a branch of knowledge thereby was brought into serious question.

After the death of Ritter there was no professor of geography in any German university and the rapid subsequent growth was largely the work not of the historical geographers. Those who followed Ritter were trained as geologists and tended to specialise in the study of non-human features of the Earth, i.e. physical geography as the term stands for. With the rise in academic status of geography and the productive work of this period, especially in the development of geographic thought, its major problem was to overcome the apparent disunity in the methodology of the field and thus definitely establish its position as a single field of science (Hartshorne, 1939, 1976, 86).

The new viewpoint of a materialistic scientific philosophy in science which came to dominate the contemporary scientific thought appeared to have been marked by the increasing specialisation of sciences, by an increasing emphasis on the development of 'scientific laws' and 'causality', and by a conscious isolation of science (and specially of geography) from any particular Weltanschauung. This latter point of view, however, often represented an equally definite philosophical presumption of a materialistic, mechanistic, universe in which humankind was to be considered as a 'Thing'.

The *emergence of new geography in Germany* appears to have been primarily due to the works of Peschel, Richthofen, Ratzel, Penck and Hettner. However, the impact of Darwinism in this methodological innovation cannot be ruled out. It should be emphasised that Darwinism did not represent a complete break with the major ideas upon which geography was founded. The study of development over time continued to be regarded as very important and, although it is not clear whether Darwin supported it or not, a deterministic explanatory framework was retained. The development was especially characteristic of the natural sciences which were making major advances. Research workers, beginning from inductive arrangements of their observations or from intuitive insights, attempted to devise for themselves *a priori models* of the structure of reality. These were used to postulate a set of hypotheses which could be confirmed, corroborated or rejected by testing empirical data through experiment. A large number of confirmations led to a verification of the hypothesis, which was then, for the time being, established as a law.

Germany became the leading nation for the *development of academic geography.* Although there had been a number of chairs in geography before 1874, a decision by the Prussian government in that year to set up permanent chairs in geography at all Prussian universities was an event of major importance. It is reasonable to assume that Prussia took this step in the belief that geographical knowledge could be used to further its political expansion. At any rate, this decision meant that geography was firmly established as an academic discipline in one of the leading European nations, and by 1880 there were professors in Geography in 10 of the Prussian Universities. However, the lead in the new direction in the field of academic geography was taken by Oscar Peschel and Ferdinand Von Richthofen, and the direction showed by them seemed to have been followed by the subsequent academicians.

SYSTEMATIC STUDIES

Geographers trained in the Darwinian tradition, in the late nineteenth century made significant contribution to systematic studies, both in physical geography and human geography. The contributions of Peschel, Richthofen, Ratzel, Gerald and Partsch are studied below.

Oscar Peschel

Oscar Peschel initiated a movement which sought to give the contemporary German geography a new outlook, orientation and dimension that was opposed to the geographic heritage founded by Humboldt and Ritter. Peschel was the first great geographer in Germany before the full impact of Darwinism was felt; and although he worked after the enunciation of Darwin's theory of evolution, its implications in the interpretation of Earth features and human societies were not yet recognised. To

Peschel geography was to be a systematic, empirical science, its method, observation, drawing induction from those observations, and correcting these by still new observations. His enormous respect for natural law led him to attempt causal classification of the relief features of the Earth's surface.

At the age of 45, Peschel was appointed professor of geography at Leipzig in 1871. His book on the history of geography 'Geschichte der Erdkunde' was published in 1865 and that created a stir in the contemporary academic circles of Germany: In his studies of the history of geography, Peschel attempted to use the term 'Vergleichend' to designate the method of investigation which compared all forms of one particular relief type with one another, so that by recognising intermediate stages one could arrive at an understanding of their mode of formation. He studied the much-indented fjorded coasts which he found on the western sides of continents in higher middle latitudes. He offered the hypothesis that the fjords were fissures in the Earth's crust that had been occupied and gouged out of glaciers. The causality Peschel sought was purely mechanical; it had nothing to do with purpose.

His systematic studies of fjords together with lakes, islands, valleys and mountains were published in 1870 in *Neue Problem der vergleichenden Erdkunde als Versuch einer Morphologie der Erdoberflache*. In this work, Peschel attempted to imbue the scientific spirit into geography. Through this work, he led geographers to study primarily the morphology of landforms. The *scientific morphology,* the idea he developed in his two volumes of papers, entitled 'Physische Erdkunde' was given its foundations by geographers who had been trained as geologists. His geographical work was not only simply confined to the study of the morphology of land forms but, on the contrary, he was concerned also with the study of the influence of landforms on human history (Hartshorne, 1976, 88). In his systematic studies, Peschel appeared to have concentrated more on the Earth's physical features and did not treat the elements of human geography in a similar systematic manner.

His morphological research created an academic stir in physical geography of contemporary Germany. Interest was moved from 'Landerkunde' and tended to focus on systematic studies in a manner reminiscent of Humboldt, though there was no revival of Humboldt's broad interest in all aspects of physical geography nor of his idealistic stressing of the unity of nature. Systematic studies in their new guise were highly specialised, and prosecuted by scientists who tended to be geomorphologists and climatologists first and geographers only secondly. Peschel's genetic classification of relief types provided the concept from which the formulation of geomorphology could and did proceed.

He seemed to have recognised the dualism in geography. Peschel excluded the study of man from it, but devoted his scientific energies and his teaching to both. He made outstanding contributions to the study of the science of physical features of the Earth's surface and also made serious investigations of the races and cultures of humankind, which he, like his contemporaries, defined as Volkerkunde. His early death, at the age of 49, caused irreparable loss to the emerging 'new geography' in Germany, as it cut his work short just at the time when his university position would have enabled him to be most productive.

Ferdinand von Richthofen

Ferdinand von Richthofen was one of the forerunners of 'new geography' in Germany vis-a-vis Europe as he carried forward the scientific spirit which was imbued in the contemporary geography by Oscar Peschel. Richthofen was an experienced field observer. He began his researches in the Alps and then carried on geological studies in the Carpathians. He studied the granites and dolomites in the south Tyrol, and interpreted the latter as coral formations. This view was later used by Albrecht Penck in his studies of Alpine geology.

In 1860, he was selected to undertake an expedition to eastern Asia for the study of land resources. After working in China, he sailed across the pacific to California, where he spent six years in active geological studies. He showed interest in the nature of the relations between volcanic rocks and the occurrence of gold which he studied in Hungary. He reported from San Francisco on the famous Comstock lode. For sometime he worked as a journalist providing information about the gold resources of California. Later, he was financed by the Bank of California and the Chamber of Commerce of Shanghai to provide information about the economic conditions of China, especially the occurrence of coal. In his subsequent survey of China, he was the first to report and map the Chinese coal fields. He is credited to have discovered the major resources of China.

Richthofen is known for his valuable work on China which was published in five volumes, between 1875 and 1912. The work also contained a large atlas of China. The work on China was a landmark in the contemporary geographical literature. It is based on his personal experience and observation, and as such it is a representation of the Humboldtian tradition. The work deals with the structure and the physiography of the mountains of central Asia and their influence on the outward migration of peoples. Richthofen recognises the loess deposits of northern China as wind-blown dusts from the steppe. He also noted that in this part of Asia the loess, and also the stratified rocks, were deposited over a relatively level or gently rolling surface and this more-or-less level surface cuts across ancient rocks of varying degrees of resistance to erosion. He believed that the only force strong enough to erode such resistant rocks must be the ocean.

Richthofen also discussed the geology and landform of North China. He provided information, based on his observation, about the economic and social activities of the people of North China. He provided a detailed account of the land and people of South China. However, he did much more than locating minerals and fuels. His generalisations on land/human relationships, particularly in the context of China, were primarily based on the empirically observed facts, and he attempted to identify the reciprocal relationship between the physical and the non-physical phenomena.

For a brief period, Richthofen was appointed in the University of Berlin by the Prussian government in 1875, but in 1877 he became a professor of geography in the University of Bonn. He moved to Leipzig as Professor of Geography in 1883, where the chair had fallen vacant after the death of the first professor of geography, Oscar Peschel, in 1875. In 1886, Richthofen moved to a new chair in physical geography in Berlin. It was long before he found his due place in the Akademic der Wissencheften in Berlin, but his personality and abilities gradually prevailed. He raised questions in his lectures that had never been touched in the circles of the university. His pronouncements on the geography of settlement and transport were published posthumously. His Kolloquium had a number of pupils who later became distinguished in their fields of research.

Richthofen's *Fuhrer fur Forschungsreisende* (1886), as a guide to scientific explorers, was a morphology of the Earth's surface feature. It was a masterpiece in the field of physical geography. The *first part* of the book deals with the techniques of field observation, especially in its relevance to physical geography and geology. The *second part* provides a systematic interpretation of the processes involved in the shaping of the surface of the Earth. In this part, the landforms are classified in terms of the processes. The *third part* deals with observation of the soils, rocks and mountain structure. The landforms are classified as to type under the headings of the various processes. The differential development of a river system and the associated kinds of relief are examined in areas of unconsolidated strata, horizontal strata, and abrasional surfaces. Also, in relation to processes involved (endogenetic and exogenetic processes and deposition), lakes, coasts and islands are classified accordingly. In the book, Richthofen attempts to distinguish between mountain and lowland with suitable criteria. He also distinguishes between the mountain ridges and mountain masses. Mountain ridges have long axes, while the mountain masses do not have such a form. It appears better to use the genetic factor and to subdivide mountains into types according to the basic forces in the development of their outer forms.

Richthofen would be long remembered for his pioneering statement on the scope and method of geography which came to be regarded as the programmatical statement of modern German geography. In his inaugural address at the University of Leipzig in 1883, he took over Marthes' chorological conception and made it a fundamental basis of his concept of geography. In this address, Richthofen proved himself to be the actual one to inherit and carry forward the ideas of Humboldt and Ritter, and also, in a consistent sense, to fulfil Peschel's research. In contrast with the latter, he had the sound, unprejudiced historical sense to fit himself into the course of development and to determine his position exactly (Hartshorne, 1976, 92).

To him, *geography is the science of the Earth's surface and the things and phenomena that are causally interrelated with it.* It is not Earth Science (Erdkunde). For this would be comprehensive. Its methods are measurement and observation of phenomena in the field. Geography may be pursued through the most detailed investigation of the smallest areas, and through comparative study of larger areas. The distinctive purpose of geography is to focus attention on the diverse phenomena that occur in interrelation on the face of the Earth.

To reach useful and reliable conclusion, a geographical study of any part on the face of the Earth must begin with a careful description of the physical features and then must move on to an examination of the relationships of other features of the Earth's surface to the basic physical framework. He distinguishes the study of the processes creating the surface features (geology) from the description of the surface features themselves as the frame of reference to which other elements on the face of the Earth (including the works of humans) are to be related. *The highest goal of geography is the exploration of the relationship of humans to the physical Earth* and to the biotic features that are related with the physical features. This became the basic pattern for geographical studies, not only in Germany, but also in other parts of the world.

Richthofen attempted to distinguish between the 'general geography' and 'special geography'. *General geography studies earth-bound phenomena* on a four-fold base the forms; the material; forces and causes of change; and movement. These four points of view lead to the morphologically material, dynamic and genetic modes of approach. General geography is not progressive. It is rather regressive; since it passes from the particular to the general, from the condition to the cause. It is analytical. Richthofen was very much concerned by the contemporary problem of geographical methodology: 'should geography be concerned only to describe the unique features of particular regions or should it concern with the formulation of generalisations or theory'? He attempted to simplify the problem without being controversial.

Special geography is primarily descriptive and synthetic. 'Every area on the Earth', says Richthofen, 'no matter how large or small, whether a continent, a small island, or a naturally bounded inland area, an artificially bounded state, a mountain, a river basin or a sea, is examined as a grouping of smaller unit areas, as well as in the perceptible appearances, among which are included the works of human culture' (Richthofen, 1877–1922, 46). To him, the essential observation on which any framework of concepts must be built had to be made in the field in particular areas where the features are unique.

There may be *two approaches to the descriptive, synthetic or special geography*: *chorography and chorology.* Chorography does not go beyond the systematic assembly of all the appearances of the individual land areas. The chorological approach is made possible by the advancement of various contributing sciences. It is concerned not only with registering the areal facts that are there, it also attempts to explain the areal distribution of these phenomena through the introduction of causative and dynamic interrelationships of the phenomena on every single portion of the Earth's surface. This approach has been facilitated, since Ritter's time, by the growth of the special disciplines that deal with the explanation of the spatial distribution of particular sets of phenomena. This leads to the field of general geography. The more deeply a study goes into process rather than distribution, the less is its use to a chorological treatment of a particular area.

Thus, Richthofen attempted to distinguish between chorography (which is non explanatory description, providing material for systematic geography), and chorology, a final step, the explanatory study of regions, based on systematic geography. The actual purpose of systematic geography is to lead to an understanding of the causal relations of phenomena in areas, an understanding which may be expressed in principles that can be applied in the interpretation of individual region, chorology or regional study. Regional study first must be distinctive, but it must also go beyond the description of unique features to seek regularities of occurrence and to formulate hypotheses that explain the observed characteristics.

Richthofen *attempted to revive the concept of unity of the Earth's surface* and to bring the analytical approach into a closer relationship with chorological study. To him, geography attempts to link the problems of the various sciences in a unity, the basis of which is the Earth's surface and the things and phenomena a really arranged on it. Though Richthofen set the guidelines for the systematic study of the Earth's physical features, but he also put emphasis on the geography of humans. This was concerned with the distribution of races and tribes, peoples and nations, languages and religions. He felt the need to develop what he called the dynamic anthropogeography which would aim at the understanding of the influence of the nature of lands on Humans, and the influence of Humans on the transformation of the nature of lands. He favoured the application of a genetic approach to deal with the study of human because it is concerned with the relation of human to the natural environment.

Richthofen seemed to have felt the need to begin the geographical study based on observation from the smallest segments of the Earth's surface. For it would be rather easier to describe the unique features of the units, and at the same time, might help the formulation of generalisation or theory. He distinguished the different methods of study in areas of different size (which he termed the order of decreasing size), the major divisions of the Earth, major regions, landscapes or small regions and localities. It has been rightly observed that *Richthofen followed the precedent of Humboldt, attempted to revive the close connection of geography to the natural sciences, and at the same time restored the Ritterian tradition and sought to fulfil the methodology of Peschel.* He was probably the greatest and most effective contributor to the development of modern geography.

Friedrich Ratzel

Modern geography's first encounter with structural explanation began with a debate between Friedrich Ratzel and Emile Durkheim at the end of the nineteenth century. Durkheim attempted to assimilate human geography to his own conception of morphologie sociale while *Ratzel provided the guidelines for a comparative systematic study of human geography* (or to use the word he introduced, anthropogeographie). This reflects Ratzel's background in the natural sciences, particularly zoology. It suggests the geography of humans in terms of individuals and races—*anthropological geography.* The major objects of Ratzel's concern were the works of human, particularly the products of human's social life in relation to the Earth (Gregory, 1978, 81). He was the greatest single contributor to the development of the geography of human. He made significant contributions to ethnography that are today highly esteemed by cultural anthropologists.

He came to geography from a quite different background. He did his graduation at Heidelburg in 1868, working in zoology, geology and comparative anatomy. Ratzel appears to have been *much influenced by two of Datwin's major themes:* 'struggle and natural selection', and 'association and organisation'. His major works thus demonstrated that cultural as well as natural phenomena could be subjected to systematic studies. He published in 1869 a commentary on Darwin's work that showed strongly the influence of Ernst Heinrich Haeckel, the famous Jena zoologist. His dissertation especially dealt with the significance of Darwin's ideas. Like his predecessors, Ratzel also felt that generalisations must be drawn on the basis of the empirically observed facts. Thus, he was basically interested in the observation of plants and animals out-of-door than in laboratory studies.

In 1869, he went on a trip to the countries around the Mediterranean. He came in contact with a French naturalist with whom he visited the area and learnt much about the bio-diversity of the region. He showed keen interest to write popular accounts about the people of German origin living outside Germany, especially the way they lived and made use of resources. After the Franco-Prussian war, Ratzel resumed study at the Munich University for a brief period. Here he met Moritz Wagner, the noted naturalist and ethnographer, who introduced him to his theories of the importance of the migration of species. Many of his writings were subsequently published in a book, *Travels of a Naturalist*.

Ratzel had the opportunity to visit the United States and Mexico in 1874–75. This American experience appeared to have made a deep life-long impression on him and was the source of many of his ideas. He attempted to compare the way of living of the German living in Eastern Europe whom he visited earlier and the German living in the middle-west and south-west of America. He was impressed to see the ways in which the German in America had been making use of the local resources. He was sad to see the deteriorating condition of the minorities such as the Negros, the Red Indians[1], the Africans and the Chinese in California.

He began to formulate some generalisations about the 'genre de vie' (way of living) of different communities, especially about the geographical pattern resulting from 'the westward expansion of the American Frontier'. These generalisations were based on his observation and travels. He recognised and described examples of the destructive use of land, which, he hoped was a characteristic of land settlement at an early stage and would be remedied as settlement progressed and matured.

He published a book on Chinese emigration in 1876, especially that which occurred in California. Later in 1878 and 1880, he published two volumes on the United States (*Physical and Cultural Geography with Reference to Economic Condition*). In 1893, he wrote another book on the United States, in which he described the contemporary political conditions with special reference to natural and economic conditions of the States. It was this experience in the interpretation of an extensive region in 1874–75 that turned his attention specifically towards the study of geography.

Ratzel is often described to have broken from what he called the integrative morphological approach of Humboldt, who was concerned with the interdependence rather than the origins and spread of things and ideas over the Earth. Through his biological approach, Ratzel also departed from Ritter's much criticised approach to geography, but he was far more sympathetic towards it than such contemporaries as Peschel. Ratzel was critical of the mechanistic evolutionary approach of both Darwin and Haeckel, yet the influence of their works was profound on his writings. In his later years, he became increasingly influenced by philosophical rather than biological thinking.

It was in Munich that Ratzel first began to publish his ideas concerning the study of systematic human geography. The first volume of his main work was published in 1882. It was entitled *Anthropogeographie*, or *Outline of the Influences of the Geographical Environment upon History*. In this volume, Ratzel attempted to develop the new method of natural science within human geography; he *stressed on the extent to which humans live under nature's laws.* He regarded cultural forms as having

[1] The words 'Negros' and 'Red Indian' were mentioned in Ratzel's original account of his visit to the United States.

been adopted and determined by natural conditions. *Ratzel saw humans as the end-product of evolution, an evolution* in which the mainspring was the natural selection of types, according to their capacity to adjust themselves to the physical environment.

Thus, whereas Ritter wrote of the reciprocal relation of human and nature, a relationship which was part of a harmonious whole serving the creative purpose of God, Ratzel tended to see humans as the product of his environment, moulded by the physical forces that surrounded him, and succeeding only in so far as he made the correct adjustment to their demands. The ultimate purpose of these adjustments, if any, lay outside the core of his inquiry. It must be recognised that Ratzel *initiated new ground in demonstrating that cultural and natural phenomena could be subjected to systematic study.* Before his time, human geography had largely confined itself to regional studies.

However, in Germany itself some of his contemporaries, notably Kirchhoff, held the opposite approach to the study of human geography. Instead of describing the influence of the physical Earth on human affairs, Kirchhoff attempted to focus attention on human himself. *He studied human geography by the reverse method—by considering human condition in relation to natural condition.* He gave more attention to the culture of human groups than to the physical aspects of the Earth. This was the approach Ratzel adopted in his second volume of *Anthropogeographie* which was published in 1891. The first volume was a reworking of the theme developed by Ritter in his *Erdkunde*, and Ratzel himself stressed the fact that he was developing the ideas of Ritter in conformity with the newly established scientific method.

The modified second volume of *Anthropogeographie* (1891) attempted to discuss the concentration and distribution of population, settlement forms, migration and the diffusion of cultural characteristics. *Ratzel did not merely explain phenomena in human geography in terms of natural conditions, but also stressed the significance of the historical development and cultural background of population.* But many of his followers, and Ellen Churchill Semple in particular, laid great significance on the deterministic opinions of Ratzel. In scientific circles, the first volume of *Anthropogeographie* had much greater impact than the second volume. The second volume appears to have attempted to revive the Kantian concept of the chorological science which 'studies things that are mutually coordinated, not subordinated in space'.

Dickinson has remarked that the first volume (1882) appears to treat *the cause of human distributions* and refers to the dynamic aspect of geography, and the second volume seems to deal with *the facts of distribution* which reflects the static aspect of geography. The first volume is an application of geography to history, and the second the geographical distribution of man. Implied in *Anthropogeographie* (vols I and II) are the influences of the Darwinian concepts of struggle and survival which emphasised on the method of natural sciences, or more precisely the hypothetical-deductive method, that sought prime causes by deduction.

The reaction against too much orientation towards environmental determinism (as implicit in his second volume of *Anthropogeographie*) prompted him to clarify that *cultural differences were far more pronounced than differences in the physical character of the land.* He pointed out the great contrast in the way people make use of the land in two places that are physically very much alike. He attempted to give emphasis on the *genre de vie* of the people. He observed the contrast between the French and the German cultural traditions in the physical resemblance between the two places of the low mountain regions on either side of the middle Rhine Valley (Ratzel, 1898).

Ratzel is also *called the 'founder' of political geography* and is often recognised as the father of modern political geography. Undoubtedly, it was he who brought a radical change in political geographical thinking. Although following logically after Ritter, he represented a break with the largely impressionistic writings of the past by concentrating on the *growth of political entities as living organisms* and by evolving, through an organic analogue model, a set of laws by which he attempted to explain the processes of the spatial growth of states.

In 1897, Ratzel published his work on political geography. This reshaped the old topic along lines conformable with the principles he had introduced in his *Anthropogeographie*. He discusses the state as an organism attached to the land, a piece of humanity and a piece of Earth. Ratzel, like many other scholars of his time, was greatly influenced with the ideas of Herbert Spencer regarding the similarity of human societies to animal organisms. *He attempted to show that a state, like some simple organisms, must either grow or die and can never stand still.* When a state extends its borders at the expense of another state, there is a reflection of internal strength. Strong states must have room to grow in order to survive. Every state is a piece of humanity and a portion of the Earth. Out of this develops the *concept of living space* (*lebensraum*). He contended that the superior people must have the right to enlarge their living space at the expense of inferior neighbours. Ratzel appears to have *believed in what may be called 'racial superiority'*. This was in marked contrast to Humboldt's assertion that 'all races of man had a common origin and that no race was necessarily inferior to the other' (Kellner, 1963, 83).

Although Ratzel derived this organism analogy from the natural sciences, its roots may also be traced in the earlier idealist philosophy and the idea of ganzheit or 'whole' as used by Ritter. Ratzel also appears to have adopted Ritter's notion of evolution from youth to maturity as a correlation between growing old age and diminishing size. He, like Ritter, *attempted to comprehend 'the Earth as an integral whole, an interacting unity'*. But his works differed in two important respects from Ritter's. He treated Human Geography systematically and from a Darwinian point of view. His last great work attempts to present the interrelation of the phenomena of the Earth's surface.

Ratzel's attempt in *Anthropogeographie* (1882 and 1891) and *Politisctic Geographie* (1897) *to change geography from an inventory to an explanatory science* invited a reduction of geography to social morphology. The only difficulty one could see in this was that Ratzel had provided two conceptions of geography, and only one of them was admissible. At times, he clearly seemed to propose as the object of political geography, the settlement forms assumed by societies on the surface of the Earth, and this, properly speaking, is social morphology. Sometimes, he assigns it the goal of establishing the effects of physiographical features on the political development of these peoples. However, Durkheim regarded the first of these as the concern of any properly constituted social science, whereas the second accorded primacy to what he always insisted were entirely environmental constraints.

Ratzel's systematic studies of human geography led him to discard a number of ancient concepts. He seems to have provided an opposite view with regard to various stages of human's use of the land, a view that continued to be accepted since the time of Dicoearchus. His criticism of the concept of stages was recognised by Edward Hahn, who in 1892, prepared a new map of economic systems of the world, based on Ratzel's viewpoint.

Dickinson points out that Ratzel's concept of geography seems to be based on two approaches—the measurement of the consistent interrelation of environment and humans, and the measurement of the interrelation of human phenomena over the Earth that are a really coincident. The former focuses on relationships while the latter is concerned with the coincidence, correlations, and interrelations of distribution. He also focused attention on a third aspect—the visible human-land landscape. (Dickinson, 1969, 72).

In the contemporary German geographical scholarship and thinking, there was a tendency to overstress the physical aspects of geography, but it was Ratzel who attempted to establish a balance that equal importance be given to both physical geography and human geography. But it was the period when dualism in geography acquired a professional dimension. The question was: are the concepts and methods of physical geography so utterly different from those of human geography that these two branches cannot be properly included in a single discipline? However, it cannot be disputed that Ratzel, who attempted to incorporate the Darwinian themes into geography, tried to give the discipline a new paradigm, based on hypothetical-deductive method, that set in a new horizon in the contemporary geography not only in Europe, but also in North America.

Ratzel had attempted to rescue the study of humans from its earlier subordinate position as a part of Landerkunde, but he had not thereby affected the dualism in geography. Indeed, his work served to fix interest still more firmly on systematic studies, and regional geography continued to receive scant attention.

Georg Gerald

It was at the time when Ratzel's systematic but scientific human geography had been making inroads into the contemporary German geographical thinking, that Georg Gerald (1833–1919) came out with a different argument which tended to focus on the study of the physical aspects of the Earth. He was professor of geography at Strassbourg from 1875 until his retirement in 1910. He attempted to develop the views of Kant and Peschel in an effort to develop geography as an exact science. He favoured the complete exclusion of humans from the field of geography. He maintained that geography should be exactly what the word Erkunde implied—study of the whole Earth without reference to humans. He concluded that physical sciences could formulate exact laws, but no such laws could ever be formulated which could account for the behaviour of human groups.

To him, the Earth was like millions of other units in the universe, a great complex of materials in a state of change, bound together and interrelated by various forces to form a unit whole, though acted upon also by external forces, notably from the Sun. The problem and object of geography is therefore to study the interrelations of these forces and the resultant changes in Earth's material. Though he advocated in favour of the study of physical science with major emphasis on non-human organism, but all the same he attempted to include the geography of humans in his courses at Strassbourg. He kept the contemporary dichotomy between human geography and physical geography alive for a longer period.

Joseph Partsch

One of the distinguished scholars in the contemporary German geographical scholarship was Joseph Partsch (1851–1925) who was often ranked with Hettner and Penck in the growth of modern geography in Germany. On account of his close associations at Breslau with Carl Neumann and H. Kiepert, he attempted to revive the Ritterian tradition. He *made significant contributions to the advancement of regional geography,* notably in the areas of Silesia and Central Europe. Some of his generalisations on regional geography of Silesia and Central Europe seemed to have been based on his travel accounts. He developed a particular interest in the study of glaciers. Partsch visited the Alps several times and attempted to study the glacial landform in the Riesengebirge, and paticularly observed moraines and granels that spoke of extended glacial activity in a particular period of time. His book on glaciology was published in 1882.

He completed and published the unfinished work of Neumann on the physical geography of Greece in 1885. On the request of H. J. Mackinder of Great Britain, Partsch prepared a regional geography of Central Europe in 1899. It was published in 1903, almost simultaneously with the first volume of his work on Silesia. These works undoubtedly enhanced the reputation of Joseph Partsch not only in Germany, but also outside Germany. He succeeded Ratzel at Leipzig in 1905, and there he completed his work on Silesia. The second part on Mittelschlesien and the third part on Niederschlesien were published in 1907 and 1911 respectively.

It is interesting to observe that Joseph Partsch also got involved in the contemporary dichotomies and dualism in geography because he *stressed more on the study of physical geography rather than on human geography.* It is academically and professionally more interesting to know that he was unmoved by the tradition of Ratzel (*Antropogeographie*) even though he succeeded Ratzel at Leipzig which had long remained a centre for the study of the systematic human geography, a tradition that Ratzel had founded there. The development of the discipline from its overemphasis on classical geography to a newly conceived and applied field of regional geography was evident in the career of Partsch, a great scholar of his time whose contributions to the development of new paradigms in the study on the glacial landforms and applied regional geography had remained a source of inspiration to the succeeding generations in Germany and outside Germany also.

Albrecht Penck

A versatile genius and a great scholar of his time, Albrecht Penck is long remembered for his valuable *contribution to the development of modern physical geography.* He was a phenomenon in himself as the philosophy he developed in the realm of physical geography had not only struck the German geography alone, but also the geography outside Germany. He was *the forerunner of the modern 'geomorphology'* for he is credited with the first use of the term to refer to the origin and development of the Earth's landform. Penck was without doubt one of the scholastic giants of the last generation. His greatest contributions to knowledge were his studies of the Ice Age and its fluctuations, but he was, throughout his life, a professional geographer (Dickinson, 1969, 100).

Albrecht Penck was born in Leipzig in 1858. He studied natural sciences at the University of Leipzig, beginning in 1875. He went to Munich in 1880 to work under Karl Zittel, and there he became a private doyen at the University in 1883. He was Professor of Geography at the University of Vienna from 1885 to 1906. During his stay at the University of Vienna, Albrecht carried out his greatest work on the *fluctuations of the Ice Age,* based on the observation of the deposits in the valleys of the Alps. Here he collaborated with Edward Bruckner in the identification of four separate ice ages in the Alps. He was associated at the University with Eduard Suess, who prepared maps of the major geological regions of the world, outlining the great crystalline shields of ancient rocks. He succeeded Ferdinand von Richthofen at the University of Berlin in 1906, and remained there until his retirement in 1926.

Penck's first paper on the boulder clay of the lowland of Germany appeared in 1879; Penck seemed to have supported the 'ice sheet theory' in place of the 'drift theory'. He *further developed the ice-sheet theory* in a book which was published in 1882. In collaboration with Edward Bruckner in Vienna itself, Penck attempted to identify four separate iceages in the Alps, and a book to this effect was published in three volumes from 1901–09. In the book, he supported the theory of climatic fluctuation in the ice, with the help of the collection of numerous evidences from the field observations. His division of the quaternary Ice Age in the Alps into three inter-glacial and four glacial periods (named after Alpine rivers Gunz, Mindel, Riss and Wurms) became the classic basis of quaternary geology and the springboard for the interpretation of human pre-history. This is the great contribution of Penck to the field of glaciology and quaternary geology.

Though Albrecht Penck belonged to the tradition of Oscar Peschel, but he *adopted the comparative method of field study* practised by Alexander Von Humboldt (which included observation, location and comparison) that must be applied in respect to a particular problem. Penck's objective was thoroughly geographical, but his method was geological. He attempted to show how the systematic study of physical features can be approached from the chorological point of view.

In his monumental work *Morphology of the Earth's Surface*, Penck attempted to introduce new paradigms and concepts as well as methods which could be applied to analyse and study the origin and development of landforms. The method he introduced in the book was hypothetical-deductive, which sought prime causes by deduction. It attempted to set up inferences and hypotheses that led to the formulation of certain physical laws to explain the reality of the landforms.

The above book was published in two volumes. The first volume deals with the metric measurement of the Earth forms and processes involved in the formation of the Earth's surface, and here Penck discusses at length the agents of erosion and deposition. The second volume deals with the individual geographic complexes of similar and related forms, classified and examined in terms of their mode of development. This involved a careful empirical analysis of the various types of land surface, and their relation to various forms of composite relief. The forms of sea are discussed in a separate section under the headings of coastlines, sea floors and islands. He attached greater importance to the processes of formation of landforms and there he made suggestion for the physical laws. *His system of classification was based on forms, not on process.*

Penck also appears to have been influenced by the Darwinian theme—change through time or evolution—which he attempted to incorporate into his new concept of the morphology of the Earth's surface. He sought the origin of landforms in genetic processes through time. He seems to have held the view that the climatic features are so dominant that they have direct impact on the landscape. He used the terms 'arid', 'humid' and 'nival' with specific pluviographic connotations, as the basis of a world classification of climates. He hypothesised that the climate of a region so impresses itself on the observable features of the landscape that a classification of climates could be made even where instrumental records were lacking. To him, the effective rainfall of a place reflects a balance between rainfall, run-off and evaporation. He also held the view that higher the temperature, greater is the evaporation.

Though Albrecht Penck overemphasised physical geography and geomorphology, he did not underestimate the works of humans which he thought essential to an understanding of the varying characteristics of the face of the Earth. He studied the population capacity of the Earth, in terms of the carrying capacity of specific land units (Fisher, Campbell and Miller, 1967, 99–106).

In his 'new geography', Penck argued that the unity of geography should be based on the spatial arrangement of phenomena on the Earth's surface in respect of both 'physical' and 'human' processes in so far as such processes help to characterise the unique areas of the Earth's surface. His new geography seemed to be a revolt of the younger geographers against the old leaders (Valkenburg, 1967, 105–06).

Though Albrecht Penck offered new paradigms and concepts in physical geography and geomorphology, but *his views on political geography attempted to revive Ratzel's basic concept of lebensraum.* The concepts of Reichsboden, Sprachboden, Volksboden and Kulturboden were all expounded by Penck with a view to explain the spatial range of German culture in Central Europe. These concepts also reveal strong influence of Ratzelian tradition on the spatial growth of the state vis-a-vis lebensraum. At a seminar in the University of Berlin on the concept of natural frontiers, Penck maintained that 'natural boundaries are those which correspond to the economic needs of the nation'.

Some of his politico-geographic concepts seemed to have stimulated the German geographers of the inter-war period to concentrate on the German need of more living space vis-a-vis geopolitics. He wrote of 'Zwischeneuropa' in 1916 to explain a major transcontinental zone of change in the build of Europe; of the 'Grossgau' in the heart of Deutschland (1921); and of the concept of Deutschland as a geographic form (1926 and 1928) which speaks of the limits of German culture that stretches from early times beyond the contemporary German concentration in Europe.

He was of the opinion that the study of geography would remain incomplete unless there is the provision of accurate maps showing the major features that are associated with areas. *Systematic study of geography cannot proceed without accurate maps.* He favoured the idea of large maps, full of physical and non-physical information. Thus, he put forward the idea of topographical maps which would show the shape of relief, water bodies, vegetation and the cultural features. The preparation and compilation of such accurate maps would require detailed survey. When he raised the issue, a very small part of the world was covered by surveys detailed enough to permit the preparation of topographic-scale maps.

At the International Geographical Congress in Berne in 1891, Penck proposed that the nations of the world must cooperate in the compilation and preparation of an International Map of the world at a scale 1 : 1,000,000. He suggested that there should be agreement about the accuracy of the maps, the categories of things to be shown and conventional symbols to be used and the projection. Thus, he is credited with having encouraged the nations to cooperate on the issue of compilation of topographic-scale maps, and this was achieved in the Paris conference in 1913.

Though Penck's specialised contribution lay in the field of physical geography and geomorphology, and especially Pleistocene geology, his conceptual and substantive contribution to political geography made him one of the distinguished architects of modern German geography. It is significant that many of the distinguished German geographers of the last generation who studied under Penck began as geomorphologists, but made lasting contributions to the geography of humans. This change is evident in the professional growth of Penck himself, in his change-over in the later period on the imprint of the works of historic humans on the landscape. His imposing figure, actual and figurative, and his tremendous research ability made him the most influential geographer in Germany.

CHOROLOGY AND LANDSCAPE MORPHOLOGY

Chorology and landscape morphology developed side by side in Germany in the early part of the twentieth century. It was Alfred Hettner who revived the Kantian philosophy of chorology, while Otto Schluter that of landscape morphology.

Alfred Hettner

Much of the present German geography owes its indebtedness to Alfred Hettner, who by his methodological innovations in geography, created considerable stir in the professional circles of Germany of his time. And even after the Second World War, German geography tended to show a natural orientation towards the heritage and tradition which was long established by Hettner.

Hettner was one of the most influential figures in Germany in the development of geography. He was the first professor in a German University after Cad Ritter (to be) trained as a geographer. He also studied contemporary philosophy and had the opportunity to come in contact with Kirchhoff who introduced him to geography. He worked with Theobald, Fisher and Richthofen, who infused in him the concept of chorology. He did his doctorate dissertation on the climate of Chile and Patagonia at the University of Strasburg with Professor Gerland. He submitted a thesis on the geomorphology of the Saxon Highland for the 'habitation' at Leipzig under Ratzel. He made extensive field studies in Germany and in the Andes of Columbia that led to a multitude of publications in geomorphology. He became a professor of geography at Tubingen in 1897. However, in 1899 he was appointed professor at Heidelberg where he remained till his retirement in 1928. Though he came much closer to Ratzel while at Leipzig, yet he was not influenced by Ratzel's approach to geography of humans, but *emphasised on the physical basis in geography.*

Hettner made significant contributions in publications of geographical materials. His first publication entitled *Travels in the Columbian Andes*

(1888) was based on his field observations of the region. In 1895, he started his own periodical, the *Geographische Zeitschrift* of which he was the editor till 1935. It usually dealt with various discussions on the methodological exposition. The periodical was the forerunner of modern German geographical heritage and tradition. In 1907, Hettner published a large volume on the regional geography of Europe and in 1924 published the continuation of the earlier work on extra-European states. These two books on regional geography later came to be known as *Foundations of Regional Geography*. In 1933–35, he published *Comparative Regional Geograhy* in four volumes.

His other important books were on Russia (1905), *England's World Domination and the War* (1915), *Surface Forms of the Continents* (1921 and 1928) and *The Spread of Culture over the Earth* (1928 and 1929). His *Handbuch der Geographischen Wissenschaft* contained many ideas on the contemporary regional geography. The work was published in 11 volumes and it was a monumental work in geographic literature.

Alfred Hettner's essays on the concepts, paradigms and methods of geography, published between 1895 and 1905, including additional section, were published in a single book entitled *Geography: Its History, Character and Methods* (1927). However, some of his works were published posthumously which included: *A Geography of Man* (3 vols); *Transport Geography*; *Economic Geography*. These works of Hettner gave contemporary geography a firm philosophical and scientific basis.

Concepts and Methods of Geography. Hettner defined *geography as the chorological science of the Earth's surface.* It is concerned mainly with the interplay between human and nature, an evaluation of spatial (Raum) relation. Its aim is primarily to study areas or regions. Such a study should contain description and explanation derived either analytically or synthetically. Or, in other words, *geography is the study of the Earth (Erdkunde) according to the causally related differences—the science of areal differentiation of the Earth's surface.* Though Hettner criticised Richthofen for not being entirely successful in finding the sharp methodological expression for his chorological concept, yet he attempted to elaborate Richthofen's concept of chorology.

It is also a fact that Alfred Hettner did his utmost to redevelop and elaborate Kant's thought. He admitted that for a long time he had not paid sufficient attention to Kant's exposition of geography, but had later rejoiced to discover a close correspondence between himself and Kant. It has been rightly observed that Hettner in his writings had revived the Kantian definition of geography, and within this framework welded the systematic studies of Humboldt, Peschel, Ratzel and the regional studies as defined by Ritter, Marthe and Richthofen into a coherent whole. However, Doring concluded that in Hettner's synthesis of the geographic thought of the two founders of modern geography, Humboldt stands not beside but before Ritter.

Hettner considered that the biggest difference between history and geography does not lie in that geography sets out to study a given time (namely, the present), but that in geography the time aspect recedes into the background. *Geography does not study development in time as such,* although this particular methodological rule is often broken and geographers cut through reality at a distinct point in time and only consider historical developments in so far as they are necessary to explain the situation at that chosen point of time (Hettner, 1927, 31–32).

Hartshorne outlines Hettner's concept of chorology as presented in 1927: 'The goal of the chorological point of view is to know the character of region and places through comprehension of the existence together and interrelation among the different realms of reality and their varied manifestations, and to comprehend the Earth's surface as a whole in its actual arrangement in continents, larger and smaller regions and places' (Hartshorne, 1959, 13).

According to Hartshorne, he had traced the repeated appearance of the concept of chorology in the place of geography among the field of learning in the writings of different scholars since the eighteenth century.

In considering the *logical position of geography among the sciences,* Hettner, like both Kant and Humboldt before him, proceeds not from the consideration of particular branches of science, but from a view of the whole field of objective knowledge. He has developed the concept most fully as compared to Kant and Humboldt. It is clear from the following excerpts from Hettner's book published in 1927:

> Reality is simultaneously a three-dimensional space, which we must examine from three different points of view in order to comprehend the whole; examination from but one of these points of view alone is one-sided and does not exhaust the whole. From one point of view we see the relations of similar things, from the second the development in time, from the third the arrangement and division in space. Reality as a whole cannot be encompassed entirely in the systematic sciences, sciences defined by the objects they study, as many students still think. Other writers have effectively based the justification for historical sciences on the necessity of a special conception of development in time. But this leaves science still two dimensional, we do not perceive it completely unless we also consider it from the third point of view, the division and arrangement in space.
>
> The systematic sciences ignore the temporal and spatial relationships and find their unity in the objective likeness or similarity of the subjects with which they are concerned. With the same justification as the development in time, the arrangement of things in space demands special study. Together with the systematic or material, and the chronological or historical or time-sciences, there must develop chorological or space sciences. There must be two of these – one has to do with the arrangement of things in the universe; that is astronomy, the other is the science of the spatial arrangement on the Earth, or since we do not know the interior of the Earth, we can say on the surface of the Earth.
>
> Geography as the second chorological science, is the study of the spatial arrangement on the surface of the Earth. If no causal relation existed between the different places on the Earth, and if the different phenomena at one and the same place on the Earth were independent of each other, no special chorological conception would be needed; since, however, such relationships do exist, which, by the systematic and historical sciences, are comprehended only incidentally or not at all, we need a special chorological science of the Earth or of the Earth surfaces (221–24).

Alfred Hettner seemed to have been *opposed to Otto Schluter's conception of 'landscape morphology'* which was a form of regional geography, developed in Germany in the inter-war period. Schluter asserted as early as 1906 that geographers should consider the form and spatial structure created by visible phenomena on the surface of the Earth as their unifying theme. *Hettner was basically concerned with the uniqueness of areas, whether this uniqueness was evident in the visible landscape or not.* He recognised the focal interest of landscape, but 'refused to recognize the limits set by it on the study of the human facts in space. In spite of this disagreement, landscape morphology can be seen as an attempt to study regions on the basis of Hettner's principles. The supporters of the landscape morphology agreed with Hettner on the main philosophical issue that geography is a chorological science which should focus on regional synthesis and only delve into historical developments to the extent necessary in order to explain the contemporary situation' (Holt-Jenson, 1981, 31).

Because of his strong inclination towards the concept of chorological science, Hettner *came out strongly against environmental determinism.* He was of the opinion that a geographical synthesis is distorted when nature is regarded dominant and human as subsidiary, that the unity of geography could only be maintained through the concept of a chorological science, i.e. a subject which studies things which are mutually coordinated, not subordinated in space. Thus, Hettner appeared to be *a significant contributor to further the development of Possibilism.*

Hettner made a distinction between *'systematic geography'* which seeks to formulate empirical generalisations or laws, and the study of the unique in *'regional geography'*, whereby generalisations are tested so that subsequent theories may be improved. In other words, the general geography follows systematically the distribution of the various geographic phenomena over the Earth's surface, and special or regional geography (Landerkunde) elicits the concept of geographic region.

It is mainly due to Hettner that dualism, which so long hampered geography, has been successfully overcome. *He rejected the view that geography could be either idiographic[2] or nomothetic[3] but not both.* He attempted to make it clear that geography is both idiographic and nomothetic, as indeed almost all other fields of learning must be. Both Hettner and Hartshorne made this view of geography very clear. However, some writers have accused Hettner of defining geography as essentially idiographic, thereby obscuring the underlying continuity of geographic thought. Both Hettner and Hartshorne *considered region to be a functional unit —an organism which was more than the sum of its parts.* Functionalism affected much geographical research in the late nineteenth and early twentieth centuries and philosophy was strongly interwoven with method. Hartshorne emphasised the need for a functional approach to political geography.

Hettner is often accused of giving more emphasis on physical environment rather that on cultural environment (Husain, 1988, 130), but he seemed to have *focused attention on the theme of humans relation to his physical and biotic surroundings.* The traditional outlines of a regional study started with location or position and then proceeded through chapters on geology, relief, climate, vegetation, natural sources, settlement, population, the form of economy, communication and political divisions. The outline was based on the notion that this formed a kind of sequence of cause and effect; in dealing with each topic the relationships with the physical base were discussed, but not the relationships with other topics (Dickinson, 1961, 122).

Hettner made no significant contribution to political geography, but his periodical *Geographische Zeitschrift* was put on a war basis and gave much space to articles on the areas of battle and the countries involved. The most significant of these was his editorial statement published in 1919 on the Treaty of Versailles. 'Our work in the world is destroyed, our land is dismembered, our national wealth has been taken away from us, our economic life is burdened with numerous mortgages; and what is worst is that the German nation, which had broken its back through the destitution of war and the insolent revolution, and thus a dishonourable peace, is thrust on us' (Hettner, 1923, 5). This statement is significant in the sense that it added philosophical basis towards the development of geopolitik in Germany.

This editorial statement on 'Peace and Political Geography vis-a-vis the Treaty of Versailles' was given much political propaganda by Karl Haushofer. Despite this, Hettner was often accused of being a 'positivist' by the Nazi people in Germany. This was because the positivists were opposed to everything that smacked of metaphysics and unverifiable phenomena. They, therefore, became bitter opponents of Nazism, which they saw as a mixture of irrational prejudice and ideological dogma. Positivist became a term of abuse in Nazi Germany.

Harvey has criticised both Hettner and Hartshorne for having relied too much on Kant's concept of a chorological science because it was developed on the assumption of 'absolute space' which was tied to Euclidean geometry. As the science has accepted Einstein's theory of relativity,

---

[2] *Idiographic:* It is concerned with the unique and the particular.

[3] *Nomothetic* : It is concerned with the universal and the general.

it has rejected the concept of absolute space. It is, therefore, ironic that such influential geographers as Hettner and Hartshorne took guidance from Kant, rather than from Gauss who also directly made major contribution to the science of geography in his work with map projections. As a consequence, the main current of philosophical opinion within geography in the first half of the twentieth century was based upon concepts which other scientists have already rejected as untenable (Harvey, 1969, 209; Holt-Jenson, 1981, 63).

A number of contmporary geographers in Germany expressed scepticism about the identification of geography as a chorological science and, hence, defined it by its method rather than by its subject matter and its concepts. Others were also concerned about the overemphasis on the significance of the physical features of an area that resulted from following the scheme. By tying back everything to the physical features, other important relationships were over looked—such as the relation of population density to the economy, or the economy and the routes of circulation, or even the relation of all of those to the political units. Many of the interrelations observed in regional studies were in the process of change through time; and only by examining the past geographies and the changes taking place could the idea of processes, or sequences of events, be introduced to geographical work.

Hettner-Schluter Controversy

Hettner-Schluter *controversy on the 'content' and 'purpose of geography'* in the early twentieth century seems to have had significant impact on the contemporary geographic tradition not only in Germany, but also outside it. The controversy sharply divided the German geographers into two broad groups—one adhering to the Hettnerian tradition, the other to the heritage being established by Otto Schluter (1872–1952). The controversy appeared to have developed with the publication of J. Wimmer's *Historische Landschaftskunde* in 1885, and it was Otto Schluter who in 1906 applied the term 'Landschaftskunde' (landscape science) to designate the concept of geography.

Schluter insisted that geographers look first at the things on the surface of the Earth that could be perceived through the senses and at the totality of such perceptions—the landscape. He objected to the chorological definition of geography, and noted that accepting the landscape as the subject matter of geography would give the field a logical definition. Schluter's landscape morphology was a distinct form of regional geography and he insisted that geographers should consider and emphasise the form and spatial structure created by visible phenomena on the surface of the Earth as their unifying theme. All human distributions of non-material character, such as social, economic, racial, psychological and political conditions, should be excluded from the study as ends in themselves, but they could be considered while explaining the observable landscape.

He raised objection to both the environmental emphasis on land-human relationships of the current schools—Ratzel, Davis, Mik and Vidal Blache—and to the view of areal associations of terrestrial—phenomena as expounded by Hettner. (Neither, it seemed to Schluter, provided a distinctive field). The regional concept (Landschaftbegrift) of Schlutcer, like that of Hettner, is a geographical position of the Earth's surface which showed a certain degree of homogeneity within boundaries that could be defined.

However, *Hettner emphasised upon the 'Wesen'*, a personality of an area as based upon the similarities of contiguous places and then phenomena, spatial cohesion, and the causal cohesion of the various natural realms in this area. *Schluter emphasised upon the 'Bild'*, the association or assemblage in space of landscape elements as the essence of his landscape unit. The morphology of landscape is related with the ecology of landscape, and it interprets the non-material character of human distributions only in so far as they are relevant to the landscape.

Hettner was opposed to Schluter's limitation of geographical study to the visible landscape.

He was concerned with the uniqueness of areas, whether this uniqueness was evident in the visible landscape or not. He refused to recognise the limit set by it on the study of human facts on space. Hettner's approach provided an electical structure for geography as a distinct discipline whereas Schluter's concept of cultural geography—the landscape concept—has stood the test of time and has continued to grow by increasing clarity in concept and practice, and in the shaping of new problems relevant to the rapidly changing society.

Schluter was very influenced by history and it was under the influence of Kirchhoff that his interest from Germanistic studies to geography was made possible. He applied historical methods in geographical studies, especially in analysing landscapes. In his three-volume work on the settlement areas of Central Europe in early historical times published shortly before his death, he *distinguished between 'Urlandschaft' and 'Kultarlandschaft'*, the former refers to the landscape that existed before major changes were introduced through activities of humans, and the latter means a landscape created by the human culture. He attempted to reconstruct the 'Urlandschaft' of Central Europe. He discovered that the approximate date when the major movement of new settlement into the forests began was 500 AD By using the evidence of place names and reading the descriptions by ancient Greek and Roman geographers, he reconstructed the pattern of forested land and open land for that date. He then traced the process of settlement that created the 'Kultarlandschaft'.

With regard to the content and purpose, Hettner focussed attention on terrestrial areas while Schluter stressed the terrestrial landscape. Hettner's approach resulted in the orderly presentation of categories of data on the Earth surface viewed in relation to a physical framework. Schluter's approach stressed on the use of land by man in terms of his heritage, purpose and technology. Penck recognised that the essence of geography lies at the fact that it deals with terrestrial areas, and that the visible content of the landscape determines the content of modern geography. He opined that the smallest region (Landschaft) is an area which has unity of form and function.

Schluter is often credited with having founded 'modern cultural geography' on a base comparable to that of geomorphology. He attempted to draw a distinction between *genetic analysis* (that is, the process of change in form) and *causal or dynamic analysis* (that is, the causes that lie behind the process of change). To him, human geography should aim at the recognition of the form and arrangement of the earth-bound phenomena as far as they are perceptible to the senses.

He claimed that his method was morphological and its procedure parallel to the study of landforms. All human distribution of non-material character, such as social, economic, racial, political, etc., are excluded from this study as ends in themselves and are considered only is so far as they are contributory to an understanding of the evolution and character of the landscape. This mode of approach is parallel to that of physiography and is concerned not with the exclusive study of the processes (i.e. mechanics of evolution), but with the facial expression of these processes on the landscape. The cultural landscape embraces both immobile and mobile forms. This morphological approach is clearly a fundamental departure from the Darwinistic environmental approach.

The geographic concept of landscape seemed to have made deep impact on Jean Brunhes in France, and the ideas of Schluter and Brunhes became the research objectives of the continental geographers. However, the British geographers showed lack of interest in the concept of landscape morphology. In the USA, Carl Sauer gave a basic evaluation of the new direction in his *Morphology of Landscape* (1925), but apart from this, landscape morphology attracted little interest in the new world. Carl Sauer attempted to represent geography as a science that finds its entire field in the landscape. Its concern is only to establish the connection of phenomena in the visible landscape and these connections are the ones of spatial associations and not of some hidden causality. In fact, the 'morphologic method' of Schluter, and later of Carl Sauer, was simply a restatement of Comte's designation of 'la reel' as the only appropriate domain for scientific enquiry (Gregory, 1978, 28–29).

It may be noted that Schluter's concept of landscape morphology represents an approach to regional geography, but his approach to the geography of cultural landscape appears to belong to the branch of systematic geography. There has been an increasing interest in the cultural landscape in recent years, perhaps as a reaction against the uniformity of most modern buildings and other landscape features.

Other Geographers

Among the other geographers we study Passarge, Koppen, Philippson, Gradmann, Schmieder and Carl Trole.

Seigfried Passarge (1866–1958) was a medical doctor and geologist turned into a geographer. He was a contemporary of Hettner and a pupil of Richthofen, but his geographic ideas were often complex and contradictory. Despite this, he was one of the makers of modern geography. He succeeded Partsch when the latter left the University of Breslau in 1905, but in 1908 he was appointed professor of geography in the newly established Kolonial Institute in Hamburg where he remained until his retirement in 1936. He attempted to blend together the ideas of Penck, Hettner, Schluter and Vidal la Blache, but he rejected the view of Richthofen that regional geography was an orderly presentation of a large variety of spatial distribution.

In his *Physiologische Morphologie* published in 1912, he rejected the Davisian approach which attempted to explain the areal variables of landscape in terms of the genetic processes. As regards the Davisian formula of 'structure, process and stage', Passarge substituted 'region, place and form' which refer to the region, the local forces, and the resultant landforms. He thus *proceeds to an analysis and classification of the form elements grouped in terms of their genetic determinants.* If one genetic force is dominant, the resultant form of the land is called a 'monodynamic form'; and if two or more are dominant, it is called a 'polydynamic form'. The Davisian approach provides an apparent explanation of landforms, based on abstraction, but it is often obscure and complicated and raises difficult problems of surface that the Davisian theory cannot properly explain.

His first major work 'Grundlagen der Landschaftskunde' (1919–20) published in four volumes deals with unit elements of landscape, the causes of the landscape phenomena, the origin of landforms, and the imprint of Human's works on the landscape.

In another work, *Vergleichende Landschaftskunde*, published in five parts during 1921–1930, Passarge attempts a heirarchy of the unit area of landscape. 'The smallest areas are the landscape elements such as slopes, meadows or valley bottoms, ponds, dunes, etc. A group of contiguous small areas together form a section and contiguous sections make up a region. A group of contiguous regions leads to Landschaftsgebiete such as the North German plain, and a number of these make up a region such as the Central European forest. It is a constituent member of the world's great regional belts which are essentially climatically determined.

His *Landschaftsgurtel der Erde* published in 1923, was based on major categories of vegetation, and here he suggested that it would be possible to identify certain regularities in the arrangement of the world's major landscape zones in relation to latitude and position on the countries.

Passarge *developed the idea of four spatial forms* (*raum, mensch, kultur, geschichte*), and the idea that cultural geography is concerned with the influence of humankind and its works in the creation of the cultural landscape or the man-made landscape: This view is identical with Schluter's idea of a cultural landscape of perceptible phenomena. Passarge analysed the impact of the landscape on the psychology of the human occupants, and introduced the concept of Stadtlandschaft. He seemed to have held the view that the ideas of milieu and lebensraum as embracing the totality of the spatially operating forces contribute to the individuality of areas and along these lines, developed the distinctiveness of regional geography (Landerkunde).

Albrecht Penck's suggestion about a classification of climates based on observable features of the landscape appeared to have made great impact on the Russian-born climatologist **Wladimir Koppen.** Between 1884 and 1918, Koppen made several attempts to provide a satisfactory classification of climates, using temperature distinctions only. It was only after the publication of Penck's ideas concerning the relation of rainfall effectiveness to temperature and seasonal changes that Koppen offered a new system of classification in which he made use of annual variations of temperature and rainfall. His new classification first appeared in 1918 and its final version came out in 1936.

With the help of these climatic parameters, Koppen attempted to explain the causes of deficiencies of moisture throughout the year on the west coast of continents between 20° and 30° latitudes in both the hemispheres, and the abundance of moisture on the west coast of continents about 40° and along all the east-coast. He demonstrated these regularities by drawing a geometric figure for a generalised continent and plotting on it the hypothetical position of his classification of climates (Koppen, 1936).

It was largely due to the efforts of **Gerhard Schott** that a comprehensive world map of oceanic regions was prepared. This map revealed certain regularities of arrangement on the Earth. It was made posible to prepare a diagram of a generalised ocean similar to the generalised continent of Koppen. Gerhard Schott initiated a tradition of the geography of oceans in Germany, the heritage of which continued to dominate the realm of oceanography in Germany.

The period from 1905 to 1914 is described as the 'Golden Age of German Geography'. This was a period of rapid growth and increasing productivity. The methodological discussion made working in this professional field more exciting; and each new proposal regarding the scope and method of geography resulted in new field studies to provide examples. A number of new professional periodicals appeared as the influence of German geography spread throughout the world. This period witnessed the consolidation and nucleation of the Germanic heritage of modern geography, initially founded by Alexander Von Humboldt and Carl Ritter, but nourished and cultivated by Richthofen, Preschel, Ratzel, Hettner, Penck and Schluter.

However, the rise of Nazism in Germany made it difficult to proceed with further studies in objective scholarship. The Nazi interference with the academic works, often followed by intimidation of the geographers, resulted into what may be called the academic and professional recession in geographic tradition. **Alfred Philippson** was deported to the concentration camp by the Nazi Government in 1942. He was a pupil of Richthofen at Leipzig. He published his work on general geography in three volumes in 1921, 1923 and 1924. He was the distinguished author of the study of the Mediterranean region. His monumental work *Greichische Landschafter*, published in the 1950s, stands as an exemplary scholarly exposition of the concept and procedure of regional geography (Landerkunde) in the light of the ideas and teachings of Richthofen.

Robert Gradmann introduced the *concept of the harmonic landscape* into geography. He made outstanding regional study on southern Germany in two volumes (*Suddeutschland*) published in 1931. It closely followed the current idiographic tradition in German Landerkunde. Gradmann was the supervisor of **Walter Christalier**, who was the first German geographer to make a major contribution to location theory in 1933. Christaller noted that the 'crystallisation of mass about a nucleus is part of the elementary order of things, and human settlement obeys this principle as well as physical elements'. The foci nodes around which settlement tends to cluster, he called central places, each surrounded by a complementary area with which the central place is functionally related. His attempt to explain the pattern and hierarchy of central places by a general theoretical model was not acceptable within the reigning paradigm.

Oskar Schmieder (1955) attempted to focus on the idea of a focal theme of geography as the transformation of the original landscape to the humanised habitat or the cultural landscape, a conceptual procedure that obviously calls for historical interpretation. **Hermann Lautensach** worked on both physical and cultural geography, and was encouraged to this end by his pursuit of regional geography. He completed Kreb's unfinished work on comparative regional geography whose third edition was published in 1966. His theoretical framework of geography was based on a sound knowledge and appraisal of the natural sciences, mathematics, philosophy and pedagogy. He arrived inductively at a hierarchy of world units comprised within a major framework of 40 unit areas.

Carl Troll, a leading world-geographer of the postwar decades, initially showed deep interest in glacial morphology. His view on glaciology was identical with that of Penck's concept. Though he was not a pupil of Penck, but he was one of his sincere followers. He developed the concept of regional ecology in which geomorphology, soil science, climate and hydrology are all interrelated. He was deeply concerned with the relations of vegetation to altitude and climate in the mountains of the Tropical Andes. As a professor of colonial geography at Berlin in the 1930s, Troll busied himself on the possibilities of white settlement in East Africa. He also contributed to the methodology of geography.

One of his distinct achievements was the foundation of the periodical *Erdkunde* in 1947. He made a comprehensive survey of the progress of German geography from 1933 to 1945. Though Carl Troll's survey does not mention about Christaller, but it certainly points to some other positive advances made by German geographers during the Nazi period. There was a conservation movement to which geographers contributed field studies as a basis for landscape planning . There were studies of the relation of people to living space that were not politically oriented. Troll also reported in his survey that it was largely due to the effort of Max Eckert that German geographers founded the Deutsche Kartographische Gesellschaft in 1937 for the purpose of promoting cartography as a separate discipline. In the words of Dickinson, the works of Carl Trou clearly manifest the imprints of von Humboldt and Albrecht Penck, the two makers of modern geography (Dickinson, 1969, 166).

GEOPOLITICS

Apart from the professional development of geography in Germany during the inter-war period, there developed a distinct discipline loosely connected with geography, called geopolitik (Geopolitics) which focused on *the application of geographic concepts to politics.* It was different from political geography. It was **Rudolf Kjellen** (1846–1926), a Swedish political scientist who is credited with having coined the word 'Geopolitics'. Geopolitics was defined as 'the study of the theory of the state as a geographic organism or a phenomenon in space' (Kjellen, 1917). Basic to geopolitik's ideas were Ratzel's organic state theory, its refinement by Kjellen and Sir Halford J. Mackinder.

It was largely due to the ideas of Ratzel and Kjellen on the one hand, and that of Mackinder on the other, that **Karl Haushofer** (1869–1946) founded the Institute for Geopolitics in Munich. He also founded a periodical which dealt with the problem of political strategy. Karl Haushofer was not a professional geographer. Although Geopolitik collapsed with the fall of the Third Reich, many of its associated notions survived in the general field of geopolitics or applied political geography.

CONTEMPORARY TRENDS

The division of Germany after the Second World War had a significant impact on the contemporary German geographic scholarship and thinking. The geographers of the Federal Republic of Germany (FRG) continued with the traditional concept of geography as Landschaftskunde, but with the use of new and more precise methods of analysis. The idiographic tradition in German Landerkunde continued to be the centre of interest of the great majority of FRG geographers in the 1960s. In the post-War period, the philosophy of possibilism in the form of 'cultural determinism' replaced the earlier physical determinism. The geographers emphasised the need for the study of 'culture' which came to be defined as social geography. The purpose is to interpret the cultural landscape with the clear recognition that the major force for landscape change is the human group—'the attitudes, objectives, and technical skills' that are part of human's culture. This is a manifestation of the Schluterian tradition (Hajdu, 1968).

Geographers of the Democratic Republic of Germany (DRG), on the other hand, rejected the idiographic tradition in geography, but for sometime sided with Schluter's idea of a landscape unit of perceptible phenomena, natural and human, that together form a single association. The East German geographers have made significant contribution to the field of landscape science and in their attempts they got help from the Soviet geographers. East German geographers attempted to distinguish between physical geography and economic geography as separate disciplines.

8

Development of Professional Geography in France

HIGHLIGHTS

- Early Geographical Studies
- Deterministic Trend
- Development of New Geography
- Antithesis to Environmental Determinism and Regional Studies
- La' Tradition Vidalienne
- Political and Historical Geography
- Contemporary Trends

It was during the Age of Explorations that there was much speculation by the historians and political philosophers in France regarding the *'influence of natural environment' on human behaviour.* The philosophy of geographical determinism seems to have been inherited from the Greeks through Ptolemy and also from the writers of the Middle Ages. One of the earliest was Jean Bodin, a French political philosopher, who lived in the sixteenth century. In his major work published in 1566, he attempted to establish 'a cause and effect' relationship between Humankind and his Environment. His conception regarding the effect of environment on human behaviour reflected the implicit influence of the Greek and Roman traditions. Abbede Bos, writing in 1719, found that weather had a definite effect on suicide and crime rates in Paris and Rome. He also attempted to show that works of art are produced only in the zone between latitudes 25° and 52° N.

EARLY GEOGRAPHICAL STUDIES

One of the principal advocates of the philosophy of determinism was de la Brede et de **Montesquieu**, the noted French political philosopher who lived in the eighteenth century. His great work *De l' esprit des lois*, published in 1748 contained certain ideas related to the *environmental determinism,* that people develop different qualities and characteristics in cold climates and become more energetic than in hot climates where the people tend to be lethargic. However, Kriesel pointed out that Montesquieu also recognised the importance of other factors than climate alone, which included religion, the maxims of government precedents, and customs. In any country, as some of these factors act with stronger force, the others are weakened. Kriesel, therefore, describes him as a 'possibilist rather than an environmental determinist' (Kriesel, 1968, 557–74). Nevertheless, Montesquieu was very persuasive in his discussion of the effect of differences in climate on behaviour.

Comte de Bufton (1707–88) was director of the botanical garden in Paris between 1739 and 1788. He had access to a vast number of specimens of plants and animals, and description written by travellers and explorers from all over the world. His *Historic Naturelle* in forty-four volumes, published between 1749 and 1804, was completed after his death by La Capede. In this book, Bufton made repeated reference to the *relations between humankind and the environment.* He attempted to emphasise the change on the face of the Earth by the action of humans, in the process of developing a civilisation. He insisted that humans could adjust to any climate depending upon his capacity to adapt. Bufton also speculated on climatic change and suggested that the removal of forests and the drainage of marshes might lead to temperature increase.

It was not until 1756 that **Philippe Buache**, the French scholar, initiated a movement leading to the establishment of the *'Reine Geographie'* (pure geography) in France. He postulated that the Earth was divided into drainage basins by mountain ranges which even continued under the oceans so that the whole surface of the Earth was divided into naturally defined compartments. He favoured a river basin as the best kind of natural region. The Danish-born French scholar, *Malte-Brun* founded the universal geography in France. He offered a carefully considered refutation of the idea that river basins are bordered by ranges of mountains or hills. But so strongly implanted was Buache's concept of the river basin that Malte-Brun's refutation of it was ignored and overlooked. Some of the generalisations Malte-Brun made seemed to have been inherited from the tradition founded by Strabo 2,000 years earlier.

J. J. d' Omalius d' Halloy prepared a geological map of France and the low countries in 1823. The map also contained a description of the relation of landforms to the soils and underlying rocks. A French army officer prepared the first of the two works on geographical methodology in 1829, in which he attempted to show all the drainage basins bordered by ridges and mountain and that appears to be identical with that of Buache.

DETERMINISTIC TREND

The academic movement for the professional geography in France in the later part of nineteenth century appears to have been largely influenced by **Frederic Le Play** (1806–1882). He had a profound effect on the development of geography in France and Britain, though his influence was negligible in Germany and the United States. Basically, he was a sociologist. He introduced the ruling deterministic paradigm into his studies of families which reflect environmentalism.

Le Play's classic work appeared in 1855. It was based on his first-hand investigation of social life and organisation of the places of Europe where he travelled. His method of investigation was elaborated in France by his two most well-known disciples, Henry de Tourville (1843–1903) and Edmond Demolins (1852–1907). Like his contemporary historians, political philosophers and scientists, La Play also inherited the philosophy of geographical determinism, which he applied to study the development of social structure of the European people under different physical conditions. He was also greatly influenced by the 'le reel'[1] methodological precept of Auguste Comte. Le Play felt the lack of both a unit of measurement and a scientific inductive method for the systematic examination of social phenomena.

He *recognised three main family types,* each of which appeared to have developed under three distinct geographical environments. The Asiatic steppe was the home of stable nomadic families under the control of patriarchs. The particularist family is the characteristic of the maritime shore of Europe, and is especially localised in the fjords of Scandinavia. This is the home of the famille-souche. Forested land, covering great areas of much variety, with grass openings, heath, and varied soils, was the birth place of the 'unstable family' that had also developed in the urban environments of Europe and had spread to America. Le Play's

[1] *'Le reef* refers to the scientific status of knowledge which has to be guaranteed by the direct experience and observation of an immediate reality which requires a particular conception of causality.

studies of the development of social groups in a particular geographical (physical) milieu had been a basic objective of geographic research in France and was given special emphasis by Paul Vidal de la Blache as he developed his idea of the genre de vie[2].

Edmond Demolins in his book published in 1901 and 1903 demonstrates that the primary and decisive cause of the diversity of peoples and races is the route which has been followed by the peoples. It is the route (the environment) which created race and social type. The environment rigidly controls the structure and development of peoples. This is a most extreme form of geographic (environmental) determinism.

Henry de Tourville particularly studied the geographic impact on the development of the patriarchal society which usually developed in the fjords of Scandinavia. It is in the mountain environment that the development of self-reliance and initiative takes place. The married children in the region left the family and took to fishing and cultivation on another site, for they were dependent entirely on self-help. This led also to the private ownership of land and government by election and contractual associations on a voluntary basis. The 'genre de vie' that appeared to have been developed on the maritime shore of Scandinavia seemed to have spread in two directions, one into Germany and the other moved across to Britain.

DEVELOPMENT OF NEW GEOGRAPHY

Apart from the influence Le Play and his two most prominent disciples had on the development of new geography in France, the work of Ratzel on human geography (or his *Anthropogeographie*) had a far greater influence on future developments of the discipline than in his own country. Otto Schluter's works also had significant impact on the development of geographic tradition in France. Vidal de la Blache's method of regional description on a local basis also revealed the explicit impact of Hettner's methodological contributions which he made in his first major essay of 1905. The views of W. M. Davis made a widespread impact before World War I in France and Germany as well.

The awakening of interest in geography among the general public in France dates from their losses in the Franco-Prussian War of 1870–71. Smarting under this defeat in Europe, France sought compensation in new lands in Africa and South-East Asia. The newly founded geographical societies in the provinces of France were the leaders in this movement in favour of colonial expansion. Mekay attempted to trace out the development of France as a colonial power after 1871 to the influence of the French geographical societies, all of which urged France to bring the benefits of French civilisation of the less developed parts of the world.

The *French have successfully merged the historical and geological schools and have achieved a regional synthesis* unsurpassed elsewhere and for which they are justly renowned. It is this breadth of vision and a fine literary style that made French regional studies pleasurable to read. Scientific analysis sustains their descriptions, and their expository art has survived. The 'region' that particular field of study of geographers, has been regarded as a *unique functional complex* which, despite a steady stream of material and 'energy', is in apparent equilibrium and constitutes a 'whole' which is more than a sum of its parts. This understanding of the region is found in the French School of geography, which we shall see in the discussion on the professionalism of French geographical heritage.

ANTITHESIS TO ENVIRONMENTAL DETERMINISM AND REGIONAL STUDIES

French geographer, Paul Vidal de la Blache had the clearest insight into the weakness of deterministic

[2] *genre de vie* stands for the complex of institutions, tradition, attitudes, purposes, and technical skills of a people. It is a basic factor in determining which of the various possibilities offered by nature a particular human group will select.

arguments, realising the futility of humankind's natural surrounding in opposition to his social milieu and of regarding one as dominating the other.

Vidal de la Blache

The growth of professional geography in France, unlike that in Germany with 'its several distinct schools of thought, has been shaped by the work of one man, namely, Vidal de la Blache. His disciples spread throughout France the point of view and the method that came to be known as 'la tradition Vidalienne' (Dickinson 1969, 208; Buttimer, 1971).

Blache graduated in history and geography in 1865 at the Ecole Normale Superieure in Paris. He found his way into geography through the study of ancient history and classical literature. He spent some time at the French School in Athens and returned to France in 1872 where he completed his doctoral dissertation. He taught geography at the University of Nancy from 1872 to 1877 and then returned as Professor to the Ecole Normale Superieure.

While at Nancy, during 1872–1877, he prepared the manuscript of the *La France de L' Est* (published in 1917). This is in many ways Blache's most original work. It studied the development of the landscape and agricultural societies in Alsace and Lorraine over a period of 2,000 years. A considerable portion of the book, which is arranged chronologically, is concentrated on changes which preceded the French Revolution of 1789. The revolution produced a great ripple in the picture, but afterwards the mainlines of development reasserted themselves (Vidal, 1917). In this book, Blache attempted to develop the concept of 'landscape chorology'. It is from Richthofen whom he met at Leipzig (Germany) in around 1876–77 that he learnt about the causal interrelation among the diverse things in particular areas, to which the term 'chorology' was applied.

The period between 1877 and 1898 was significant for the future development of geography in France, in the sense that during this period Blache devoted himself to improving the training of teachers of geography and to making up-to-date materials and ideas available to them. He was exposed to the ideas of 'new geography' of Germany, and the apparent dichotomy in contemporary German geographic tradition. In 1891, in collaboration with Marcel Dubois, he founded a new professional periodical, the *Annales de geographie*, with the purpose of publishing the new methods, concepts and paradigms of geography. In each number of the periodical, he included a bibliography of published materials where additional information and new geographic concepts could be found. In 1894, the first edition of his *Atlas generale Vidal Lablache* was published.

In 1898, Blache was appointed professor at the Sorbonne, when the chair of the professorship became vacant. He was the first professional geographer to have been appointed to the chair of geography of Sorbonne since the chair was established in 1809. His appointment as the first professor of geography in France in 1898 marked the beginning of professionalism in the French geography, or in other words, it could be said to have laid the foundation of 'new geography' in France.

Geographic Concepts. Vidal de la Blache seeks to *establish geography as a distinct discipline.* Its field of study is the ensemble of phenomena that occurs in the zone of contact of the solid, liquid, and gaseous masses that make up the planet. These phenomena are studied in relation to place, localisation and distribution. Thus, the field of geography has a double aspect—Nature and Man. He urged upon his disciples to focus attention on the close relationships between humankind and the immediate surroundings *(milieu)* by studying small homogeneous areas which are popularly known as *'pays'* which roughly resemble the German 'Landschaft'. The area over which an intimate relationship between humans and nature has developed through the centuries constitutes a region. *Neither humans nor nature is subordinated to each other,* because the very 'synthesis' would be disturbed if any one of them is regarded as dominant at the cost of the other. He felt that the peculiar obligation of the geographer lies in the correlation of physical and human conditions in their spatial interrelations.

Blache pointed out the inherent weakness of the geographic concept of 'environmental determinism' which was prevalent at the time as a Darwinian heritage. He realised the futility of setting humankind's natural surroundings in opposition to his social milieu and of regarding one as dominating the other. From Ratzel's second volume of *Anthropogeographie*, Blache *formulated the concept of 'possibilism'* which was fully developed by the historian Lucier Febvre.

If chronology is to be believed, then it can be said that the concept of possibilism was long developed in an amateurish way by Montesquieu, but Blache treated the concept within the framework of a scientific professionalism and precision in which he incorporated the Comtean positivism. 'Nature sets limit and offers possibilities for human settlements, but the way man reacts or adjusts to these given conditions depends on his own genre de vie (traditional way of living).' He points out that through his occupance and imprint on the land, Man creates distinctive countries, be they states or minor unit areas (pays). It is unreasonable to draw boundaries between natural and cultural phenomena; they should be regarded as united and inseparable. In an area of human settlement, nature changes significantly because of the presence of man and these changes are greatest where the level of material culture of the community is at the highest level.

Blache points out that *man establishes relations with the environment not as an individual, but through the heritage and objectives of the group to which he belongs.* The same environment has different meanings for the people with different heritage and objectives (genres de vie). The genre de vie is a basic governing factor determining the man-milieu relationships because it lays down which of the various possibilities being offered by nature are to be selected. The physical environment provides a range of possibilities which Man turns to his use according to his needs, wishes and capacities, in creating his habitat.

The 'paradigm of possibilism' appears to have resulted on account of an academic encounter between Durkheim, Ratzel and Blache at the turn of the century. In fact, Blache mediated between the two protagonists. He rejected Durkheim's proposed reduction of geography to social morphology by insisting that the human being 'joins nature's game', and that the 'milieu externe' was a partner not a slave of human activity. While he shared Ratzel's belief that society ought not to be left 'suspended' in the air, he was quick to dispel any lingering determinism by insisting that nature is never more than an adviser, and that the 'milieu interne' revealed the human beings as at once both active and passive.

These mediations were doubly important because *Blache's scheme led neither to a radical possibilism nor to environmental determinism.* His was essentially a neo-Kantian philosophy that embraced the organising freedom of the human being, bounded by the realities of the mechanisms of the natural realm. Blache believed that whereas society and nature were usually represented as two adversaries in a duel, the human being was in fact 'part of living creation' and was the most active collaborator who acted upon nature only through her and by her. It is this dialect which he subsumed in the concept of the genre de vie, wherein the moments of a recursive creativity could be both powerful factors and yet so vital agents (Johnston, 1986, 360–61).

In his article, *The Geographical Conditions of Social Facts* (1902), Blache considers the physical milieu as the 'natural' or 'geographic' environment of human occupance. It is indeed a reflection in a modified way of La Play's study of human societies in their environmental settings. In another article, 'Modes of Life in Human Geography' (1911), Blache points out that social conditions find their expression in distinctive modes of life (genre de vie), which in simpler societies show a close adjustment to the natural environment.

In contradiction to Buache's concepts of drainage basins, Vidal de la Blache *supported the idea of small natural regions* ('pays'). Such small natural regions are manifestations of intimate relationship between man and nature that developed through the centuries. The study of such small natural regions, each one of which is unique, should be the task of the geographer. Blache, therefore, *argued for regional geography* and against systematic geography as the core of the discipline. Man-nature relationship cannot be studied along systematic lines.

His method of regional geography was *inductive and historical*, and best suited to small regions ('pays') which were 'local' in the sense of being somewhat isolated from the world around them. The idea of identifying the regions of France as proposed by Blache was picked up by Lucien Gallois, who provided a meaningful review of the history of the regional idea in France.

Blache *supported the concept of chorology* as the study of things associated in area, mutually interacting, characterising particular segments of Earth space. He asserted that geography must have a place among the natural sciences and in human sciences (Vidal, 1913). He laid down six main principles :

1 The unity of phenomena, by which he meant the interdependence and interconnections of physical forces.
2. The variable combination and modification of phenomena, especially as seen in the study of world climates.
3. Geography is concerned with all phenomena at the surface of the Earth.
4. The need to recognise the force of the environment in its various forms and types, such as the vegetational belts of the Earth and man's adaptation within them.
5. The need for a scientific method in defining and classifying phenomena.
6. The recognition of the great part of man in modifying his basic environment.

Geography is the science of places and not that of men; it is interested in the events of history in so far as these bring to work and to light, in the countries where they take place, qualities and potentialities that without them would remain latent (Vidal, 1913). *Blache was truly the founder of modern scientific geography in France.*

Vidal Blache is regarded as *one of the founding fathers of modern human geography.* He had developed human geography as an entity, incorporating into it the positivist legacy of Comte. His *Principles de Geographie Humaine*, which was published posthumously in 1922 (through the efforts of de Martonne), is regarded by geographers as a classic. The historical evolution of each phenomenon in human geography is followed by a study of its present setting, localisation and correlation. In the Introduction, he analyses the principle of terrestrial unity, concept of milieu, the significance of environmental factors and of man's work in it.

The first part of the book is devoted to studying population density, clusters, major agglomerations in Europe and movements of population in Europe. Part two deals with the man-milieu relation, and the method man has used to develop his environment and of his several civilisations. Part three deals with transport and communication. Emmanuel de Martonne completed the book, as he added few aspects in part three relating to the origin of races, diffusion of inventions and cultural regions as well as cities. The arguments in the book are closely reasoned and the book has had great success in English-speaking world through its translation. *Blache regarded human geography as a natural science* (Johnston, 1986, 207). With his death in 1918, France lost a doyen of the subject. It was largely due to his geographic skill that the ongoing dichotomy between physical geography and human geography in Germany could not succeed to motivate the contemporary French School of geography.

In the words of Joerg, 'nearly all the occupants of chair in geography in France are pupils or pupils of pupils of the late Paul Vidal de la Blache. In no other country, it may be said, has the development of geography centred about one man as in France' (Joerg, 1922, 431–84).

Gregory writes, 'although Febvre had the greatest admiration for Vidal Blache himself, although he later said that la geographie vidalienne had played an important part in the formation of the *Annales* School of history, his interest in the structuration[3] of specific societies set him some way apart from the mainstream of French geography. The Vidalian School continued to be preoccupied with a socialised nature; even vidal Blache had described geography as a natural rather than a social science, and many of the pupils who followed his teaching took this to mean that 'the nearer man is to the brute, the more geographical he is'. The complexities of his social structure received little attention, and where they were examined at all the typology of genres de vie what resulted was a resolutely empiricist one with few concessions to any kind of structural explanation. This dismayed Febvre too, because although he was no structuralist, he was undoubtedly influenced by several precursors of the structuralist school' (Gregory, 1978, 90).

Lucien Gallois

Vidal Blache was succeeded at the Sorbonne by Lucien Gallois who was one of his first pupils. At this time, the historical tradition in the geographical studies was much strong in France, which was because of the influence of Himly who preceded Blache at Sorbonne. Gallois attempted to evaluate the relations between Man and Nature. His doctorate dissertation was an important contribution to the history of cartography (Church, 1967, 73–74).

He collaborated with Vidal Blache in the direction of the *Annales de geographie*. Gallois prepared a book in 1908 very much on the lines of the Vidalian tradition of regional studies. He attempted to show how the idea of the region naturelle emerged at the end of the eighteenth century in the preparation of the first geological maps of France. Gallois concluded that it is to the land entities, large or small, all of which are of a physical order, that we should reserve the name of regions naturelles. 'Of the variety of facts in which the activity of Man intervenes with Nature, those which bear the influence of the environment form the distinctive field of human geography' (Gallois, 1908). This dictum had a profound influence on the purpose and direction of geographic work in France.

After the death of Vidal Blache, Gallois took over the organisation of the Geographie Universelle. This remarkable series shares the joint editorial names of Vidal de la Blache and Lucien Gallois and the latter was, in fact, responsible for editing the series until the time of his death in 1941. Undoubtedly, he carried forward the concept of landscape chronology, with a major emphasis on the development of regions over time.

LA' TRADITION VIDALIENNE

Paul Vidal de la Blache, who is regarded as the founder of the Modern Geography of France, appears to have established a new tradition in geography which his followers and pupils, and their pupils carried forward.

Jean Brunhes. He was one of the outstanding pupils of Vidal de la Blache. He not only elaborated Blache's ideas about regional and human geography and spread them throughout France, but also transmitted the Vidalian tradition to other countries (James and Martin, 1981, 193). He was an outstanding scholar of France and a maker of modern geography. Brunhes is best known for his presentation of the essential facts of geography, from which one can readily quote statements which specifically limit the field. He sought to give the geography of man a clear disciplinary basis. It can be said of Brunhes that he tried to blend together the Vidalian tradition of 'genre de vie' and the Schluterian heritage of 'Kulturlandschaft', in an attempt to give geography a more dynamic and professional outlook.

[3] *Structuration* : It is an approach to social theory concerned with interaction between knowledgeable and capable human agents and the wider social systems and social structure in which they are necessarily implicated.

He studied at the Ecole Nonnale Superieure in Paris, like so many of the contemporary French geographers, with the intention of becoming a teacher in the discipline. He worked under Vidal and took the 'agrege' in history and geography in 1892, and continued his studies in Paris until 1896. Brunhes, by his training in history, the natural sciences, law and finance with his final specialisation in human geography, was particularly well equipped to assess rightly the varying parts that physical factors on the one hand, and social, economic, political and historical factors on the other hand, played on various human groups now and in the past (Church, 1967, 80).

The French geographers of the late nineteenth and early twentieth centuries were not unnecessarily motivated by the contemporary dichotomy between physical geography and human geography, as were their counterparts in Germany. This may be attributed to Jean Brunhes who led in the development of human geography, and Emmanuel de Martonne who nourished the physical geography. Both carried forward the 'la tradition Vidalienne' not only to Franc, but also outside it. An important part of la tradition Vidalienne has been a relative freedom from the dispute about whether geography is one field or many.

Brunhes's major work appeared in 1910 with the title *Geographie Humaine* : *essai de classification positive*. The book was the first large-scale attempt in France to examine fully the facts in human economy. The tradition he laid down in the book suggested an inevitable shift of emphasis from La Play's heritage of 'geographical conditions' and 'social groupings' to the study of the surface features of the Earth. He elaborated a system of human geography with its raw material in the visible and tangible facts of human activity on the Earth's surface. These are *the essential facts of human geography* which he grouped into three categories:

1. Facts of unproductive occupation of the soil—houses and roads (including rural habitations, urban agglomerations, and circulation patterns).
2. Facts of plant and animal conquest—the cultivation of plants and the raising of animals.
3. Facts of destructive exploitation—plant and animal devastation, mineral exploitation.

These are examined in particular type of areas or regional studies to show the interdependence of these phenomena in contrasted physical and social environment. Brunhes admitted that the study of these three sets of observable phenomena does not set the limit to geography, a fact that is not generally appreciated. Beyond the essential facts, but in association with them, is an area of wider study. This is described as *'the geography of history'* that includes five major aspects: distribution of population; economic geography; geography of political groupings; social geography where these are associated with the 'essential facts'; and 'regional geography', which is a synthesis of the foregoing.

Brunhes' methodological considerations are perhaps more suggestive than definitive, but his detailed portrayal of concepts, in *La geographie humaine, presents essentially a modification of the dualistic concept of Geography* in which a large part of human geography is defined in terms of relationships (Hartshorne, 1939, 1976). Brunhes seemed to have a strong orientation towards the Schluterean tradition of landscape morphology. He illustrated this view when he said that we should study the Earth as if we were sitting in a balloon and looking down upon it. We should analyse the landscape and the characteristic interplay of observable phenomena there. The visible landscape is a result both of natural conditions and forces and a manifestation of the work of man. The landscape itself creates a synthesis. There is no longer a gap between physical geography and the geography of man, both have the same object—the visible landscape—and there is also close contact in terms of method.

Brunhes was of the opinion that the physical and cultural phenomena are in a state of perpetual change and they must be studied in the temporal change, instead of taking them as static in the time scale. Therefore, in order to understand the interrelationship between physical and cultural components of small regions *(landschaft* or *pays),* the principle of activity is to be kept in mind within the framework of landscape morphology, which would help to arrive at a just synthesis. He further pointed out that our efforts are based fundamentally on the great geographical principle of interaction between man and environment like plants and animals, and therefore, the concept of interaction should dominate every complete study of geographical facts. The forces of physical nature are bound together in their consequences, in relations and in the consequences of these relations (Husain, 1988, 140–41).

In collaboration with C. Vallaux, Brunhes wrote a book in 1912. In 1928, he translated I. Bowman's work on political geography. The translation of Bowman's work into French, and Brunhes' work on human geography into english is one more example of the close rapport, both professional and personal, between the leaders of American and French geography in this period (Dickinson, 1969, 213). He had a personal contact with the German scholar Otto Schluter also who had a direct impact on his geographical writings.

Camille Vallaux. The effect of the Vidalian tradition is also apparent in the writings of Camille Vallaux. It was for this heritage that there was no problem of dichotomy in the French School of geography even after 1920. Vallaux's book, *Les sciences geographiques* (1925), clarifies his understanding of the la tradition Vidalienne with regard to the dichotomy, that geography is both a unitary and autonomous field of study and also an auxiliary aspect of many fields. Not only does geography have a philosophy of its own, he writes, but also, 'it is almost, in itself a philosophy of the world of man'. Vallaux's book thus presents the most thorough methodological study in French geography, *probably the ablest presentation of the concept of human geography as the study of interrelations.* It was, therefore, assumed that the French geographers could continue to make important contribution to systematic or topical studies and at the same time continue to produce regional monographs. As regards the frequent usage of the term 'paysage', Vallaux has commented on the confusion because of the double meaning of the term that results in its undefined use in the writings of Brunhes and others.

As Sauer puts it, Vallaux stated: 'to describe human landscape without knowing how these landscapes are constructed is to put the cart before the horse. The first solid basis to establish is, therefore, the physical geography which supplies it. These are to be reconstituted therefore in the natural landscape, in which the activity of the living world is comprised, such as nature made them, as though there had not been a living person on the Earth' (Saur, 1925, 97). Undoubtedly the method which Vallaux and Sauer recommend, and which many geographers have followed, is in accordance with the logic of nature (in the broadest sense of the universe), assuming that we can impute a logic to the universe (Hartshorne, 1976, 357).

On the contemporary controversy of whether or not the concept of region be included into the scientific geography, Vallaux concluded: 'We are not even sure that for many parts of the world, the regional synthesis is anything more than a logical artifice and a method of instruction, that is to say, after all, an actual deformation of truth, excused but not legitimized by the arguments which one could put with validity in its favour' (1925).

Franz Schrader. He was a contemporary of Vidal de la Blache. He was not his pupil. He was not an official, engineer or academician, but he was *known for his cartographic skill.* He had been a successful field observer, surveyor, artist and draughtsman. He prepared new maps of the Pyrennes, notably of the central section. With the aid of the collaborator, Schrader prepared three Atlases, the last of which was a masterpiece in cartography and was also recognised by Vidal de la Blache. He had also prepared the *Annee Cartographique,* and the first fascicule appeared in 1891 and the last one in 1931. His major work, entitled *The Foundation of Geography in the Twentieth Century* was published in 1919, in which he defined geography as the study of 'the areal interdependence of terrestrial phenomena'.

Emmanuel De Margerie. He was a great modern French scholar in the natural sciences. Neither he held any university post nor did he have any university degree. Nevertheless, he was an ardent supporter of the Vidalian School of French geography. He pioneered the study of landforms in France, and in collaboration with G. de la Noe, he published his famous work *Formes de Terrains* in 1888. He reviewed a large number of works on landforms in the *Annales*. He was actively associated with Vidal Blache and was fully conversant with the Vidalian heritage. He translated into French the classic work of the Viennes geologist Ed. Suess which was published between 1883 and 1909. The work was appreciated and endorsed by the German geographer Richthofen. He produced a comprehensive geological and morphological study of the Jura in two volumes, the first of which appeared in 1922 and the second in 1936. He was actively associated with Albrecht Penck in launching the International Map of the World at the International Congress in 1913.

Emmanuel De Martonne. He was the *first French geographer to concentrate on physical geography.* He was the recognised leader of French geography from 1918 to 1945. Martonne entered the Ecole Normale Superieure in 1892 to study geography under Vidal Blache. He was trained in the Vidalian tradition and carried forward the heritage of geography founded by Vidal to different parts of France. When he succeeded Gallois as the Director of the Institute de Geographic at the Sorbonne, he was able to secure the maximum possible recognition for his specialisation. Besides his outstanding works on physical geography, he was actively engaged in various International Geographical Congresses and International Geographical Union of which he was secretary and then became its president (Church, 1969, 87–88).

De Martonne's first published work dealt with the coastal morphology of Brittany and then he concentrated on the studies of physical geography in the Carpathians. His regional monograph on Wallachia (1902) manifested the explicit influence of the Vidalian tradition. His monumental work was the publication of *Traite de Geographie Physique* in 1909 which was extended to three volumes in 1925–27. This classic work was rearranged in shorter edition. This work showed an inevitable orientation towards the Davisian idea. He was an ardent supporter of the American geographer W. M. Davis, and carried forward his ideas to the French-speaking world. In 1943, he edited the magnificent *Atlas de France* which, in nearly 90 sheets and some 200 superbly coloured maps, covered every conceivable aspect of France capable of cartographic representation.

To him, physical geography was an essential part of the whole geographical study of an area. He attempted to demonstrate this fact in his regional monograph of the Wallachian plain and also in his studies of the Carpathians. During all of his professional carrier, he maintained a regional interest in Central Europe, and a systematic interest in geomorphology and climatology. His attempt to identify regions by using an aridity index was a major contribution to the systematic study of climate. Martonne was one of the leading physical geographers of the world and became the most influential geographer in Europe between the two World Wars.

It was indeed a major achievement of the la tradition Vidalienne that it successfully withheld the influence of apparent dichotomies in geography on the French School as had already worried the German School of geography. It was largely due to Brunhes and Martonne, the two forerunners of the Vidalian tradition, that human geography and physical geography developed separately and distinctively. Martonne, being a physical geographer, sought to combine the physical and human aspects in one framework which is evident in the two volumes on Central Europe in the *Geographie Universelle* (1930 and 1931). He also *attempted to blend together regional geography and systematic geography so as to give geography a united and unitary outlook.*

Albert Demangeon. He was one of the forerunners of 'La tradition Vidalienne'. He studied under Vidal Blache. He was a contemporary, student, friend and colleague of de Martonne. He started his career as a school teacher in Picardy. He was appointed to the University of Lille, and there he remained till 1911 when he was appointed to the chair of geography, alongside de Martonne, at the Sorbonne in Paris. He remained there until his death in 1940. Demangeon made outstanding contribution to human geography, and produced a manual of human geography.

His monograph on La Picardie (1905) reflected the conceptual framework of the Vidalian tradition. Demangeon moved across every lane in Picardy before publishing the monograph. In its preparation, he also sought help from the archives in the national library in Paris and also other written sources. The skill with which he analysed the data collected from personal observations and the archives, and prepared the monograph, seemed to have approached the inductive method very closely.

Demangeon, in appreciation of Vidal Blache, wrote: 'Every region (sic) has its unique character to which contribute the features of the soil, atmosphere, plants and man. The aim of all (geographical) research consists in the analysis of these features. The aim of description is to synthesise these and to show the interlocking of all the phenomena which comprise regional type' (Demangeon, 1905).

While at Lille, he published two important articles in the *Annales*, one of them related to the human use of the 'montague', and the other related to the relief and its development through a chequered series of cycles of erosion, rather than one uninterrupted process as postulated by W. M. Davis. Though he contributed a number of research articles to the *Annales*, but his two articles published in 1920 and 1927 justify his stand of a minor shift from the contemporary methodological approach in human geography. The first article provided a study of the regional variations of the farmstead, and the other dealt with the distribution of the varied forms of nucleated settlements in rural areas over the Earth.

Another book published in 1920 was a major *contribution to colonial geography.* It dealt with the territorial losses of France and the potentialities of recuperation thereafter. In 1923, be published another book on Britain, which dealt with the spatial pattern of its overseas colonial possessions. Demangeon's classic works on British Isles and also on the Low countries, appeared in the *Geographic Universelles*, in the mid-1920s. These two works further enhanced his academic position among the contemporary professional geographers in Europe. His philosophy of human geography is contained in his posthumous work *Problems de Geographic Humaine* published in 1942. Demangeon's study of the economic geography of France was fully published in 1946–48 in the *Geographic Universelle.*

He pioneered the study of rural settlements in France, and a number of articles on this subject appeared in the *Annales* between 1920 and 1939. He took the interior plan and agricultural function of houses as the basic differentiating factor. On the question of settlement, he regarded the degree of concentration or dispersion as the vital matter. In the later 1930s, he prepared questionnaires concerning rural habitat, agrarian structure and methods, and the part of foreigners in French farming. He revised his classification of farmsteads and produced remarkable areas of their distribution.

The *Problems de Geographie Humaine* of Demangeon reveals the nature of his conceptual framework of 'human geography'. *Human geography, to him, is the study of human groups and societies in their relationships to the physical environment or geographical milieu.* He emphasised the work of man in modifying his environment by means of communications, artesian wells, the control of rivers, and the evolution of new plants for human food (reminiscent of the Vidalian tradition of genre de vie). The study has four main aspects: (i) the influence of the geographical milieu on modes of life; (ii) the changes in the genres de vie under the impact of the human milieu; (iii) the distribution of human groups as the result of the natural milieu, and the degree of civilisation; (iv) the establishments in the landscape, due to the impact of the human groups on the land.

In these aspects, Demangeon invoked three guiding principles. 'The first is the Vidalian idea of "possibilities" of human occupance of land than determinants of human occupance by the geographical milieu; the second, never to lose sight of the terrestrial substratum (land), and build up always inductively from a regional base; and the third, to recognise always the fundamental importance of historical perspective so as to explain the present modes of human occupance of the Earth' (Dickinson, 1969, 233).

Raoul Blanchard. He is often credited with having established the Ecole de Grenoble, which was in many respects independent of Vidal's Ecole Normale Superieure in Paris. He was in fullest sense a maker of modern geography. Although he was one of the pupils and followers of Vidal, he always had a sense of academic distinctiveness. He was appointed to the chair of geography at Grenoble and spent almost 50 years there. At Grenoble itself, Blanchard founded the Institute de Geographie Alpine in 1908 and started a periodical.

His regional monograph on Flanders, which was published in the early years of 1906, revealed the impact of the Vidalian tradition in its approach. His first paper on rainfall in northern France was published in the *Annales de geographie* in 1902. Blanchard *followed the empirical method* in his geographical studies which could be known from his great work on the French Alps which he started in 1937 at the age of 60. The work continued until the publication of the final and the twelfth volume in 1956. It was a monumental work in the field of geomorphology and seemed to be independent of the Davisian tradition which had already made inroads into the contemporary French School of physical geography.

Blanchard opposed the cycle of de Martonne in its relevance to the Alps by pointing out the rule of structural and lithological condition and glacial actions. This seemed to have developed into a permanent 'de Martonne-Blanchard controversy' in the French School of physical geography.

Blanchard made substantial *contributions to urban geography.* He organised a regular school of urban geographers at Grenoble. To the study of urban geography, he made his classic contribution on Grenbole in 1910. The work, since then, has passed through a number of revisions, and it also witnessed incorporation of newer conceptual frameworks which he brought from the United States. He produced two volumes on the geography of eastern Canada containing long geographies of the cities of Quebec and Montreal. Blanchard published urban studies of Lille, Nancy, Lyous, Marseilles, Nice and Bordeaux. He was also interested in rural house types in the Alps, and published a monograph on it.

The purpose of his urban geography was to provide description of the environment of man and its activities on the transformation of the landscape. His method was based on personal observations and recordings, interviews and enquiries, and at the same time, he also used the materials of the historian and archivist, and official records. Thus, his works seemed to have approached the inductive method very closely.

Maximilien Sore. He began his career as a school teacher, but he joined the military service in 1914 and was seriously wounded and demobilised in 1915. He worked with Henry Baulig and Lucian Febvre at Strasbourg. He was appointed rector of the University of Clermont Ferrand and then at Aix Marseilles in 1934. He succeeded Albert Demangeon to the chair of geography in Paris and there he remained until his retirement. He was fully conversant with the concept of the genre de vie, as developed by Vidal de la Blache, but he had a deep sense of orientation towards a *biological (ecological) approach,* and that makes his human geography distinct from that of Albert Demangeon.

His monumental work '*Les Fondements de la Geographie Humaine*' was published in four volumes during 1947–1952. It was a classic work on human geography. His last work was published in 1961, containing new ideas, but it was essentially a resumption of the larger work. The approach throughout is ecological and provides a basic conceptual framework for the geographer. Sore became the leader of a valuable new tendency in French geography which sought closer contacts with the biological sciences.

Sore *considered physical geography as a framework for the study of human groups.* He was especially concerned with the physical milieu only in its relevance to human occupance. The first volume of Sore's *de la Geographic Humaine* dealt with the biological foundations of man's use of, and life on, the Earth. The second volume was concerned with the technical foundation of human occupance of the Earth and the development of communication that embraces the techniques of social life. The third volume dealt with the technical foundations of production and the transformation of raw materials; and the fourth volume discussed the 'human habitat'. In fact, the work seemed to be a reflection of the Vidalian tradition on the one hand, and the ecological approach on the other.

Henri Baulig. Although he was a student and a follower of Vidal de la Blache, but he was trained in the Davisiari tradition. He went to Harvard (USA) and there came under the strong influence of Davis. *He carried forward the Davisian School of physical geography into France.* He was appointed to the chair of the University of Strasbourg, where he remained until his retirement in 1947. He share with de Martonne the distinction of being a leader of geomorphological research.

His work on the morphology of the Central Massif, particularly its several erosion and peneplane surfaces, seemed to be a reminiscent of the Davisian heritage. He had a remarkable gift for observing and interpreting physical landforms and his work on the role of eustatic equilibrium in the formation of the Central Massif was one of rare distinction. In 1950, all his published works were assembled in one volume as *Essais de Morphologie* and several years later he published two long articles which were a critique of the specialised methods and claims and trends of the new geomorphology in France as expounded in the work of P. Bivot. Baulig's work on North America was perhaps the finest work on general regional geography of North America in any language. For this magnificent work, he was awarded the Daly award by the American Geographical Society.

POLITICAL AND HISTORICAL GEOGRAPHY

Apart from emphasis on regional studies, French geographers also made significant contributions to the philosophical foundation of political and historical geography in France.

Jacques Ancel. He was one of the forerunners of the French School of political geography. Before his tragic death in the Draney concentration camp, he had completed three volumes of the *Manuel Geographique de Politique Europeenne*. He wrote several other books. Ancel was a staunch opponent of German geopolitics as developed by Karl Haushofer. He had attempted to rescue the term geopolitics (geopolitique) for international science. He denied that language could be a sure guide to true national feelings.

Andre Siegfried. He is credited with the development of *electoral geography* in France. For long, he held a Research Chair in Political and Economic Geography at the College de France where he undertook work on the geographical factors in French political parties and elections. In his electoral studies, he attempted to measure the influence of the physical milieu on the electoral pattern. In fact, he incorporated the 'deterministic tradition' rather than the 'possibilist tradition' into his electoral analyses.

Siegfried's *political geography* revealed the impact of the tradition of Tocqueville in his first book published in 1904. His important work in political geography was his study of political attitudes in Western Europe which was published in 1913. His books on England, the United States, Canada, Latin America, New Zealand, the Suez and Panama Canals and the Mediterranean have been widely acclaimed by the geographers and general readers in France and outside. Though Siegfried had never produced any general theoretical system, either in politics or in human geography, yet in many ways he founded a school of thought, though not formally recognised. He was the sole geographer to have the distinction of becoming a member of the Academic Francaise in 1945.

Pierre Deffontaines. He was the pupil of the pupils of Paul Vidal de la Blache. His first teachers were de Martonne and Lucien Gallois, but later he came under the influence of Jean Brunhes. He obtained his doctorate from the Sorbonne. His thesis published in 1932 appeared to be *a departure from the traditional regional study.* There was an increasing tendency in the contemporary regional studies in France to concentrate on single central themes or single problems, and irrelevant materials were to be omitted. Deffontaine's study was thus organised in this way around the changing impact of human society on the landscape.

Deffontaines carried forward the tradition of Jean Brunhes. He actively helped Brunhes in the publication of the second volume on the human geography of France in I928. He edited books on 'Geographie Humaine' of which he had written volumes on colonisation, man and the forest, the geography of religion, and man and the winter (in Canada). He also directed a *Geographie Universelle Larousse* in three large volumes (1959–60). He founded a new periodical 'Geographie Humaine et Ethnologie' in 1948, to further the research in human geography and ethnology. He was definitely a maker of modern human geography in France in the post-War period.

On *human geography,* Deffontaines pointed out that it is a young subject and it is still in the exploratory stage. It proceeds by the sampling method. One aspect of the study is the struggle of man against his raw environment and the other aspect deals with the results achieved and classifies the types and limits. Brunhes's study of irrigation belongs to the first aspect, and Demangeon's work to the second. He differentiated two types of regions of human occupancy: one in which there is an individual pattern of human economy and the other where through intensive and long contact, present-day life is the result of many complicated influences.

Roger Dion. He was a contemporary of Pierre Deffontaines, and was trained according to the Vidalian tradition by the immediate pupils of Vidal Blache. He was appointed to the post of professor at the University of Lille in 1934, but from 1945 to 1948, Dion was professor of political and economic geography at the Sorbonne. In 1947, he served one session as professor at the University of Sao Paulo in Brazil. In 1948, he was appointed professor in the historical geography of France at the College de France.

The publication of his thesis *La Val de Loire* brought academic credit to him. The monograph revealed a change of approach from Vidal's initial framework. The monograph was divided into three units. The first unit dealt with the natural characteristics of the Val de Loire. The second unit dealt with the history of man's transformation of the river plain and his struggle with the vagaries of the flow of the river. The third unit dealt with the 'genre de vie' of the Val.

His another work on the formation of the rural landscapes of France (1934), showed his ability and skill in the interpretation of documentary records. The essay on the formation of the rural landscape of France was entirely based on the case studies on documentary records available from different places. He attempted to trace out the differential development of agrarian systems in northern and southern France, and their role in accounting for difference in the emergence and present forms of field patterns, road nets, and the grouping of farmsteads and villages. He also published a historical geography of viticulture in France (1959), and another work on the history of levees on the Loire in 1961. *Dion's works reflect the continuity of the historical tradition in geographic scholarship in contemporary France.*

Dion, writing on historical geography in the middle of the twentieth century (*L 'Information geographique*, 1957, 185), pointed out many errors of interpretation that resulted from attempting to demonstrate strict controls. Man's economic life must be seen as in the process of transition and makes no sense at all unless it is viewed in historical perspective.

Andre Cholley. He was a renowned *physical geographer.* Though he belonged to the second generation of geographers in France, but his late entry into the academic profession put him in the third generation. Martonne is believed to have been the greatest influence in moulding Cholley's development. He pointed out that Cholley carried regional geography a big step further by revealing 'une morphologie rationnelle' in his classic work on physical geography. Cholley was the director of the Institute de Geographie in Paris from 1944 until his retirement in 1956.

On Cholley's morphological expertise, as applied in the Pre-alps, Dickinson points out: 'Cholley *sought to determine the phases of evolution* revealed by the correlation between surfaces of evolution and plains formed during marine transgressions of the same epoch. From many studies in the Paris Basin, he discovered a series of peneplained surfaces resulting, at the end of Miocene, in the establishment of a "polygenic surface" that effects the whole ensemble of the Paris Basin. Following on the tectonic movements of the Pliocene period, the existing surface forms are to be attributed not to such extensive structural and erosional surfaces, but to slopes (pentes)—valley slopes, structural talus slopes, that were imprinted by climatic oscillations of the quaternary period' (Dickinoon 1969 [1973]).

Regarding the post-War trends in geography in France, he pointed out that quantitative methods have entered into geomorphology, phytogeography and more recently into the field of human geography. Problems required to be sharply defined in terms of content and method and practicability, as well as precision. On account of increasing emphasis on the quantitative explanation of phenomena, it became difficult to pursue the geographical study of an area in all its aspects according to the definition Vidalienne.

Pierre George. He was appointed to the chair of geography at the Sorbonne in 1948, and made substantial contribution to the field of modern French geography. His doctorate thesis was published in 1936 and another work in 1938. George's geographical approach appears to have been built upon the conceptual frameworks of Albert Demangeon and Maximilien Sore. His *approach was ecological* with a strong emphasis on the motifs of characterisation, action and organisation of communities as spatial complexes.

In his book on *Sociologie et Geographie* (1966), George regrets the habitual claim of the French geographers about the unity of geography which, he believes, does not exist. He points out that the skilled analysis of natural processes, pursued in the laboratory or the field, has no relation with geography as a human science. *The central concern of geography is with the study of human societies,* and this is the essence of the French approach to geography as formulated by Vidal de la Blache. As a 'human science', its object is the study of the globe as a whole and of its segments, with respect to all that conditions and all that is relevant to the diverse human collectivities that make up the population of the globe. The task of geography, as a scientific study, involves what George calls *'horizontal synthesis rather than 'vertical analysis'.* Its basic concern, therefore, is with the location and measurement of types of correlation of spatially arranged phenomena over the face of the Earth.

Jean Gottmann. He is one of the world's leading contemporary geographers who earned forward the geographic tradition of France to the United States. He was appointed to the chair of geography at the University of Oxford, England in 1967. He lectured in a number of US universities. Initially he was specialising on human geography of France and the Mediterranean countries, but after the Second World War he concentrated on political geography. His contribution to the field of political geography inevitably surpassed that of J. Ancel and Andre Siegfried. He has given the professional status to the French School of political geography.

Gottrnann's first publication on the commercial relations of France appeared in 1942. A textbook on America in French was published in 1949 and another on the geography of Europe in English was published in New York in 1950. His book on political geography of States appeared in 1952. He published an article on 'French Geography in Wartime' in the *Geographical Review* (1946). Gottrnann edited a small collaborative work on regional planning for the International Geographical Union in 1952. His other publications include: *Virginia at Mid-century* (1955), *Megalopolis* (1961), *Planning of Occupied Land* (1966), and *Metropolis on the Move* (1966). He also published a short article on political geography in the book on 'Geographie Generale' in 1966.

The *fundamental aspects of political geography,* according to Gottrnann, are functional structure (cohabitation); distribution (repartition); transport network (reseau des acces); circulation and group policy or system of beliefs (iconographie). To him, circulation and iconography have always been associated in the formation of public authority and in the political life of peoples.

In contrast to Hartshorne's *Functional Approach in Political Geography* (*Annales of the Association of American Geographer*, 1950), Gottrnann developed his own methodological conception (or it may be called 'a paradigm') in his article: 'The Political Partitioning of Our World: An Attempt at Analysis' (*World Politics*, 1952). *On the basic assumption that the space is limited, Gottmann defines accessibility as the major variable in political partitioning,* giving emphasis to the uniqueness of position. Each political territory is under a distinctive iconography—a set of symbols in which people believe and with which they identify in contrast with those in neighbouring areas. While this iconography is the stable factor (Hartshorne opposed it), circulation —the movement factor—is constantly acting as a dynamic force to change existing iconography, there is a corresponding change in the spatial pattern of political partitioning. In another article, published earlier, 'Geography and International Relations' (*World Politics*, 1951), Gottrnann examines the relationship between environment and the behaviour of states and its role in international relations. This methodological conception appears to fit closely with the chorological conceptual framework of Kantian philosophy.

CONTEMPORARY TRENDS

On the changing trends of French geography, it has been pointed out that 'there has been a weakening of 'La tradition Vidalienne' and loss of the feeling of coherence as French geography reacts to contemporary currents of thought that are international in range. The old unity of physical and human geography is being lost as geographers draw closer to scientific geomorphology on the one hand or to biology or sociology on the other hand. Though regional studies of high standard have continued to appear but systematic work has expanded impressively, particularly in urban geography and economic geography. This has both encouraged and been encouraged by what has been referred to as the quantitative revolution. In part this late arrival of numeracy owes to the grand literary tradition in French geography exemplified in regional studies, and in part to a disciplinarily conservative epistemology' (Gottmann, 1946, 80–81; Gallais 1967, 79–80; James and Martin, 1972, 198).

The fundamental mathematical techniques have now established themselves in most academic curricula, and this has manifested itself in the more recent publications of French geographers. 'There has been increasing involvement of the geographers in the planning processes. After much debate in the 1960s and 1970s, the French geographers have ultimately shown their interest in the la geographie appliquee (applied geography). Interest in the history of geographical thought has also developed in French geography in the 1960s and 1970s'. *At the present time critical theory with phenomenology*[4] *and hermaneutics*[5] *dominate the French school of geography* (McDonald, 1964, 21; 1975, 17).

[4] *Phenomenology:* It is a philosophy that seeks to disclose the world as it shows itself before scientific inquiry, as that which is pre-given an pre-supposed by the sciences.

[5] *Henneneutics:* It is the theory of the interpretation and clarification of meaning. In geography, hermeneutics has been formally developed to contest the epistemology of positivism and hence to reject the exclusive claims of spatial science.

9

Modern British Geography

HIGHLIGHTS

- Early British Geography
- Development of Professional Geography
- Regional Studies
- Historical Geography
- Current Trends

Development of professional geography in Great Britain took place much late as compared to Germany and France. This is because it lacked people like Alexander von Humboldt, Carl Ritter and Vidal de la Blache who laid the foundation of modern scientific geography in their respective countries. As a result, *there was no conceptual revolution in geography in Great Britain in the nineteenth century, as had occurred in Germany and France.* Delayed development of professional geography may also be attributed to the lack of 'academic entrepreneurship' among the British scholars in the nineteenth century. The extension of knowledge of the Earth's surface through exploration and research appeared to have continued to be a major concern of British geographers and of British geographical societies. A large part of the papers published in British geographical periodicals dealt with reports of explorations of the still relatively unknown parts of the Earth. Thus, too much of dependence on reports of voyages, expeditions and explorations in the nineteenth century prevented individual attempts to give the subject an organised and professional outlook.

EARLY BRITISH GEOGRAPHY

Nevertheless, the collection and classification of information about 'unknown' parts of the Earth was undertaken by geographers. Many of their expeditions were financed through the geographical societies established during the nineteenth century. They obtained money from commercial and philanthropical sources, for the information gathered was of great value to the mercantile world. The Royal Geographical Society of London took special initiative in the development of teaching geography at England's two oldest universities, Oxford and Cambridge. The importance of exploration declined as geography matured in its new academic-discipline status during the early twentieth century, although in 1899, Halford Mackinder felt it necessary to establish his credentials as a geographer by becoming the first person to climb Mt. Kenya (Freeman, 1980).

One of the earliest amateurist geographers was **Mary Somerville,** who read widely and was in close touch with the leading scholars of her time. Her book on 'Physical Geography' appeared in 1848. It discussed the surface features of the land, the oceans, the atmosphere, plant and animal geography, and one of the most revealing themes about *man being an agent of change of the physical features of the Earth.* It may be noted that the identical theme was long developed by Comte de Buffon in his *Histoire Naturelle* which was published in four volumes between 1749 and 1804. Somerville's 'Physical Geography' does not make any reference to Humboldt's *Cosmos* which was published in 1845. However, she made little impact on geography in Britain.

Another amateurist geographer was **Francis Galton** who served on the council of the Royal Geographical Society from 1854 to 1893. He was the first to make a weather map of Britain in 1861, based on reports from 80 stations. He was the first to point out that weather patterns could be revealed by plotting lines of equal air pressure on a map (isobar), and pointed out the nature of air circulation around a centre of high pressure. He also prepared a map of the world showing lines of equal travel from London (an isochronous map) in 1881. He made a map showing the spatial distribution of female beauty in Great Britain. In his 'Notes on Modern Geography', Galton defines geography as 'a peculiarly liberalizing pursuit, which links the scattered sciences together and gives to each of them a meaning and significance of which they are barren when they stand alone'.

At the time when Sommerville and Galton were making their contributions, there was no professional organisation of scholars to carry forward their ideas. Physical geography was very much a part of geology in Great Britain and it was within the domain of the geologists to carry on studies regarding the influence of physical features on human beings. However, it is with the appointment of a geographer at Oxford in 1887, and at Cambridge in 1888, that the era of professionalism in British geography began.

DEVELOPMENT OF PROFESSIONAL GEOGRAPHY

In Britain, the first personal chair of geography was held by Captain James Machonochic at University College, London, from 1833 to 1836, but permanent university teaching dates from the appointment of Halford J. Mackinder to a readership at Oxford in 1887 and the establishment of the first British department there in 1900. Yule Oldham was appointed to the chair of geography at Cambridge in 1893. In 1889, the Oxford School of Geography was founded by Halford J. Mackinder and A. J. Herbertson who became an assistant to Mackinder.

Halford J. Mackinder. He was both a scholar and a practical man of affairs. He held many important positions. While at Oxford he also held the post of Principal of University College at Reading (1892–1903). He was the director of the London School of Economics and Political Science. He was a member of the British Parliament between 1910 and 1922. Between 1920 and 1945, he was chairman of the Imperial Shipping Committee, and from 1925 to 1930 he acted as Chairman of the Imperial Economic Committee.

Mackinder was trained in the contemporary tradition of natural science and history. The Darwinian themes of 'association and organisation' (man as part of a living ecological organism), and 'struggle and natural selection' appeared to have had a profound effect on Mackinder. He pointed out that history without geography was a mere narrative and that since every event occurred in a particular time at a particular place, *history and geography, which dealt respectively with time and place, should never be separated.* The Darwinian paradigm of determinism was so much overriding that Mackinder, in one of his extreme statements, stated that 'no rational (human) geography can exist which is not built upon and is subsequent to physical geography' (Mackinder, 1887, 143).

Mackinder's first major work, *Britain and the British Seas,* published in 1902, was an example of a regional study in a global context. In the work, he sought to state the unity of geography in terms of natural regions based primarily on the hydrosphere rather than on lithosphere—a conception that undoubtedly owes much to Herbertson (Hartshorne, 1976, 100).

Mackinder's major contribution to the geographic paradigm lies in his *spatial political model,* which he developed to explain the world political patterns in the light of the declining relative strength of the sea power, and was primarily based on the sequence of events which occurred between the fifth and the sixteenth centuries. In 1904, he delivered his famous lecture at the Royal Geographical Society on 'The Geographical Pivot of History' which was published in the geographical journal.

Mackinder argued that the European civilisation was an outcome of the secular struggle against Asiatic invasion. Through the steppe, there came from the unknown recesses of Asia, by the gateway between the Ural Mountains and the Caspian Sea, in all the centuries from the fifth to the sixteenth, a remarkable succession of the Turarian nomadic peoples. The region in the centre of Asia from which these invaders came 'is on the whole a steppeland supplying a wide spread of scanty pasture, and the horse and camel mobility was relatively easy over the steppe, and the ruthless and idealess horse-men could turn in any direction they wished from their pivot area. The *geographical pivot area of history* was that vast area of Eurasia drained either to inland lakes or to a frozen northern ocean and thus inaccessible to ships and to sea power' (Mackinder, 1904).

'Enclosing the pivot area on the west, south and east, was the *inner or marginal crescent,* consisting of those areas of Eurasia from Scandinavia to Manchuria which are drained into the ocean and presumably, accessible to sea power. Outside this again, and separated from it by seas of varying width, are the lands of the *outer or insular crescent.* Sea power gave a kind of unity, especially after the opening of the Suez Canal, to the lands of the inner crescent. All the settled margins of the old world sooner or later felt the expansive force of mobile power originating in the steppe. Russia, Persia, India and China were made tributaries or received Mongol dynasties. There was a certain persistence of geographical relationship' (Mackinder, 1940, 430).

Mackinder further argued, 'if Turkic and Tartar nomads, equipped only with the horse and camel, could so shape the course of history in the peripheral lands of the inner crescent, what might not a modern power, its mobility enhanced by the railroad, achieve this respect? This might happen if Germany were to ally herself with Russia; it could also happen, were Chinese organised by the Japanese to overthrow the Russian Empire and conquer its territory. In such an event, France, Italy, Egypt, India and Korea would become so many bridgeheads where the outside navies would support armies in order to prevent the powers of the pivot area from expanding down to the sea' (ibid).

In the light of some events; such as the failure of the British navies to force entry through the Turkish Straits into the Black Sea, and their exclusion from the Baltic Sea by the German mines, Mackinder made some changes in his original thesis. Fifteen years later since his concept first appeared, he published a small book entitled *Democratic Ideals and Reality* (1919). In the book, he emphasised the relative ease of movements through the South Russian Steppe from the *pivot area (here for the first time called the heartland)* into East Europe.

The heartland of 1919 was not quite same as the pivot of 1904. 'The heartland as a region to which, under modern conditions, sea power can be refused access, though the western part of it lies without the region of Arctic and continental drainage' (Mackinder, 1919). Nevertheless, Mackinder no longer regarded the heartland as completely inaccessible to power of the inner marginal crescent. Indeed the heartland had its 'Achilles heel', that region of the south Russian steppe through which the German armies had advanced in 1917. He summed up his view of the global strategy with the famous dictum:

> Who rules East Europe commands the Heartland.
>
> Who rules the Heartland commands the Worldlsland.
>
> Who rules the World Island commands the world.

In 1943, Mackinder offered yet another modification of his own theory, which was strongly influenced by contemporary events. He separated off from the heartland that part of the Soviet Union lying east of the Yenisei River, which he called 'Lenaland'. In opposition to 'heartland Russia', he established the 'midland basin' consisting of the North Atlantic Ocean, the Eastern United States and Western Europe. The East Central Europe, thus separated from the Midland Basin and the Russian heartland, came to be known as the 'shatter belt' (Mackinder, 1943).

Commenting on the spatial-political model, as developed by Mackinder to explain the world political pattern, James and Martin observe: 'the paradigm that Mackinder set before the British geographers called for basic work in physical geography to provide scientifically accurate description of the stage-setting on which the human drama was to be played out. In this point of view, he was close to the concept of Richthofen and Hettner, but not Schluter, and close to Vidal de la Blache and de Martonne, but not Brunhes. He himself was concerned with the global view, but most of his contemporaries and his followers turned their attention to the analysis of man-land relation in small areas' (James and Martin, 1981).

Mackinder's heartland theory is an example of a simplified model in which too much truth has been sacrificed for too much simplicity (Chorley, 1964). It can be further said that the application of this historical natural model involves a translation of the simplified model into a different time and/or place, on the assumption that what has happened before will happen again, or what happens here will happen there. The assumption well fits into the paradigm set forth by Mackinder in his successive modifications of his earlier thesis regarding the world political pattern vis-a-vis the global strategy. It appears that Mackinder's thesis made little impact on British geographers, but it certainly formed a symbiosis which seemed to have a special relevance to German 'Geopolitik'. He had a group of followers in Germany who strictly adhered to the Nazi propaganda.

George G. Chisholm. He was a contemporary of Mackinder. He also contributed to the development of British geography. He pioneered the field of commercial geography in Great Britain. Chisholm seemed to have maintained the colonial trend in his geographical writings that appeared in the closing years of the nineteenth century. Organisation of the commercial world required a great deal of information about the various countries concerned, the provision of which became a major task of geographical research whilst its propagation was the keystone of geographical education. Much of the material put together was about commercial activities and infrastructure as in *Chisholm's Handbook of Commercial Geography* (first edition, 1899). He attempted to introduce Hettner's concept of chorology in Great Britain (Hartshorne, 1976).

Hugh Robert Mill, a contemporary of Mackinder, seemed to have been largely influenced by the Darwinian tradition of environmental determinism. He stressed the environmental emphasis on land-man relationships and extended his ideas of physiography to include the geography of man, though limiting the latter to what was consistently related to the physical environment and derivative from it. Mill conceived of Richthofen's system of geography of spatially distributed phenomena as a pyramid, with a physical base and an economic rubble heap at the top. Although he was trained into the Darwinian environmental approach, but Otto Schluter's morphological approach also appeared to have deeply affected his thought.

He conceived of water as an important environmental element governing the terrestrial phenomena. The study of water was the major part of interest and research. However, he could not undertake empirical study on water because of his ill-health. Wrigley sums up Mill's view on the study of water: 'the study of the part played by water in the economy of the world through the action of solar heat and terrestrial gravitation raising vapour from the sea and carrying the condensed moisture back over the land, sustaining all forms of life and furnishing hydroelectric power, the only inexhaustible supply of energy' (Wrigley, 1950, 657–60).

The study of *water as part of the 'realm of nature'* had fascinated him as early as 1891, but his interest increased when he became the joint director of the British Rainfall Organisation in 1900 and director the following year. He directed the preparation of rainfall maps of Great Britain and the maps were prepared on the basis of fifty-year averages. It was a monumental work that enriched the contemporary heritage of British School of Geography.

In 1896, Mill suggested using the sheets of the Ordnance Survey (I inch to the mile) as basis on which to plot categories of land, quality arid landuse for all of the British Isles, and four years later provided an example of how such mapping of relevant information could be done and could be used. Though the idea was initially not approved, but it was furthered by L. Dudley Stamp during the 1930s (James and Martin, 1981, 205). Undoubtedly, Mill's suggestion reflected the dominance of some form of positivism or naturalistic pragmatic trend which was current in the contemporary social thinking, and at the same time that called for the solution of practical problems. His first book, *The Realm of Nature* was published in 1891, in which he emphasised the study of water. He wrote biographies of polar explorers which contained accurate recordings of events. His major works on polar explorations were *The Siege of the South Pole* (1905) and *The Life of Ernest Shackleton* (1923), which attempted to focus on 'man's effort of manoeuvring with nature.'

REGIONAL STUDIES

Regional studies in Great Britain acquired a distinct professional dimension since the beginning of the present century, which might be attributed to the increasing emphasis on studies of 'Landschaft' in Germany, and on studies of 'pays' in France. Regional studies in Britain were greatly influenced by Vidal de la Blache and other French regional geographers, but to an even greater extent by the French sociologist Frederic Le Play through his influential Scottish follower Sir Patrick Geddes (1854–1932).

Patrick Geddes. The idea of field study—observation and recording in the field—was in Geddes' teaching basic to the diagnosis and treatment of maladjustments of social groups in each country and town. From this idea, there developed what Geddes called '*regional survey*', which embraced the triology of 'place', 'work' and 'folk', alternatively defined as environment, function and organism or, again, as geography, economics and anthropology (Dickinson, 1969, 204). He saw an important application of regional survey in regional planning. To him geography should be an 'applied science', which should state what ought to be and not what is. Geddes also influenced the study of regionalisation in Great Britain.

There are at least *three different meanings to regional studies* as a characteristic of British geography. (1) There are regional studies which amount to descriptions of segments of the Earth's surface, broadly synonymous with regional geography or Landerkunde. (2) There are regional studies which seek to divide the surface of the Earth into either homogenous or functional areas of varying size, which may be called 'regionalisation'. (3) Regional studies may denote regional specialisation, when an individual geographer devotes a large part of his life to study different aspects of some part of the world.

Andrew J. Herbertson (1865–1915). He is believed to have carried forward the tradition of Sir Patrick Geddes. He was an assistant to Geddes in Dundee, and later became an assistant to Mackinder at Oxford. He succeeded Mackinder in 1905. He was a *determinist in true sense of the term.* He attempted to combine the Geddesian tradition of regionalisation with that of the Darwinian heritage of deterministic approach. *He introduced the term 'natural regions' in England.* He suggested that the natural region should be identified in terms of associations of surface features, climate and vegetation. Herbertson considered his regions as specimens of types, thus associating widely separated areas in the same type.

His system was therefore a classification of region regardless of location and association in space, in contrast with an areal division of the world into major parts, each sub-divided into sub-divisions which are contiguous and together form an associated whole. For his major division of climate, Herbertson heavily relied on Supan. He divided the world into 15 major natural regions, and his natural regions revealed the regularities of the climate because the same regions appeared in similar positions on each of the continents. However, in his later work Herbertson attempted to extend the term to include human activity. His concept of major natural regions on a world basis also appeared to have been influenced by Hettner.

Grigg observes that Herbertson's regional system appears to have been obtained by beginning with the world, and dividing it into sub-classes on the basis of a number of principles. This is a quite different process from synthetic regionalisation and has been called *analytical regionalization.*

Stoddart (1965, 248–49) identifies the explicit influence of the Darwinian theme of 'organization and ecology' in Herbertson's regional system. Herbertson (1905) used the term *macro-organism* for the 'complex unity' of physical and organic elements of the Earth's surface. To him, 'natural regions are definite associations of inorganic and living matter with definite structures and functions, with as real a form and possessing as regular and orderly change as those of a plant or an animal.' He further pointed out that 'if the Earth be regarded as an individual, and the geographical region, districts, and localities, as representing organs, tissues and cells, then we perhaps get nearest to a useful comparison' (Herbertson, 1913–1914).

His theory of natural region appeared to have revived what may be called 'reine geographie' or pure geography in the contemporary school of British geography. Watson sums up that Herbertson seems to have used 'natural' to distinguish those regions which, because of a certain natural homogeneity of physical and human relationships, were more or less self-evident from regions, such as political ones, drawn by man. Yet, using vegetation as a basis for his scheme of major natural regions, he appeared to stress the material rather than the social factor, and left himself open to misinterpretation that a natural region is primarily physical (Watson, 1967, 467).

Herbertson's idea was taken up by his own students in England and by various followers in England as well as by Gradmann and Passarge in Germany. Herbertson's works appear to be the *precursor of the 'ecosystem concept'* (Johnston, 1983, 43), which in later periods of the twentieth century became an important paradigm.

Marion I. Newbigin. Like their French counterparts, the British geographers also developed keen interest in regional studies and devised different ways of classifying regions. One of the earlier geographers was Marion I. Newbigin who in her little book on *Man and His Conquest of Nature* (1912) suggested an approach to the definition of regions that was a clear reflection of "La tradition Vidalienne". The focus of the book was on the *relations between groups of men and their surroundings.* The central geographical question, according to her, was why it is easier for men to make their living at some places than at others. She insisted that the answer could be found by examining the relationship between the genre de vie and the productivity of the land. She pointed out: 'It is no part of the work of geography to lay down rules as to the best solutions of particular problems of land utilization, but it is her work to set forth as clearly as may be the conditions which reign at different parts of the Earth's surface, for a consideration of these furnishes the new raw material upon which all political and social adjustments must be made'.

The book, in fact, is a presentation of Vidal's idea of the mode of lite (genre de vie) and is a direct descendant of Le Play's classification and treatment of the world's societies. The presentation embodies an approach that was much more mature than the theoretical statements of her contemporaries. The classification of region, therefore, should be based on the bonds of relationships observed between human communities and their natural surroundings.

H. Fleure. The idea of modes of livelihood (genre de vie) in relation to the environmental setting was being developed at this time by Roxby and Fleure, the professional leaders of geography in Britain, and both of them seemed to have been deeply influenced by the tradition of Blache, Geddes and Le Play.

Fleure offered a regional classification of the world and the basis of the classification was the problems and difficulties posed by the condition of the physical environment. He assumed that *all human activities were primarily directed towards the accomplishment of three functions: nutrition, reproduction and the increase of well-being.* In seeking a criteria for the delimitation of human regions, he eliminated the first two because without them 'a race would perish'. He classified human regions according to the measure of the Earth's response to man's effort in the pursuit of well-being. Thus, Fleure offered a classification of **seven types of regions**—regions of hunger; regions of debilitation; regions of increment; regions of effort; regions of difficulty; regions of wandering; and industrialised regions (Fleure, 1917). Fleure's regional classification seems to reflect a combination of the deterministic standpoint with the most critical possibilist precision. He emphasised on the human factors rather than on the physical factors. He recognised that the new technology and finance of industrialised regions had so modified the relation of man to his environment that such regions could replace any of the others.

The experiments by Newbigin and Fleure were imaginative efforts to formulate illuminating empirical generalisations; they were offered at a time when there were neither the statistical data nor the electronic equipment with which to store and process such data. However, the regional classification of Fleure does not refer to the description of unique features of particular region, and the trend therefore does not appear to be idiographic in the true sense of the term.

John F. Unstead. He proposed a classification of geographic region in which the *physical and human factors were to be given equal weight.* He argued that the smallest unit area of the Earth's surface is the 'feature', and that features (individuals) can be grouped on resemblance to give first order region called 'stows'. Such classes can be successively grouped into higher order categories called 'tracts', 'subregions', 'minor regions' and 'major regions' in a hierarchy with five orders or categories (Unstead, 1916). Unstead recognised the problem and called the individual (the feature), but gave little attention to the recognition or definition of such individuals. Nevertheless, he recommended that the conditions of the past periods should be drawn upon only where necessary to interpret present features that represent relict forms of the past (Hartshorne, 1976, 179–80).

Unstead wrote of *regions as organisms,* comparable with biological organisms (Darwinian heritage). 'He spoke of the evolution of regions, in the sense of increasing complexity and of their pathology, in the sense of conditions harmful to man. He admitted that regions, unlike organisms, cannot be said to die, but he compared continuity of existence in a region with that of the "germ plasm" of organisms through the successive generations, thus endowing regions at one time with the properties of individuals and at other times of populations' (Stoddart, 1966). Similar organismic views were developed by Bluntschli, Stevens and Swinnerton.

Unstead *defined the term 'geographical region'* to describe regions where there was a high degree of homogeneity in both physical environment and human activity. The belief that an area can be logically divided into a system of such geographical regions is one that did not gain currency in England until recently. It is analogous to the belief in the biological sciences that all aspects of the objects studied can be incorporated into one natural classification (Unstead, 1916). Unstead is believed to have introduced '*synthetic regionalization*', a procedure which was frequently used by him. This was in contrast to Herbertson's analytical approach to regionalisation.

There were two difficulties with Unstead's (1916) scheme of geographic regionalisation that tended to weaken the scheme. 'The *first difficulty* was associated with the concept of a uniform unit of area, a stow, that could be used as the basic building block in erecting a structure of world regional division. The idea of a unit so homogeneous that it could not be further sub-divided. It is clear that the stow is only indivisible because it is so conceived by the observer, and actually it is necessary to start with the observable fact that no two microscopic points on the face of the Earth are identical and that any area enclosed by a line and described as homogeneous is only homogeneous. With respect to selected features the difficulty arises when the region is identified as a unit area, an indivisible segment of space. The *second difficulty* with the scheme was Unstead's announced intention to recognise homogeneous associations of physical and human factors. The French geographers recognised these difficulties, and following Vidal, defined regions in terms of the way people live' (James and Martin 1972).

Charles B. Fawcett. The revival of Geddesian tradition again occurred in the works of Fawcett who attempted to avoid regional schemes that involved the association of too many diverse factors. In his book *The Provinces of England* (1919), he proposed a federal structure of the 12 autonomous provinces for England. All six principles of the division were clearly inspired by Geddesian regionalism, particularly so in the organisation of each province round a 'definite capital which should be the real focus of regional life', that provincial boundaries should be 'drawn near watersheds rather than across valleys', and that 'the grouping of areas must pay regard to local patriotism and tradition'. Fawcett translated the somewhat obscure ideas into a workable form, and made one of the first identifications of '*functional regions*'.

It is also interesting to note that the use of the catchment area of a stream, as a basis for regional divisions, is still regarded as rather versatile by modern geographers. Fawcett mapped the continuously built-up urban areas of Britain (conurbations, as Patrick Geddes had called them). Fawcett's regionalisation tended to strengthen the idiographic tradition in British geography.[1]

Regional Trend

The regional trend continued in Great Britain quite for some time, however, in the form of an applied approach (in planning) and that involved non-geographers also. It is often assumed that stimulus to the development of regionalism in Britain during the inter-war period was provided by broadcasting which gave impetus to the growth of a regional consciousness, quite distinct from the old county spirit. In 1939, England and Wales were divided into 11 Civil Defence regions, and Scotland formed a twelfth region. The first kind of regional study leads to the *Civil Defence regions* which were of much geographical interest and this sort of regionalisation was devised to meet the threat of invasion. When the war ended, the Civil Defence Regional system was soon withdrawn. This form of regionalism may have been the ideal divisions in war-time, but that does not mean that they are also the best possible areas for planning in time of peace (Gilbert, 1967).

E. G. R. Taylor attempted to divide England and Wales into *planning regions* regardless of the existing administrative boundaries (second kind). But the division was based on geographical conditions because it was done after careful consideration of all geographical factors. Gilbert

[1] Fawcett's regionalisation (1919) was an expansion and modification of his earlier paper to the Royal Geographical Society on the 'Natural Division of England', published in *Geographical Journal*, 1917. A somewhat similar scheme was advocated by Professor G. D. H. Cole. His books *The Future of Local Government* (1921) and *Local and Regional Government* (1947) both argue in favour of changes in the boundaries of local government. Cole's divisions are based on political and economic regions and his divisions are not designed to be so strictly in accordance with geography as those of Fawcett.

offered a planning regionalisation for England and Wales in 1941. It was a modification of the earlier Civil Defence regions which usually consisted of groups of counties and that tended to infringe the county network. It was, therefore, decided to devise a sounder grouping of the counties than that of Civil Defence and at the same time, to rationalise the counties by dividing them in certain cases where it was inevitable. In the regional scheme, the changes in county boundary were reduced to a minimum, and had been adopted only where they would greatly help to produce more rational areas.

Third kind of regional study leads to *regional specialisation,* when an individual geographer devotes a large part of his professional life to studying different aspects of some part of the world. This appears to be a form of applied study, because each of the publications deals topically with some aspects of the area.[2]

One of the concepts derived from the work of Geddes was that of the regional survey of potential land quality and land use as a basic input to plans for economic development. It was **L. Dudley Stamp** who organised and directed the first British land utilisation survey during 1930s employing school children in the mapping of landuse on a scale of 1/2500 in their home districts under the supervision of school and university teachers. The importance of these maps was quickly recognised as essential data for the programme of food production necessitated by the German blockade. The first Land Use Survey eventually formed the basis for an official agricultural regionalisation of Britain. It was largely due to the effort of Stamp that the International Geographical Congress in Lisbon in 1949, resolved for the establishment of a World Land Use Survey, and for that an international commission under the International Geographical Congress was· set up to supervise the work.

In Britain, applied geography—as information gathering and synthesising—has a long and substantial record. Stamp's land utilisation survey of Britain in the 1930s, his involvement in the preparation for post-War land use planning, and the use of geographical material in that preparation, provide a major example of a tradition of 'liberal contribution' by a geographer (Johnston, 1983, 167). Logical positivism seemed to be inherent in the conceptual framework of Stamp's land utilisation survey because it was concerned with empirical question (i.e. with a factual content)—the economic depression of 1930s that largely accentuated the effects of long-term economic changes which created marked divergences of economic development and distress in Britain.

The regional paradigm was dominant in the years before and just after the Second World War. By the 1950s, disillusionment with the empiricist philosophy of the regional paradigm led to the dominance of the topical specialisms and the regional synthesis was ignored. British geographers seemed to have been less concerned with philosophical and methodological debate than their American counterparts during the 1920s, 1930s and 1940s. They did not elevate the regional doctrine as much as did their American counterparts. Nevertheless, 'the aim of regional geography is to gather up the disparate strands of the systematic studies, the geographical aspects of other disciplines, into a coherent and focused unity, to see nature and nurture, physique and personality

[2] Great Britain produced a number of regional specialists, but David G. Hogarth and Percy M. Roxby deserve to be mentioned-here. David G. Hogarth built up a formidable expertise on the Near East. His book especially dealt with the genre de vie of the people of the Near-East. Percy M. Roxby attempted to develop the idea of modes of livelihood in relation to the physical milieu, a tradition developed by La Play and further nourished by Blache. Roxby built up his expertise on China. His publications on China dealt topically with a number of aspects, such as terrain, economics, population and people, etc. His regional classification reflected the continuity of the traditions of La Play and Vidal as well as of Geddesian regionalism. He also made significant contributions to the methodological studies.

as closely related and interdependent elements in specific regions' (Woolridge and East, 1951,53).

HISTORICAL GEOGRAPHY

Only since World War I has a movement appeared to enhance the emphasis on historical geography in Great Britain. Mackinder is credited with having laid the foundation of historical geography. He pointed out that history without geography was a mere narrative and that since every event occurred in a particular time at a place, history and geography, which deal respectively with time and place, should never be separated. He also insisted that a geographer should find out suitable historical paradigms to recreate the past geographies and show how sequences of change have led to the present observable features. Though large number of British geographers have made many important contributions to historical geography,[3] but special mention must be made of Darby, Kirk and Brookfield .

H. C. Darby. Darby has made significant contributions to British school of historical geography (1936, 1951 and 1952). The historical approach which was closely associated with the work of Darby involved the detailed study of past geographies. This was done in a series of cross-sections, whose location in time were almost always determined by the available source material, such as the *Domesday Book of C.* 1086 which was analysed in great depth by Darby and associates. These cross-sectional analyses, complete with their regionalisations in many cases, were linked together by a narrative outlining the changes between the periods studied, most emphasis was placed on the cross-sections for which data allowed analysis rather than interpretation (Johnston, 1983, 46). The positivist method in historical geography implies objectivity, but the geographer in describing a landscape is subjective.

W. Kirk. Kirk and Brookfield appeared to have proposed alternative philosophies to that of positivism, philosophies which were humanistic in their orientation, and which could be applied in historical geography. Kirk emphasised that the environment is not simply a thing, but rather a whole with shape, cohesiveness and meaning added by the act of human perception. He recognised two separate, but not independent, environments: a phenomenal environment which is the totality of the Earth's surface, and a behavioural enviroment which is the perceived and interpreted portion of the phenomenal environment. He urged that sequence of historic events or re-creation of past geographies must be studied within these frameworks though appreciation of the behavioural environment should be central to the study of historical geography (Kirk, 1963).

H. C. Brookfield. With a field experience of South Africa, Mauritius, New Guinea and the Pacific Islands, Brookfield also took up the *concepts of behavioural or perceived environments.* He pointed out that the concept of the perceived environment has a considerable pedigree in historical geographical scholarship. *Historical geographers must study a trilogy of worlds*—the real world, as recorded in documents, and in the landscape; the abstract world, as depicted by general models of spatial orders in the past, and the perceived world; past world seen through the eyes of contemporaries, perceived according to their culturally acquired preferences and prejudices, shaped in the images of their assumed works. Brookfield also surveyed the literature which showed that decision-makers operating on an environment based their decision on the environment as they perceive it, not as it is (Brookfield, 1964, 1969). Though Kirk and Brookfield pioneered the historical behavioural geography in Great Britain, but still the reconstruction of past environments seems to be extremely difficult.

---

[3] They include Marion I. New Bigin (1916), E. G. Taylor (1930, 1934), E. W. Gilbert (1933), W. G. East (1935, 1951).

CURRENT TRENDS

The idiographic method and the humanistic way of thinking began to be strongly criticised in the 1950s. A new and dynamic school was developed and new methods brought into use. There was much talk of paradigm crises and revolution in the contemporary British school of geography. By the early 1960s, the *quantitative and theoretical revolutions* were having considerable impact on the British geography, and overlapping the regionalism of the post-World War second period came geography as spatial science, followed sequentially by overlays of the behavioural and radical structural approaches.

Regional paradigm seemed to be inadequate in the light of the nature of problems that accumulated, which raised serious doubts about the relevance of the ruling paradigm leading to a crisis phase in the contemporary British school of geography. Quantification, statistical description of pattern and statistical manipulation and testing of hypothesis tended to inaugurate a revolutionary phase leading to a new paradigm.[4]

The acceptance of *statistically-based new paradigm* was revolutionary because it attracted the allegiance of the younger research workers, such as Richard J. Chorley, David Harvey and Peter Hagget in opposition to established scientists. Richard J. Chorley was a geomorphologist and had spent some time studying in the United States. Harvey, a graduate of the Cambridge, joined the Johns Hopkins University in 1969, and became an Anglo-American geographer. He was one of the forerunners of modern American geography and made significant contributions to both positivist geography and critical geography.

Peter Hagget was trained in human geography and had visited the United States and experienced the development there. Chorley and Haggett worked on the adaptation of certain statistical techniques to geographical (both physical and human) ends, but their most lasting contribution was probably in editing two collections of papers which resulted from courses that they directed, aimed at introducing the new geography to teachers—*Frontiers in Geographical Teaching* (1965) and *Models in Geography* (1967).

Frontiers in Geographical Teaching is a collection of articles that tend to focus on *methodological innovations and revolution* in the field of geography. In one of the articles, Wrigley discusses the changing philosophy of geography and the increasing use of statistical techniques as the contemporary development of singular importance. Pahl introduces the *models* of the Chicago School of Urban sociologists and suggests a social geography in which the prime factor is distance. Hagget writes on the use of models in economic geography. His chapter on scale problems illustrates methods of sampling and map generalisation from samples. Timms demonstrates the use of certain statistical techniques for the analysis of social patterns within cities.

'The sciences concerned with the study of social variation have as yet produced few models which can stand comparison with the observed patterns or which can be used to predict those patterns. Prediction rests on accurate knowledge of the degree and direction of the interrelationships between phenomena. This can only be attained by the use of techniques of description and analysis which are amenable to statistical comparison and manipulation. If the goal of geographical studies be accepted as the formulation of laws of areal arrangement and of prediction based on those laws, then it is inevitable that their techniques must become considerably more objective and

[4] It is interesting to note that a small group of Cambridge graduates (Haggett was among them, but worked within the quantitative revolution) remained relatively unconcerned with the quantitative-based geographical paradigm that developed in Britain as a reaction to regional paradigm. The group included Michael Chisholm, Peter Hall and Gerald Manners, all of whom were taught by A. A. L. Caesar. Chisholm worked with the theoretical developments within systematic fields, but with a relatively little concern for detailed quantitative analysis. Both Hall and Manners were more concerned with geographical analysis of contemporary issues, although in some cases the analysis led to attempts at theory derivation.

more quantitative than heretofore' (Chorley and Hagget, 1965, 262).

Models in Geography (1967) presents a synthesis of most of the work completed before the mid-1960s by adherents to the quantitative and theoretical revolutions. In this book, Hagget and Chorley used the Kuhnian concept, in a normative sense, as part of an argument for, a revolution in geographical methods. They identified an inability of the then dominant paradigm to handle both the explosion of relevant data, for geographical research and the increasing fragmentation and compartmentalisation of the sciences. They proposed a new '*model-based' paradigm.*

Model is a widely used term, being given a variety of meanings (Chorley, 1964). There are two functions of a model: (a) as a representation of the real world, such as a scale model, a map, a series of equations and some other analogue; and (b) an ideal type, a representation of the world under certain constraint conditions. Both are used in the positivist method to operationalise a theory, as a guide to the derivation of testable hypotheses.

Haggett and Chorley (1967, 20) presented the model as 'a bridge between observational and theoretical levels, concerned with simplification, reduction, concretization experimentation, action, extension, globalization, theory foundation and explanation'. Individual authors were asked to discuss the role of model building within their own special fields of research. Haggett and Chorley distinguished between the varieties of models: those that are deterministic in that if specified conditions exist at an earlier time or will exist at a later time; and stochastic, which specify sequences of events within a certain range of probability. The approach in the book is strongly nomothetic and that attempts at changing British geography towards a more scientific approach.

Chorley has made another significant contribution to the new scientific geography in Great Britain. His *Geomorphology and General System Theory*, published in 1962, represents a great step forward in the development of *paradigm-based modern scientific method* in geography, for it was the first major paper exclusively on a system theme and on the *framework of general system theory.* By interpreting the Davisian cycle of erosion in system terms, Chorley provides a valuable example of these points and incidently demonstrates that knowledge of these frameworks of thought is as important to the interpretation of the theoretical deductive system of enquiry as to any empirical inductive one. He distinguished between open and closed systems and the application of these ideas to geomorphology. He concluded that an open system framework is more applicable to geomorphology and would lead to several major advantages. The most important of these would be to focus attention on the multivariate nature of most problems, and on the relationships not only between form and stage, but between process and form (Chorley, 1962).

Stoddart reintroduced the *ecological approach.* He justified the ecological approach not by empiricism, but on methodological grounds, on the creation of a framework of concepts. Recognition of the ecosystem implied that man, the environment and the flora and fauna are brought into one monistic framework. This framework is structured in a particular way and it is the elucidation of this structure, together with the functional relationships of the whole system, that represent the focus of enquiry.

It is admitted that a multitude of structures and relationships can be recognised in any ecosystem, but the potential value of the concept depends on the correct selection of the components to deal with the situation. Moreover, particular attention must be paid to the appropriate scale of investigation of the problem when these components are analysed. Stoddard pointed out that the *ecosystem is a form of general system* and its properties may be compared with the system derived by information and communication theorists and from these comparisons fruitful lines of further investigation may be suggested.

British geographers in recent periods have also shown interest in the studies of *public policy, welfare and the social good,* especially that of a social action organisation. The positivist based spatial-science methodology, according to them, may be inappropriate for such studies. House (1973) has reviewed the tradition of involvement in public policy by British geographers, and Stoddart has identified the late nineteenth century views of Reclus and Kropotkin as the origins of a socially relevant geography—the latter was discovered by the 'radicals'. Santos (1974) has reminded English geographers of the Marxist inspired works of Jean Dresch on capital flows in Africa and of Jean Tricart on class conflict and human ecology, both prior to the Second World War. Keith Buchanan is known for his left-wing opinions in New Zealand during the 1960s, and called for the need for studying geographical primacy of the state, especially in the non-West world.

The British journal 'Area' seemed to have specialised in studies of major societal problems during the early 1970s. Chisholm attempted to identify differences between governments, with their interests in cost-effective research and their primacy in decision-making and academies. Eyles pointed out that the focus of relevant research should be some of the social and spatial inequities in society, and the challenge is to study the distribution of power in society, which is the mechanism that allocates scarce resources. Research would then identify the disadvantages of relative powerlessness and would provide the base for policy which redistributes resources.

10

Professional Geography in the Former Soviet Union and Contemporary Russia

HIGHLIGHTS

- Beginning of the Tradition
- German Influence on Early Russian Geography
- Pre-Revolutionary Russian Geography
- Determinism and Indeterminism
- Marxist/Socialist Geography
- Dichotomy in the Former Soviet Geography
- Geographical Synthesis
- Contemporary Trends

Geographic heritage in pre-revolutionary Russia appears to have been founded much earlier than in France and Great Britain and even in Germany, and it was here that the establishment of the tradition received royal patronage. Thus, the Russians had a long history of geographical work, including the preparation of maps and atlases and the writing of regional monographs. It was the colonial expansion in the late seventeenth century and in the early eighteenth century which necessitated the accurate geographical information and mapping of the area being colonised. Several expeditions were carried out along with the colonial expansion, with the purpose of mapping down the location of rivers, coasts and mountains, and to identify places where fur and precious metals could be found.

BEGINNING OF THE TRADITION

Peter the Great, who ruled Russia between 1682 and 1725, recognised the vital importance of having accurate geographical information to guide the eastward expansion of the empire. In 1719, all official Russian map making activities were placed under the direction of Ivan Kirilov, the first Russian to be appointed as the head of the cartographic office. He supervised the preparation of an atlas of Russia which was published in 1734. In the preparation of the atlas, he took technical assistance of French cartographers. The great Russian encyclopaedist M. V. Lomonosov insisted that the exploring parties be asked to make systematic collection of information about the physical character of the land, the population and the conditions of the economy. In 1758, Lomonosov was appointed the head of the world's first officially named department of geography, which was in the Russian Academy of Sciences. It was not until 1768, three years after the death of Lomonosov, that the Academy of Sciences organised the first expedition for the specific purpose of gathering information and reporting on the physical and economic geography of a part of the national territory. The southern part of European Russia and the Ukraine were surveyed in the late seventeenth century and the maps on the basis of the survey were published in Amsterdam (Bagrow and Skelton, 1964, 170–76).

The establishment of the department of geography in the Russian Academy of Sciences in 1758, much before the birth of the founders of modern classical geography, Alexander von Humboldt and Carl Ritter, showed that *the tradition of geography in an organised way first developed in Russia rather than in Germany.* Because of the language barrier, as most of the earlier works were written in the Russian, the contemporary Russian geography could not make its impact felt on Germany and France.

GERMAN INFLUENCE ON EARLY RUSSIAN GEOGRAPHY

Although the Russian tradition of geography was older than that of Germany, but it was the German tradition whose impact on the Russian geography was more pronounced. The works of Auton Friedrich Busching had a profound effect on the Russian geographic heritage. It was in 1766 that the part of his *Neue Erdbeschreibung* dealing with Russia was translated into Russian, in which he described the division of European Russia into latitudinal zones of differing natural conditions—north, middle and south. Busching's proposal was given due academic recognition by the Russian geographers who initiated a movement calling for the division of the national territory into natural regions for the practical purposes of administration.

Large number of regional descriptions, very much on the pattern suggested by Busching, were prepared by the Russian geographers. Most of the regional descriptions followed political units as regional base. Thus, the geography that developed in Russia in the later part of the eighteenth century, as a result of Busching, was not 'pure geography' (reine geography). The concept of *Staatenkunde* seemed to have prevailed in the contemporary geographical methodology in Russia, in contrast to the concept of *'Landerkunde'* in Germany and France.

James and Martin identify certain persistent characteristics of Russian geography in the early nineteenth century which included: emphasis on regions as the basis of organisation for geographical works that the regions are real entities that can ·be objectively defined; the study of these regions was undertaken for practical purposes, i.e. the emphasis on the social relevance of these regions, a form of applied study. While classical geography was undergoing analysis in Germany as each academic discipline sought to establish its separate existence, in Russia the tendency was for scholars with diverse interests to come together as geographers. In 1840, this created need for some kind of institution to provide a forum for the presentation and discussion of different kinds of studies dealing with the physical world and its human habitants. In 1845, was founded the Imperial Russian Geographical Society by Arsenyev (James and Martin, 1981, 225). The society was known for its diverse specialities, though the promotion of geographical studies was its prime concern. The diverse specialities represented in the Imperial Russian Geographical Society were known collectively as 'the geographical sciences'.

One of the earlier writers, **K. I. Arsenyev** who founded the Imperial Geographical Society, published a *Short Universal Geography* in 1818. The book had strong economic emphasis, reflecting concern with improving the quality of life of the peasants. This book was probably the oldest contribution in *'liberal geography'.* Arsenyev prepared a monograph on historical geography of Russian cities in 1832 in which he offered an economic classification of cities. There is no evidence to show whether Arsenyev met Alexander von Humboldt when the latter visited the city of St. Petersburg in 1829. The Russian Tsar, like Peter the Great, also recognised the vital importance of having accurate geographical information about the virgin land islands of Siberia across the Urals, and under his royal patronage several expeditions to Siberia and the Far East were organised.

Alexander von Humboldt was entrusted by the Russian Tsar to explore the virgin lands across the Ural Mountain. He explored the virgin lands of Siberia as far as the borders of China. He visited the shores of the Caspian Sea. In course of his expedition, Humboldt observed that temperatures varied at the same latitude in accordance with the distance from the ocean. The Russian Tsar was so influenced by Humboldt that he agreed to set up weather stations where weather data could be recorded. By 1835, the Russian network of recording stations extended all the way from St. Petersburg to an island off the Alaskan mainland. On the basis of the weather data collected at the recording stations, Humboldt deduced the *concept of continentality.* This heritage which Humboldt founded, had a profound effect on the contemporary Russian geography, and empirical studies of areal phenomena with the positivist approach formed an inevitable symbiosis for the conceptual framework.

PRE-REVOLUTIONARY RUSSIAN GEOGRAPHY

Most of the Russian geographers of the pre-revolutionary period seemed to have had direct or indirect contact with the professional German geographers; some of them studied in the German universities, and were pupils of Ritter, Richthofen, and Ratzel. But the impact of German geographical ideas had produced quite different results in Russia. Russian geographers of the pre-revolutionary period always *believed in the mechanical explanation rather than in teleological explanation,* where the phenomena and observation were understood as outcomes of prime causes.

Pre-revolutionary Russian geography seemed to have been centred around Petre Petrovich Semenov, Alexander Ivanovitch Voeikov, V. V. Dokuchaiev, and D. N. Anuchin. They formed the true heritage of the Russian geography which continued its existence even after the October Revolution of 1917.

Petre Petrovich Semenov is often regarded as the 'grandfather' of the pre-revolutionary Russian geography who acted as a bridge between the scholars of the classical period, such as Lomonosov, Busching and Arsenyev, and the scholars of the modern period. He attempted to build his own conceptual framework. In 1853–54 Semenov attended Ritter's lectures at Berlin and worked with Richthofen to prepare for exploring work in Central Asia. He was more influenced by Richthofen than by Carl Ritter. Ritter's teleological philosophy made no appeal to him.

Sernenov is credited with a distinct humanistic approach in the contemporary Russian geography which emphasised questions of social inequality. He developed a *'welfare paradigm'* for social relevance in geography that stressed the quality of life. The empirical identification of social inequality in terrestrial distribution involved developing social indicators such as poverty in rural areas, hunger and crime. His empirical generalisations on social relevance appeared to have a naturalistic pragmatic trend which had a profound effect on the future paradigm of Russian geography.

His regional monographs, including a five-volume work on Russia (*Geographical and Statistical Dictionary of the Russian Empire, 1863–68*) dealt with 'the causal interrelations among the diverse things in particular areas'. The work has been described as a perceptive blend of natural, historical and economic phenomenon (Hooson, 1968). In 1871, Semenov also published a general work on the historical geography of Russian settlement. He was also a member of the committee that planned and directed the first Russian census of population in 1897. He is often credited with having set a distinctive stamp on Russian geography, giving it unity in spite of the variety of its parts and pointing it towards practical and remedial objectives.

Prince Petre Kropotkin was a contemporary of Semenov. Though he never held a university post, he was a geographer of the Anarchist School to which belonged Elisee Reclus. He is believed to have introduced radicalism into the contemporary geographic scholarship. From 1876 to 1917, while Kropotkin was in exile from Russia, he worked for the Royal Geographical Society in London. His radical views were in contrast to what may be called the teleological explanation and appeared to be more humanistic in nature. Kropotkin's geography seemed to have been *deeply rooted in idealism.* Most of what Reclus had to say about Russia was based on Kropotkin's views.

Professional Geography

The development of professional geography in pre-revolutionary Russia took place between 1880 and 1914 when the ideas of Richthofen, Ratzel and Hettner made rapid inroads in the contemporary methodological discussions in Russia. It is interesting to note that the Darwinian tradition which had already consolidated in Germany, failed to make impact in Russia. This was largely because of the fact that similar studies of evolution had already been done in Russia by the biologist N. F. Rulye whose evolutionary ideas did affect a group of Russian historians who specialised in the geographical historical studies,

Professionalism in pre-revolutionary Russian geography seemed to have been initiated by three outstanding followers of Semenov who left distinct imprints on the contemporary methodologies. They were Alexander Ivanovitch Voeikov, V. V. Dokuchaiev and D. N. Anuchin. The first two are credited with innovative studies of climate and soils, and the last one established geography as a major university subject and drew up the curricula for the primary and secondary schools.

A. I. Voeikov studied at Berlin, Gottingen and Heidelberg universities of Germany. His doctoral dissertation, 'Direct Insolation in various Parts of the Globe' was accepted at Gottingen University. Voeikov was *essentially a physical geographer.* He studied the Earth's heat and water balances which he continued for the rest of his life. His studies of climatology were directed to the improvement of agriculture. He is credited with the snow-science. He laid down the foundation of systematic agro-climatic geography through elaboration of the *comparative method* as expounded by Cad Ritter. His concern with the improvement of Russian agriculture led him to compare the farm practices in places similar to those of European Russia.

On man-environment relationship, Voeikov followed the tradition of Comte de Buffon that *man was an agent of destruction.* The positivist approach was inherent in his report on the destructive effects of man's use of land. He pointed to the overgrazing of some of the Russian steppes with a consequent acceleration of gully erosion. The cleaning of forests in the north could produce a change in the climate towards increasing drought. He was always enthusiastic about what irrigation could do to improve the productivity of arid or semi-arid lands.

V. V. Dokuchaiev was a distinguished geographer of pre-revolutionary Russia. He became the first professor of geography at St. Petersburg in 1885. His doctoral dissertation dealt with the Russian chernozem or black Earth which was published in 1883, before his appointment to the chair of geography. In 1889, he published his concept of soil-forming processes and of the natural zonation of *soils according* to climate. He made outstanding contributions to the soil geography. His innovative studies of soil within a systematic framework led him to formulate certain generalisations regarding the formation of soils.

Dokuchaiev was the first geographer to realise that soil is not just disintegrated and decomposed rock. He observed that different kinds of soil could be identified by looking closely at the layers or horizons, which differed because of differences in the soil-forming processes. Soil is formed by water percolating through the loose material at the surface and carrying away soluble minerals, and soil is also formed by the mixture of organic matters from plants and animals with the upper horizon. The soil, according to him, reflects the extraordinary complex interaction of climate, slope, plants and animals, with the parent material derived from the underlying geological formation. A soil which was exposed to all these conditions for a long time, would more closely reflect the complex of climate and vegetation than it would the patent material. His generalisations were quite different from those of German and French soil scientists who conceived of the soil as a reflection of the underlying geological formations. The generalisation of Dokuchaiev was essentially based on empirically observed arrangement of climatic zones which could produce observable soil differences over a large tract of plain.

While defining the genre de vie of the people that brought about the transformation of natural zones into agricultural regions, especially in the steppe region, Dokuchaiev attempted to adopt *holistic approach* to explain the very concept of *man as a major agent of change* on the surface of the Earth, which Voeikov did earlier. Dokuchaiev's concept of natural zones transformed by man resembles Schluter's concept of the landscape type. In fact, geography to him appeared as 'landscape science' (Landschaftovedenie).

Undoubtedly, he was one of the leading professional geographers of Russia and deserved major place among the world's leading geographers of his time because of his contributions to systematic soil studies. Since all his works were written in Russian, his major ideas could produce little impact on west European geographers. It was only through the translation of his works into German and French after his death that he came to be known beyond the frontier of Russia as one of the founders of applied soil geography.

Dokuchaiev is believed to have developed a distinct heritage of soil geography at St. Petersburg in pre-revolutionary Russia and the tradition he established was carried forward by his pupils.[1] His innovative soil studies received wide recognition in the United States. This was because of the presence of large expanses of plain on which zonal soils could be observed and which resemble 'the Russian conditions of large expanses of plain.' Dokuchaiev seemed to have employed the hypothetical deductive method in his empirical studies in which he sought prime causes by deduction. The positivist approach was inherent in the works of his pupils who dealt with empirical questions which had factual content.

D. N. Anuchin was an able administrator and organiser, and primarily an educator. In 1887, he was appointed to the newly created chair at the department of geography and ethnography at Moscow State University. He studied anthropology and anthropogeography at Heidelberg and was trained in the Germanic traditions of geography. It was he who developed geography as a major university subject. His textbooks provided a new kind of geography to teach in school. It was due to his efforts that geography was recognised as a major field of landscape (in enquiry) and geography also became a popular subject throughout the country.

There was one **V. P. Semenov Tyan-Shanki** (the son of Petrovich Semenoy, the grandfather of pre-revolutionary Russian geography) who published a work on geomorphic regions of the Russian plains with special reference to man-land relations. His 19 volume semipolar regional geography of Russia was published in St. Petersburg between 1899–1914. In 1928, he contributed a report on the Russian census of 1928 to the *Geographical Review.*

DETERMINISM AND INDETERMINISM

Geography in pre-revolutionary Russia and in Soviet Russia seems to have been badly hit by the contrasting paradigms of 'determinism' and

[1] L. I. Praslov, one of his pupils, became the editor-in-chief of a soil map of the Soviet Union compiled from a variety of detailed studies on a scale 1/1,000,000–a project finally completed by Gerasimov. N. M. Sibirtsev, who died in 1900, developed the concept of zonal soils as distinct from azonal soils, the former reflecting climatic pattern while the latter more closely related to underlying structure. In 1908, K. D. Glinika published a major work on the zonal soils of the world, *The Great Soil Groups of the World and Their Development.* The work specifically incorporated the concepts of Dokuchaiev to types of zonal soils not found in Russia. The work was translated into German in 1914, and into English in 1927, by an American soil .geographer C. F. Marbut.

'indeterminism', creating an apparent dichotomy in the contemporary geographic methodology vis-a-vis the tradition. It is interesting to note that neither in Germany nor in France, such a situation of contrasting paradigms of determinism and indeterminism has ever existed. It was the deterministic paradigm that dominated the German geographical thinking, and in France the possibilistic tradition was dominant.

Determinism

The concept of struggle and selection is reflected in geographical writings of both pre- and post-revolutionary periods. Similarly, the concept of genre de-vie is inherent in the views of both Russian and Soviet geographers. It was the Russian historians who carried forward the environmentalist views, and the geographers followed them. The famous historian Sergey Solovyev pointed out that the nature of a country has important significance in history, in the influence which it has on the national character. Similar environmentalist view was expressed by Vasily Klyuchevskiy who explained the effect of the forests, steppes and rivers on the history of Russian people. He said that each of these separately, by itself, took a lively and original part in the formation of the life and the ideas of the Russian man.

Lev Mechnikov carried forward the idea of Klyuchevskiy that the river as a synthesis of all physical geographical conditions was one of the factors which determined the development of society, but at the same time he also recognised that man had played a significant part in the formation of the geographical environment. He pointed out: 'Far from geographical fatalism, which the theory of the influence of the environment is often accused of being, in my estimation one should seek the principle of the rise and character of primitive institutions and their subsequent evolution not in the environment itself but in the relations between the environment and the capacity of the people inhabiting a given environment for cooperation and solidarity' (Mechnikov, 1924). Thus, Mechnikov spoke of both determinism and indeterminism—a sort of contradiction which seemed to have a profound effect on subsequent geographical works.

G. Plekhnov. The works and ideas of the pre-revolutionary historians are of interest mainly because of their effect on early Marxist scholars and hence on a later generation of Soviet geographers. It is interesting that most pre-revolutionary Russian geographers seemed to have been free from environmentalist thinking.[2] It was the historians and not geographers who were largely responsible in influencing the thought of the roost eminent of early Russian Marxist philosophers, G. Plekhanov.

Plekhanov elaborated the idea of Lev Mechnikov in a different way. He asserted that in order to assess correctly the influence of the geographical environment on the historical destiny of humankind, it was necessary to trace how the natural environment acted upon the type and nature of the social environment which in the closest way determined the character and proclivities of man. His idea was to introduce the theory into the framework of the Marxist analysis of society. Plekhanov further claimed that the 'peculiarities of the geographical environment determined the development of the productive forces, the development of the productive forces determined the development of the economic forces and directly after them also all the other social relations' (Plekhanov, 1957). Marxism for Plekhanov appeared to be the application to social development of the Darwinian theory of the adaptation of biological species to the conditions of the environment.

[2] A. Voeikov showed few traces of environmentalism in his works. D. N. Anuchin saw man as one of the productive forms of nature and regretted that the development of geography in the past had led to the placing in the foreground of the influence of geographical condition on man and his culture — the influences for the exact study of which there were neither means nor methods (Anuchin, 1892).

The *mixture of geographical determinism and 'economic deviation'* is further reflected in Plekhanov's conclusion that the character of the surrounding natural environment determines the character of man's productive activities and the character of his means of production. However, the means of production determine just as inevitably the mutual relations of people in the process of production (Plekhanov, 1956, 54).

N. N. Baranskiy, the noted *economic geographer* of the Soviet Union, carried forward the basic ideas of Plekhanov. In a statement in 1926, he said: 'the influence of natural conditions on man is taken into account in the Marxist scheme of social development to the extent in which these natural conditions form a natural basis for the material productive forces which determine the productive relations, and through them the legal and political superstructure and, finally, the forms of "social conscience"' (Baranskiy, 1960, 53).

Baranskiy also *ridiculed 'geographical nihilism'* (referring to an insignificant role to the geographical environment in the development of society) as being theoretically incorrect, as, denying any significance for natural conditions and removing human society from the material environment of its existence and development, it inevitably leads to Menshevik idealism. The Marxist-Leninist position is as far from this view as it is from that of environmentalism, or 'geographical fatalism' (Matley, 1966; Baranskiy, 1960). He, however, managed to build up within Marxist framework a strong case for a more realistic approach to the natural environment by Soviet geographers. Baranskiy's ideas contain an echo of Ratzel as well as a rejection of the ideas of Buckle, Chernyshevskiy and Pokrovskiy.

Plekhanov and Baranskiy appeared to have held a positive approach to the geographical environment. Baranskiy's views on the relationship between man and the geographical environment seemed to have influenced other Soviet geographers, and Moscow University witnessed the establishment of a distinct heritage of Baranskiyan tradition of 'geographical fatalism'.

Adapting Baranskiy's basic idea, **Anuchin** attempted to develop a concept of *'neo-detenninism'*. To him the nature of geographical environment included not only pure nature, but also man and the results of his activities. By 'geographical environment', he meant that part of the landscape envelope of the Earth in which direct links are found between human society and the rest of nature. It is the part of nature which has been changed by man's activities, and which in turn exercises certain influences on the development of human society (Anuchin, 1957).

In his book, *Theoretical Problems of Géography*, published in 1960, Anuchin realised that the influence of geographical conditions on the life of a society is beyond all doubts. It would be wrong to see in the assertion of this influence a manifestation of geographical determinism. The scientific worthlessness of the latter does not rest in the fact that it recognises the influence of the geographical factor on the development of society as a basic condition of this development, but in the fact that this influence was elevated to the position of a basic principle of development.

He identified that the main fault of geographic determinism lay in its mechanical nature. The environmentalists saw certain influences as producing certain consequences; whereas in fact the problem of cause and effect is a very complex one. It is this mechanical aspect of determinism which Marxism rejects and not determinism in the broader sense of the word. To him, indeterminism appeared to be anti-scientific, an idealistic view of the world, arising from subjective idealism. Indeterminism sharply separates and contraposes human society to the rest of nature, and so rejects geography as a science.

Anuchin found an ardent supporter of his views in **Saushkin** who attempted to elaborate Anuchin's basic idea in his *Introduction to Economic Geography* published in 1958. In the book, he emphasised that the geographical environment is not a passive and uniform witness of human history. The history of nature and the history of society are linked in the closest manner and influence the development of one another. This becomes even clearer the more the society changes the surrounding geographical environment. Man at every stage in his history has become more capable of changing nature according to his own interests, yet he can never change the laws of nature or formulate new laws of nature. 'The whole mastery of man over nature is confined to understanding her laws and being able to apply them in his interests' (Saushkin, 1958).

Indeterminism

In contrast to the concept of geographical determinism as one of the ruling paradigms, a group of Russian scholars of the pre-revolutionary and also of the Soviet period, attempted to focus on the concept of indeterminism. Indeterminism (i.e. opposition to geographical determinism) appeared to have been built upon the conceptual framework of 'possibilism' and that developed into a 'state-sponsored' ruling paradigm during the Stalin era. **Chernyshevskiy** was one of the earliest pre-Marxist revolutionary scholars who, in his article in 1887–88 attempted to trace out the difference of character between peoples and their national character, and the general character of elements promoting progress without raising the bogey of the geographical environment.

Pokrovskiy stated in 1931 that 'Man can master nature and it is not nature that is the foundation of his economic activity. Nature is the only material for this activity. The foundation of the economics of man is man's labour' (Pokrovskiy, 1931). Both Chemyshevskiy and Pokrovskiy were of the opinion that at an early stage of human development, the environment has a more direct and decisive influence than at a later stage when improved tools and techniques enable man to control nature to an increasing degree. At a later stage of development, however, geographical and climatic factors are no longer the determining forces. The concept of geographical determinism was criticised as being a scientific weapon of the bourgeois revolution.

It was in 1938 that 'indeterminism' became an officially approved ruling paradigm when **Stalin** in his official statement clearly stated: 'The geographical environment indisputably is one of the constant and necessary conditions of society, and of course, influences the development of society; it accelerates or retards the speed of development of society. However, its influence is not a "determining" in as much as the changes and development of society proceed incomparably faster than the changes and development of the geographical environment. Geographical environment cannot be the chief cause of development of that which undergoes fundamental changes in the course of a few hundred years' (Stalin, 1952).

Voskanyan also raised doubt over the role of the geographical environment, and emphasised the activities of man in changing the natural features. He believed in the ultimate control of nature itself by man. A Canadian geographer Sebor felt that a strong reaction to geographical determinism might lead Soviet geographers along the road to possibilism. However, I. Gerasimov rejected the assertion of Sebor on the grounds that possibilism is simply a form of geographical determinism. Sebor's claim that Soviet geographers are possibilists is not borne out by the evidence.

Saushkin also felt that possibilism is just as incompatible with Marxism as environmentalism, although he admitted that the idea that geographers should be interested in nature only from the point of view of the possibilities that it offers to the economy, is rather widely held among Soviet economists. *He agreed that possibilism completely rejects determinism,* but nevertheless considered that it errs in regarding nature as only an object for exploitation. His explanation of the origin of possibilism was, to say the least, novel. After pointing out that relations between man and nature are not similar to the relations between the exploiter and the exploited, he went on to say that in the capitalist world it frequently happens that the bourgeois scientists try to apply the social relationships of their system, based on exploitation, to natural phenomena, hence the idea of geographical possibilism.

However, a group of prominent Soviet geographers such as Kalesnik, Konstantinov and Semevskiy, strongly attacked the neo-determinist concept of Anuchin as being too impractical to be considered. On the man-environment relation, they insisted on the validity of the concept of 'anthropocentrism' which considers 'man as the master and exploiter of nature and of nature as only an object of exploitation.' The controversy of determinism and indeterminism has led to open debate on the nature of geography in Russia/Soviet Union. While Anuchin, Saushkin and their supporters held the view of a unified geography, Gerasinov, Kalenisk and others (mainly physical geographers) called it a segregated science. Almost on the same line developed an apparent dichotomy between positivism and critical theory in the Soviet geography.

MARXIST / SOCIALIST GEOGRAPHY

Marxist/Socialist geography, that developed in the Soviet Union after the Great Revolution of 1917, tends to emphasise on *spatial patterns of the determining influence of class forces* operating through the laws of historical and dialectical materialism. It stresses on social justice and equality in contrast to class-based inequalities and attempts to offer explanations, seeking for the laws to analyse the societal processes. The socialist revolution in Russia, as a result of the class-based struggle, seemed to have inevitably created what may be called *'spatial dialectics'* which included the two most intrinsic elements: 'social process' and 'spatial form', which formed the very basis of a dialectical unity.

It was as early as 1818 that K. I. Arsenyev realised the need to provide explanations to study the genesis of the rural poverty and its spatial form. The same tradition of giving emphasis on the study of class-based inequalities in their spatial dialectic seemed to have continued even after the Great Revolution, and formed the major ruling paradigm in the Soviet geography. Russian/Soviet geographers are more concerned with the material conditions of the class-based society. Despite the conflicting approaches, they all focussed on the material life of the society and the character of the social system.

Commenting on the new direction of geography after 1917, James and Martin point out: 'Most of the distinctive characteristics of Soviet geography can be traced back to the pre-revolutionary period. The tendency *to look at natural landscape as systems of interrelated parts is typically Russian* and derives from men like Semenov, Voeikov and Doukuchaiev. The continued interest in such physical processes as the heat and water balances goes back to Voeikov. The pre-occupation with the drawing of regional boundaries began in the eighteenth century, as did the concern with practical problems of economic development. And to find that geography is a focus of diverse specialities rather than a remnant of the separation of the disciplines is in line with long-standing Rusian tradition' (James and Martin, 1981, 232). The geographers in the Academy of Sciences made practical use of the large amount of information collected during the preceding half century under the guidance of Semenov and Anuchin. Geographers were asked to make an inventory of natural resources.[3]

One of the distinctive aspects of the Socialist geography in the Soviet Union immediately after 1917 was *continuing emphasis on economic regionalisation.* The State Planning Commission

[3] To further geographical researches for practical purposes, the Institute of Geography at Leningrad University was founded in 1915. In 1925, the Institute of Geography at Leningrad was recognised as the Faculty of Geography. The Commission for the study of Natural Productive Forces which was established in the Academy of Sciences in 1918 was first reorganised as the Institute of Geomorphology under Gregoriev in 1930, and then in 1934, it became the Institute of Physical Geography, and finally it was reshaped as the Institute of Geography in 1939.

which was established in 1921, appointed a special commission on regionalisation which was authorised to recommend a national plan for the division of the national territory into functional units. The commission formulated 'the concept of the economic region' in 1922: 'A region should be a distinctive territory that would be economically as integrated as possible, and, thanks to a combination of natural characteristics, cultural accumulation of the past, and population trained for productive activity, would represent one of the links in the entire chain of the national economy. This principle of economic integration makes it possible to construct further, on the basis of a properly selected complex of local resources, capital assets brought in from outside, new technology, and the national plan of economic development, a regional development plan that would make optimal use of all possibilities at minimal cost. This will also help achieve other important results: the regions will specialise to a certain extent in those activities that can be developed most fully in them, and exchange between regions will be limited to a strictly essential amount of purposefully directed goods' (Sanshkin, 1966, 12).

'Regionalization will thus help establish a close link between natural resources, working skills of the population, and assets accumulated by previous cultures and new technology, and yield an optimal productive combination by insuring a decision of labor among regions and, at the same time, organizing each region as a major economic system, thus evidently insuring optimal results' (Saushkin, 1966, 12). The State Planning Commission divided the Soviet Union into 21 such regions and then proceeded for a detailed study of each of them. L. L. Nikitin, a distinguished geographer, attempted to formulate generalisations by combining older data with latest information regarding natural resources, physical condition, population and types of economy for each of the 21 regions.

The concept of economic region vis-a-vis the productive activity also motivated a group of geographers not directly related to the State Planning Commission, to think on problems of industrial location and resource development. The plans for the organisation of inter-regional combines were the natural outcomes of such thinking. N. N. Kolosovskiy prepared such a plan for the great Ural-Kuznetsk Industrial Combine which called for the movement of raw materials and finished goods between regions. A theoretical model for an integrated industrial region, combining basic resources with steel production and with industries using the steel was prepared in 1927 for the Dneiper Basin.

It is interesting to note that the positivist approach was very explicit in the works of geographers who were associated with the State Planning Commission, and who worked on practical problems of economic planning and sought for the formulation of explanation to real problems. University geographers were involved in the methodological and philosophical discussions whether geography should be 'idiographic' or 'nomothetic'. Baranskiy attempted to revive the Geddesian heritage of regional approach within the framework of the dialectical unity because this appeared to him the only way that Socialist geography could make a useful contribution to practical problems of economic planning. (*Dialectic must not be confused as an abstract, model or theory but it is a method of analysis*).

A group of Soviet geographers who attempted to revive Dokuchaiev's tradition of geographers as 'landscape science' favoured a nomothetic character of geography because geographical laws, based on hypothetic-deductive method could be developed to study spatial relations of social processes arising out of the class-based inequalities. Another group of Soviet geographer led by N. N. Baranskiy insisted that Socialist geography must concentrate on the unique characteristics of spatial relation of social processes.

This methodological-philosophical debate in the 1940s and 1950s appeared to have given rise to an apparent dichotomy in the Soviet Geography, breaking the very essence of the unity of geography.

DICHOTOMY IN THE FORMER SOVIET GEOGRAPHY

The dichotomy, that physical geography and economic geography are clearly differentiated branches of the discipline, with separate concept, methods and laws seemed to have 'exposed' the inherent inner contradictions of the Soviet Geography of the 1960s. The argument is that laws governing the physical world are not at all the same as those governing man's economic behaviour and, therefore, these two fields of study cannot logically, or even practically, be included in the same discipline.

Physical Geography. Voeikev and Dokuchaiev were the two most prominent pre-revolutionary Russian geographers who advocated in favour of physical geography, that geographers should concentrate on the physical aspects of the Earth's surface. In fact, Dokuchaiev's 'landscape science' makes geography a nomothetic science, in which generalisations may be formulated to explain the physical nature of the Earth. The pre-revolutionary tradition of physical geography was carried on by I. P. Gerasimov, who was the director of the Institute of Geography at the Academy of Sciences in 1963. Physical geography attempts to seek the laws which lie behind the observable physical features of the Earth's surface, and that the development and testing of theories and models of universal application appear to be intrinsic to the genuineness of physical geography. On the various specialities of physical geography and biogeography, Gerasimov pointed out: 'The particular physical geographic disciplines made extraordinary rapid progress throughout the Soviet period, both in the development of theory and in the formation of new approaches and research methods.[4] Although the achievements of the particular disciplines provided the main elements of modern theory of physical geography, the formation of such generalisations was also complicated by the growing differentiation of the physical geographic disciplines, in which scholars concentrated their attention increasingly on specific components of the natural geographic environment' (Gerasinov, 1968, 247).

Systematic studies of climatographic analogue, soil geography glaciology, geomorphology and biogeography, accompanied with the universal and generic concepts, appeared to have given a distinct status to physical geography in the Soviet Union. The Soviet geographers of the 1970s proposed a new model-based paradigm for physical geography. From Dokuchaiev's 'law of zonality' has evolved the '*landscape science*', which ultimately developed into a 'model-based' or predictive physical geography during the Soviet period. The landscape, according to Isachenko, is a dynamic system in which matter and energy are circulating and in which there are rhythmic (seasonal) changes of heat and water balance and biological productivity. A classification of landscapes with a hierarchy of scales or degrees of generalisation prepared for the whole Soviet Union to be mapped on a scale of 1/4,090,000. (Isachenko, 1968).

Gerasimov, the forerunner of the modern Soviet physical geography, believing in the nomothetic

[4] In climatology, for example, Soviet scholars developed the theoretical principles of forecasting and typology of climatic phenomena, based on dynamic meteorology: they developed the concept of an integrated climatography, and in recent years gave rise to the promising study of the radiation budget and the moisture, cycle and their role in the formation of climates. In the field of hydrology, Soviet scholars worked out the theory of the water budget, the relationships between components (surface water, soil water and ground water) and methods of transforming one into the other. Glaciology saw the development of the physical theory of glaciation processes, based on the study of heat and mass exchange in various types of glaciers. Geomorphologists established the dynamic character of many exogenous processes (erosion, deflation, abrasion, etc.) and, on the basis of a general theory of external and internal forces and the study of recent crystal movements, developed the morphotectonic or morphostructural approach to geomorphology. Soil scientists identified many new soil type characteristics of taiga, desert, and mountain areas, and worked out new approaches (for example, physical and biogeochemical) to the study of the dynamics of soil forming processes and the circulation of substances in the natural environment. Biogeographers gave emphasis to the ecological and biogenetic approaches in the study of plant groupings and of animal populations.

character of the discipline, *defined a landscape as a combination of interrelated environmental components* (local climate, landforms, soils, plants and animals) occupying a discrete territory. It exists objectively in the natural environment. The landscape science appears to incorporate the general systems theory as it emphasises on a set of objects together with relationships between the objects and their attributes. The approach to landscape science, therefore, is to study them as an assemblage, in which the various parts may be related in an areal and time sense. The landscape science or the modern Soviet physical geography attempts to set out to discover universals, to build and establish theoretical structures into which geographical realities of the physical Earth might be fitted.

Gerasimov insisted that the study of landscapes just for the purpose of describing them is not enough: *constructive geography*[5] must use this knowledge for the effective transformation of nature. He offers four examples of constructive geography:

1. The study of the geophysics of natural and cultural landscapes of the wooded steppe in the central chernozem zone. The purpose was to investigate the heat and water budget on the Earth's surface in virgin land and in cultivated land and then to experiment with various technical devices for controlling natural processes and so increase agricultural productivity.
2. The study of the irrigated lands of central Asia. The purpose was to find ways to control salt accumulation, to use water more efficiently, and to increase the irrigated area. Questions investigated include the fate of the Aral Sea and what the drying up of this body of water would mean for economy of the region.
3. The study of means of reclaiming the swamps of the valley by using properly placed dams and diversion canals. The study also includes the hydroelectric potential.
4. Studies of the water in Lake Baikal for the specific purposes of reducing pollution, regulating the flow through the Angara River and finding new ways of putting this natural resource to better use (Gerasimov, 1968).

The naturalistic-pragmatic trend appears to be explicit in the constructive geography vis-a-vis the landscape science.

Economic Geography. The protagonists of economic geography, on the other hand, seemed to have believed in the idiographic tradition of the discipline, which is exemplified as follows: '... economic geography teaching should concentrate on a large body of concrete economic geography material, systematically represented on maps. The specifics of economic geography, the location of productive forces, and economic regionalization should be of central concern to teaching personnel. Most of the content of economic geography should be taught on a regional basis'. (Saushkin, 1966, 30–31).

It was during the initial period of Soviet Geography that the dichotomy between physical geography and economic geography became more pronounced. In 1926, O. A. Konstantinov pointed out: 'We reject the possibility for economic geography to be simultaneously part of two quite different systems of science (the geographic and economic sciences). We hold that ours is not only an economic discipline, but a purely economic discipline. In other words, we are for a complete break with geography, in the sense of complete outlawing of geographic approaches' (Saushkin, 1966, 23). This very statement provided a normative character to the discipline. The Soviet economic geography sought for the empirical generalisations which could be valid for specific time and place. Spatial laws governing man's economic behaviour are not universal and at the

[5] *Constructive Geography* refers to geography which is applied to the practical problem of building the socialist economy.

same time economic geography could not be a law-seeking discipline, as that of the physical geography. Empirical generalisations, prepared within the time-frame by the Soviet economic geographers, were built upon the uniqueness of economic phenomena.

Unity of Geography. This dichotomy in the Soviet geography, that physical geography and economic geography being clearly differentiated branches with separate concepts and methods seemed to have continued for a long time until the publication of V. A. Anuchin's book *Theoretical Problems in Geography* in 1960. In the words of Hooson, 'the book set out to investigate the theoretical basis of geography as a whole, through historical and philosophical analysis' (Hooson, 1962, 469).

In the book Anuchin elaborated his ideas with the aim of demonstrating the *unity of geography.* He devoted considerable space to a review of the history of geographical determinism, and analysed the theory from a Marxist point of view. He realised that the influence of geographical conditions on the life of society is beyond all doubt. It would be wrong to see in the assertion of this influence a manifestation of geographical determinism: the scientific worthlessness of the latter does not rest in the fact that it recognises the influence of the geographical factor on the development of society as a basic condition of this development, but in the fact that this influence was elevated to the position of a basic principle of development.

He saw the main fault of geographical determinism as lying in its mechanical nature. The environmentalists saw certain prime influences as producing certain consequences, whereas, in fact, the problem of cause and effect is a very complex one. It is this mechanical aspect of determinism (according to Anuchin) which Marxism rejects and not determinism in the broader sense of the word. On the basis of this assertion, Anuchin ruled out the dichotomy in the Soviet Geography. He attacked both an 'inhuman' physical geography and an 'unnatural' economic geography.

In fact, *the movement for the unity in geography* was initiated by N. N. Baranskiy in the 1920s when Stalin's policy of strong centralised control of the economy made the study of regional synthesis less relevant and seemed to negate the whole idea of an economic region 'as a major territorial production complex with a specialization of national significance'. Baranskiy successfully managed to buildup within the Marxist framework a strong case for a more realistic approach to the natural environment by Soviet geographers. He also demonstrated a very positive approach to the geographical environment vis-a-vis the unity in geography (Matley, 1966).

In the 1930s, the number of geographers who believed in the Baranskiy-led movement for 'the unity in geography vis-a-vis the idiographic tradition in geography', increased rapidly so as to give a legitimate and professional recognition to it. In 1934, two decrees gave official recognition to the professional support that had been given to Baranskiy. Though Baranskiy carried forward the movement for the unity in geography, but his belief in the unity seemed to be somewhat inclined towards economic geography, and implicit in his belief was some form of a dichotomy. He himself organised several seminars on economic geography and in one of the seminars, a discussion on the economic geography of the USA was held, in which several specialists were trained on this subject.

In spite of this, Baranskiy is credited with the movement for the unity in geography, and his idea was carried a stage further by V. Anuchin. He successfully challenged those who believed in the dichotomy. Later, Anuchin, for his methodological and philosophical stand on the unity of geography, had the full support of Baranskiy and Saushkin. Commenting on Anuchin's book, Baranskiy (1960, 107) pointed out: 'Anuchin comes out, on the one hand, against geographical determinism and on the other hand against a nihilistic underestimation of the environment, which he aptly calls geographical indeterminism. The geographical environment in itself does not determine the historical process of social production, but in a number of cases it exerts, in intermediary form, a decisive influence on the development of the economy'. Saushkin had, for some years before Anuchin's first methodological article appeared, been a supporter of a unified geography and of a more realistic appraisal of the role of the natural environment in economic geography. Saushkin's book *Introduction to Economic Geography* (1958) revealed his thinking on the major issues of methodological and philosophical aspects as being very close to those of Baranskiy and Anuchin.

GEOGRAPHICAL SYNTHESIS

The view that geography is a synthesising subject has always been basic to the philosophy of the discipline. *A belief in synthesis has been the teleology of geography,* the purpose that justified the activities of geographers (Halt-Jenson, 1981, 110). The critical geography in the Soviet Union has not, in fact, abandoned the idea of synthesis which is exemplified in the following statement of Anuchin: 'It becomes self-evident that production can only fully develop when it is based on geographical forecasting. Man's means of influencing nature have so increased that their application cannot continue without a study of their possible consequences. Geography, however, has not shown itself ready to solve this problem.... For this purpose one must have synthetic general geographical studies, the results of which would provide practical forecasts of the consequences of interference with natural processes, which inevitably are taking place. It is necessary to have a science concerned with the utilization of nature, a science which will connect natural science with the group of social sciences—what is needed is a geography without adjectives' (Anuchin, 1973, 61–62).

Anuchin is largely repeating the old view that geography must be a synthetic science. What is interesting is the justification for the Soviet geographer in asserting that geography can only add something useful to research in the future if it can make the geographical synthesis work (Holt Jenson, 1981, 140). Anuchin further stated: 'the society requires synthesis at different geographical levels. The synthesis needed, especially in economic planning, provides a basis for the development of the discipline and possibilities for new jobs for geographers, other than those employed by Universities (Anuchin, 1973, 45).

James and Martin point out that *'1970s is the beginning of a period of synthesis and unity',* but they foresaw a return of the tendency of split before the year 2000. It seems that geographers are brought together because of their concern with the geographical landscape and with specific segments of the Earth's space. The processes going on in a landscape are interconnected spatial systems, and the study of them separately is not the same as the study of them as systems. In the study of the economy, there is need for the full consideration of physical factors and there is also a need for the study of the changes in the landscape resulting from human action, whether planned or unplanned. The basis of physical geography is landscape science. The general and the particular complement and attempt to enrich each other. Especially in the field study does the essential unity of the geographical approach become apparent.

CONTEMPORARY TRENDS

Soviet geographers made significant contribution to urban geography. Though Baranskiy (1960) felt the need to develop a method of classifying 'cities very much on the line adopted by American geographers, but it was not until the 1960s that the Soviet geographers showed interest in urban studies. O. A. Konstantinov is credited with having played a leading role in defining the methodology and philosophy of Soviet urban geography (Konstantinov, 1968, 419). C. D. Harris, who reviewed the works of the Soviet geographers on urban studies, reported that the Soviet urban geography, during the 1960s, became highly 'model-based' as it involved the application of sophisticated quantitative techniques, followed by the development and testing of theories and models through hypothetic-deductive methods (Harris, 1970, 400).

Soviet geographers developed model-based paradigms to be adopted in the study of the management of natural resources. They showed considerable interest in the concept of territorial industrial complexes to further their thinking on national planning. An ecological point of view seems to have emerged in the Soviet Union in the 1970s that justifies the ecological paradigm not by empiricism, but on methodological grounds on the creation of a framework of concepts. Soviet geographers devoted more time to environmental and regional problems, and by referring to Anuchin's model, they attempted to demonstrate the practical value of an integrated geographical approach in their solution.

One of the contemporary trends is to emphasise the study of the theoretical foundation of recreational geography, known as constructive research. The constructive research requires to be based on a comprehensive understanding of the fundamental laws underlying the organisation and functioning of natural system. It is a model-based geography applied to the practical purposes of building the socialist economy within the framework of Marxist-Leninist tradition.

11

New Geography in the United States

HIGHLIGHTS

- Exploratory Geography
- Development of Professional Geography
- Challenge to the Davisian Model
- Ontography
- Other Professional Geographers
- Environmental Determinism and Possibilism
- Delimitation of Regions and Regional Geography
- Systematic Geography
- Current Trends

If tradition is to be believed, then it can be said that the heritage of geographical studies and teaching was established by Thomas Jefferson in the late eighteenth century. He made important contribution to the study of geography in the United States of America. Count Buffon's conception of the occupied space in America sparked off a dispute which led Jefferson to visit Buffon in Paris to protest at such a conception. However, Buffon agreed to recognise his error. It was this misconception of the occupied space in America that motivated Jefferson to publish his geographical study *Notes on the State of Virginia* in 1787.

The expedition of 1803–1806 to explore the Missouri river and its tributaries and to find the best route to the headwaters o f the Columbia River and thence to the Pacific Ocean, was organised under Jefferson's official patronage who directed Lewis and Clark to prepare a systematic record of observations concerning the nature of the country and its inhabitants. Thomas Jefferson asked them to prepare a specific report on the genre de vie of the Indian, and also to provide information about the ecological condition of the regions to be explored. This marked the beginning of an era of an exploratory geography in the United States.

EXPLORATORY GEOGRAPHY

An ecological viewpoint of traditional nature seemed to have been developed by **George Parkins Marsh.** Marsh, who had never studied geography as such, developed a novel approach to the study of the relations of man to the land: *he turned attention to man's effect on nature,* to the modifications of the organic and inorganic parts of the habitat that resulted from human action. This conception of Marsh was, by and large, based on Buffon's paradigm of 'man's destructive use of land'. He believed in induction from observation rather than deduction from theory, and his great work 'Man and Nature' (1864) was based on his observations of the damage done by human action in Turkey. He had been working on the theme of the modification of nature by human action long before he left for Turkey. He warned against interference with the spontaneous arrangement of the organic and inorganic world.

About Marsh's book Glacken writes, 'the technical ideas of the nineteenth century scholars converge with the seventeenth century conception of nature as a divinely designed balance and harmony, and they sought the restoration of disturbed harmonies.' Marsh's plea for conservation was not really effectively heard until the twentieth century. Glacken again writes, 'Our greatest debt to him is that he studied the technical works of European foresters, meteorologists, agronomists, drainage engineers and hydrologists, botanists and plant geographers and scientific travellers and for the first time placed the results of their investigations where they belonged in the forefront of human history' (Dickinson, 1969, 74).

Matthew Fontaine Maury is *credited with his generalised model of atmospheric circulation,* prepared in 1850. He was a naval officer and collected data on the winds and currents of the ocean. With the help of this data, he sought to develop a generalised picture of the surface winds of the Earth by identifying the prevailing winds and eliminating the temporary and local interruptions. He also devised new instruments for sounding ocean depths and was, therefore, able to produce the first map of the floor of the North Atlantic Ocean (Maury, 1855). Maury also supported the idea of an open polar sea, and explained his belief in the existence of such an open sea at great length. He believed that the warm Gulf Stream flowed under colder surface water and emerged again at the surface in the vicinity of the pole. However, much of his creativeness was sabotaged by Joseph Henry, director of the Smithsonian Institute in Washington who was trying to set up a network of weather reporting stations in the United States and Canada for the study of storms.

Maury seems to have established the tradition of the climatic studies[1] in the United States. Apart from climatic studies and generalisation, based on the availability of weather records and data, the US scholars also continued to concentrate on exploratory geography by organising large-scale surveys of the western territories.[2]

During 1867 and 1868, **John Wesley Powell** travelled to the Rocky Mountains and climbed Pikes Peak and Longs Peak. In 1869, he sailed down the Green River and the Colorado River through the Grand Canyon. In 1871, he carried out the second canyon trip, and reported on the Indian customs and language. He saw for the

[1] It was in 1870 that an official United States weather service was established. In 1891, it was transferred to the Department of Agriculture and in 1940 it was moved to the Department of Commerce. Based on the tradition of Maury, Samuel Forry prepared one of the earliest maps of the United States showing the distribution of temperature by making use of isotherms. In 1853, Lorin Blodget prepared a controversial temperature map for the United States. In 1854, James H. Coffin compiled a wind map of the Northern Hemisphere, and in 1875 the map was extended to include the whole world. His paper on the laws of atmospheric circulation went far ahead of Maury. In 1882, Elias Loomes of Yale prepared the first rainfall map of the Earth. William Ferrel provided a mathematical explanation of the way moving bodies must behave on a rotating sphere. His was the first explanation on the subject.

[2] Four such expeditions were organised during 1866 to 1879, under the supervision of Ferdinand V. Hayden, Clarence King, George Wheeler and John Wesley Powell. Hayden's team surveyed the mountain country of northern Colorado and Wyoming. He believed in the possibilities of settlements and defended the notion that rainfall tended to increase with settlement clearance. King's team surveyed the Uinta and Wasatch Mountains. He also identified and named the now-dry lake basins of the Great Basin in Utah and Nevada—Lake Banneville and Lake Lahonton. George M. Wheeler headed the US geographical surveys west of the one-hundredth meridian. He was directed not only to make topographic maps, but also to report on a land classification. He identified a four-fold classification of the landuse which included land suited for agriculture; land suited for timber; land suited for grazing; and arid land. Between 1874 and 1879, some 175,000 square miles were classified and mapped.

first time how a river could cut down through a mountain that was being raised up across its course. He *coined the term 'antecedent river'* for such a river. He also contradicted the hypothesis of Charles Lyell about the evolution of landforms in the western part of the United States which, to him, appeared to be the result of the action of the running water. Powell also developed the concept of the base-level—that there is a base-level below which rivers cannot cut a level determined by the surface of the body of water into which a stream flows. Later, this base-level concept was taken up by W. M. Davis.

It is interesting to note that a man, not trained in geographic tradition, could offer such an explanation which seemed to have given a professional touch to geographical heritage in the United States. It was in 1878 that Grove Karl Gilbert, who was recruited by George M. Wheeler, but preferred to work with Powell, developed his famous 'concept of grade', i.e. the equilibrium reached between slope, volume of water, velocity of flow and load of detritus (Gilbert, 1878). Thus, Powell and Gilbert set forth new paradigm in physical geography.

Powell also had an *ecological viewpoint* and his observations led him to believe that the only way to maintain the flow of water from the mountains to the bordering lowlands was to preserve the forest cover. He warned that the clearing of the forests on the watersheds would bring ecological disaster. In 1880, he became the second director of the US Geological Survey, and remained at the post till 1894. His monograph on the arid lands was a landmark in the contemporary tradition of physical geography.

The 'new' geographical tradition in the United States in the nineteenth century was further enriched by two European geographers, Agassiz and Guyot. **Jean Louis Rodalphe Agassiz,** a Swiss scholar, was appointed Professor of Natural History at Harvard in 1848. He developed the hypothesis of universal glaciation which had a profound effect on the young scientists of the USA. He turned attention of his younger colleagues to the study of glaciers and the effects of glacial actions in producing distinctive kinds of mountain landforms.

Another European geographer was **Arnold Guyot** who was appointed professor of physical geography and geology at the College of New Jersey (Princeton), and there he remained from 1854 to 1880. Guyot was an ardent supporter of the Ritterian tradition of teleological geography, which he attempted to establish in the United States. He remained unshaken by the philosophical revolution and the biological paradigm set forth by Charles Darwin. When Guyot retired in 1880, the very philosophical foundation of 'new geography' he taught was outdated and it had been largely discredited because of the shift in emphasis on the method of natural science, or to the hypothetic-deductive method that sought 'prime causes by deduction'. The teleological explanation was replaced by the mechanical explanation, and the Darwinian paradigm became the ruling paradigm in the United States. However, **William Libbey,** who succeeded Arnold Guyot as professor of physical geography, carried forward the tradition of Guyot in his studies of oceanography, especially of the relationship between the Gulf stream and the Labrador current.

Shaler was influenced by the Swiss scholar, Agassiz, especially his conception of a universal glaciation. Shaler learnt much from Agassiz about the observation of landforms and developed the habit of seeing man and his works as part of the landscape. Shaler attempted to elaborate the concept of man's role in changing the face of the Earth, which Count Buffon in the eighteenth century and George Perkins Marsh in the middle of the nineteenth century had developed. Shaler transmitted to Davis a vision of the Earth as the resource base on which the human habitats were dependent. He is often regarded as a geologist by training, but was a geographer by instinct in that he was always conscious of studying the Earth as the home of humankind. Shaler's was a positive view of geography that is reflected in his book *Nature and Man in America* published in 1893.

DEVELOPMENT OF PROFESSIONAL GEOGRAPHY

The 'new' geography in the United States, as founded by **Marsh, Maury, Powell, Gilbert, Agassiz** and **Guyot**, attempted to establish the tradition of field survey, with greater emphasis on induction from observations rather than deduction from theory. The year 1880 is a watershed in the history of American geographical scholarship because it marked the beginning of professionalism in the discipline, leading to the development of academic geography and making the subject a permanent independent university discipline.

The pioneer, who introduced professionalism in geography, was the geologist **William Moris Davis**, who was appointed instructor of physical geography in the Department of Geology at Harvard in 1878. He set forth the early paradigm for geographical study and helped in establishing some of the professional institutions. Davis was much influenced by the Harvard geologist; **Nathaniel S. Shaler**. Davis is believed to have acquired the habit of careful observation and logical argument from N. S. Shaler, when he was working with him as an assistant.

W. M. Davis pioneered 'professionalism' in the US geography. He helped in the establishment of some professional institutions. It was due to his effort that geography acquired an independent status as a discipline. It was usually associated with geology, but soon independent departments for geography were created in a number of universities, notable among them were Harvard, Yale, Chicago and Pennsylvania. These centres became the chief sources of the geographical ideas which were evolved in the scholarly competitive discussions from 1904 to 1914.

Davis was appointed assistant professor of physical geography in 1885 at Harvard and later promoted as professor. In 1889, he was named the Sturgis Hooper Professor of Geology at Harvard, and he occupied the chair until his retirement in 1912. He was one of the founders of the Association of American Geographers in 1904. He recognised that if geography was to become established as a professional field, it would be necessary to organise a professional society in which members could present their ideas. He was a visiting professor at the University of Berlin in 1909, and also at the Sorbornne from 1911 to 1912, and learned much about the methodological debates which were going on in Germany and France.

He learned *three distinctive habits of thought* from Shaler: (a) habit of careful field observation and making use of it in logical and impersonal argument; (b) habit of seeing man and his works as part of the landscape, not separate from it; and (c) clear appreciation of the importance of processes of change in explaining the varied features associated on the face of the Earth.

Davis was influenced by the Darwinian theme of change through time, and he *took evolution as his inspiration in the idea of the geographical cycle.* In his first paper (published in 1884) on the development of landforms, Davis referred to a 'cycle of life', and used such terms as birth, youth, adolescence, maturity, old age, second childhood, infantile features, and struggle to emphasise the analogy of an organism undergoing a sequence of changes in form through time (Stoddard, 1966).

The concept of *the cycle of erosion* was first formulated in 1884, and in 1899 it was presented in a revised form. Davis appeared to be indebted to Powell and Gilbert, because his model of the cycle of erosion included in it the base-level concept and the concept of graded profile. Davis preferred to use the word *'the geographical cycle' in place of the cycle of erosion.* This was in the nature of a model, an ideal representation of the development of landforms that would take place during the erosion by the rivers of a newly uplifted portion of the Earth. He postulated that after the upheaval no further up-or-down movements would occur, and that during the resulting cycle there would be no essential change of climate. To him, the geographical cycle refers to a period, during which a newly uplifted landmass is reduced to its base-level, or developed into a low featureless plain called the peneplane. The development of landforms proceeds through three stages—youth, maturity and old age. The whole cycle may begin again with another uplift, resulting in rejuvenation. The low mountains that stand above the general level of a peneplane, Davis called them 'monadrocks'. To him, the *landscape is the function of structure, process and stage* (Davis, 1899).

Davis also applied the evolutionary concept to the ideal sequence of forms in mountain regions sculptured by glacier, in islands bordered by coral reefs, and in a limestone region with solution caverns (Davis, 1928, 1930). However, he insisted that his ideal sequence was not to be considered as rigid, but rather that it could provide a theoretical framework in reference to which actually observed landforms could be described. This is what he called the *'explanatory description of landforms'*. He realised clearly that there would be an infinite variety of disturbances to the ideal sequence because almost every region observed in specific detail would constitute a special case.

Throughout his working life, Davis *emphasised the theme of orderliness and development through time, which he termed evolution;* but it is perhaps significant that he took his illustrations not from the species or the population, but from the individual. So successful was Davis in promoting this view that in his hands, geomorphology became more the study of the origin of landforms than of landforms themselves, and was thus readily channelled into the restricted field of denudation chronology. His geomorphology was deductive, time-oriented, and imbued with mechanistic notions of causation, deriving its uniformitarianism from Lyell and its theme of change through time at least partly from a simplified view of evolution (Stoddart, 1966). The geographical cycle of Davis provides a scientific procedure for describing the landscape phenomena in relation to their development over time.

Chorley (1962, 282–300) has studied the Davisian model of landscape development within the framework of the general systems theory. The Davisian model of geographical cycle, leading- to the landscape development, contains certain elements of closed system thinking—including, for example, the idea that upliftment provides initially a given amount of potential energy and that, as degradation proceeds, the energy of the system decreases until at the stage of peneplanation there is a minimum amount of free energy as a result of the levelling down of topographic differences.

The Davisian peneplain, therefore, may be considered as logically homologous to the condition of maximum entropy, general energy properties being more or less uniformly, distributed throughout the system with the potential energy approaching zero. The positive change of entropy[1] and connected negative change of free energy implies the irreversibility of events within closed systems. This again bears striking similarities to the general operation of the geomorphic cycle of Davis. In closed system, there is the inherent characteristic that the initial system conditions, particularly the energy condition, are sufficient to determine its ultimate equilibrium condition. This inevitability of closed system thinking is very much associated with the view of geomorphic change held by Davis.

CHALLENGE TO THE DAVISIAN MODEL

The so-called Davisian system of 'explanatory description of landforms' was subject to bitter criticism by certain German geographers in the years following his lectures in Berlin. Alfred Hettner, Walther Penck and Siegfried Passarge challenged the accountability and validity of the concept of geographical cycle. Hettner's criticisms first appeared in 1911 and his major attack

[1] Entropy: A measure of the amount of uncertainty in a probability distribution or a system subject to constraints.

appeared in his book published in 1921. Walther Penck's criticisms appeared in his posthumous work edited by Abrecht Penck.

Hettner criticised the *Davisian model as being an arbitrary scheme.* It neglected other contributory local factors operating in an area, such as weathering, variations in structure and lithology, and climatic factors. These factors not only account for climatic change during the cycle of erosion, but also account for differences in the combination of the processes of erosion and deposition in different milieus of the world. The Davisian concept is applicable only to the humid areas in Europe and Germany. The German scholars made extensive scientific observations of landforms in the arid regions of the Earth. They claimed that the development of landforms through the operation of the wind and rain and weathering, and the resultant processes of erosion and deposition, vary in different climates and with changes in climate in one and the same area. The soil and vegetation cover have repercussions on surface changes.

Hettner and Passarge *rejected the generalised system that was deductively applied* so as to portray, on this limited explanatory base, the surface of an area. They argued that the geographic problem was to characterise landforms in their locally repetitive forms and to explain them in terms of all the forces operating on them. Davis did not study the landforms as they occurred in reality, but defined and explained them deductively within the framework of a theoretical system which, it was argued, contained fundamental errors. Hettner found the Davisian model too rigid and too specific to fit real world conditions as they exist under unique situation, and also found them to be too misleading.

Passarge also criticised the method of explanatory description of Davis, and insisted on a purely empirical treatment of the landform base. But Davis pointed out that Passarge's model would create numerous inconsistencies. For Davis, the natural history of landforms was the core of geographical work to which other elements of landscape could be related; for Passarge the goal was the treatment of landscape, including the many elements of diverse origin associated with areas in which the landforms were basic, but not necessarily important.

As early as 1912, Passarge had emphasised the description of the present landforms as functioning factors in the region (or landscape). Lecturing at the Sarbornne, Davis stated that 'if it proves very difficult to work out the development of the landforms, so that, in the problems of the past one loses sight of the landscape, the geographer will go better to describe the land simply with the help of the older orographic presentation' (Hartshorne, 1976, 398–90).

ONTOGRAPHY

W. M. Davis is credited with the *term 'ontography' to denote 'the organic half of geography*—a conception he took to a large extent from Ritter and which had to be contrasted with and derived from physiography. He points out: 'In its present modern phase, geography is essentially concerned with the rational correlation of the items that fall under its two parts: on the one hand, the items of inorganic conditions that constittite the physical environment of living forms, and on the other hand, the items of organic response made by living forms to their environment. Thus understood, geography is concerned with the combination of physiography and ontography—that is, with the correlation between inorganic environment and organic response' (Davis, 1902, 1903).

Davis began to seek an even larger conceptual structure for geography. He *attempted to seek cause and effect generalisation,* 'usually between some elements of inorganic control and some elements of organic response'. He pointed out: 'Any statement is of geographical quality if it contains a reasonable relation between some inorganic element of the Earth on which we live, acting as a control, and some element of the existence or growth or behaviour or distribution of the Earth's organic inhabitants serving as a response. There is, indeed, in this idea of a causal or explanatory relationship the most definite, if not the only unifying, principle that I can find in geography' (Davis, 1906). The positivist approach appeared to be explicit in Davis' ontography, as it concerned itself with factual content or empirical questions.

The Darwinian idea of natural selection and adaptation formed the basis of statement regarding environmental determinism by Davis (1906), who identified the core of geography as the relationship between the physical environment (as the control) and human behaviour (the response).

Davis was one of the greatest geographers of the late nineteenth century and early twentieth century, whose works had already influenced the European scientific thought, totally in Germany and France. He set forth a tradition which continued to dominate the American geographic scholarship for several decades. His ontography almost remained unchallenged in America and outside. His ontography paradigm, which was subject to criticism in the earlier period, developed into a major paradigm in the 1960s when major emphasis was attached to the ecosystem as a geographic principle and method.

He had a number of colleagues and friends at Harvard who became his followers and carried forward his ideas and tradition to different parts of the country and outside. The Davisian tradition of physical geography was carried forward by **Douglas W. Curtis F. Marbut** and **Robert Dec Ward**. Johnston made significant contributions to the evolution of shorelines. Marbut applied the Davisian paradigm in his soil studies, and classified soils as young or mature. He is credited with the English translation of Dokuchaiev's major work on the zonal soils of the world. Ward taught climatology at Harvard for 40 years, and became the president of the Association of American geographers in 1917. In 1908, he published his book on the effect of different kinds of climate on human life. Thereafter most of his efforts were directed to the organisation of material or regional climatology.

OTHER PROFESSIONAL GEOGRAPHERS

It would be relevant to discuss the works of the two most important geographers of the contemporary United States, who made significant contributions to the development of professional geography in the country and carried forward the 'heritage' of new geography, founded by **W. M. Davis**. They were **Mark Jefferson** and **Isaiah Bowman**.

Mark Jefferson. 'None of Davis' students did more to promote and improve the teaching of geography in the United States than Mark Jefferson, who was professor of geography at the Michigan State Normal College in Ypsilanti from 1901 to 1939. Jefferson deserves a special place in the history of geography not only because of the enthusiasm he kindled among his students but also for the many contributions to the conceptual structure of geography that came from his pen' (James and Martin, 1981, 296). He was at Harvard during 1897–99 where he studied geography with Davis. He was given the responsibility to teach new geography to some 1,300 Cuban school teachers.

He was opposed to the systematic approach in geography, but favoured some form of regional paradigm. He prepared reports, based on field surveys, on European colonies in Chile, Argentina and Brazil in 1921, 1924 and 1926. *He wanted geography to emphasise on 'man on the Earth', not 'the Earth and man'*. In fact, he was not influenced by the Darwinian paradigm of determinism, but favoured the concept of 'indeterminism', attaching much importance to man's creative ability. Thus, he was opposed to Davis's assertion of 'environmental control and man's response'.

On the *nature of geography,* Jefferson pointed out: 'Some one has said that anything that you can put on a map is subject matter of geography. That is what I would call locational or distributional geography.... But geographers are contemplative persons who cannot be satisfied with so meagre an account of the subject. The nature of geography is the fact that there are discoverable causes of distributions and relations between distributions.

We study geography when we seek to discover them.... But there is an art of geography—the delineation of the Earth's features and inhabitants on maps—cartography, and a science of geography, which contemplates the fact delineated and seeks out causes of the form taken by each distribution and its relationship to others' (Martin, 1968, 319–21). Jefferson's was apositive view of geography which is reflected in his much quoted statement that *geography concerned itself with man:* 'where they are, what they are like, and why they are there'.

He was the pioneer, with such papers as that of 1909 which dealt with the *Anthropogeography of Some Great Cities.* It was followed by numerous other papers in 1913, 1916, 1917, and 1920, dealing with the distributions of Japanese, British and American urban peoples. The early writers, Ratzel in Germany and Reclus in France, gave many plans of cities in their publications, but they rarely gave much attention to the principles underlying urban agglomerations, which Jefferson successfully referred to in his papers. According to him, 'a country's leading city is always disproportionately large and exceptionally expressive of national capacity and feeling'—which he called the *'Law of the Primate City'.*

In the period between 1909 and 1941, he had 31 papers published in the *Bulletin of the American Geographical Society* and its successor, *The Geographical Review.* This is by far the largest number of professional articles published in these prestigious periodicals during this time by any one scholar. Some of the articles were major contributions to the concept of geography. As one of the first American students of population distribution and of urban structures, he was an innovator and trail-blazer.

Isaiah Bowman. He was one of the outstanding geographers of the American geographical scholarship of the early twentieth century. He made significant contributions to the conceptual structure of geography, which reflected explicit *positivist approach.* He believed that the scientific status of knowledge had to be guaranteed by the direct experience of an immediate reality, which required a particular conception of causality in which causal relations amounted to regular associations with phenomena. This philosophical assertion very often inevitably belonged to one of the methodological precepts of Comte's 'le reel'. With this approach, he moved on to work for the 'new geography' initiated by W. M. Davis. To some extent, Bowman's contribution to the conceps of geography was far greater than that of his predecessors.

Bowman was graduated at Harvard and Yale. After completing his undergraduate work at Harvard with Davis in 1905, he received an appointment in the Department of Geology at Yale. He led a number of scientific expeditions to various parts of South America from 1907 to 1913. In 1907, he landed at Iquique, Chile and made his way across the Atacama Deserts to the Bolivian Altillano and thence to the forested eastern slopes of the Andes. He returned by the way of Peru, through Cuzco, and Arequipa, and Mollendo. Based on this scientific expedition and observations, Bowman prepared his doctoral thesis entitled 'The Geography of the Central Andes'. It was a landmark in the contemporary conceptual structure, of geography in the United States. In 1911, he went on the Yale Peruvian Expedition headed by Hiram Buigham. In 1913, he went for a third Peruvian expedition, which was financed by the American Geographical Society.

These expeditions, in fact, led him to discard the Davisian paradigm which was the ruling paradigm for the American geographic community. The results of these expeditions were published in two books, *The Andes of Southern Peru: Geographical Reconnaissance Along the Seventy-third Meridian* (1916) and *Desert Trails of Atacama* (1924). In these books, he sought for an effective way to generalise the many detailed observations regarding the terrain and people which he had collected in his expeditions. His imaginative innovation was his application of 'regional diagrams' in his books, which in itself set forth a distinct paradigm for the research community.

Bowman remained at Yale until 1915 and from 1915 to 1935 was the director of the American Geographical Society. He was engaged on research after the First World War at the Peace Congress at Paris. From 1935 to 1948, he was President of Johns Hopkins University. It is equally important to know that Bowman during this period made significant contributions to the conceptual structure of geography. He attempted to apply geographical methods to the study of practical problems during and after First World War that tended to strengthen the conceptual structure of 'Applied Geography' not only in America, but outside also.

At Paris, Bowman was appointed as the chief territorial specialist of the American Commission to Negotiate Peace. There he applied his geographical skill to study the politico-territorial problems of Europe, arising out of the territorial disorganisation. He treated the territorial problems within the framework of the regional paradigm, as reflected in his *The New World: Problems in Political Geography* (1921). The book dealt with particular problems of particular regions, with adequate description of the local setting and historical backgrounds so that people could know about the resultant effects of the territorial reorganisation. Extraordinary effective use of maps in the book has made it a unique contribution in the field of political geography (Hartshorne, 1976, 248). The basic notion of the book, concerning the post-First World War boundary problems of Europe, and the application of regional paradigm to find out remedial measures, inevitably pointed towards the explicit nature of the positivist approach in political geography. It was a book on 'applied' political geography. He was probably the first geographer to have introduced empiricism in politico-geographical studies in the USA.

Bowman *pioneered boundary studies* in the United States. He was asked by the Secretary of State to arrange for a study of Guatemala-Hondorus boundary problems and suggest a solution. The survey of the disputed region was made in 1919–1920 and the solution to the problem was submitted to the Secretary of State, which was accepted only in 1933.

From boundary studies, Bowman moved on to the *study of pioneer belts.* In 1925, he turned attention to problems relating to the thinly populated areas on the margins of settlement. Pioneer belt studies are not solely concerned with the possibilities of new settlements; they may also be concerned with the need to withdraw from less favourable places. He proposed to study the pioneer movements all around the world and to identify certain general conditions, not only the physical conditions considered favourable, but also the attitudes and objectives that led people to become pioneers and the economic, social and political institutions that could best support pioneers. He also proposed to investigate the particular and unique conditions in specific pioneer areas, knowledge of which would be essential to the formulation of policy. Bowman's proposal ranged widely over all the fields of the social sciences and was essentially multi-disciplinary in approach and concept. His book *The Pioneer Fringe* was published in 1931, which studied the nature of problem and offered examples from the western United States, Canada, Australia, southern Africa, Siberia, Mongolia, and South America.

Bowman, who was earlier a follower of the doctrine of environmental determinism, an ardent supporter of the Davisian heritage, ultimately championed the cause of the 'possibilist philosophy' and *became a possibilist geographer.* The change was a result of his Peruvian expeditions and the pioneer belt studies. He offered structural explanations on possibilism, based on his empirical generalisations. On human activity, he pointed out: 'As knowledge of the world spread, the associations of events or conditions with place widened, they became more complex, they. had less or more significance with respect to mankind. The potato and maize plants were unknown to pre-Columbian Europe. Their discovery raised the question, are they useful to the rest of humanity and where can they be grown? The whole known world was in a sense resurveyed by the rough processes of trial and error and the result has been astounding. These two. plants largely changed the economy of Europe. The soil had not changed; man had gained a little more knowledge of it through a new plant. An element of one environment had been added to the elements, long fixed, of many other environments' (Bowman, 1934).

On the role of environment, he said: 'The geographical elements are fixed only in the narrow and special sense of the word. The moment we give them human associations they are as changeful as humanity itself. That is why modern geography has so definitely steered away from determinism and towards a study of types of actually working regional combinations of human and environmental conditions.... Earth facts do not determine the form and nature of human society in development. They condition it. New Earth facts are continually being discovered and old Earth facts given new significance as human knowledge, thought and social action develop. The relations are reciprocal'. However, he also pointed out: 'While the physical laws to which mankind responds are variable in their application and in degree of effect, yet this is also true that all men everywhere are affected to some degree by physical conditions' (Bowman, 1934, 106).

Bowman's possibilist philosophy appears to have incorporated the neo-Kantian philosophy that sought for a symbiosis for the organising freedom and capacity of man bounded by the realities of the mechanisms of the environmental realm. In fact, he laid the foundation of a new professional geography in the United States which was markedly different from that of Davis and others.

ENVIRONMENTAL DETERMINISM AND POSSIBILISM

These competing paradigms appear to have had greater impact on the contemporary school of American geography. Davis was an ardent supporter of the paradigm of determinism, while Isaiah Bowman championed the cause of the paradigm of possibilism. Mark Jefferson spoke in favour of 'indeterminism' which, epistemologically, formed a part of the possibilist philosophy.

Ideas of natural selection and adaptation formed the basis of statements regarding environmental determinism that tended to identify the relationship between physical environment as the control, and human behaviour, the response. Davis' ontography belonged to this category of the paradigm of determinism.

Ellsworth Huntington. He studied with Davis at Harvard and was associated with Bowman at Yale. He was one of the forerunners of the deterministic concept vis-a-vis the Davisian tradition. During his field studies in Asia between 1903 and 1906, he found much evidence to support the idea that there had been a world-wide progressive desiccation since the glacial period. However, he recognised that there were cycles of cool-wet and hot-dry periods of varying length. He *stressed the importance of climatic determinism.* In his book *The Pulse of Asia* (1907), he attempted to correlate the periods of drought with historical dates. He concluded that the migration of nomadic peoples from Central Asia, that led to the Mongol conquests of India and China and the invasions of eastern Europe in the thirteenth century, could be explained by the drying up of the pastures on which the nomads were dependent. He formulated empirical generalisations based on these deductions.

Huntington's generalisations on climatic determinism seem to be identical with those of Montesquieu, especially the one which he developed in his second book *Civilisation and Climate* (1915). The generalisation he formulated in the book is that man's civilisations could only develop in regions of stimulating climate and that the monotonous heat of the tropics would forbid attainment of the higher levels of civilisation. Obviously, he referred to the cool temperate oceanic regions because he found these regions highly civilised and developed.

Though Huntington (1915, 68) emphasised the importance of climatic determinism, but urged that it be 'put into relation not only with the results of other factors of the physical environment, but also with the opposite side of the shield, i.e., with the purely human factors'. He even suggested that diet was as important as was climate as an explanation of human energy. His was clearly a normative theory and his generalisation might be called *'pragmatic determinism', or some kind of 'scientific' determinism.* Huntington's pragmatic determinism appears to be very much in line with G. Taylor's 'stop-and-go-determinism', and V. Anuchin's 'neo-determinism' as well.

Ellen Churchill Semple. One of the most important protagonists of the paradigm of determinism was Ellen Churchill Semple who studied with Ratzel at Leipzig in 1891–92 and again in 1895, and carried forward his ideas into the United States. In fact, she was greatly stimulated by Ratzel's different approaches to the paradigm of determinism, reflected in his *Anthropogeographie* where he attempted interpretations of so-called geographic influences on the course of history, as well as his modification emphasising the culture of human groups rather than the physical Earth.

Ellen Semple published her first professional article in 1891 dealing with the Appalachian Barrier in American history, and in 1901, she published a paper based on her own field observations on the highlands of eastern Kentucky regarding the results of isolation on the settlers of that area. Her first professional book *American History and Its Geographic Conditions* appeared in 1903. With its publication, Semple's professional status in the horizon of the US geographic scholarship came to be recognised. She offered a mechanical approach to explain the so-called geographic influences on the course of American history. In the preparation of the book, Semple seems to have relied on pre-existing causes to explain the observed phenomena of American history.

Semple is known for her book *Influences of Geographic Environment* (1911), which is an explicit bowdlerisation of Ratzel's own lectures. About the method in the book, she points out: The writer's own method of research has been to compare typical peoples of all stages of cultural development, living under similar geographic conditions. If these peoples of different ethnic stocks but similar environments manifested similar or related social, economic, or historical developments, it was reasonable to infer that such similarities were due to environment and not to race. Thus by extensive comparison, the race factor in these problems of two unknown quantities was eliminated for certain large classes of social and historical phenomena (Semple 1911, 69).

Then she pointed out *man to be 'a product of the Earth's surface', 'a child of the Earth, dust of her dust',* which had 'entered into his bone and tissue, into his mind and soul'. But at the same time, she insisted that 'geographical factor and influence' shuns the word determinants and speaks with extreme caution of geographic control. Her book revealed numerous examples which sharply contradicted her principles. The examples she used to explain her propositions were frequently so crude that they effectively constituted 'a metaphysical assertion' rather than 'a basic hypothesis'. It may be argued that claims of this sort failed to recognise that determinism was not a universal hypothesis capable of empirical validation, but rather a logical structure of cause and effect which needed to be translated into the scientifically credible vocabulary of necessity and sufficient conditions (Harvey, 1969).

However, James and Martin (1981) point out two basic characteristics of Semple's viewpoint contained in the book. (a) First, that under certain circumstances there is a tendency for people to behave in predictable ways—which is a v,erbal approach to probability theory. (b) Second, her 'islands of ethnic expansion and islands of ethnic retreat' offer an important modification for the contemporary theory of innovation dispersal. *Positivist determinism* appears to be expressive in her later works on the geography of the Mediterranean, entitled *The Barrier Boundary of the Mediterranean Basin* and *Its Northern Breaches as Factors in History* (1915, Annals) and

The Geography of the Mediterranean Region: Its Relation to Ancient History (1931).

Albert Perry Brigham. Another protagonist of "the deterministic concept of the Davisian tradition" was Albert Perry Brigham who worked with Shaler and Davis at Harvard in 1891-92. He was on the faculty of Colgate University until his retirement in 1925. He held several posts in the Association of American Geographers and in 1914 he became the President of the Association. His book 'Geographic Influences in History' appeared in 1903, the same year in which Semple's book on the same subject was published. Both of them had identical views with regard to the geographic control. In fact, Brigham had a pragmatic approach towards geographic influences and control, but he rejected the climatic determinism.

He sought for empirical generalisations to test the scientific validity of the paradigm of determinism He pointed out: "Our goal is broad generalisation. But the formulation of general laws is difficult, and the results insecure until we have a body of concrete and detailed observations. Detailed investigation of single problems ... must for a long time prepare the way for the formulation of richer and more fundamental conclusions and general principles than we have yet been able to achieve" (Brigham, 1915, 24-25). He was of the view that under certain favourable conditions, the inhabitants tend to behave in a predictable manner, and there the environmental control gradually becomes ineffective (Brigham, 1903). Hartshorne (1976) observed that in spite of Brigham's caution with regard to the environmental control and human response, and the need for empirical generalisation to test the alleged control and influences, many geographers continued to draw plausible, but unverified, conclusions from their studies.

Rollin D. Salisbury, who was the Chairman of the Department of Geography at Chicago from 1903 to 1919, was sceptical about the paradigm of determinism. He rejected the idea of simple cause and effect relation between the physical Earth and the human response. To him, physiography was the scientific study of the stage setting on which the human drama unfolded, but, he said, the relation of the stage setting to human action was not a causal one.

Harlan H. Barrows, on the other hand, *opposed the paradigm of determinism,* offering a very careful statement, reversing the usual form. He speaks of geography as "the study of man's adjustment to the natural environment" or 'human ecology'. He considered that geography should concentrate on the study of man's associations with his natural environment, and that it should also concentrate on such themes as lead towards synthesis. Adjustment was not caused by the physical environment but was a matter of human choice (Barrows, 1923). He insisted that the physical conditions must be studied in relation to man. Such a view was first developed by Kirchhoff in Germany and Ratzel incorporated it in the second volume of his Anthropogeographie. Barrows' was essentially *"a pragmatic possibilism"* that sought to embrace the organising freedom of man, bounded by the realities of the physical environment. His was a normative theory.

Carl O. Sauer (1924). He was an *activist of "the possibilist movement"* in the United States during the inter-war period. He was influenced by the concept of "Kulturlandschaft" of the German geographer Otto Schluter who focussed on the study of the human habitat as the creation and work-place of human groups, whose relations are conditioned by their group habitat and heritage. Sauer called for such field studies which aimed at exposing the areal expression of man's activities – Man, behaving in accordance with the norms of his culture, performs works on the physical and biotic features of his natural surroundings and transforms them into the cultural landscape. The design of the landscape includes the features of the natural area, and the forms superimposed on the physical landscape by the activities of man, the cultural landscape. To him Man is the latest agent in the fashioning of the landscape.

Sauer insisted that instead of going into the field with a set of *a priori* principles concerning the effect of the physical environment on man, one should seek to observe the facts and then draw conclusions from them. He insisted that no field of study can be defined in terms of a single causal hypothesis that would commit the students to a particular outcome of an investigation in advance. To go into the field to look for influences or evidences of control exerted by the physical Earth (condition), is to accept a single dogma. He did not deny the possibility of environmental determinism in specific cases, but insisted that the concept of influences should be exposed to objective testing (Sauer, 1927).

Carl Sauer (1925) *represented geography as 'a science that finds its entire field in the landscape'.* According to him, the systematic organisation of the content of the landscape proceeds with the repression of *a priori* theories concerning it, so that 'geography relies on a purely evidential system, without prepossession regarding the meaning of its evidence'. It is concerned only to establish 'the connections of phenomena in the visible landscape, and these connections are ones of spatial association and not, definitely **not,** of some hidden causality. Geography could now be established as a positive science'. In effect, his *'morphologic method'* was simply a restatement of Comte's designation of 'Ie reel' as the only appropriate domain for scientific inquiry (Gregory, 1978, 28–29).

Sauer's landscape paradigm appears to be an intrinsic part of the possibilist philosophy, the focus of which was the study of on-going processes leading to landscape change up to, and including, the present and the beginning at the pre-human stage of occupance. The human geographer must be obliged 'to make cultural processes the base of his thinking and observation' (Sauer, 1941). Physical environment is conceived not as the principal source of explanation for the condition of the humanised cultural landscape, but rather as the setting and environmental patterning upon which agents sculpt their peculiar ideas in ways conditioned by theif cultural make-up. He approached the human use of the Earth ecologically and while he was critical of the paradigm of a determinism represented in the United States by Ellen Semple, he was himself strongly influenced by ecological concepts of culture traceaple to the same biological sources (Johnston, 1986, 87).

Derwent Whittlesy. His concept of *'sequent occupance'*—'the view of geography as a succession of stages of human occupance which establishes the genetics of each stage in terms of its predecessor'—represents anti-thesis of environmental determinism, or sharply contradicts the paradigm of determinism. In a sense, *the concept of sequent occupance represents a distinct form of 'pragmatic possibilism',* as it explicitly refers to the French concept of genre de vie. 'Sequent occupance' tends to recognise the fact that any change in the genre de vie (way of living) of a region inevitably leads to the reappraisal of the significance of resource base. Sequent occupance studies the ways in which each culture uses a region in its own way. This is demonstrated in America where most regions experienced a sudden change from the Indian to the European culture, and in many parts of Europe which progressed from agrarian to industrial cultures. Sequent occupance stresses the stages in the development of a region, not through studies of local differentiation as a result of the long continued and largely undisturbed interplay of man and nature over the centuries (as in the France of Blache), but by emphasising how easily shifts in regional character can take place.

A large number of studies that appeared during 1920s and 1930s made use of the method of 'sequent occupance', and in this regard special mention must be made of **Robert Platt.** He was a member of the Department of Geography at Chicago from 1919 to 1957. In his works *Problems of Our Time* (1946) and *Environmentalism vs. Geography* (1948), he rejected the paradigm of determinism. In fact, he became one of the advocates of the possibilist. philosophy, and his generalisations were based on empirical observations.

Geography in the United States, during the inter-war period, however, witnessed the publication of large number of studies which emphasised over the paradigm of determinism. It was probably because of the changing 'genre de vie' of American people that brought about changes in the landscape through 'innovation diffusion' in the field of technology and science.

James and Martin (1972, 326) observe: 'Out of these studies of sequences of development of cultural landscape vis-a-vis settlement, certain principles began to emerge. One was the principle that same physical conditions of the land could have different meaning for people with different attitudes towards their environment, different objectives in making use of it, and different levels of technological skills. In agricultural areas it was clear that slope had one meaning for the man with a hoe and quite another for the man with a tractor-drawn plow. It might be that the introduction of machinery could reduce the arable area of a country or change the kind of soil that was to be considered desirable. People with one kind of culture might concentrate their settlements on flattish uplands, whereas another people in the same area might concentrate in the valleys. Water power sites that were useful for the location of industries before the advent of steam lost that attraction when power came from the other sources.'

DELIMITATION OF REGIONS AND REGIONAL GEOGRAPHY

In the early years of the twentieth century, a group of American geographers, trained in the Davisian tradition, were attracted by the British approach to the delimitation of regions. Many concluded that the highest expression of geographic research was regional geography. The tradition of delimitation of regions, in fact, started long back when John Wesley Powell in 1896 divided the country into physiographic regions. He divided the nation's territory into 16 regions, some with several sub-divisions. Davis also prepared a similar map in 1899. But the professional touch to the approach of delimitation of regions began with W. L. G. Joerg in 1914. It was he who used the term *'natural regions'* which he defined as 'any portion of the Earth's surface whose physical. conditions are homogeneous'.

Joerg's concept of natural region was highly acclaimed in the United States. N. M. Fenneman also published a study of physiographic divisions of the United States in 1914. His map incorporated such features which were not found on the 21 maps that Joerg compared. However, the map of the physiographic regions with a detailed statement of the characteristic of each region was published in 1916 and included a folded map on the scale of 1/7,000,000. C. R. Dryer moved a step further when he incorporated into his scheme the concept of physical control and human response with a view to attempt at drawing of regions. He proposed that the best way to identify natural regions is to measure the economic functions of each. Economic activities fit into these regions, and a study of the economic activities provides the best guide to the delimitation of natural regions. He called these regions as 'natural economic regions'.

For some little more than one-and-a-half decade since 1915, the American geographers were involved in different kinds of studies, other than regional studies. The ideas and methods of regional geography reappeared a little later in the United States in the late 1930s. Like environmental determinism and possibilism, the regional study too was an attempt at generalisation, but at generalisation without structured explanation, and thus of a different type from the increasingly discredited law-making attempts of the previous paradigm (Johnston, 1983, 42). In the late 1930s, two non-geographers published a major survey of American regionalism (Odum and Moore, 1938).

But it was **Richard Hartshorne** who is *credited with a definitive statement of the 'regional' paradigm.* His monograph *The Nature of Geography: A Critical Survey of Current Thought in the Light of the Past* (1939) published under the patronage of the Association of American Geographers, was a landmark in this direction. Hartshorne pointed out that there was much debate among American geographers during the 1930s about the nature of their discipline. He was concerned about the tone and content of the debate and in 1938 submitted a paper in the *Annals* as a contribution to the philosophical discussion. He went to Europe for field work on boundary problems, but it was interrupted by the political situation, and he concentrated on reading the works of German geographers on the nature of geography. He used this field work and his study to extend his 1938 paper, adding the subtitle to it. The result was a 'paper' of 491 pages which became the major philosophical and methodological contribution to the literature of geography in English then available.

To Hartshorne, the *focus of geography is on areal differentiation.* It is the mosaic of separate landscapes on the Earth's surface. To him, geography is 'a science that interprets the realities of areal differentiation of the world as they are found, not only in terms of the differences in certain things from place to place, but also in terms of the total combination of phenomena in each place, different from those at every other place'. It provides 'accurate, orderly and rational description and interpretation of the variable character of the Earth's surface'.

Geography seeks to acquire a complete knowledge of the areal differentiation of the world, and therefore discriminates among the phenomena that vary in different parts of the world in terms of their geographic significance, i.e. their relation to the total differentiation of areas. 'Phenomena significant to areal differentiation have areal expression—not necessarily in terms of physical extent over the ground, but as a characteristic of an area of more or less definite extent'. The principal purpose of geography, according to this tradition, is synthesis, an integration of relevant characteristics to provide a total description of a place—a region—which is identifiable by its peculiar combination of those characteristics.

There is a *close analogy between geography and history;* the latter provides a synthesis for temporal sections of reality, whereas the former studies the spatial sections of the Earth's surface of the world. Geography, therefore, studies the world, seeking to describe, and to interpret, the differences among its different parts, as seen at any one time, commonly the present time. This field it shares with no other branch of science, rather it brings together in this field parts of many other sciences. These parts, however, it does not merely add together in some convenient organisation. The heterogeneous phenomena which these other sciences study by classes are not merely mixed together in terms of physical juxtaposition in the Earth surface, but are causally interrelated in complex areal combination. Geography must integrate the materials that other sciences study separately (Hartshorne, 1976, 460).

Hartshorne also indicated the methodology to be used for this integrating science aimed at orderly description of the Earth's surface. To him, the ultimate purpose of geography, the study of areal differentiation of the world; is most clearly expressed in regional geography, and so the accepted procedures were necessary for regional identification. 'Regions are characterised by their homogeneity on prescribed characteristics, selected for their salience in highlighting areal differences' (Hartshorne, 1959).

Two types of regions were identified: the *'fomal region' (or uniform region)* in which the whole of the area is homogeneous with regard to the phenomenon or phenomena under review, and the *'nodal' or 'functional region'* in which unity is imparted by organisation around a common node, which may be the core area of a state or a town at the centre of a trade area. 'Preparation of the materials for a regional synthesis required collection both from other sciences specializing in certain phenomena (though usually not their areal patterning) and from the topical systematic specialisms which complemented, but which were eventually subsidiary to, regional geography' (Johnston, 1983, 44–45).

Regional paradigm was at the forefront of the American geographic scholarship in the 1940s and systematic studies were the procedures of information for that enterprise. 'Regional geography in the traditional sense seeks to bring together in an areal setting various matters which are treated separately in topical geography'. Urban geographers studied towns because they 'constitute distinctive areas', in line with the regional concept; political geographers studied the functions and structures of an area, as a region homogenous in political organisation, heterogeneous in other respects (Hartshorne, 1954). In defining the 'new' field of social geography, Watson (1953) saw it 'as the identification of different regions of the Earth's surface according to associations of social phenomena related to the total environment'.

Each of these topical specialisms, therefore, sought to produce its own regionalisation. It had its link with relevant systematic sciences—social geography with sociology, for example. The very differentiating factor between the two was the geographer's focus on the region, the single-attribute region of his specialism and multi-attribute region in the synthesis of his work with that of the other to produce regional geographies. Given this focus on the region, it is not surprising that the literature of the paradigm contained many contributions discussing the nature and delimitation of such homogeneous areas, for virtually every region was in effect a generalisation, complete homogeneity being very rare over more than a small area' (Johnston, 1983, 45–46).

Hartshorne *emphasises 'generic principles' and avoids mention of causal mechanisms.* In the minds of many geographers, causality was still uncomfortably close to the discredited thesis of environmental determinism. Hartshorne was therefore anxious to present geography as a science concerned with the *functional integration of phenomena* rather than with the processes of particular kinds of phenomena. This reaffirmed its commitment to the discovery of spatial associations: geography was a 'naive science' which looked at 'things as they are actually arranged and related'. Such a prospectus echoed the empiricist intentions of the German School of History which sought to reconstruct the past, and this was hardly surprising in an account which relied heavily on an exigesis of the German intellectual tradition.

Although geography could now operate at a non-phenomenal level, therefore it continued to be confined to the demonstration of regularities—or 'simple correlation'—and was not permitted to disclose causalities. This obliged systematic geography to provide clearly-defined generic principles. Hartshorne conceded that to this extent geography had to be 'a nomothetic science', but its essential task remained the idiographic one of locating these principles in specific regional contexts and describing their interlocking configurations (Gregory, 1978, 30–31).

Schaefer-Hartshorne Debate

The Hartshornean tradition of regional paradigm was subject to a bitter criticism in the United States itself, and it was raised by F. K. Schaefer in his paper published posthumously in 1953 in the *Annals*. The paper is often referred to by those who seek the origins of the 'quantitative and theoretical revolution'. Schaefer was originally an economist; he joined the group of geographers teaching in the Economics Department of the University of Iowa after his escape from Nazi Germany.

In this paper, **Schaefer** attempted to criticise the exceptionalist claims made for regional geography and to present the case for geography adopting the philosophy and methods of the positivist school of science. He elaborated the nature of science, and then defined the peculiar characteristics of geography as a social science. He argued that 'to claim that geography was the integrating science which put together the findings of the individual systematic sciences was arrogant, and that in any case its products were somewhat lacking in … startlingly newer and deeper insights' (Schaefer, 1953, 232). A science is characterised by its explanations and explanations require laws. To explain the phenomena one has described, means always to recognise them as instances of laws.

In geography, the major regularities which are described refer to spatial patterns and hence geography has to be conceived as *the science concerned with the formulation of the laws governing the spatial distribution* of certain features on the surface of the Earth. It is these spatial arrangements of phenomena, and not the phenomena themselves, about which geographers should be seeking to make law-like statements. Geographical procedures would then not differ from those employed in the other sciences, both natural and social. Observation would lead to a hypothesis—about interrelationship between two spatial patterns, for example—and the hypothesis would be tested against large number of cases, to provide the material for a law if it, were thereby verified (Schaefer, 1953).

Schaefer observed *exceptionalism in Hartshorne's regional geography* which, according to him, claims that geography does not share the methodology of other sciences because of the peculiar nature of its subject matter—study of unique places or regions. Using analogies from physics and economics, Schaefer argued that geography is not peculiar in its focus on unique phenomena; all sciences deal with unique events which can only be accounted for by an integration of laws from various systematic sciences, but this does not prevent the development of those laws. It is, therefore, wrong to maintain that the geographers are distinguished among the scientists through the integration of heterogeneous phenomena which they achieve. There is nothing extraordinary about geography in that respect.

Schaefer traced the exceptionalist view in geography back to an analogy drawn by Kant between geography and history, an analogy repeated by Hettner and Hartshorne. In a rejoinder to Hettner's argument that both history and geography deal with the unique, and thus do not apply the method of science, Schaefer argued that this is a false position, for in explaining what happened at a certain time period, historians must integrate laws of social sciences. Time periods, like places, are undoubtedly unique assemblages of phenomena, but this does not preclude the use of laws in unravelling and explaining them. History and geography can both be sciences, for what scientists do is that they apply to each concrete situation jointly all the laws that involve the variables they have reason to believe are relevant.

The last part of Schaefer's paper reviews some of the problems of applying his nomothetic (law-producing) philosophy to geography as a spatial, social science. He attempts to recognise, for example, the problems of experimentation and of quantification, and suggests a methodology based on cartographic correlations. A major point concerns the difference between laws produced in geography and those from other 'maturer' social sciences. The former are morphological laws; the latter are process laws. In order to fully comprehend the assemblages of the phenomena described in geographer's morphological laws, therefore, it is necessary to derive process laws from other social sciences. *Geography, then, is the source of the laws on location, which may be used to differentiate the regions of the Earth's surface* (Johnston, 1983; Schaefer, 1953).

In three of his major publications (1955, 1958 and 1959), **Hartshorne** expressed his reaction to Schaefer's attack and criticism to his regional paradigm. In his paper (1955), he attempted to indicate certain inconsistencies in Schaefer's paper, and claimed that it ignores the normal standards of critical scholarship, and in effect offers nothing more than personal opinion, thinly disguised as literary and historical analysis. Hartshorne also pointed out that in coming to the conclusion that geography should take process laws from the.systematic sciences and use them to produce morphological laws, Schaefer came very close to preaching the sort of exceptionalist claim that he sought to destroy. In his second paper, Hartshorne (1958) attacked Schaefer's claim that Kant was the source of the exceptionalist view. In fact, both Humboldt and Hettner reached the same view independently, although unaware of Kant's views when they were writing.

Hartshorne's most bitter rebuttal of Schaefer's criticism was published in his monograph entitled *Perspective on the Nature of Geography* (1959), in which he defined geography as that discipline which 'seeks to describe and interpret the variable character from place to place of the Earth as the world of man'. He considered that human and natural factors do not have to be identified separately—any prior insistence on this was a function of the arguments of environmental determinists. He said that a division into human and physical geography is unfortunate, because it limits the range of possible integration in the study of reality. He made an important *distinction between 'expository' description and 'explanatory' description.* 'Geography is primarily concerned to describe the variable character of areas as formed by existing features in interrelationships ... explanatory description of features in the past must be kept subordinate to the primary purpose'. Thus, 'historical geography should be the expository description of the historical present but the purpose of such dips into the past is not to trace developments or seek origins but to facilitate comprehension of the present'.

On the question of *'dichotomy between systematic and regional geography',* Hartshorne accepted that studies of interrelationships could be arranged along a continuum, from those which analyse the most elementary complexes in areal variation over the world to those which analyse the most complex integration in areal variation within small areas. The former are called topical studies and the latter are regional studies. Whereas every truly geographical study involves the use of both the topical and the regional approach, there is no argument that one is superior to the other (1959, 120–21).

On the question raised by Schaefer, 'Does geography seek to formulate scientific laws or to describe individual cases?', Hartshorne replied in the negative and pointed out following difficulties in establishing such laws through geographical investigation: (a) scientific laws must be based on large number of cases, but geographers study complex integrations in unique places; (b) scientific laws can be best established in laboratory experiments which allow only a few independent variable to vary, but such work is impossible in geography; (c) interpretation requires skills in the systematic sciences which are beyond the capacity of geographers; and (d) scientific laws suggest some kind of determinism, but this is inappropriate to the human motivations which are in part the causes of landscape variation. For these reasons *the search far laws is irrelevant to geography.*

Hartshorne upheld the Hettnerian tradition of geography as a chorological science with history as a chronological science. This is valid because it describes the way in which geographers have worked, on both topical and regional subjects, with reference to interrelationships and integrations within areas. *Hartshorne's was a positive view* of geography: geography is what geographers have made it. *Schaefer's was a normative theory* of what geography should be, irrespective of what it had been (Johnston, 1983).

By the 1950s, especially in the United States, disillusionment with the empiricist philosophy of the regional paradigm appeared to have developed, which was followed by a gradual dominance of the topical specialism. As a result, a full revolution was launched with the presentation and ultimate acceptance of a new paradigm.

SYSTEMATIC GEOGRAPHY

There is no doubt, that within about a decade of the publication of Schaefer's paper, most geographers had adopted at least part of his paradigm, with growing concern for quantification and law-making. The younger geographers in the United States in the 1950s had a choice before them whether to accept the regional paradigm or to search for quantification and law-making. It is pointed out that most geographers opted for geography as a law-seeking science.

During the 1950s, systematic studies became much more important in the research and teaching of American geographers. The growing popularity of topical specialisms is revealed in the review chapters in James and Jones' edited work *American Geography: Inventory and Prospect* (1954) and also in the *Journal Literature* of the 1950s. Very few of the investigations reported were aimed at the generation of laws in any sense; indeed some could almost be categorised under the exploration paradigm in that major purpose seemed to be the provision of new factual material. Schaefer, Ullman and Ackerman were the precursors of this new paradigm of systematic geography in the United States. However, most of the methodological changes and discussions in systematic studies in geography during the 1950s occurred at Iowa, Wisconsin and Washington.

At Iowa, it was **Harold McCarty** and his associates who made considerable contributions to the methodological aspects of the systematic studies in geography. Their intent was to establish the degree of correspondence between two or more geographical patterns, akin to the morphological *laws of accordance* discussed by Schaefer. It is interesting to note that though Schaefer was at Iowa until his death in 1953, he is not referred to by McCarty and his associates in their works.

McCarty *attempted to give the morphological laws a definite theoretical shape.* The purpose of theory is to provide explanations and he recognised two types of explanations. The first is based on a search for the causes of observed locational pattern and the second type focuses on association. Such laws of association are built up in a series of stages: (a) statement of the problem and of the necessary operational definitions; (b) measurement of the phenomena (with attendant problem of sampling in time and space); and (c) statement of the findings in tabular or graphical form. Thus, geography, in its search for morphological laws, is to a considerable extent a consumer of the laws of other disciplines. These laws may be theoretically rather than empirically derived.[1] The method that McCarty attempted to develop, with its focus on the testing of simple hypothesis derived either from observation or from theoretical deduction, became the model for much research in the ensuing decades.

Earlier pioneers in the department of geography at the University of Wisconsin, Madison, were John Weaver who used multiple correlation and regression to identify the influences of climatic variables on barley yields (1943), and J. W. Alexander & G. A. Zahorchak who focussed on the measurement of population pattern (1943). A. H. Robinson (1961) later on attempted to combine these two interests in his research works, whose main interests were in cartography and cartographic correlations. His concern was to develop statistical methods of map comparison, and he paid particular attention to the problems of representing areal data by points and of using correlation methods in the comparison of isarithms maps.

The pioneers of the systematic tradition in geography of the Schaeferian heritage in the University of Washington, Seattle, were W. L. Garrison and his associates, and E. L. Ullman who moved to Seattle in 1951. A large group of graduate students worked with Garrison, and some of them became leaders in the new methodology during the subsequent decades. The group was benefited from a visit to Seattle by the Swedish geographer Torsten Hagerstrand, who was developing methods of generalising spatial patterns and processes. The dominant thrust of the works of **Garrison** and associates involved the derivation from other systematic sciences of relevant normative theories, mathematical methods and statistical procedures with which to develop morphological laws. They had catholic interest in urban and economic geography, and also in regional planning.

[1] In a major demonstration, McCarty and others discussed several statistical procedures for measuring spatial associations and adopted the well-known technique of multiple regression and correlation.

In one of his articles, 'Spatial Structure of the Economy', published in the *Annals* in 1959, Garrison urged upon the geographers that to advance their locational analysis and theories they would have to depend much on the works of economists, because their works are specifically devoted to the question 'what determines the spatial arrangement of economic activity'? Garrison also worked on the possible application of the mathematical procedures of linear programming, which produce the optimal solutions to problems of resource allocation in constrained situation. In this, he illustrated how the procedures of neoclassical economic analysis could be adopted in order to investigate ideal solutions to the problems of where to locate economic activities and how to organise flows of goods. Garrison and his team undertook an empirical work with a planning orientation on the impact of highway developments on land-use and other spatial patterns.

One of the outstanding associates of Garrison was **W. Bunge,** whose *Theoretical Geography* published in 1962 established the scientific credential of geography, and contradicted the Hartshornean paradigm of regional geography. In the book, he claimed that Hartshorne confused uniqueness and singularity. He also contested Hartshorne's claim that geography cannot formulate laws because of its paucity of cases and argued in favour of even more general laws. He countered the argument that geographical phenomena are not predictable with the claim that science does not strife for complete accuracy, but compromises its accuracy for generality. He also favoured the application of descriptive mathematics in preference to cartography as a more precise language. The rest of the book is devoted to the aspects of the substantive content of the science of geography beginning with 'a general theory of movement' and then a chapter on 'central place theory'.

B. J. L. Berry and M. F. Dacey, two other distinguished associates of Garrison, also made significant contributions to the field of systematic studies in geography. Berry's work (1967) has always had a very strong empirical and utilitarian base, whereas Dacey (1973) continued to work on the mathematical representation of spatial, especially point, patterns. In total, the work of this group of scholars influenced the research and teaching of a whole generation of human geographers throughout the world, and has undoubtedly been the most significant in the development of the first post-regional paradigm.

Quite distinct and different from any of the above three, was the publication of an article by **J. Q. Stewart** (1956), an astronomer at Princeton University, in which he developed the 'concept of social physics'. He observed certain regularities in various aspects of population distribution, regularities which were akin to the laws of physics, such as a tendency for the number of students attending a particular university to decline with increasing distance of their residences from its campus. From these observations, he developed the concept that 'the dimensions of society are analogous to the physical dimension and include numbers of people, distance, and time. Social physics deals with observations, processes and relations in these terms' (Stewart, 1956, 245). One of his associates was William Warntz (959) who worked on the investigation of distance as one of the basic dimensions of society. The wide range of empirical regularities which Stewart and Warntz observed was used to develop **the** *concept of macrogeography.*

It was not until the publication of D. Harvey's *Explanation in Geography* in 1969 that the conceptual revolution in the new scientific geography initiated by Schaefer remained inconclusive. A key concept that Harvey developed in the philosophy is that *laws must be proved through objective procedures,* and not accepted simply because they seem plausible. A valid law must predict certain patterns in the world, so that having developed an idea about those patterns, the researcher must formulate them into a testable hypothesis, a proposition whose truth or falsity is capable of being asserted.

CURRENT TRENDS

Current geography in the United States is characterised by three distinct approaches, each with a separate epistemology or theory of knowledge. (1) First, there is the *'positivist philosophy',* with belief in the objectivity of scientific description (empiricism) and analysis of the world, its goal of formulating laws about that world, and its asumption that explanation can be derived by studying the outcomes of the laws. It believes that the laws of spatial organisation can be revealed by analysing spatial pattern. (2) Second, there are *'the humanistic philosophies'* based on a belief that men live in subjective worlds of their own creation, within which they act as free agents. (3) Finally, there are the *'structuralist' or 'realist'* philosophies which argue that explanation for observed pattern cannot be discovered in the analysis of those patterns themselves, but only by the development of theories of the underlying processes that generate the conditions within which human agents can create those patterns. Foremost within this group is Marxism which argues that the processes are themselves changing—and can be altered by concerted political action so that no laws of spatial organisation are possible.

12

Geography in Modern India

HIGHLIGHTS

- Geography in Modern India
- Use and Application of Geomatics: Remote Sensing, GIS and GPS

Modern Indian Geography is a blend of different 'schools' of thought as of British, French, Russian (erstwhile Soviet Union) and the American. However, the influence of the British School of thought was more expressive in Indian Geography. It has been rightly observed that: 'Ideas similar to those developed in Britain, France, former Soviet Union, United States, and Germany were presented in many different languages and in many different national and regional settings, and these ideas had shaped the geographical scholarship of different nations around the world' (Deshpande, 1983). Modern geography in India, therefore, had its development 'not independently'; rather, its development was 'dependent in nature', lacking in indigenous 'inputs'.

To quote James and Martin (1972, 258–59): 'The new geography was transmitted from Great Britain to the former British colonies and dominions. In fact, many of the universities in these areas have close administrative connection with a university in Britain. Many of the distinctive characteristics of British geography were passed on to universities throughout the Commonwealth'. The mid-1920s is a watershed in the institutionalisation of the discipline of geography in India. It was in this time that India's first geographical association was formed and under-graduate classes were first started in colleges affiliated to Lahore (now in Pakistan) in 1920, Aligarh in 1924 and Patna in 1927. The maiden seeds for the discipline had been planted. To gain root in a sterile landscape, the pioneers faced enormous hardships. All these happened because of British initiatives.

The British were official couriers of the seeds of a segregated and defined discipline of geography in India. For over a hundred years, the British used the tools of this science to measure, collect, organise and document the geographical facts of India. But the study of geography in India did not find its expression till the early twentieth century. The knowledge of geographical horizon increased successively with the inputs of geographical information provided by the institutions, founded by the British for the purpose:

The establishment of these institutes by the British in the eighteenth and the nineteenth centuries necessarily provided inputs for geographical studies in India, but it was not until the early twentieth century that the fruits of geographical information provided by these institutes were realised. The Survey of India was set up in 1767 with the purpose of knowing the geography of this vast subcontinent through the map-works. The first map of Hindustan on a scale of 1:3 million was prepared in 1788 by James Rennell, the first Surveyor General. Following this success, the Great Trigonometrical Survey resulted in geodetic investigation of marked importance. By producing an interrelated triangulation series and using a theodolite, the heights of 79 major Himalayan peaks were determined, including the height of the highest peak.

Sl. No.	Institutes established	Place	Year	Information
1.	Survey of India	Dehra Dun	1767	Maps
2.	Royal Botanic Garden	Calcutta	1787	Vegetation
3.	Geomagnetic Observatory	Bombay	1840	Tectonics
4.	Geological Survey of India	Calcutta	1851	Rocks
5.	Census of India	Calcutta	1861	Population
6.	Archaeological Survey of India	Calcutta	1861	Past Cultures
7.	Indian Meteorological Department	Pune	1875	Climate
8.	Forest Research Institute	Dehra Dun	1906	Flora

Source: Kapur Anu., Ed. (2002) *Indian Geography: Voice of Concern.* Concept Publishing Company. New Delhi.

towering to 8,848m and was named Mount Everest in honour of the Surveyor General of India.

The Geological Survey of India was set up in 1851 with the purpose of discovering the mineral resources, especially the strategic minerals and coal beds as fuel for the British-laid railways. By the end of the nineteenth century, most of the major geological formations were identified and located throughout the subcontinent that largely helped to map the distribution and/or locations of occurrence of different mineral resources. The horizon of geographical knowledge witnessed an upward increase with the activities of the Survey of India and the Geological Survey of India, but the study of geography in India continued to be academically inexpressive and confined to primary-level teaching.

Geographic information and knowledge, so far collected, were used to further British colonisation and influence to the inaccessible regions of the subcontinent. Geography was used as a tool for political expansion. Surveys and maps were not only the tools for storing geographical information, but describing and compiling statistics were additional supports. The geographer with the East India Company, John Walker, undertook the responsibility of compiling the *Atlas of India*, on a scale of 4 miles to one inch. The compilation work began in 1882, and the *Great Atlas of India* was completed only by the early twentieth century.

The gazetteer, a geographical dictionary in which names and descriptions of places were given in alphabetical order, was an important invention of Johnn G. H. Hassel, a German geographer in 1817. By 1881, nine volume sets of the *Imperial Gazetteer of India* were compiled by W. H. Hunter, but the number of volumes increased to 26 in 1907. A retinue of provincial and district gazetteers followed soon. The. British had left a legacy of 500 gazetteers on India when they left in 1947 August. Nevertheless, the geography teaching in the nineteenth century British India continued to confine to the primary level only, where pupils were asked to memorise names of the places, their location, etc. The horizon of geographical knowledge about the subcontinent was politically and strategically motivated, but its academic relevance continued to remain redundant. The British neither promoted geography teaching at the college/university level nor training for the school teachers at the graduate level. The British established a university in Calcutta, Bombay and Madras in 1857, followed by one at Lahore (now in Pakistan) in 1882 and Allahabad in 1877, but without a department of geography. Even at the college level, there was no facility for geography teaching. The Indian Universities Act, passed in 1904, was designed to embark upon post-graduate teaching and research in both the humanities, and the physical science, but there was no mention 'of the discipline of geography to be introduced in the college/university curriculum.

It was after 16 years since the Indian Universities Act of 1904 was passed that the geography teaching at the undergraduate level was started and/or introduced in colleges affiliated to Lahore in 1920, Aligarh in 1924 and in Patna in 1927. However, credit must be given to the Indian Science Congress Association, formed in 1914, for promoting the geography teaching at the under-graduate college level, particularly, where infrastructure was available for beginning teaching, although geography was not in its priorities at its initial phase. It was geology, not geography; which was earlier given an academic recognition by the Indian Science Congress Association.

Geographical Society and Association in British India

The first geographical society ever established in India was the Bombay Geographical Society, Bombay in 1832. However, it was not an indigenous society by birth or by its focus of interest; rather, it was established under the patronage of the Royal Geographical Society, London. The Bombay Geographical Society was later merged with the Royal Asiatic Society, in 1837, which was founded in 1823, with its headquarters in Calcutta. The East India Company had showed its special interest in the establishment of the Bombay Geographical Society for the simple reason of acquiring vital geographical information of the land and people of India. The Society had no Indian members. Since the Bombay Geographical Society was merged with the Royal Asiatic Society, Calcutta in 1837, it was not known what contribution did it make with regard to India's land and people. The Society had political ambition, rather than academic interests.

However, the geographical societies and associations that sprangup in the early twentieth century were founded where the undergraduate teaching of geography was started:

1. Curzon Geographical Society, Aligarh 1925
2. Madras Geographical Association, Madras 1926
3. Patna College Geographical Society, Patna 1929
4. Calcutta Geographical Society, Calcutta 1933
5. Bombay Geographical Association, 1935

The Bombay Geographical Association, however, was the joint effort of school teachers of geography, but all the other four societies/ associations were creation of incipient number among scholars and experts who had realised the relevance of the discipline in academic and applied fields. All the four societies and associations were founded by non-geographer British scholars. Advocating the cause for geography within the sphere of education and research, all the societies played a key role in their own region of influence. All the four geographical societies/associations are still active in their academic pursuits, promoting geographical research and studies of international standard, except the Patna College Geographical Society that has remained dormant for the past several decades.

The discipline of geography at the time of the end of the British rule in 1947 was still in a pre-embryonic stage, because of the lack of interest shown by the indigenous scholars in the discipline. Geography was not a fascinating discipline that it would attract many people for academic purposes and works. Geology was more popular and fascinating because of its applied nature. Those interested in the study of the Earth's surface and features were necessarily inclined towards geology, rather than towards geography. 'While geologists had greater recognition, they played an important role to popularize geography in the coming decades. The focus in physical geography, the development of which Calcutta was to be proud of later, is rooted in its allegiance with geology. In fact, the Calcutta Geographical Society owes its birth in 1933 to geologists who constituted a sizeable group among its founder members' (Kapur, 2002).

GEOGRAPHY IN MODERN INDIA

Although the roots of the development of geography in India lay in the British period, but the development was 'intermittent', 'discontiguous', and 'isolated' in nature, because the discipline had few takers in British India. The possible reason might be the lack of professionalism of the discipline vis-a-vis the continued relevance of geology, rather than in the studies of the earth-bound phenomena that the British encouraged and supported. The status of geography on the eve of the British departure was too low to be mentioned, given the size of the land and people of the country.

However, the departments of geography grew in number from 17 in 1947 to 48 in 1970. The post-graduate teaching was started in quite a number of universities, and the discipline in 1950s of the last century started experiencing professionalism. Good number of geographical societies/associations came into existence in different parts of India with the intention of promoting geographical teaching, research and training. One of the significant historical aspects with the Indian geography in the 50s was its gradual recovery from the moribund waters. As the discipline underwent conceptual and methodological changes at an unprecedented fast rate in the world as a whole, so was the case with the Indian geography in the late 1950s but the rate was not 'unprecedently fast', rather it was quite slow, but it did initiate changes in the status of Indian geography in general. Nevertheless, the non-geographers headed most of the important departments of geography during their formative phases. They mostly belonged to the discipline of geology and commerce. Later on, the professionals trained in geography teaching, research and training took over the reins of these departments, and/or succeeded the non-geographers as the heads of the departments. This change-over had its impact on the contemporary Indian geography from geology/commerce-centric geography to 'geographi-centric geography', of course with British flavour. Most of the people who occupied the chair of the various departments of geography of contemporary India were British-trained, and had a long tenure of more than decade(s) as the head, or chairman of the departments. They are popularly called the 'pioneers' or 'doyens' of Indian geography. However, in 1956 a government agency, National Atlas Organisation, started work on the 'Atlas of India', and that played an important role in popularising geography teaching and research in independent India.

Geography in India in the 1950s and 1960s of the last century proliferated under the leadership of first generation geographers which included S. P. Chatterjee, P. Dayal, R. L. Singh, H. P. Das, G. S. Gosal, B. Arunachalam, N. B. K. Reddy, R. Vaidyanadhan and their role cannot be overlooked in building the department at Calcutta, Patna, Banaras, Gauhati, Chandigarh, Bombay, Tirupati and Waltair respectively. Most of them, however, had obtained their PhD degree from Great Britain, particularly from the University of London. Similarly, I. R. Khan, the founder of the department of geography in the Aligarh Muslim University had also obtained his PhD degree from the University of London. Unique inter-disciplinary institutes like the Centre for Regional Development, at the Jawaharlal Nehru University, New Delhi, the Institute of Development Studies at Mysore, and the Centre for Area Studies at Osmania University were the major achievements of Moonis Raza, R . P. Mishra and M. Alam. Those who were trained in the Universities of Great Britain, particularly, in the University of London, for their higher studies, and who founded and built the departments of geography on their return to India, had introduced curriculum for geography teaching, research and training necessarily based on the British pattern. Only those branches of human geography were introduced in the geography curriculum for the under-graduate and post-graduate levels which were taught in the contemporary British universities, and to which the Indian geographers were trained.

Since political geography/geopolitics was not in the curriculum in the contemporary British Universities, this branch of human geography was not introduced in Indian universities. Though historical geography was very much a part of the tradition and heritage of British geography, but this branch of geography did not get a place and/or expression in the curriculum, prepared by the leaders of the first generation, who pioneered geography teaching in India. Thus, at the embryonic stage of Indian geography, British geographical ideas, tradition and heritage and methods were propagated, popularised and set forth for higher studies. Despite a rich heritage of geographic knowledge, there was no attempt to revive and/or reconstruct the indigenous methods of the historic past, which could have given, if attempted, an 'independent' identity to Indian geography, completely free from the colonial legacy, of which the contemporary Indian geography was accused. However, Indian geography, in the 1950s and part of the 1960s of the last century, having been influenced by the British geographers of the regional and/or chorological heritage, was drawn towards 'regional approach' in geographical studies as being evident by the contemporary publications.

Some Indian geographers such as G. S. Gosal, C. R. Pathak, R. P. Mishra and Mahadev, who could be put into the category of the second generation leadership, had the distinct opportunity to do their higher studies in the United States, at a time when the discipline had been undergoing theoretical methodological changes, and experiencing 'paradigm-shift', at the cost of the 'regionalist' geographer, like Richard Hartshorne. Indian geographers who had gone there, however, could not remain isolated from the changes that the discipline had been undergoing on account of the positivist-led quantification. On return, they attempted to introduce, propagate, popularise the American School of Spatial Science at their respective areas and/or departments. Indian geography was, thus, brought at the cross-roads of British heritage of regional geography and the newly developed American heritage of spatial science. Nevertheless, there was nothing like a dichotomy in the contemporary Indian geography, but Indian geographers, by and large, showed their preference to the American School of spatial science, as being evident from the publications of that period.

In 1964, the Indian Science Congress Association, to which Indian geography was associated since 1934, published a review entitled *Fifty Years of Science in India: 1910–1960,* in which there was a separate discussion on the status and dimension of Indian geography, based on the available and published literature, pertaining to the nature of geographical studies conducted and/or undertaken during this period, by S. P. Chatterjee. Three categories of studies were identified to be divided into: (1) Physical Geography emphasising studies of the physical and geological formation of historic past, studies of different rock series, studies of deserts and arid features, studies of the Himalayas, studies of the great plain and other river-plains, studies of coastal areas, etc; (2) Economic Geography, in which major emphasis was found to be on agriculture, irrigation, mineral and power resources, and industries, particularly, with special reference to the Bihar-Bengal-Orissa mineralised zone; and (3) Cultural Geography, in which major emphasis was given on emerging urban scenario, rural morphology, migration, density of population and human geography.

In 1968, S. P. Chattrjee prepared a report on geography, entitled: *Progress of Geography in India, (1964–1968).* It was based on the review of work done in the discipline during this period. It was more of a nature of country report. The study, however, revealed that Indian geographers had concentrated their work on geomorphology, agricultural geography, soils, urban geography and cartography. There was also a report on Geography in Indian Universities, published by the University Grants Commission, New Delhi, on the eve of the 21st International Geographical Congress of the International Geographical Union, in 1968. It was held in New Delhi.

In 1972, the Indian Council of Social Science Research, New Delhi, published a project-report, entitled *Survey of Research in Geography,* with regard to the contemporary state of geography in India. Eminent geographers, mostly belonging to the first generation leadership, like V. L. P. Rao (Delhi School of Economics), R. L. Singh (Banaras Hindu University), P. Dayal (Patna University), M. Shafi (Aligarh Muslim University), G.S. Gosal (Punjab University), S.M. Alam (Osmania

University), C. D. Despande (Bombay University), S. P. Das Gupta (National Atlas Organisation), Moonis Raza (Jawahar Lal Nehru University) and P. Mishra (Institute of Development Studies, Mysore) were the major contributors to the project report. They studied and surveyed most of the publications, and researches done in geography until 1969, and divided the research into eight areas and/or categories: Economic Geography; Geography and Planning, Historical Geography, Human Geography, Political Geography, Regional Geography, Methodological Review, Research Methods.

The Indian Council of Social Science Research, New Delhi, published the *Second Survey of Research in Geography* in 1979 under the editorship of Moonis Raza. This survey report was concerned with the development of the discipline during 1969–1972. This period was identified as being the most important in the history of the development of modern geography in India. It ushered in an era of 'New Geography' in India. The survey report attempted to identify the salient characteristics of contemporary Indian geography: (1) concern with regard to the problems of development and planning; (2) growing tendency of moving beyond the boundary of the discipline into the realm of other discipline of social sciences with result, the research frontier in geography being widened; (3) growing adaptability of quantitative techniques in research methodology; (4) lack of generalisation despite the adoption of quantitative techniques and models in geographic studies; and (5) growing tendency to adopt and apply western models in geographical studies under Indian context.

The second survey report identified five areas of research during the period under review that included: Economic Geography; Social Geography; Historical and Political Geography; Population and Settlement Geography; Regional Geography and Planning. Each of the five broad areas of research was further divided into several sub-fields with academic and professional relevance. Works in the area of economic geography included resource geography, agricultural geography, land-use review, industrial geography, and transport and commercial geography. In social geography, the survey report revealed that the major sub-areas of research work were in the field of caste and ethnicity, tribal communities, religions, language and sub-language, geography of social change, neglected areas of social geography, and application of quantitative techniques in social geography. Works in historical and political geography included the studies of the past historical geography, construction and mapping of historical landscape of the past, political geography and studies of political phenomena in areal/spatial context, and geography of international relations. In population and settlement geography, the major areas of research during the period were in the fields of population geography, rural settlement geography, and urban geography. Similarly, in regional geography and planning, the research trends were towards regional geography, regional planning, regional disparities and regional hierarchies.

The Indian Council of Social Science Research published the report of the *Third Survey of Research in Geography* in 1984 under the editorship of S. M. Alam. The report of the survey dealt with the period of 1972–1975. The report identified six major areas/sub-areas of research in contemporary Indian Geography, that included: economic and resource geography (agricultural geography, industrial geography, commercial geography, and resource geography), social, historical and political geography (social geography—tribe, community and ethnicity, caste, religion, historical geography of the ancient and medieval period, contemporary political geography), population geography, settlement geography (rural settlement geography, and urban geography), regional and theoretical geography with emphasis on regional geography) geography of development (regional planning).

In 1999, the Indian Council of Social Science Research published the report of the *Fourth Survey of Research in Geography* under the editorship of G. S. Gosal. It covered the period of 1976–1982. Apparently, it was a short period for identifying any long-term trends in geographic research in the country, like the previous three survey reports of the Indian Council of Social Science. But when considered in conjunction with what was happening in this field of study before 1976, it did throw up some important clues relating to developments in concepts and methodology in this discipline. Though the period of this survey is short, it is marked by the publication of a large number of papers in the major Indian journals, and these papers necessarily, indicated the contemporary trends in the development of research methods in various branches of geography. It was on the basis of the research publications in the various branches of geography that the following sub-fields were identified:

1. Geomorphology;
2. Economic geography (agricultural geography, resource geography, marketing and transport geography);
3. Regional and Theoretical geography;
4. Social, historical, political and medical geography;
5. Population and settlement geography;
6. Urban geography, regional planning and development;
7. Environment and ecosystem.

Along with the new approaches in methodology were also seen studies which had an applied value. Besides the usual trend, geographical research was noted to be more diversified, particularly so in agricultural, marketing and transport geography, political (including electoral geography), medical, settlement and development geography and finally the studies of environment and ecosystem. Broadly speaking, the contemporary geographical research focused on urban geography and regional planning. The inter-disciplinary research in geography was a new methodological development/phenomenon, and it appeared that the contemporary Indian geography had a major inter-face with regional science (Reddy, Singh, Routray, 1999, 8–29).

The fourth survey report identified the research trends that indicated: (1) acceptance of quantification (statistical methods and techniques), (2) growing emphasis on model building, (3) progress in the observation and handling of data, (4) progress in the analysis of processing research, (5) emphasis on social process, (6) data integration process with space and cultural phenomena, and (7) implication of research to spatial development.

1. *Quantification.* The application of statistical methods and quantitative techniques became very common in most of the on-going research in physical, human and social geography. The process of quantification is well-perceived and adopted in more than one-half of the research papers during the period under review. Indian geographers appear to have fully realised that without the aid of statistical and quantitative techniques, it is difficult to present facts and describe phenomena precisely and scientifically. As a result of this conviction, adequate emphasis was laid on the quantitative methods in the curriculum of geography in many universities, particularly at the post-graduate level.

The application of quantitative techniques became more common in fluvial geomorphology, morphometric analysis, agricultural efficiency, crop-association, productivity, agricultural regionalisation, transport studies, hierarchy of markets, assessment of domestic market potential, identification of marketing region in marketing geography, and the study of settlement patterns. Hardly any sub-branch of geography had remained untouched of quantification; even political geography, particularly the electoral geography, experienced quantification during the period of the review. However, in some cases, the application of quantitative techniques has proved to be disastrous because of: (i) lack of knowledge of statistics and mathematics, and (ii) wrong choice of the statistical/mathematical techniques for the analysis. It cannot, however, be denied that very few of the Indian geographers possessed sound knowledge of statistics and mathematics and as a result some of the publications were found to be ill-conceived.

2. *Model Building.* The growing emphasis on theoretical geography during the past decades led to work on model-building. The use of

model became quite common in the field of agricultural, marketing, transport and urban geography as well as in regional planning. But, it is true that Indian Geographers could not develop any indigenous 'Indian' model, and they increasingly applied the models developed by the Anglo-American geographers. The application of such 'western-developed' models by the contemporary Indian geographers for their studies under Indian situation did not yield much desired results.

Nevertheless, it is true that the contemporary geographical literature in India became flooded with application of various models during the period under review, and the process got diversified at a 'reasonably' faster rate. It became clear that theoretical studies were no longer confined to either economic or urban and transport geography, rather, sub-branches like electoral/political, medical, education and geography of tourism seemed to have employed analytial methods using models and quantitative techniques to facilitate the appreciation of development of precise relationship between the theoretical approach.

3. *Progress in the Observation and Handling of Data.* The main fields of geographic research lay partly in physical and natural sciences and partly in social sciences. The availability of right kind of data—qualitative as well as quantitative—was crucial in Indian context. For a very long time research in geography in India was confined to empirical studies though field observation was not ruled out. But, during the period under review, besides empirical and field observation, major sources of geographical data and information had been the topographical maps of the Survey of India and administrative and thematic maps of the Census of India and National Atlas and Thematic Mapping Organisation (NATMO). The secondary sources of data were found very much in use for research in geography. The primary survey through questionnaire, schedules, field survey and observation, however, got an impetus in the various kinds of research works, including in MPhil dissertations, doctoral theses, and in research projects as in other social sciences. However, these field-oriented research involvements were relatively fewer.

The trend of research during the period under review was found to be very much development-oriented, the quality of research was dependent on the quality of available data in all sub-branches of economic, social and cultural geography. However, the organisations of data-collecting were not found to be relevant in the context of spatial orientation. There was a need for up-to-date data at national, state, district, block and village-level for quality and predictable research work. The development-oriented research work was reflected in the fields of agriculture, land use, transport, marketing and urban and regional planning. Studies and researches in these fields could be strengthened by improving the nature and quality of information. Some geographers and non-geographers, specialising in geographic research, have used aerial photographs and satellite imageries, particularly in the fields of geomorphology, landforms, land-use, resource evaluation and transport network studies. Even those working in the field of regional and urban planning also used aerial photographs and satellite imageries in their studies. But this research work, involving the use of aerial photographs and satellite imageries was confined to a few specialised geography departments of the country.

4. *Analysis of Processes.* The cultural processes, altering the continually changing Earth's space contents, have attracted significant attention in fundamental geographic researches. Working within a pattern-emphasising process, research in geography expanded in several new directions since the beginning of the professional geography in India. Many such developments were initiated due to growing interest of geographers in public policy issues, like the problems of environmental control, resource management, human relations, poverty and nutrition, housing, transportation and urban renewal and re-development. One

of the many directions was the ecological process and the ecosystem. Many of the studies have dealt with the perception of adjustment to the spatial and temporal dimension of floods and drought. Attention of the research to environmental perception and its effects on the behaviour of urban and rural inhabitants was found to be getting focused/attracted by a section of Indian geographers.

5. *Social Process.* During 1970s and 1980s, an increasing number of geographers recognised the need for studying the social, as well as political, economic and physical processes that affect spatial structure and interaction of human societies. This is a new field for social geography to study the relationship between social pattern and natural environment. 'The scope of social geography is to study the spatial patterns and functional relation of social groups in the context of their social milieu/ environment, the internal structure and external relation of the nodes of social activity, and the articulation of various channels of social communication' (Raza, 1979). The major subfields of research in social geography were found to be political, medical, education, tourism geography, economic geography as well as the geography of religions.
6. *Integration of Spatial Phenomena.* The systems analysis or systems approach has provided a new dimension to the studies of integrated problems. There are three major sub-systems in overall geo-spatial (geographical) systems which include: (i) human sub-system, (ii) physical sub-system, and (iii) an activity sub-system. The geographer's role is to understand the spatial structure and spatial interaction of the components of the several sub-systems in a given time or through a period of time. In this context, regional geography reflects the interaction and synthesis of the three sub-systems in a given area. However, at present, the traditional regional study was being replaced by regional planning and regional development studies which were at times methodological and at times developmental. This aspect of research attracted the attention of the Indian geographers not only during this period, but also a decade or two decades earlier. The present trend of consciousness in this line of research in geography shows that it further expands and will intensify both methodological and developmental studies.
7. *Implications of Recent Developments.* Space is the first and most important fundamental element in geographical research. The study of spatial development was emphasised in different sub-branches of economic geography along with urban and regional planning and regional development. In the present context of development processes in our country, the trend of research, as evidenced from the various geographical studies published during this period, was very much directed to match with the local, regional and national requirements. It may be noted that this was by far the most positive development in contemporary research in geography in India which had immense applied value.

In economic geography, the spatial development approach was adopted in studies related to agricultural regionalisation and development, transport and economic development, hierarchy of market centres and identification of service areas, identification of drought-prone areas and development perspectives. Similarly, the urban and regional development perspectives went on from micro-level to macro-level studies pertaining to development problems. In urban geography, the important studies were related to functional classification of towns and hierarchy of urban centres.

All the four successive surveys of research in geography, conducted by the Indian Council of Social Science Research, New Delhi, reveal one significant aspect that Indian geography since independence has moved forward in multi-directional ways without any 'expressive' dichotomies. But, at the same time, it is equally true that Indian geography, during the period of its formation phase (1947–1982), was heavily dependent on Anglo-American geography, particularly with regard to the philosophical and methodological bases, and the discipline progressed along the Western model of the 1950s and 1960s in the country. Indian geography passed through a 'positive' phase of its development in the late 1970s and early 1980s, when law-based geography overtook the tradition-based regional geography with the discipline of geography being conceived of a spatial science, rather than a chronological science; however, the Indian geographers never quarelled over this binary issue. They continued to treat the discipline as being unified.

Geographical Socities and Geographical Journals and Periodicals

Majority of the leading departments which were accorded special treatment, started their own geographical societies and associations to further geographical research and training. By 2000, India had about 50 societies and associations. However, majority of them were tied up with the colleges where there were departments of geography. No doubt, geographical societies have been playing a very significant role in promoting geographical teaching, research, training and studies in the various sub-branches of the discipline, particularly at the centres and/or departments, where geographical societies have come into existence. Although the societies were mostly manned by the non-geographers, but they flourished over time, under their dynamic guidance. These societies made significant contributions to the development of geography, especially after independence, by publishing journals and periodicals, and organising local and regional seminars, workshops and summer schools. However, these geographical societies are more regional in character rather than national. Bulk of the papers so far published in the journals and periodicals of these societies necessarily reveal local and/or regional bias. Nevertheless, the regional bias in the research publications has nothing to do with regard to the development of professionalism in Indian geography.

As the teaching of geography spread to different universities and colleges, both at the under-graduate and post-graduate levels, so came up a few more national- and state-level geographical societies attached to the new centres of geographical teaching and learning. The most important among them are the Indian Institute of Geographers, Pune, Association of Punjab Geographers, the Rajasthan Geographical Association, and the Association of Geographers of Bihar and Jharkhand. The latter three associations are not associated with any universities, rather they came into existence primarily because the local and regional geographers felt the need of such associations to express their views and talents. These associations have their own journals and periodicals. Most of the state-level geographical associations/societies, however, have caused more harm than development of the discipline in the country since they run on money and politics.

However, the theme-based associations, particularly, the Association of Population Geographers of India, Chandigarh, has done commendable work in promoting population geography in India. It is attached with the Department of Geography, Punjab University, Chandigarh, and its contributions to the field of population geography in India are in no way inferior to the international standard. The Regional Science Association of India, and the Indian Society of Social Survey and Land-Use Planning, Nagpur, are known for their achievements in their respective areas/fields. These two associations have made significant contributions to enhancing the methodological dimension of regional science, and soil and land use planning, giving a 'nomothetic' character to their respective areas of specialisation. The Indian National Cartographic Association, Hyderabad, has been doing commendable work since its inception by organising periodic workshops in cartography, making it more relevant in geographic teaching and research.

All the above-mentioned societies and associations have been working individually without any interaction between them. So, the lack of togetherness among these societies and associations has badly hurt the very spirit of Indian geography and its development. Each unit of societies/associations developed its own geography based on localism and parochialism. Realising this inherent weakness in Indian geography, a group of dedicated geographers, trained in the Anglo-American universities, laid the foundation of the National Association of Geographers, India (NAGI) in 1978. 'Echoing the need of the time, NAGI was not created, but was born: A voice of concern for Indian geography'. '... To capture the voice of concern the architects of the National Association of Geographers, India wanted to reach across the country. With a natural verve for region so that all areas are contained, it (NAGI) carved out five territorial divisions in India. Differing in size, area, population and even the number of departments of geography, the homogenous criteria for regionalization seems to be "common location". Hence the names North-Western, Northern, Western, Southern and East Regions.... Hailed as a learned society since its very inception, NAGI was keen to publish a biannual journal. Named 'Annals', a word that the dictionary defines as narrative of event year by year, or a historical record' (Kapur, 2002, 27–30).

The reach of NAGI expanded over years across the country, and the number of members, specially the life-members, has continued to swell every year since its inception. Soon it became truly 'national'. Holding of annual meets and/or the Indian Geographical Congress at different locations across the country, since its foundation in 1978 has brightened its national identity.

The growth and development of the geographical society/association largely depends on its literacy excellence that it achieves through its publication of journal and periodical. Geography developed and flourished in classical western and Anglo-American nations through the society-journal association and combination. Each society in the West has its own journal. A society is, therefore, identified by the quality. In India, the pre-independence geographical societies which were re-named after independence have started their own journals and periodicals, except the Patna College Geographical Society. However, the *Indian Geographical Journal* of the Indian Geographical Society, Madras (Chennai), and the Geographer of the Aligarh Geographical Society, Aligarh, are the only two pre-independence geographical journals in India, otherwise, the *Geographical Review of India* of the Geographical Society of India, Calcutta, *National Geographical Journal of the National Geographical Society*, Varanasi, *Transaction of the Institute of Indian Geographers*, Pune, are some of the earlier post-independence journals of the country which are of international repute, and are known for their valuable contributions to the development of professional geography in India. *Annals* of NAGI also emerged as one of the finest geographical journals of international repute. The *Indian Journal of Regional Science of the Regional Science Association of India*, Kharagpur, Population Geography of the Association of Population Geographers, Chandigarh, have also made valuable contribution to the development of geography in India. The first two journals are known for their contribution to enrich the law-abiding systematic geography in India. These are the few Indian geographical journals and periodicals which are recognised by the Indian Council of Social Science Research, New Delhi, and the University Grants Commission, New Delhi, hence they are called 'referred' journals.

Annals of the Association of Rajasthan Geographers, Bhilwara, and the Punjab Geographers, Patiala, are the fast-emerging state-level geographical journals in the country, while the *Uttar Bharat Bhoogol Patrika* used to be the only geographical journal in Hindi in the country, being published from the Department of Geography, DDU Gorakhpur University. *North East Geographer*, Gauhati, is another regional geographical journal, specifically dealing with the geography of the North-Eastern region of India. Its contribution to enhance geographic research and training in the region cannot be overlooked and undermined, given the quality of research papers being published in this journal.

At present, there are more than 200 geographical journals and periodicals in circulation in India, including those journals initiated by individual geographers. These journals and periodicals have caused more harm to the discipline rather than its development, by publishing sub-standard papers. In spite of the circulation of these journals and periodicals, geography in India has moved forward in the past 60 years or more because of the scholarly efforts by some distinguished geographers of the country, belonging to the first, second and third generation of leadership.

Geography and Scientific Research Institutes

It is mentioned earlier that during the British period, development of geography was very much dependent on the information provided by the Survey of India, Royal Botanical Garden, Geomagnetic Observatory, Geological Survey of India, Census of India, Archeological Survey of India, Indian Meteorological Department and the Forest Research Institute. On the basis of the information these institutes provided that the map of the sub-continent could be prepared, natural vegetation of the land together with the flora and faunal species of the land could be known, tectonics of the sub-continent could be known, the distribution of rocks and minerals were mapped and identified, demographic character of the sub-continent could be known, past cultures and cultural heritage identified through excavation, and the nature and character of the monsoon became known. These institutes, in fact, came up with multi-dimensional geographical information, concerning South Asia, which in the later periods, particularly in the twentieth century, provided the necessary inputs for making geography a relevant discipline, at par with other disciplines of human sciences in the country. The beginning of geography teaching at the under-graduate level at Lahore, Aligarh, Patna, Calcutta and Madras in the 1920s and 1930s of the last century was a natural outcome of the achievements of these institutes with regard to the data-base they prepared and provided to authorities for further analysis that, in turn, formed the basis for 'academic' geography in British India.

Independent India also established the following illustrative Research and Scientific Institutions:

Sl. No.	Year	Illustrative Institutions
1.	1947–1949	Indian Institute of Agricultural Research
2.	1949	Dept. of Architecture and Regional Planning, IIT, Kharagpur
3.	1950-1959	Planning Commission, Centre for Rural Studies, Indian Institute of Public Administration
4.	1960-1969	Institute of Applied Manpower Research, Indian Council of Social Science Research
5.	1970-1979	Institute of Development Studies, Institute of Social and Economic Change, National Institute of Urban Affairs
6.	1980-1989	Institute of Himalayan Studies
7.	1990-1999	National Institute of Waste Lands and Rural Development

Much of the growth and development of geography in India depends on the above-mentioned institutions because of their data-base, which necessarily helps in geographic research. No research in geography can be undertaken without seeking help from these institutions. Over the years, geographers have been taking help from these institutions for their studies and research. These institutions are also providing job accommodation to geographers.

One of these illustrative institutions was the Planning Commission, where geographers found their relevant expression. It was the approach towards balanced regional planning of the Planning Commission that made the role of geographers quite relevant. It became quite clear that no attempt at planning was going to be successful unless the spatial data-base and its modalities were significantly enlarged. This created willingness to explore the role of geographical participations in regional aspects of national planning. Planning Commission invited O. H. K. Spate of the University of Canberra, Australia and A. T. A. Learmonth from the University of Liverpool to spend a couple of months to report on the possibility of geographical contribution to the national planning.

Under the Technical Cooperation Program of the Colombo Plan, Learmonth put together a team of two economists, one statistician and four geographers to undertake a regional survey for planning purposes. It was efforts like these which served to redirect the attention of geographers in India to problem-solving by using the frame of regional development. This greatly helped the geographers in India to engage themselves in efforts of initiating urban and regional planning at various centres and localities. A number of urban and regional planning organisations came in existence all over India during 1955–1975. Geographer-planners could cut a profession in their ambit. As nutrition to grasp the complexities and problems facing development, the Government of India launched several centres for social science and policy research. They, too, served as possible havens for geographical works. The blueprint of planning and research institutions relied on a team of social scientists and the scope of involvement by geographers was an inherent part of this design (Kapur, 2002, 21–22).

The Planning Commission in 1959 had sanctioned many projects to various geography departments of the country, like the Damodar Valley Diagnostic Survey to the department of geography of Patna and Calcutta Universities, and the department of Regional Planning, IIT, Kharagpur, manned by a geographer. Similarly, the Terrain Analysis Cell of the Ministry of Defence had approached several departments of geography to undertake projects of terrain evaluation in areas of specific needs. The Terrain Evaluation study in Bikaner and Jaisalmer Districts (1967–69), by the department of geography, University of Delhi, and Terrain Evaluation of Punjab Plains, Punjab Himalayas, and Shipki Tract, Rajasthan border, 1965–68, by the Department of Geography, Punjab University, Chandigarh, manifested the efforts to fulfil the defence needs of the country, and, at the same time, these studies represented the professional characteristics of Indian geography vis-a-vis the Indian geographers.

Considering the professional skill and quality of geographers, various institutes sanctioned projects to them for meaningful studies, and the projects and studies proved their applied relevance, particularly in the policy-making fields. Thus, professionalism in Indian geography got further enhancement and impetus with the successful completion of projects and that, in turn, sustained the effectiveness of geographic research in the country. Some innovative studies by geographers in the field like regional planning with emphasis on hill areas, flood-prone areas, drought-prone and tribal and industrially backward areas and like during the past few years, have made geography 'an applied science' making the application of geographic concepts and skills to the study of practical economic, social, or political problems more relevant than before. Geographers have been found to be best field surveyors, so more and more geographers are being appointed to conduct field surveys and to prepare detailed notes based on the survey, by the different institutions, government and non-government, and other private institutions. In doing so, geographers have developed new methods of study, new techniques of observation and analysis, and new cartographic methods.

Perspective of Indian Geography

To quote N. B. K. Reddy, 'It is indeed paradoxical that although geography is a subject of great antiquity, as a modern academic discipline and applied science, it is only 150 years old. Indian geography, being hardly 50 years old, is much younger than modern geography' (Presidential address at 5th Indian Geography Congress, Tirupati, 1982).

Commenting on the present status of geography in India, the report of Indian Council of Social Science Research Review Committee observed: '... geography, despite the late start it had in the Indian universities, has manifested considerable potential for growth in recent times. Geographers have prouced volumes of research literature over a wide range. Geographical research is still dominated by traditional methods of cartographic compilation of history, archaeology and of qualitative descriptions. A noteworthy trend in geographical research today is the increasing use of quantitative methods to spatial structure, spatial organisation and spatial process' (Indian Council of Social Science Research, 1972 First Survey Report).

Moonis Raza (1979), who edited the survey report in 1979; had commented: 'Indian geographers, by and large, have shown little interest in theoretical discussion on the nature of geography. The reasons are partly historical. When the controversy between the exponents of environmental determinism and possibilism was raging in the west, modern Indian geography had not been born. It did not get involved in the ideological turmoil through which modern geography acquired its philosophical foundation and intellectual rigour. By the time geography emerged on Indian campuses, a storm had already taken place.

Having 'improved' and uncritically accepted a complete system of ideas without participating in the making of it, Indian geography tended to develop at best, as a dinosaur. The tradition continues. Concern for theory is still considered futile pedantry and search for logical system is equated with esoteric cause. This has led to a situation where new developments do not grow inside the organism as the inevitable consequence of an evolutionary process in the realm of ideas. They are just mechanically aggregated to the amorphous mass in response to the latest intellectual fashion in the west. This has given to Indian geography a highly fragmented character'.

To quote C. D. Deshpande: 'It is not surprising that those who have been trained in the United Kingdom, the United States of America and more recently in the Union of Soviet Socialist Republics and East European countries have influenced our research activity. Lamentably, the influence of France and Germany, which significantly continues more rigorously in the natural aspects of geographical science, is lacking. But apart from these exposures, it is time that we review our problems through our own perspective' (Reddy, 1982).

It is, no doubt, true that the quantitative and conceptual revolutions as the problem-oriented and applied approaches have elevated the status of Indian research. There has been an increase in the technological sophistication, modeling and communicative jargon. But despite these developments and large volume of output, Indian geography is yet to make impact on the international scene. It is still weak in its conceptual, empirical, analytical and inferential aspects. There is a lack of 'in-depth' studies of a quantitative nature. While qualitative assessment has its own place, it is important to lay greater stress on quantitative methodology. Quantitative methodology fixes the seal of exactness on the data. It is, therefore, necessary to strengthen the classical tradition retaining the gains of the quantitative revolution.

It is necessary to understand the complex geographical problems of India with reference to the international setting. Panikkar (1955) in his pioneering geographical study of India and the Indian Ocean brilliantly summed up the situation as follows: 'Geography has been at all times one of the great and most regrettable gaps in Indian knowledge. Worse than our lack of interest in things historical has been our utter neglect of geography. Geography constitutes to be the permanent basis of every nation's history. It is one of the major factors that determines the historical evolution of a people. Today, free India has to realize her position in the world where geography has placed her in relation to other countries. She has also to understand the strength and weakness arising from her geography. A nation can neglect its geography only at its peril.' Therefore, it needs hardly be stressed that Indian geographers should take greater interest in studies on the politico-geographical and/or geopolitical aspects of the country. But this branch of geography is yet to get its expressive recognition in India.

As mentioned earlier, there has been a growing tendency to apply foreign models to the Indian setting. If foreign models are uncritically accepted and applied, they become blind imitations. Such applications stifle incentives for indigenous approach towards original and creative thinking. It is needless to state that the Indian context is unique and as such the methodology and constructs have to be validated in an Indian frame of reference so that the interpretation would carry a sense of realism.

Social, cultural and historical geography are some of the neglected fields in geographical research. They are almost untouched by geographers. India with its hoary antiquity, a rich and diversified cultural tradition provides a fertile ground for research in these fields. Such studies go a long way in understanding the complete socio-cultural problems of the country. However, particular emphasis requires to be given on awareness of humankind's immediate environment and regional geography. The discipline has to be made more relevant to provide a broad base for an interdisciplinary superstructure at higher academic levels. Also, studies with greater bias relevant to a region and studies that are more problem-oriented need to be introduced without neglect of theoretical issues.

The modern era of scientific and technological developments has thrown up a great challenge to geography in India. The need for wider application of geographical approach to various practical problems in India is increasingly felt, especially in the field of regional planning, resource development and management, and environmental planning. With rapid growth of population and urbanisation, besides increasing regional disparities, the need for regional planning and scientific land-use planning has acquired greater urgency. Technological developments such as remote sensing, aerial photography, and computer technology have opened up new vistas for speedy data collection, data storage, data retrieval and data analysis. The implications of these technologies in geography are far-reaching and their utility in geographical research is immense. However, the success of our planning mainly depends on the development of agriculture and its stability. To achieve this end, research has to be carried out on various aspects of agriculture, cropping pattern, optimisation of land use, development of irrigation and supply of necessary inputs. Therefore, agricultural geography will have to play a prominent role in the economic development. Efforts should be made to develop new methods to carry on in-depth researches in agricultural geography.

Geography in India has witnessed a tremendous growth in a short span of time and the people have now realised the importance of the subject as an academic discipline and as an applied science also. There is virtually no sphere where geographer cannot play a significant role.

To quote G. S. Gosal: 'An examination of the overall performance of geography as a major discipline in India brings one to the encouraging conclusion that it has come a long way in establishing its place among the sciences dealing with reality and that is making steady progress in various directions. Geographers have been showing concern for problems of development and regional planning. They have also responded well to collaborative work with colleagues in allied disciplines and inter-disciplinary projects on the strength of their own knowledge and ideas. What discipline is better fitted for such an integrative role than the one concerned with spatial arrangement and organisation as changing through time? Consistent with the requirement of the times and the nature of problems facing the society, they have increasingly taken to quantitative methods of analysis and synthesis, although a lot remains to be learnt and applied appropriately. Thus far they have been preoccupied with studies of specific problems in specific areas or places ... now they should subject these studies of the specifics to comparative analysis and draw generalisation of wider validity. It is their distinctive function to develop theories regarding spatial relation on the face of the Earth. The theories must grow out of the facts that make India' (1st Indian Geography Congress, Chandigarh, 1980).

In the sphere of the Geography of Development, there is a need to develop the themes: People, Resource, and Government. Whereas state plans for socio-economic development are the same within the administrative units, there are wider spatial inequalities in the achievements within the units. The political patronage in areas and differing receptivity of the people to plan within the region, even having a similar resource base are potent factors in this regard. Geographers in India should come in a bold way in studying these problems of spatial organisation within their local areas and make prescriptive recommendations (Ibid).

B. Arunachalam observed: 'the initial impetus to Indian geography has come through the testing of the western models in the Indian context, though this exercise at times has been carried too far: it has, however, not logically led to this innovation of Indian models. The deductive Central Place Theory of Christaller, the ecological models of urban structure, studies to examine stages in the growth of a Primate City, Growth Pole Theory and the like have all found their echoes in the Indian realm. Many of the achievements of geography in India over the last three decades have resulted from its ability to address itself to range of Indian problems, especially those of urban development and industrialisation. Many of these problems are common to the developed countries of the West—the only difference is that India is passing through a phase of development that they have already gone through—and Indian studies have positively helped in building a bridge in the international forum through such comparative studies' (Third Indian Geographical Congress, Bombay, 1980).

Expressing his reaction to too much importance being given to urban geography, Arunachalam commented: 'The "urban" bias in Indian geography is being overdone; every university, whether located in an urban metropolis or a rural interior, deems it necessary to specialize in urban geography. In a country, predominantly rural and agricultural, why has Indian geography developed intellectual indifference to the problems of rural development? Environmental hazards like floods and droughts, and the permanent threats of food shortage also pose compelling interest to Indian geography. If the western world has no parallels to such Indian situations, the field is rich enough to provide ample ground for indigenous virgin researches through primary field investigations. There is a need for Indian geographers to work out strategies for planning rural development' (ibid.).

Given the professional edge through intensive training, the geographers in India can become important and useful members of any planning cell. Look at the immense prospects for the professional utility of the geographer in a set-up where planning units are being visualised at the grass-root level in every district of the country and the market for his cartographic skills when every planning cell contemplates a local/regional/national atlas. The question is: can and will Indian geographers build themselves into national/regional and local grids of co-operating and interacting teams committed to these specific goals of national priority? Can geographical training in the schools, colleges and universities be restructured as to give the right orientation and consciousness of local, regional and national environmental problems of conservation and wise use of resources? Can our geographers pledge their research interests to the priorities of the National Data Bank and Resource Inventory so that the geographers too can contribute in pulling the 'Juggernaut' of Developing India? These were some of the querries put forward by Arunachalam. He expressed his anguish with regard to the unexplored frontier of evolution of Indian thought. The history and evolution of Indian geographical thought is an unexplored frontier inviting challenges.

C. R. Pathak, a regionalist par excellence, while elaborating the role of geographers, regional scientists and regional planners, has noted that 'geographers have been observers of the phenomena and their interactions that characterise the landscape. Such characterisation of landscape varies from place to place and geographers throw light on the characteristics of such places and give a clear description of the phenomena and their interaction. However, no phenomena or place to place variations can be analyzed and evaluated under some tools and techniques of analysis and developed to ascertain the relative role of different factors, and bring out relevant facts which need proper attention. Those who develop such tools and techniques for analysis are identified as regional scientists. Based on their analysis, regional planners formulate strategies and find solution for tackling such problems. Hence, the geographers, the regional scientists and regional planners have demarcated their own domain of study and research.... While it is very difficult to distinguish geographers, regional scientists and regional planners, particularly when they work on the field and contribute their might to the problem solution, nevertheless, they necessarily maintain their identity' (7th Indian Geographical Congress, Kharagpur, 1985).

On the relationship of geography and regional science, C. R. Pathak said: '... in treating spaces, regions and related areal and distance concepts both geography and regional science have an orientation different from that of other disciplines. If, for example, we view economics, sociology, political science, anthropology, geology, oceanography, meteorology, botany, etc. each as having a vertical orientation and alignment, then geography and regional science can be said to have a horizontal orientation and alignment, cutting across many of the fields with vertical alignments.'

If the combined scope of geography and regional science and their total ramifications and numerous procedures and methods are beyond the absorptive capacity of one person, it seems appropriate that the overall body of subjects and techniques should subdivide, much as the human cell. Subdivision, along methodological channel, is a logical, and ... a most rational procedure ... that (a) the social science segments which are theoretical and involve model-buildings, advanced statistical techniques and comprehensive statistical frameworks oriented to general theory be viewed as one distinct field, a cell in itself, i.e. regional science, (b) the rest of the field, traditionally designated geography, either continues as it has in the past, or subdivide into two or more parts (example social geography, including political, cultural, historical and economic and physical geography), including bio-geography and related areas; and (c) one of the chief meeting grounds (zones) of regional science and social science geography be at the intermediate statistics level where the geographer with his feet well-rooted in the facts of reality scrutinises and appraises the abstract concepts, operational models and theories of the regional scientists and where the regional scientist with his logical schemata and broad framework suggests to the geographers new types of data which may be usefully collected and new and more efficient ways of processing and utilising data and where together utilising both the system and vectors approaches, they make a possible greater understanding of society in its diverse spatial and related forms.

The regional scientist and geographer as general spatial practitioners over the several sciences, one at the abstract theoretical level, the other at the more empirical, grass-roots level. Each attempts a partial (horizontal-type) integration of the social sciences, an integration involving the synthesis of spatial knowledge and techniques, but only at his own appropriate level of knowledge.

Regional science theory tries to overcome such intuitive judgement by formulating guidelines for planning in dynamic setting of reality which requires publication. The regional scientists' kit of tools such as input-output, benefit-cost, industrial complex, programming or factor analysis will be handy first to employ criteria to determine the finite number of alternatives to be considered and second to ascertain criteria for choosing from among the alternatives. The planners need to optimise within the system of objective functions and constraints which are intimately related to values and goals. The regional scientists have to resolve the conflicting values and goals, policies; the regional scientists should be able to make a more positive contribution to social analysis and policy formulations. The role of regional scientists is further strengthened by their ability to develop superior tools and methods to identify the loci of decision-making in the society within which they operate.

Though Pathak makes a strong plea about the relevance of regional science, particularly, for the country like India which is confronted with the problem of regional disparities and imbalances, and requires conscious planning efforts to achieve the material and social well-being of the masses through removing regional disparities, but nowhere in the departments of geography across the country, there is a separate unit and/or cell exclusively devoted to regional science. He further says that social scientists often confuse between regional science and regional planning/development. Regional science is something different from the geographer's approach to regional planning, because of philosophical and methodological differences. Nevertheless, 'space' is common to both regional science and geography, and that often brings them 'together'.

S. A. Alam (9th Indian Geographical Congress, Srinagar, 1987) makes a very critical assessment of the 'state' of geography in India when he says: 'Geography is one of the most ancient disciplines, but in terms of development has lagged behind most of the social and physical sciences. These disciplines, besides possessing a strong theoretical background, have focused their studies to problems of life and society and have therefore made themselves more meaningful to the society. Geographical studies too have moved in this direction but not to the extent they ought to have. In order to stand on equal terms among the leading disciplines of knowledge we must focus our studies strongly to problems of life and society in order to establish firmly our social relevance … the key problems … such as the energy problems, environmental degradation, urban and industrial pollution, growing urban poverty and proliferation of slums, etc. have hardly attracted the attention of geographers in India…'.

He further says that: 'In the field of urban geography, geographers have made significant contribution in explaining growth and morphology of Indian towns or in explaining city hinterland relationship. Unfortunately, we are now overdoing it. Let us grow out of this growth and morphology syndromes and pick up the key urban issues for in-depth analysis. As for instance, we can examine and analyse the problems of metropolitan India…. The study of the location of industries is indeed important but more important are the consequences of industrial location on urban living: Industries have become major source of pollution of ground and surface water and therefore constitute a major health hazard. Every major metropolitan city of India is a potential Bhopal…. All these developments have significant spatial consequences and should constitute major concerns of geographers…. The social geographers should critically examine the social consequences of technological developments. Technological transformations are likely to influence significantly the social contours of our established social order and may lead to the emergence of a new social system which will reflect not only a new pattern of social relationship between man and man but also between man and land.'

On the trends and quality of research in geography in India, R. Vaidyanadhan (Tenth Indian Geographical Congress, Pune, 1987) observes: 'The bane of geographic research in India seems to have been waywardness of a majority of the researchers. Hardly does one equip himself in one of the upcoming trends, whether it is urban geography, quantitative geography or environmental science, another wave from the West strikes him or gets generated from within the country by scientists in other disciplines, and he is right there so as not to be left behind; we hurriedly join the stream, to make our own limited contribution. This has perhaps not given us much time to project our discipline to the extent it deserves'.

P. Dayal (17th Indian Geographical Congress, 1994, Hyderabad) while identifying the agenda for Indian geography, says: 'One of the first tasks would appear to be re-orientation of our courses of study to include more of physical geography to take care of the growing awareness about the natural environment and its sustainability. Our courses of study at the moment are too heavily tilted in favour of human geography, and physical geography occupies rather peripheral position in our programmes of study and research. This imbalance has to be corrected and a better balance struck between physical and human geography irrespective of whether we emphasise landscape-areal differentiation approach or man-environment relation type approach. Enormous changes have taken place in the methodology and content of human geography providing deep analysis of the human variables in geography, aided by a highly sophisticated form of quantitative spatial analysis. Economic, social, cultural and psychological man figures so much more prominently than 'man on Earth' in the conceptual framework of human geography that the ties between human and physical geography have received a severe jolt and have been loosened to such an extent that there appears to be hardly any need to relate human geographical research to the complexities and intricacies of the physical environment and the relevant environment variable. Physical geography as it is taught in our universities is mostly geomorphology, a little of climatology and very little of bio-geography. The internal imbalance in physical geography needs correction to include the totality of the environment and the spatial element in its make-up'.

He further observes: 'An integrated physical geography has been re-discovered under the name of environmental science largely by non-geographers. The contents of environmental science are identical with physical geography and the difference, if any, is confined to approaches rather than content. Fear has been expressed in some quarters that ... physical geography may cease to be an integral part of geographical studies ... geographers as a group have not taken the lead in the interdisciplinary study of environment and have preferred to leave the field to be dominated by biologists, ecologists, engineers, with some interest shown in by economists, political scientists, and lawyers. Whether physical geography reappears in its new garb as environmental science is immaterial so long as the problem is considered real and significant, and geographers are adequately trained in this area to be able to make a worthwhile contribution, for only then we can expect them to play dominant role in environmental affairs'.

On the use of quantitative techniques in geographical studies, Dayal comments: 'On the whole this has had a salutary effect and has made possible more accurate measurement of phenomena and more sophisticated explanatory mechanism, but by no means has led it to any qualitative revolution and at its worst it has been used as mere embellishment to conceal lack of thought or judgment behind a facade of formulae and symbols at lot of sound and fury signifying nothing. This is being said not to decry the use of quantitative techniques by Indian geographers but merely to point out that quantification by itself is not geography. It is only a sophisticated tool and need to be used wisely and properly.'

A cursory review of research publications in India would show that the Indian geographers have tended to neglect certain very vital and promising areas from enquiry. These include geomorphology, hydrology, climatology, bio-geography, historical geography, and political geography. Even though geography was guided by geologists geographers, this did not lead to a great development of geomorphology or other branches of physical geography in India, like it did in Britain and the USA, in spite of the fascinating physical diversity in India.

India provides very rich materials for the study of *historical geography* with its long tradition of peasant settlements, its ancient religious centres, and the routes of movements along the length and breadth of the country as well as the evolution of ancient, medieval, pre-industrial and industrial towns. Historical geography needs impetus not only for the promotion of tourism, but also for the widespread importance of a historical developmental approach in the study of human affairs.

In *political geography*, the geo-strategic importance of the Indian Ocean, India's relation with its neighbours and better understanding of their mutual problems, interests and conflicts, boundary disputes external and internal, the phenomena of increasing violence and crime and their relationships with caste, politics, and unemployment, the need for national vis-a-vis regional aspirations, state-union relation and electoral participation, the problems and perceptions of religions, ethnic and linguistic groups all present challenging subjects for scientific and perceptive enquiry' and are of great practical importance.

Neglect of the *industrial geography* by Indian geographers is rather intriguing, as industry has been perceived to be the main catalyst 'in our programmes for economic development. While there is a great scope for technological and organisational changes in our agriculture leading to improved productivity and employment, it is generally recognised that there are far too many people on land than it can support, and diversification of our economy by the development of large and small scale industries is the way of providing more arenas of employment and improving their economic conditions. The importance of changing tehnology, the development of new processes and new uses of raw materials for diffusion of industry have to be studied in spatial perspectives and that is lacking in contemporary Indian geographic literature. Use of alternative resource and preference for dispersal of industries also assume greater significance in locational studies. Indian geographers have, by and large, neglected the cost-benefit analysis of the large industrial units, both public and private. Intimately connected with industrialisation is the question of adequate and sustainable power supply. Industry is the heaviest consumer of commercial energy in India and used more than half of the national commercial energy consumption. This aspect also required to be studied in locational and supply perspectives.

Energy production continues to lag behind consumption, and energy consumption is heavily dependent on imports of crude oil and production of petroleum products while domestic oil production continues to stagnate. It is going to be a challenging task to meet the growing demand for energy and will call for massive efforts towards utilisation of renewable sources of energy, with a major stress on conservation and efficiency.

Human development, employment generation, population control, literacy, education, health, drinking water and provision of adequate and basic infrastructure are listed as priority requirements. All these have powerful geographical dimension which geographers must explore and bring to light for purposeful use.

The involvement of Indian geographers in planning till today has been only marginal, although they play a vital role in planning by their active participation in the preparation of various kinds of specialised maps, and, therefore, a more determined effort is needed to make concerned agencies and the government aware of the potential of geographers in planning geography can do a lot to improve the lives of the people by its concern for better knowledge and the wise use of the Earth's resources.

'Man's imprints on the surface of the Earth reflect sharp inequalities in the emerging features of the socio-cultural and economic landscape. The landscape approach as a synthesis of different but related elements that give distinct personality to it has been unique to geography. Therefore, the subject is essentially interdisciplinary in nature', says L. S. Bhatt (18th Indian Geographical Congress, Jammu. 1995). *Regional geography* is considered as the synthesis or the end product of geographer's study of individual elements or structures of the region (physical structure, resource-base-physical and human, urbanisation and industrialisation, movement of goods and people and so on). Regions thus identified are either topical (as in physical, agricultural region, resource region, etc.) or composite (integrated whole combining all the phenomena of an area, for example, economic region or region as defined in regional geography). The latter are dynamic having distinct structure and interrelationship and hence they are hierarchic. This concept of regions through 'man-land relationship' is best illustrated by the statement that a formal region (e.g. agricultural region or resource region) tends to become a functional region (market centres or nodal centres of different types and their tributary areas), and the emergence of functional region is the initial stage in the formation of economic regions. Such a process of organisation of economic activities has built-in characteristics of spatial inequalities because the impact of nodal centres decreases with increasing distance, and this has been the basis of theoretical models of development, according to Bhatt.

Progress of geography in both fundamental and applied studies is towards this aspect of the study of emergence of region and regional differentiation which seems to explain the causes of regional imbalances in development. The concept of the region has undergone changes from environmental determinism in the form and functioning of region to man-made environment for which regional differentiation and characterisation of their positive and negative factors would be only the framework of reference in the initial stage of measuring inequalities. Towards minimising the ambiguity in the definition of the region in an inter-disciplinary context, the term 'space' is used as neutral, and so that introduction of spatial dimension to the sectoral-temporal development plan would be the common objective of different fields of social science which have space affinity, e.g. geography (geographic space), economics (economic space) and sociology (social or behavioural space). The search for regional boundaries is less rigid and the regional boundaries, where necessary for analytical purpose, are made co-terminus with the nearest administrative boundary to facilitate quantification of various elements of the regional structure and to understand and evaluate the spatial processes by applying appropriate quantitative and cartographic techniques of measurement of association among different elements (correlation, regression also) and areal variations in the extent of association among different elements (regional differentiation by using factor analysis, cluster analysis, discriminate analysis). Mapping techniques have been complimentary because the maps serve as 'model of the real world' and help to understand the phenomena of contiguity as in the distribution of agricultural resources or discontinuities in the formation of 'core' areas of economic development and point to location as spatial dimensions of industry and urban centres.

L. S. Bhatt observes that Indian geographers have greatly contributed to the field of regional planning, and to the study of inequality. As an extension of the study of regional geography to planning, an attempt was made for the first time by C. D. Deshpande in 1948 to identify a scheme of planning region of former Bombay Province and to identify the problems in planning. The anomaly between regional and administrative boundaries was noted. A hierarchic scheme of regions of India (macro, meso and minor) was evolved by O. H. K. Spate in 1957 as a part of the study of regional geography of India. Since the 1950s, there have been attempts in the delineation of planning regions of different scale—National, State and District level. These studies emphasised the regional and spatial approach to planning, using the concept of regional and locational hierarchy. The pilot regional survey of Mysore State (re-named Karnataka) conducted at the Indian Statistical Institute was a milestone in a systematic evaluation of the natural environment, resource base, population, patterns of urbanisation and industrialisation. Geographers have done some pioneering works in this field. The late V. L. S. P. Rao may be considered as a pioneer in the field of regional planning in India.

Emphasis on minimisation of regional disparities in development in the context of the natural perspectives of economic development elaborated in the Third Five Year Plan served to re-orient their approach to regional studies conducted by Indian geographers. They have attempted delineation of major resource regions and sub-regions of the country using secondary data at the district level and the maps of natural resources prepared by departments of forestry, minerals and others. Agricultural regions were delineated identifying the core and peripheries of major crop associations. Geographers have attempted delineation of economic region by grouping linguistic states on the basis of complementarity in their economic regions and the resultant spatial organisation of economic activities within sub-region of such state as was assumed to serve the purpose of spatial integration of the economy and reduction of regional inequalities in development.

While regional differentiation and characterisation of regions have a bearing upon the problems of regional disparities in development, the problem of measuring the extent of regional disparities is a complex exercise of interdisciplinary nature. Indian geographers in collaboration with statisticians and economists have contributed to measuring variations in level of development and in providing regional interpretation of the observed patterns.

The contributions of Indian geographers to the problem of inequality present the fundamental view of geography as a spatial discipline, merits and limitation of the approach to the problem starting from descriptive account of regions, to delineation by quantitative mapping techniques and the application of quantitative techniques for measurement of association among several phenomena and regional differentiation. This has enabled geography to focus its attention on spatial dimensions in national development planning in a multiregional framework comprising different tiers of political-administrative work through which the planning process operates.

The number of case studies conducted by geographers on district and block level planning is large and increasing, but the evaluation of such studies brings out the inadequacy of micro-regional surveys which ought to have preceded the studies on district and block level planning. A normative physical spatial development framework is required to be developed for identifying a shelf of development programmes that could create both durable assets and employment in the rural areas. The resource inventory which is considered basic to the District Development Plan is not adequately prepared using large-scale topographical maps and other information. With access to more sophisticated and latest information through satellite imageries and computer, it should be possible to interpret these data in their spatial context.

Dichotomy between rural and urban components of Indian social and cultural and economic structure needs to be understood in a regional framework. While geographical studies focus on spatial patterns there is a need to evaluate the interrelationship in hierarchies system of spatial interactions.

K. V. Sundaram (21st Indian Geography Congress, Gorakhpur, 1998) says that: 'The new wave of democratic decentralization in the country, i.e. the Panchayat Raj renaissance, offers an unprecedented opportunity for geographers to contribute their skills to the development of the country. The recent constitutional amendments have emphasised area planning at the local levels which have become basic and fundamental. As a discipline, geography has been traditionally identified with 'Area Studies'. Physical geography and regional geography are important cornerstones in this context, as they help provide the physical and cultural base for area planning. The application of geography to area studies will provide the cultural and area-specific dimension which is essential.

Geographers with their ability to recognise various development phenomena and interrelate them in space should be able to contribute significantly to some aspects of planning at all the three levels: Village, Block and District.'

On the crisis of paradigm that modern Indian geography is faced with, P. Mishra (19th Indian Geography Congress, Shillong, 1996) comments: '... Indian geographers have not yet even begun to explore the history of Indian geography. The ecological concepts enunciated by our forefathers in regard to man-nature relationship can offer us new ideas for theory building. Symbiosism, which our ancients advocated, is now getting recognition from all over the world. It offers the base for a new paradigm. Had we explored this concept more deeply, we could have presented an alternative to determinism, possibilism, and probabilism. There are many ideas contained in our Sanskrit literature which are geographic and highly relevant to the present day world'.

Gopal Krishan (23rd Indian Geography Congress, Dharwad, 2000) identifies certain priority areas for Indian geographers to carry on their researches; and which include:

1. The regional scenes of ecologies, society and polity are going to be more complex in future. This calls for a higher level of competence on the part of geographers to comprehend, describe and explain the emerging scenario. Indian geography must sharpen its reflexes by way of immediately responding to new challenges.
2. Indian geographers must find as to why they have lost the initiatives in taking up themes or methodologies which were inherently geographical in nature. These included, in particular, ecological studies, geographic information systems and regional planning. Other disciplines and professions took the lead. We, the geographers, simply followed.
3. New millennium should be visualised as an opportune time for regimenting Indian geography. We must have a clear perspective of the spirit and purpose of our discipline. A blend of theoretical and empirical dimensions of any theme should manifest itself in our research work, and field work must be our forte.
4. Indian geography should be guided by the tenets of social relevance. It must impose its visibility and make its presence felt. It must be in a position to assert that 'This only I can do' and others should acknowledge that 'Yes, this you only can do'.
5. Indian geography should concentrate more on those themes in which we have greater competence. These include eco-development, area studies, regional development, public policy and geographic education, among others. The future professional scope of our discipline lies primarily in the promotion of geographic information system. A sound base of cartography is a must for the purpose.

'Indeed, the evolved scope of Indian geography resembles a poly-cyclic landscape. It displays a co-existence between geographies belonging to different eras. The best and the rest go together. For some, Indian geography is a saga of continuous achievements and consolidation; for others, it leaves much room for its improvement' (Krishnan, 2000).

Emerging Branches and Trends in Indian Geography

To quote Jasbir Singh (2000, viii).... 'Indian geographers must take an active interest in conducting studies in the geography of agriculture/industry/population/settlements/health/environment/politics/resource/human behaviour/administration, etc. Unfortunately, in India the achievements in planning-oriented geographical research have remained behind the expected goals. At the same time there are some branches of geography which have not been touched and have remained unattended by and large by the Indian geographers, particularly gender studies, behavioural geography, transport geography, historical geography and trade geography, or in some ones a few studies are available, such as in political geography, administrative geography, social geography, industrial geography, etc.'

An analysis of the Status Reports (Progress in Indian Geography) published by the Indian National Science Academy for the 29th International Geographical Congress (Seoul, Korea, 2000); the 30th International Geographical Congress (Glasgow, UK, 2000–2004), and the 31st International Geographical Congress (2004–2008) Tunis, Tunisia, 2008 have revealed that not much changes have occurred in Indian geography during the period of 1996–2008, particularly with regard to the research in the various branches of geography which have remained more or less the same as before. The branches of geography that have attracted more focus and emphasis and interest in the field of research are as follows:

1. ***Administrative Geography***. It is one of the emerging branches of Indian geography, though not fully expressive in the arena of geographical research and studies. Nevertheless, this branch is being greatly 'emphasised' by the geographers of Punjab, Haryana and Delhi, including western UP Administrative geography is concerned with the spatial attributes of executive processes. Administrative geography finds all variety of public affairs and their management at the local level as of greatest relevance. The public policy is emerging as the focal research agenda in administrative geography in India; and five broad themes of this branch could be identified in the research arena in India:
   (i) Study of variations in quality of governance at the level of local government, this theme of study not only concentrates on governmental organisations and institutions, but also focuses on non-government, cooperative and household sector.
   (ii) Study of inter-state differentials in devolution of powers to the bodies, a highly commendable reforms in Indian polity and governance following the 73rd and 74th Constitutional Amendments effected in 1992.
   (iii) Administrative area reform, especially the reorganisation of states and districts in India
   (iv) An examination of public policies in terms of their formulation, implementation, and impact.
   (v) Study of territorial administrative units organised by private organisations, especially the private corporate sector.

   An active interface of geography with public administration is critical. While the latter can suggest special terms such as administrative process, resource mobilisation, and development organisation, the former can conveniently demonstrate their earth-bounded character. Another imperative is to interact with the practitioners of administration and to impress upon them the relevance of spatial parameters of any public policy of administrative activity. Indeed professionalism will have to be a part of the training in administrative geography.

   Gopal Krishan of the Department of Geography of the Panjab University, Chandigarh, is credited with having pioneered research works in this 'neglected' branch of geography by his publications, and other academic achievements in the field. At the moment, Suryakant of the same department and of the same university has been carrying forward the tradition of Gopal Krishan in the field of administrative geography. Both have done some joint works and publications. In a democratic-federal institution, like ours, the development of Administrative geography appears to be an 'applied' necessity, particularly those aspects of this branch which are relevant to the problem of public policy.

2. ***Agricultural Geography.*** By contributing nearly one-fourth of the GDP and providing livelihood to around 70 per cent of the population, agriculture continues to be the backbone of the Indian economy. Spatially it is the most widespread economic pursuit

claiming more than 40 per cent of the country's total area. Agricultural geography in India is as old as the country itself. During the last 60 years, geographers specialised in agricultural geography have emphasised the need to work on the various aspects and dimensions of the agricultural scenario of the country. Geographers have raised several problems and tried to analyse them in different geographical regions of the country. Their main focus was on studies of agricultural productivity, efficiency and development, soil and water management, regional studies on crops and cropping intensity, land use and land cover and nutrition and food security.

Agricultural productivity, efficiency and development are the three interrelated aspects of agricultural performance, and have always been at the core of the studies covered under agricultural geography. Various measures and analytical frames have been evolved by Indian geographers to study these aspects from different perspectives. Achievements of agricultural geographers in these fields of study are noteworthy. However, studies on agricultural productivity and efficiency and milk production were mainly based at micro-level units of study. The main focus was concentrated on field crops grown under different agro-climatic conditions. Several studies were germinated from climatic and soil data base and their correlation was sought with crop productivity and efficiency. A section of geographers who have specialised in agricultural geography, attempted to concentrate their main focus on soil studies with respect to soil fertility, soil rating, soil erosion, degradation and its response to various crops. They derived their results with the help of modern techniques of remote sensing and GIS. Hydrological characteristics of water were also studied in relation to agriculture and water management.

A group of agricultural geographers in the country have mainly focused on planning aspect of cropping pattern for the optimum utilisation of land resources. Some have made detailed study of ecological impact of cropping pattern on the environment and on the quality of land, i.e. its fertility and potency for the future. Ecological themes and approaches are becoming more and more common in the studies of agricultural geography. Lot of works have been done by Indian geographers in this dimension of agricultural geography. Geographers in India have also focused intensively on land use and land cover. Ecological impacts of land use changes have also been studied in depth. The use of remote sensing and GIS techniques, in the study of spatial agricultural scenario has increased manifold. This has made possible to make comparative studies of spatial and temporal changes that the Indian agriculture has undergone in recent years.

Problems associated with nutrition and food security in recent years have drawn serious attention of agricultural geographers in India in line with social scientists throughout the globe. Several studies have been made to elucidate the concepts of vulnerability and food security that sought to develop theoretical frameworks to analyse the issues of food security. Studies on the nutritional pattern were done on the basis of standard nutrition units of both production and consumption in order to analyse the population-food balance for some specific region with 'depressed economy'.

However, Indian agricultural geographers have given rather less emphasis on the role of socio-cultural factors on crop diversity and dynamics. Reflections of socio-cultural fabrics in agricultural practice and perception are extremely significant for the 'landscape' analysis. There is a need for such studies in agricultural geography. Such studies require to be made within the frameworks of landscape ecology, landscape chronology, and landscape morphology.

Although a number of Indian geographers have richly contributed to the development of agricultural geography, but the contributions of the late Md. Shafi of the Aligarh Muslim University deserve special mention here, because he pioneered research in this branch of geography in such a way as to have brought global recognition to Indian geographers. He developed indigenous models and paradigms for further research works in agricultural geography. Another Indian geographer who also contributed richly to this branch of geography is Jasbir Singh of the Kurukshetra University, Kurukshetra. He prepared an agricultural atlas of India that brought him world-wide recognition as an agricultural geographer.

3. ***Environment and Resource Development.*** Indian environment and ecosystem regions are getting enormous pressure due to rapid population growth, underdevelopment and careless application of developmental technology. Urgent environmental problems helped geographers for intensive research on how human beings developed with nature. Indian geographers are examining the spatio-temporal dimensions of natural, social, and human environment taking different critical regions of the country. The major environmental issues include population growth dynamics, food and hunger, urbanisation and industrialisation, socio-economic disparities, poverty, diseases, gender, migration, extinction of plant and animals, deforestation, soil-erosion, desertification, natural disasters, toxic and nuclear waste, climatic changes, atmospheric pollution, water pollution, ozone depletion and land-use/cover change. In spite of a lack of integrated techniques and approaches to sustainable resource use in the vulnerable 'Indian' environment, a section of Indian geographers have done commendable research works in this emergent branch of Indian geography.

The major thrust areas of research in this branch of knowledge (geography), emphasised by the Indian geographers are: environmental change and development, resource survey and inventory of vulnerable environment, forest and biodiversity, urban environment and climatic change, land and agricultural resources, water resources, wetland resources and mangroves, mountain environment, disaster management, environmental impact assessment, sustainable development of wastelands. Ram Babu Singh, UGC Research Scientist 'C' (Professor) of the Department of Geography, University of Delhi, has richly contributed to the. 'identified' thrust areas of Environment and Resource Development, through his scientific and paradigm-based studies.

Geographers in India, in recent years, have studied the dimension of environmental change with the objective of analysing and understanding peoples' role as both the possible cause and target of environmental change as well as recognising the local issues and use of local field-based geographical studies supplemented by national, regional and local data. They have provided an indispensable contribution to analyse the key driving forces of land-use maintenance and change. Their studies necessarily reflect the wide diversity of economic, social, cultural, and institutional practices and traditional knowledge at the local level. This aspect of study promotes interaction between local and regional communities, conflict prevention and resolution in critical environmental situations. This has direct implications for policy development and implementation in order to develop strong links between research community, policy-makers and environmental management experts.

Vulnerability provides a basis for analysing resource and environmental pressure. It depends on exposure and sensitivity to impacts and the ability to cope or adapt. Vulnerable environment is highly sensitive and responds rapidly to anthropogenetic intervention. The Resource Information System has been considered as multi-dimensional: attribute dimension, spatial and temporal dimensions. Indian geographic studies offer such compatibilities as they integrate multi-sector, multi-space and multi-period perspective. The development of various kinds of models stimulate both the socio-economic and biophysical driving forces, interactions, feedbacks and their responses to environmental change.

In assessing environmental change and prediction of future of that environment in an integrated and holistic manner, there are a number of emerging research areas in Indian geography relating to environmental changes. Indian geography attempts to integrate physical and social aspects of surrounding, and thus plays an effective role in environmental management. A section of Indian geographers with a sound knowledge of remote sensing and GIS have attempted to reconstruct landscape processes, environmental history for ecosystem, state and condition of environment and resource use, modeling of the dynamics of present-day landscapes, and analysis of causes and consequences of anthropogenic changes.

India's metropolitan cities are experiencing rapid urbanisation at a much higher rate than the 'non'-metropolitan small cities. Most of the metropolitan cities are now confronted with the problem of urban environmental degradation. Metropolitan regions already posed not only severe environmental problems like pollution and congested housing, but also brought into sharp prominence the problems of adequate access to urban basic services and infrastructural facilities. Keeping in view these environmental problems, geographers here have emphasised studies on critical issues which are: (i) environmental criticality, (ii) environmental endangerment, (iii) environmental impoverishment, and (iv) environmental sustainability. Three types of researches are being conducted in urban environment: (a) urban planning and environment management strategies, (b) problems of integration of urbanisation into larger issues of regional development, and (c) resolution of urban ecological conflicts.

With rapid population growth, demand on land resources has been increasing, and as a result, the risk to land resource sustainability is also intensifying. This aspect has been taken by Indian geographers in their studies. As a result, agro-ecosystem research and education are being preferred by geographers in the country. A focus on agro-ecosystem in Indian geography now appears to be highly imperative, highlighting different agricultural ecosystems of India, agricultural input environment, ecologically based pest management and eco farming.

A growing research area is that which concerns itself with the availability of fresh water and conservation of surface water resources. A number of studies have been conducted to reveal the degradation of water resource in the Himalayas resulting from erosion, flooding and scarcity of water and degrading water quality. A large number of literature related to watershed management, micro watershed, integrated water resource management and different tools and techniques related to water conservation have been compiled to indicate the future problems of fresh water availability in the country.

Wetlands and mangroves are considered rich in biomass productivity. However, these are under serious threat for development of aquaculture, ponds, roads, and ports. India has a rich wealth of wetland ecosystem distributed in different geographical regions. India is estimated to have an area of about 6,000 sq. km under mangroves. Recently, a directory on wetlands in India has been published to provide information on location, area and ecological categorisation of wetlands. India is a signatory to the Convention on Wetlands of international importance, especially as Waterfowl Habitat (Ramsar Convention).

With regard to mountain environment, contribution has been made to the areas related to assessment and modeling of the interactive influence of topography and land surface heterogeneity on the spatial pattern of soil moisture, evapo-transpiration, runoff generation and erosion.

Disasters, be they natural or man-made, are the real constraints to development and a threat to the environment. The enormous economic losses from natural disasters and the massive relief expenditure make disaster reduction a condition for sustainable development. Development programmes need to be designed in a way to reduce susceptibility to disaster. Disaster mitigation is required to be made a part of national development. However, this part of disaster is very limited in Indian research.

In India, environmental impact assessment is emerging as a measure for ensuring that environmental quality is fully taken into account in the decision-making processes of any developmental programme. Geographers, here, have attempted to evolve consensus and resolve conflict coming from different interested groups. Different stages of environmental assessment have been identified, studied and analysed by geographers.

Studies have been conducted on the problem of desertification, its location and 'expanding' extent with the help of satellite imageries and GIS data base, which were designed to reveal the extent of damage likely to occur to the local ecosystem: Studies have also been made on the negative part of the dry land farming in Rajasthan particularly. Development of wastelands and their sustainable management has become an important theme/area of research by Indian geographers.

To have a better understanding, Indian geographers have focused on how physical and human ecosystems operate and interact. However, there is a lack of integrated techniques and approaches to vulnerable environment. In recent years, Integrated water management is emerging as an important tool for resource management. Understanding complex resource-environmental interactions in space provides an important base for sustainable environmental planning and management. Indian geographers, nevertheless, have richly contributed to the field of environment and resource development.

4. ***Geography of Health.*** It is a developing and an emerging branch of Indian geography, and has been making significant strides in some universities of the country. It has progressed from studies in ecological association of diseases and attempts at disease mapping, to investigations into wider perspective of health and healthcare. Researches in India have been laying greater stress on changing environmental factors and its impact on health as a system and on healthcare. The focus of most researches being human welfare, various attempts have been made to adopt the cultural and structural approaches to address the problem of health and place. During the period 1996–2008, the major areas of research in the geography of health lay confined to disease ecology focusing on diseases such as SARS, bird flu and HIV/AIDS (indicating a shift in studies from the traditional communicable diseases to more modern types of diseases); nutrition and health, emphasising the nutritional status of different societies and communities, and calorie intake patterns; maternal and child health focusing on infant mortality rate in different states, and the role of MCH services in reducing IMR, besides investigation into the reproductive and child healthcare status in different parts of the country; women health with the emphasis on health status of women in the light of malnutrition, anaemia, low birth weight of children, vulnerability to infectious diseases, lung and breast cancer due to exposure to air pollution, urinary tract infection due to poor sanitary conditions, and many more such diseases that women in India are increasingly subject to, healthcare and planning with focus on access to healthcare units (some modern quantitative techniques including the techniques of geoinformation have been applied in the studies of healthcare planning); health and social well-being with the emphasis on the applicability of the model of social well-being from an interdisciplinary perspective to the geography of health; environment and health, focusing on the impact of air, noise, water, and land pollution on human health (a characteristic aspect of this thrust area has been the impact of climatic change on human health, particularly with respect to the incidence of malaria); urban environment and health. This thrust area necessarily focuses on the increasing incidence of non-communicable diseases related with changing lifestyles in the cities and resultant epidemiological transition.

Use of remote sensing and GIS techniques in the researches in the field of geography of health has been increasing. Some studies have used and applied GIS to explore relationship between malaria incidence and mosquito habitat. The purpose of the study is to develop a predictive model for malarial incidence.

Rais Akhtar of the University of Kashmir, Srinagar, Jayati Hazra of the University of Calcutta, Kolkata, and Jayashree De of the M.S. University Baroda, have pioneered this branch of geography.

5. ***Geomorphology***. Early Indian geographers, who were trained in geology, had set forth the tradition of geomorphology in the country, and in the 1950s of the last century, a number of research papers of high international standard appeared in the various journals of geography within and abroad. H. L. Chibbar of the Banaras Hindu University, Varanasi, S. P. Chatterjee of the University of Calcutta, Kolkata, were the forerunners in promoting geomorphological studies in India. The main emphasis of geomorphological studies was on 'denudation chronology'. Enayat Ahmad of Ranchi University, Ranchi, and R. P. Singh of Patna University, Patna, (he later moved to Magadh University, Bodh Gaya) also richly contributed to the development of geomorphology in the late 1950s and 1960s of the last century. The relevance of geomorphology, however, declined in the following decades, because of increasing relevance of urban geography, economic geography and commercial geography, agricultural geography and land use, population geography, etc. Most of the geographers during the formative stage of the Indian federation had gone to Great Britain for their higher studies where they were mostly trained in different branches of human geography rather than physical geography. On return to India, they attempted to set forth the tradition and/or heritage that they learned there. Since physical geography or, for that matter, geomorphology was missing in their curriculum, the founders of modern Indian geography did not attempt to give preference to physical geography and geomorphology.

For quite a long time, geomorphology remained more or less 'hidden' in the area(s) of research in India. Nevertheless, a few scholars continued with the geomorphological studies, but their number was insignificant, given the number of geographers engaged in research in different branches of human geography. However, with the establishment of the Indian Institute of Geomorphologists in the 1990s with its centre at Allahabad, that the preference for geomorphological studies started 'resurrecting' in India. This institute also started publishing a journal, known as the *Indian Journal of Geomorphology*. The journal is exclusively meant for the publication of articles dealing with geomorphological phenomena, which are mostly case studies. Another journal which publishes a fairly good number of articles on geomorphological phenomena is named as *Indian Journal of Landscape Systems and Ecological Studies*. It is published from Kolkata.

The studies in geomorphology in India during 1996–2008 have mainly concentrated on four major themes : (i) fluvial geomorphology, (ii) desert geomorphology, (iii) coastal geomorphology, and (iv) tropical weathering and erosion. In addition, some studies are based on the use of remote sensing techniques. In fluvial geomorphology, the foundation with drainage basin and network morphometric analysis continues even today. But in recent years, there has been a shift from single morphometric analysis to the study of channel forms and processes. Many of the papers under this category are related to river hydrology, especially floods. Other themes in fluvial geomorphology include the channel characteristics of usually wide piedmont streams, temporal changes in channel morphology, identification of longitudinal profiles and large-scale geomorphic mapping.

In desert geomorphology, the major areas are: (i) wind erosivity and erodibility of terrain, (ii) mechanism of sand-drift, (iii) characteristics of sand-moving and dune-forming winds, (iv) trend and sites of old dunes, (v) characteristics of modern mobile barchan, barchanoids, and megabarchanoids dunes, (vi) resettling the former sites of streams in the Aravallies.

The study in coastal geomorphology necessarily emphasises on coastal landforms and processes. The vulnerability of the coast to erosive action, threat to the fragile mangrove ecosystems and the place deposits along the beaches are some of the topics covered in this section. The studies on morphodynamics of tidal inlets, slumping of sea cliff are largely devoted to the coastal processes whereas some studies have been conducted on mud beach or lagoon to describe the morphological aspect of the given forms. Apart from these aspects of studies, another aspect of the study of coastal geomorphology is pivoted around the changes in sea level along the Indian coast. On the basis of large-scale mapping, several relict features on the eastern coast have been identified.

The major thrust areas under the theme tropical weathering and erosion are mainly concerned with: (1) erosion rate(s) in the river catchment area; (2) measurement of run-off and erosion rate; (3) evaluation of the rate of deforestation and agricultural expansion on sediment yield; (4) impact of human activities on soil erosion; (5) the rate of erosion and badland development in lateritic terrain; (6) rate of weathering profile development and rate of denudation; (7) regolith formation, and (8) colluvial deposits resulting from mass movement and fluvial erosion of regolith on hill slopes.

Use and application of remote sensing techniques have been the focus of some studies in recent years. The major thrust areas where the use of remote sensing has been carried out in studies, include the study of abandoned channel flows over the coastal plains of northern Tamil Nadu, the study of identifying palaeo-valleys in the Luni river in the alluvial plain between Jodhpur and Pali; the study and interpretation of different types of depositional, denudation and structural landform assemblages in Assam. A geomorphological mapping of Assam was done with the help of remote sensing.

Similarly, Global Positioning System (GPS) was used to measuring the coordinates and elevation of sand dunes in Rajasthan.

Savindra Singh of the Allahabad University, S. R. Basu and S. C. Mukhopadhyay of the University of Calcutta, V. S. Kale of the University of Pune and A. Kar, and others have been pioneering geomorphological studies in India for the past 20 years.

6. ***Gender Geography***. It is a new, rather an emerging branch in Indian geography. Gender geography has witnessed a gradual yet cautious widening in the 1990s. By the year 2000, the field of this branch expanded enough to claim for itself its share of academic space in the discipline. The development of this new branch of geography is necessarily confined to the universities of West-Central India, particularly Jawaharlal Nehru University, Delhi University, Kurukshetra University, Jamia Milia Islamia, and so on. In these universities, gender-geography has been introduced as a full paper at the post-graduate and under-graduate levels. The introduction of new courses that interrogate spatial pattern with reference to gender and the incremental volume of geographical research using gender as a category of analysis are major milestones marking the development of the branch. From being almost implicit, rather invisible, gender as an analytical category has crept into almost every field of geographical enquiry. Research papers using gender include themes on health, development, workforce, participation, food security, conflict, disaster management, environmental issues, resources, micro-credit and policy planning among others.

Barring a few studies, the engagement of space with gender and vice versa remains largely glossed over by geographers. One of the major thrust areas of this branch of geography in India is that which reflects equating of site with space and sex with gender. This thrust area of research has the largest number of works and publications. However, there is a paucity of studies and works on the themes of gendered experience of space, gendered spaces and spatialities of gender. The intertwining or interweaving of space and sexuality is another optimistic thrust area that deserves mention, but research in this thrust area is not too expressive. Nevertheless, the contemporary research works in gender geography reveal a healthy methodological combination of qualitative methods along with the quantitative.

Saraswati Raju of the Jawaharlal Nehru University, Delhi, has long been associated with gender geography. She has very richly contributed to the development of gender geography in the country. It will not be an exaggeration if she is called the 'founder' of gender geography in India.

7. ***Political Geography***. As a field of higher research, political geography was relatively a late arrival. Nevertheless, this branch was very much in the syllabus curricula of a number of universities in the country. But it was a neglected branch. The reason was the lack of initiative on the part of the makers of modern Indian geography who preferred to establish those branches of geography which they learned in the universities of Great Britain.

The major thrust areas of research in political geography in India are mostly concerned with different aspects of politics of India, which have spatial expressions and attributes, both in terms of internal and external functions and relations. The thrust areas of research and studies can be categorised as: (i) geography of federation—the Indian way, (ii) geography of integration with emphasis on raison d' etre and state-idea of the Indian Union, (iii) concepts of Indian nation and nationality, (iv) geography of India's foreign policy and non-alignment movement, (v) boundary studies and problem of boundaries, (vi) Kashmir conflict and related issues, (vii) inter-state and international water disputes, and (viii) geographical interpretation of India's relations with the neighbours and with other countries.

Apart from these major thrust areas of research in political geography, a new research area in the branch has come into existence in the past few years, and that is 'Electoral Geography'; or 'Geography of Elections'. This is the only thrust area in political geography of India which has experienced quantification, theory formulation and spatial modeling and hypothesis testing.

R. D. Dikshit (M.D. University, Rohtak), C. P. Singh (University of Delhi), R. L. Dwivedi (Allahabad University), Swaranjeet Mehta (Panjab University, Chandigarh) and Ram Nandan Pd. Sinha (M.S. University, Baroda) pioneered and have been pioneering research in this field of geography. Ram Nandan Pd. Sinha of M.S. University, Baroda, is credited with having developed an alternative concept, called the concept of 'Geosheeltics'—an antithesis of the concept of geopolitics. Geosheeltic is geography and politics mediated by the Sanskrit word 'sheel' meaning 'righteousness'. Hence geosheeltic is geopolitics based on Gandhian philosophy of truth, non-violence, ethics and morality. The concept has been widely discussed in Israel, Korea, and Bangladesh.

8. ***Population Geography.*** It is a well-developed branch in India, experiencing the application of modern quantitative techniques in research methodology, besides qualitative studies and discourses. The development of this branch of geography in India took place in the 1960s when the department of geography, Panjab University, Chandigarh, started teaching and research in population geography. The Panjab University, Chandigarh, is the pioneer in population geography in the country. Now, population geography is taught as optional and compulsory paper at the post-graduate and under-graduate levels, in almost all the universities of the country.

During the period of 1996–2008, population geography in India has undergone some philosophical/conceptual and methodological changes, and accordingly the areas of research have witnessed structural changes. Concepts, theories, models and paradigms together constitute a major thrust area of population studies. Some post-modern concepts on population and space have been incorporated centre in population geography with the objective of applying Derrida's models of deconstruction and fragmentation in some of the universal population theories.

Apart from this major area of research enquiry, population studies can be grouped under the following major areas of research: (i) studies on population density and distribution in terms of overpopulation; (ii) studies on vital rates with emphasis on fertility and mortality pattern (infant mortality also). Several studies in this area have been done to identify the areal differentials, rural-urban differentials and inter-urban differentials. (iii) Studies on population growth. Such studies are based on census data to focus on the overall trend in population growth and also the decennial variation in relation to the prevailing fertility behaviour. Some regional studies have been conducted to assess the impact of population growth on the socio-economic landscape of the regions. Quantitative techniques have been applied to measure the extent of impact. Few studies have been made to evaluate and analyse the impact of growth in urban context and pressure of populations. In one of the studies, concerning the impact of population pressure, modern techniques like remote sensing, GIS and GPS have been applied and the result was highly appreciable. (iv) Studies on migration have been numerically preponderant in population geography. Most of the studies are concerned with internal migration, notably rural-urban migration. However, some studies have been made on the Bangladeshi migrants in the north-eastern India. Some studies are concerned with the absence of the government policy on migration. Few studies have been done on implications and impacts of migration, and the behaviour of migrants. (v) Studies on composition, working force and occupational structure have been conducted in the light of literacy differentials, spatio-temporal changes in the age and sex structure, trend in demographic behaviour of major religious groups in India. However, some studies have been made on occupational composition of SC and ST population, spatial pattern of primary, secondary and tertiary activities, regional dimensions of work-force and levels of social developments.

Apart from the aforesaid thrust are of research in population geography, some more areas have been identified such as gender issues, population/ethnic groups, population as resource, demographic profiles, human development, management of human resources and sustainable development where studies are being conducted, using quantitative techniques and samplings in order to arrive at concrete conclusions.

The credit of enhancing the relevance and prospect of population geography in India goes to the faculty members of the Panjab University, Chandigarh, particularly to G. S. Gosal, Gopal Krishan, R. C. Chandna, Swaranjeet Mehta and M. S. Gill of the Punjabi University, Patiala. The department of geography, Panjab University publishes a journal on population studies: *Population Geography*. Population geography is, thus, a fast developing branch of systematic geography in India.

9. ***Social Geography.*** India with a vast geographical expanse and a huge population, presents a highly complex mosaic of society and culture evolving out of long-drawn historical process. It is a land of regionally-grouped socio-cultural-ethnic diversities with hardened cleavages. The Indian society is largely plural in character and disaggregated also. A proper understanding of all the above facts of India's ethnical and social realities is not possible without a critical examination of their specific spatial identities in national and regional settings. Here lies the relevance and significance of the socio-spatial perspective at different spatial levels over time towards understanding the highly complex character of the social and cultural elements and other associated phenomena of the country both in entirety and in parts. It is in this background that the necessity of 'developing' social geography was urgently realised. Social

geography is the study of social phenomena and processes in their areal and/or spatial perspective. Social geography is a new branch, and a chronological review of its development in India suggests that it is still in its embryonic stage trying to come out of a rigid disciplinary boundary and control that stands in the way of its development.

A number of thrust areas, however, have been identified in the research works, conducted and carried out in this branch of geography, which include: (i) critical social theory and globalisation, (ii) social components and social structure, (iii) social order, cultural domain and societal process, (iv) social change in mega-urban regions; (v) tribe, displacement and ethnic conflict, (vi) caste and language, and social conflict; (vii) class struggle, (viii) gender studies, (ix) issues of social concern and social well-being, and (x) education and under-development.

Though the list of research publications is limited in the social geography, but the political happenings of contemporary social relevance in the country have attracted young geographers to be actively involved in the aforesaid thrust areas for enquiry, however, in spatial perspectives. The ongoing process of globalisation to which India is now firmly committed appears to have been the driving force behind a large number of geographers interrogating the spatial impact of the changes manifesting on the wake of these changes. Researchers in social geography in India have been alive to contemporary social problems of displacement, redistribution, class struggle, caste tension, ethnic strife, social exploitation. Mandalisation of Indian society with resultant social realignment has undoubtedly opened up a new vista for the development of social geography in the country.

10. ***Urban Geography.*** It is one of the most dynamic branches of Indian geography. Urban geography in India has been moving forward seemingly continuously in its philosophical perspectives and thematic contents. It has not merely captured the prevailing paradigms of geography, but has succeeded in injecting new paradigms. This is rightly so because cities play critical role in spatial organisation and modernisation processes by contributing to regional and global economy. Teaching and research in the urban geography in India is as old as the teaching and research in the agricultural geography and land-use. Nevertheless, this branch of geography stands on sound foundation of research methodology

A survey of literature reveals that urban geography in India has been unfolding several new dimensions of thrust areas which include: (i) urbanisation, urban issues and analysis—trends and pattern—this is the most sought after thrust area in the branch. Here, the major emphasis is given on the processes, pattern, causes and trends in the light of suburbanisation, megalopalitanisation, counter urbanisation, and re-urbanisation. Apart from these trends of study, some attempts have been made to reconstruct trend of urbanisation in the historic past; (ii) urban environment-it is one of the most popular thrust areas of research as most of the cities are plagued by environmental problems. Most of the problems, which find research expression, are domestic in nature, with emphasis on air pollution, water pollution, toxic waste, and natural and human hazards. Studies have been conducted on the urban quality of life, and slum patterns. The quality of social environment has also emerged to be a part of this thrust area of research, (iii) question of urban sustainability/eco-planning of cities—it is the most captivating theme for urban geographer because the question of urban sustainability has become a key issue in contemporary context. Sustainability has many variants and is being approached differently by different scholars. Major emphasis under this thrust area is given to the effective implementation of mission made approaches such as those designed to improve basic services to urban poor, infrastructure, integrated housing and slum development programme. Another dimension of research in the thrust area is that which concerns with the greening of city, maintaining bio-diversity and managing land degradation and eco-planning, (iv) urban governance—this area of research is necessarily concerned with water supply, conservancy, sanitation, services and other facilities and solid waste management, etc. Apart from these thrust areas of research, a few more areas of research in urban geography have emerged which necessarily include mega and million cities, small and intermediate cities, urban migration dynamics, socio-economic dimension, urban transport system, urban planning, urban political conflicts and so on.

Other Branches of Geography

Apart from these branches which have shown an emerging trend on the horizon of Indian geography, there are few more emerging branches that need to be mentioned. These branches are as follows:

1. ***Cultural Geography.*** The post-modern discourse has certainly given a new meaning to studies in cultural geography, and cultural geography in India had become a shadow of its western incarnation. India has a rich cultural heritage that gets expressed and transformed through 'landscape formation', which needs to be studied in spatial-ecological perspective. The geographical implication of cultural pluralism vis-a-vis landscape-formation had hardly received any attention in the past, but with the increasing emphasis on social geography the relevance of cultural geography has increased. Social geographers are getting drawn towards cultural geography of India, with the objective of evolving new research methods and paradigms for the analysis of cultural landscape, of course within the framework of post-modern, post-structural discourses.
2. ***Historical Geography***. Historical geography has never been a priority area in Indian geography, though its importance can hardly be over-emphasised. However, most of the studies in this branch have been done by non-geographers, and the studies are strictly 'non-spatial'. There is a need for meaningful historical-geographical interpretation of the major historical events, but at the same time, methodological developments need to be done with clearly-defined thrust areas. India has a rich heritage of history from the antiquity down to the present, and that must be treated in spatial framework.
3. ***Settlement Geography***. This has been the traditionally important area of research that had attracted good number of researchers. The development of this branch of geography was somehow restricted because of the greater emphasis being placed on other branches of geography. However, with a shift from the conventional analysis of size, form and location of settlements to studies on functional aspects and locational characteristics of human activities as well as spatial-functional organisation of economic landscape that settlement geography has regained its lost place on the horizon of Indian geography. Issues pertaining to the impacts emanating from hyper-urbanisation and diversified urban system are the future thrust areas of research.

However, research on Indian rural settlements has not touched upon a wide variety of dimensions. It is confined to the same old areas of research developed decades earlier. In fact, rural settlement geography does not appear to have received the attention it ought to have received on account of the sprawling rural landscape across the country. This branch of geography requires planning-oriented teaching because India has a large number of villages. Nonetheless, a new trend has developed in recent years, in which studies are being taken up to distribution pattern of villages and their size, ruralities, accessibility to vital socio-economic amenities, socio-economic development levels, transformation of rural landscape, rural gender issues, literacy level impact on housing patterns, and so on. This branch is likely to make its impact felt on the research horizon of Indian geography in the years to come.

The combined use of GIS and DIP provides better prospects of spatial analysis, environmental monitoring and forecasting once wider areas within limited time-span. Such geographical technology also provides the spatial data base of natural resources and basis for formulating developments. It would be easy to develop predictive model capabilities in order to achieve effective public policy in years to come. Remote sensing with its unique characteristics of synoptic view, repetitive coverage and reliability has opened immense possibility for spatial mapping together with planning and management to achieve optimisation of resource utilisation and conservation.

USE AND APPLICATION OF REMOTE SENSING AND GIS IN GEOGRAPHICAL RESEARCH

One of the emerging scenarios in Indian geography in recent years has been the use and application of *Remote Sensing and* GIS in geographical studies and researches. It has opened a new research frontier in Indian geography.

1. ***Weather and Climate Change***. New developments are taking place in estimating sea surface temperature and modeling the methane emission using MODIS, the multipurpose INSAT 3A provides vital information for a detailed monitoring of the weather and an accurate forecasting. The installation of 100 data collection platforms (DCPs) and the Disaster Warning Systems (DWSs) has been done for transmitting methodological data on the one hand and for tracking the path of cyclone on the Andhra Coast on the other hand. They also provide information imageries. Significant information has been consistently received at the meteorological headquarters and centres of global climatic change on forest diversity from the satellite signaling and imageries.

2. ***Geomorphology***. Space Application Centre is continuously monitoring glaciers in order to observe climate change impacts. Remote sensing data are being used for mapping potential glacier lake outburst floods (GLOF) in the Himalayas. The most frequent applied areas of remote sensing in India is the study of Earth's subsurface and surface features. Aerial photographs are an effective tool for geological, geomorphological, relief and hydrological studies and land-use mapping. Geomorphic units have been identified based on interpretation of the aerial photographs and the Survey of India's topo-sheets and the LANSAT imageries in the various regions like Pali district, Jodhpur district, Luni basin and Tripura state. These geomorphic units have different physical potential and provide a sound base for land-use planning. Other application includes remote sensing study in identifying wind erosion areas. Remote sensing techniques have been applied for geomorphic and land-use mapping of larger areas of Assam.

Based on aerial photographs, a study of Hiran catchment (Jabalpur district), highlights the physical and hydrological characteristics of the area, using quantitative analysis for land resource development and management. GIS has been used for soil productivity assessment and mapping. Drought monitoring is an important aspect of satellite monitoring in Maharashtra. The National Remote Sensing Agency (NRSA) carried out a survey for soil association mapping, land degradation and ground water exploration to aid drought relief in Bundelkhand region of Uttar Pradesh. A GIS has been initiated by the Geological Survey of India for locational and distributional identification of various kinds of rocks and mineral deposits.

3. ***Hydrology and Water Resources***. Targeting ground water in hard rock, using remote sensing and GIS, is a recent phenomenon in hydrological science. In India, remote sensing is used for monitoring the aspects

such as measurements of evapo-transpiration, measurement of water surface roughness, rainfall distribution and infiltration, ground water discharge and salt content of water and light absorption.

Aerial photographs have been utilised to collect information regarding water storage, season and long-term fluctuations of lake and river surface aerial extent, assessment of underground and soil moisture. Estimation of surface temperature of snow has been assumed through GIS technology. The weather satellite imagery is usually used to monitor ice and snow cover conditions providing important inputs for water management and flood prediction. In recent years, microwave sensor data is being used for flood monitoring during cloud cover condition.

The sustainable benefits from coastal marine resources require considerable improvement in the application of remote sensing imageries to improve both utilisation and management of these resources. The launching of the OCEANSAT-1 satellite has brought in significant improvement in the ocean monitoring.

4. ***Forest and Biodiversity***. Assessment of grassland and their changes with time has been greatly facilitated by using remote sensing and GIS technology. It is through remote sensing that the actual forest covers of India were known through findings of satellite data analysis. The first attempt to categorise forest cover types by computer analysis of Landsat data was done in 1978 for Nagaland. In a study conducted by NRSA, the satellite digital image covering entire Periyar-Thodupuzha drainage basin was analysed. Apart from forest land classification, stock mapping and volume estimation, remote sensing is also used for damage assessment and fire detection, which is a common feature of Indian forests. GIS is used in biodiversity conservation plan. The Forest Survey of India prepared forest cover type and land-use maps on 1:50,000 and 1:63,360 scale by interpreting medium to small scale panchromatic aerial photographs for about 4,20,000 sq.km^2 in India. The main application of remote sensing in forest management has been for timber harvest planning and monitoring of logging and deforestation: Digital Elevation Model (DEM) has been used to prepare terrain complexity map. Potential fishing zones are being mapped from the data collected through satellites.

5. ***Land-Use/Land Cover Mapping***. Land information system has been developed to assess and manage the land resources of any area. Using IRS-LISS-1 data in 274 districts through visual interpretation and 168 selected districts through digital techniques, agro climatic zones are being analysed. For the first time, two season satellite data both for Kharif and Rabi seasons are used to precisely estimate the agricuitural land in Kharif and Rabi seasons. Landsat data have been used intensively for mapping. Recently, under this programme, one of the projects refers to changes on the landuse because of urban spread, while another project emphasises on industrialisation in Ahmedabad—Vapi regions, Gujarat, and Chindwara district, M. P. A detailed land capability classification is being conducted using remote sensing and GIS technology, especially for highlands and Himalayan regions, prone to land degradation to be used for the formation of integrated land use plan. The arid zone monitoring includes agricultural improvement and desertification study. The land-use land cover change through GIS technique is very significant in analysing the dynamics of land-use change. Various remote sensing and GIS layers are also being used for mapping urban heat islands.

6. ***Urban Development***. Increasing demands in urban planning and management sectors call for coordinate application of remote sensing and GIS for sustainable development of urban area. Availability of high resolution data form IRS-lC and 1D satellite has revolutionised the process of thematic mapping and spatial-

data base creation, especially in the context of urban sprawl and regional planning. Other application includes use of remote sensing in infrastructure development, urban sprawl mapping, regional planning for air and noise monitoring network, development or road monitoring and management systems and silting sanitary landfills, etc.

7. ***Monitoring of Natural Hazards and Disasters***. The remote sensing data, particularly from satellites, have unique capability to monitor the condition of atmosphere and on Earth's surface. The images from an environmental satellite in conjunction with the conventional data help in monitoring severe connective situations, e.g. cyclones and storms, responsible for large-scale destruction. The visible, infra-red and enhanced infra-red images are utilised to estimate rainfall in sensitive areas. The hydrological and meteorological data transmission from the data collection platform (DCP) system are valuable input for planning flood control and protection measure.

Drought monitoring is an important aspect of satellite monitoring in Maharashtra. The study for the period 1986–89 provides a valuable data-base to study annual biomass production, agro climatic zoning and rain-use efficiency in addition to its use in drought monitoring. In the year 1986, the NSRA completed the survey and mapping of wastelands in India using Landsat satellite data. Flood plain zonation is a key tool in managing the disaster in the various parts of India. Satellite imageries are also being used for assessing the impact of flood on biodiversity, particularly in Kaziranga National Park.

There is no doubt that Indian geography has made steady progress since 1947. Nevertheless, branches developed appeared and disappeared, and then again re-appeared on the horizon. This has been the story of ups and downs that Indian geography has witnessed and experienced during this long period of history.' B. Thakur once remarked (1994), 'Indian geography appears to have borrowed much from the western models, and there is a general lack of tendency among the Indian geographers to develop indigenous models'. There is truth in this statement. Indian geography has progressed very much in the last 20 years, and various branches have developed, and some are on the way of emerging, with remote sensing and GIS techniques being incorporated not only in physical geographical studies, but also in various branches of human geography. But the real problem is that we still stick to western models, theories and laws, and continue with traditions that the geographers of the first generation had established 60-years ago. Problems have changed, geographical phenomena have changed in the past 60 years, but we have not changed. The tendency to rely upon western traditions notwithstanding, the development and progress of Indian geography, necessarily stands in the way of the development of indigenous Indian geography.

USE AND APPLICATION OF GEOMATICS: REMOTE SENSING, GIS AND GPS

The term *Geomatics* is of recent origin. It first appeared in 1981 in *The Canadian Surveyor*, and it was Michel Paradis, a French-Canadian surveyor, who introduced it in April 1982 while addressing the Centennial Congress Ceremony of the Canadian Institute of Surveying. However, at times it claimed that the term was coined earlier by B. Dubuisson in 1969 by combining the words 'geodesy' and 'geo-informatics'. Subsequently, it was adopted by the International Organization of Standardization, the Royal Institute of Charted Surveyor, and many other international authorities.

Geomatics is the English equivalent of the French *Geomatique*, which is a discipline of gathering, storing, processing, and delivering geographic information, or spatially-referenced information. The European equivalent of Geomatics is Geo-spatial Information (Technology).

Geomatics incorporates the tools and techniques used in land surveying, remote sensing, cartography, geographic information systems (GIS), global-navigation satellite systems (GPS [Ground Positioning System], GLONASS, Galileo, and Compass), photogrammetry, geophysics, geography and related forms of Earth mapping. The rapid progress and increased visibility of geomatics since the 1990s has been made possible by advances in computer hardware, computer science, and software engineering, as well as by airborne and space observation remote sensing technologies. Geomatics as a new technology incorporates the older field survey techniques together with many other aspects of spatial data management. It is a rapidly developing discipline focusing on spatial information. The geomatics technology handles and/or manages local, regional, national and global spatial data infrastructures.

Geomatics is, thus, defined as a systemic, multidisciplinary, integrated approach to selecting the instrument and the appropriate techniques for collecting, storing, integrating, modeling, analysing, retrieving at will, transforming, displaying and distributing spatially geo-referenced data from different sources with well-defined accuracy characteristics and continuity in a digital format. Erected on the scientific framework of geodesy, it uses terrestrial, marine, and satellite-based sensors to acquire spatial and other data.

In recent years, classical geography has undergone a structural change in its philosophy and methodology with the introduction of geomatics in geographical studies, researches, and training. Now, it is more a science rather than a *naive* science as it is used to be called 50 years back. Geography is revolutionised a second time over with the incorporation of the geomatic techniques that has also caused a theoretical-shift within the discipline. The first such revolution occurred in the early 1950s when the nomothetic search for model replaced the traditional ideographic concerned with areal association, leading to what is called the quantitative revolution in geography. Increasing use of geomatics in both geography and engineering has opened up a new vista in contemporary applied studies.

The geographers of the Atlantic Community, and the Pacific Community increasingly use and apply geomatics techniques in their studies and research works, pertaining to regional planning, natural hazards and calamities, disaster management and so on. The Chinese geographers have made outstanding progress in the application of geomatics techniques in all branches of geography, and in some areas they have surpassed their American counterparts, particularly in the hazard zonation mapping and regional planning mapping.

Coming to India, geographers here are specialised in remote sensing and geographic information systems, which are components of geomatics, whereas non-geographic institutions and engineering colleges have introduced geomatics courses in their curricula, and in some of such institutions, geomatics has been introduced as a separate discipline, where the PhD courses in the field have been initiated. The institutions, where teaching and researches are done in geomatics are the College of Engineering, Andhra University, Visakhapatnam, Civil Engineering Department, BRECM College of Engineering and Technology, Bahal, Bhiwani, Haryana, Advance Training Centre for Earth System Sciences & Climate, IITM, Pune, Jawaharlal Nehru Technological University, Hyderabad, Space Application Centre, and Physical Research Laboratory, Ahmedabad, CSRE IIT-Kanpur, IIT-Hyderabad, National Remote Sensing Agency, Hyderabad, Indian Institute of Remote Sensing, Dehradun, Institute of Remote Sensing, Anna University, Chennai.

The above institutes conduct applied research using geomatic techniques in a wide range of areas, which include coastal ecosystem, geo-informatics tools and techniques, trends in image processing, marine biology, advances in sensor technology, microwave applications, coastal processes and hazards, photogrammetry and

LIDAR for terrain analysis, planetary science and exploration, groundwater conservation and exploration, agricultural and soils, geosciences and spatial infrastructure, disaster management, monsoon dynamics and climate change, web and location-based services, geomatics application in environmental monitoring, hyper-spectral imaging and application, geomatics in urban and regional planning, geomatics in water resources, geomatics in agricultural applications, geomatics application in land use/land cover and forestry and so on. One of the emerging fields in the application of geomatics techniques in India is in the field of boundary management and strategic border studies of the vulnerable areas. For the last few years, geomatics has been increasingly used by the Indian Army in tracking down cross-border infiltrators in Jammu & Kashmir.

It is, indeed, a fact that Indian geography is not fully expressive in terms of the application of the geomatics technology in totality in the area of research and training, though a good number of geographers have undergone training in geometics technology both within and abroad. Yet, somehow they are handicapped in their research endeavours. Since, most of the geography departments are not fully equipped technologically to impart geomatics-based teaching, so the development in the field of geomatics teaching, research and training is rather poor.

Nevertheless, Indian geography has witnessed some development in the area of geomatics, techniques such as GPS, and Ground Penetrating Radar (GPR), which have contributed to the growth and development of both Remote Sensing (RS) and GIS in the country. Transformation from analogue (visual) method to digital techniques in classification of remote sensing data, from manual to semi-automated methods, digital image process imaging techniques (DIP) to knowledge-based classifiers and new algorithm of data capture are some of the developments over the years in geography. In the area of GIS, there has been a transformation from simple overlaying methods to integration to multiple sets of database, modeling, visualisation tools, simulation studies Web GIS and so on. GIS offers such capabilities as they integrate multi-sector, multi-level, multi-period database. Geographic Information Technology (GIT) is emerging as a pioneering technology and serves as a powerful aid in the development of planning and governance together with disaster management education. Spatial technology has become a key science for decision support of sustainable development and disaster management (Singh 2009).

During the last two decades, there has been increasing use of Remote Sensing and GIS techniques in geography in the country by way of publication of a number of papers by geographers in major journals at regional, national and international levels, and also in various interdisciplinary journals on Remote Sensing and GIS and cartography, geomatics, town planning and various conference proceedings and so on. These papers and reviews indicate the trends in the development of RS and GIS-based research methods and application in various branches of geography taught in the country.

Following are the major thrust areas, where Indian geographers in the past two decades have greatly contributed to enrich this emerging field of geography with their studies and research works, using the techniques of geomatics that involved the massive use of Remote Sensing and GPS:

1. Land resource assessment and management: Remote sensing provides land resource data in the form of digital magnetic tapes and in different bands, useful for delineation of land use/land cover classes distinctly in remote sensing with its multi-spectral, multi-temporal and synoptic view which has the potential to provide accurate spatial and temporal information on land use/land cover of a region in time and cost-effective manner. The data generated through remote sensing techniques can be integrated with non-spatial socio-economic data for identification of planning priority zones, for land resource development. The remote sensing technique has enabled Indian geographers to identify and delineate the various coastal geomorphic units/landforms successfully for their synoptic view. This has helped to understand the geomorphology and related sedimentary environment properly.

Visual interpretation of the satellite imagery provides an important and practical method of land-use/land covers classification and mapping. However, the challenges lie in accurate mapping at higher spatial scale. Data merging of medium resolution multi-spectral imagery (IRS LISS-III) with higher resolution panchromatic imagery (Landsat ETM+PAN) is one of the cost effective methods for generating high resolution data, which has been used by Indian geographers for land use/land cover extraction. Patil (2010) focused on the analysis of cropping intensity indices with the help of GIS techniques in the tribal region of the northern part of Nandurbar district of Maharashtra.

2. Water resources assessment and management: Using remote sensing and GIS techniques, the water resource problems of Udaipur City have been studied by Rathore (2009), while Kavita et al. (2009) studied the groundwater pollution problem of Dindigul town, using the same techniques. Recent advances in RS and GIS have put forward dimension in the analysis of the Chandraprabha drainage basin by supplementing the accurate time and real time information on various aspects of the river basin. Using the digital image processing technique (false colour composites-FCCs), the monsoon, pre-monsoon, and post-monsoon surface waterlogged areas around Patna, Bojpur and Buxar and the region of the Ganga basin have been delineated. The Peddapatippasamudram watershed management study has been done with the help of IRS-18 Geo-coded data and LISS-III data on scale of 1:50,000. The study also analysed relief, slope, landforms, soil, land use, erosion susceptibility, hydro-geomorphology, and land capability to suggest watershed management for sustainability. Similarly, wetland changes in the Vedarnyam coastal area have been assessed by using IRS-LISS-III data with ArcGIS software. Remote sensing and GIS techniques are being increasingly used for effective management of land and water resources in watershed in 44 mandals of the Mahaboobnagar district. These areas have been identified as drought-prone. Various maps have been prepared with the help of modern techniques of RS and GIS that include base map, contour map, drainage map, soil map, geomorphology map, slope map, land use/land cover map. Geo-referencing has been done with the help of blending together the SOI topographical maps and the satellite images. A comprehensive strategy has been evolved for the soil and water conservation for the areas.

3. Disaster monitoring and mitigation: A new methodology has been developed using geo-spatial technology, GIS database on landslide controlling geo-systems, integration of landslide incidences, geo-threshold and GIS vulnerability maps for strategic stand in the mountain terrain. Similarly, landside zonation maps for the vulnerable areas of Himachal Pradesh have been prepared with the help of geomatics techniques. RS and GIS database have been used for accurate flood mapping along with provisions of early warning mechanisms in the flood-prone areas of Mumbai for proper urban flood management. The study has been undertaken on the basis of the satellite images of the Mumbai floods in 2005. The flood-prone Birupa Basin, nested in the Mahanadi Basin in coastal Odisha, has been studied. Mapping of flood extent from RADARSAT images was done using visual interpretation, threshold technique; rule based expert classification and neutral network. The threshold values, internal parameters and ancillary information used in different techniques on 2006 images was applied on RADARSAT image of 2008 for the same area and it was found that parameters derived can be applied in other areas with minor modifications. This field information was used for accuracy assessment. For explaining the results of flood inundation extraction by different techniques, the merits and limitations of different techniques have been identified. Validation of results was done by RADARSAT image of the same area acquired in 2008.

Mountainous areas tend to be prone to a variety of potentially hazardous geomorphic and hydrological processes. Hazard maps were prepared and superimposed using GIS technique, for the region along the Bhagirathi River, and away from the National Highway, which is a moderate risk zone. This has been studied by Singh and Kumar in 2011, and in their study, they said that increasing intensity of anthropogenic activities might contribute to the fragility and vulnerability of flooding.

4. Urban monitoring: The investigation of patterns of urban growth is very crucial from regional point of view to provide basic amenities in the urban area. The growth pattern of urban sprawl has been analysed and studied with the help of temporal multi-sensor, multi-resolution spatial data. Spatial data have been modeled to a co-registered lower resolution MSS image without altering its spectral properties and contrast by using a ratio between higher resolution image and its low pass filtered (smoothing filter) image. The study has been conducted by Menon et al. (2008). Urban sprawl is leading to urbanisation and often fuelling dispersed development both in the peripheries, and in the rural areas, with impacts such as loss of agricultural land, open space and ecologically sensitive habitats, and creation of urban heat islands. These changes in and around urban areas, together with the rural areas, however, need to be studied using RS and GIS techniques for proper planning.

5. Borderland management: India's international borders are highly vulnerable, despite round the clock vigil. Most of the borders are porous, providing access to illegal immigrations, unlawful crossings, trans-border smuggling, and overall cross-border terrorism across the Line of Control (LoC) in Jammu & Kashmir, and Chinese infiltration across the Line of Actual Control (LAC) in Ladakh, besides infiltrations into the territorial waters. Geomatics techniques have been successful in managing the borderlands, particularly with regard to border guarding, deployment of security forces, strategic manoeuvrings, particularly along the LoC and the LAC, tracking the cross-border movements and so on. Borderland mapping has become easier with the help of RS and the GIS database. With the help of the GPS, the border outposts positioning along the LoC and the LAC have been established that has been helping in the management process of the vulnerable gaps between the outposts on the borders in Jammu & Kashmir. The Centre for Strategic Studies and the School of International School (Political Geography division) of the Jawaharlal Nehru University are specialised in the RS and GIS techniques-based border mapping. The Indian Army has a separate division of geomatics engineering for defense and strategic purposes.

Conclusion

It is now a well-established fact that remote sensing and GIS together have been proved to be a very effective tool for monitoring all kinds of earth-bound problems, related to natural resource management, water management, disaster and risk management, urban resource management, border management and so, when carefully evaluated and applied within an appropriate conceptual framework. What is needed is that the geography departments of the premier Central and state universities must be properly equipped with the tools of geomatics engineering, so that Indian geographers working in the departments could prepare themselves in modern technology, at par with their American, European, and Chinese counterparts.

13

Environmental Determinism and Possibilism

HIGHLIGHTS

- Rise of Determinism
- Rise of Possibilism
- Stop-and-Go Determinism
- Changing Perception of Human-Environment Relations: Environmentalism

Environmental determinism and possibilism are the two mutually exclusive philosophies in human geography, centred around humankind—whether humankind to be looked upon as a 'passive being' or as an 'active force', reacting to his environment and changing it. In other words, in the deterministic view of struggle and survival (i.e., the environment controls the course of human action), humans are not free to have their own choices and their achievements are to be explained as consequences of natural conditions. The philosophy of possibilism, on the other hand, reflects the view that the 'pattern of human activity on the Earth's surface is the result of the initiative and mobility of man operating within a frame of natural forces. Without denying the limits every environment sets to man's ambition, they emphasize the scope of man's action rather than these limits' (Tatham, 1953; 152–53).

RISE OF DETERMINISM

The philosophy of environmental determinism is, perhaps, the oldest surviving philosophy that can be traced back to the classical antiquity. 'Among the ancients, a people and their country were inseparable, and where unusual customs or strange physiognomies were found a cause was sought in one or other of the physical elements-climate, relief, or soil' (Tatham, 1952). **Herodotus,** writing in the fifth century BC, came out with the very old idea that 'all history must be treated geographically and all geography must be treated historically' (James and Martin, 1972, 21). This idea was an implicit recognition of determinism. Herodotus further observed that geography provided the physical background, the stage setting, in relation to which historical events occurred.

Hippocrates (c. 420 BC) contrasts the easy-going Asiatics living in a very favourable region with the penurious Europeans, who must seek through greater activity some amelioration of their poor environment. He also contrasts the tall, gentle, brave folk of the most windy mountain lands with the lean, sinewy, blond inhabitants of the dry lowlands.

Plato (428–348 BC) insisted that the observable things on the Earth were only poor copies of ideas, or perfect predicates from which observable things had degenerated or were in the process of degeneration. However, he was not an extreme determinist like his predecessors; rather, he missed the chance to change the whole history of speculation concerning man-land relations by

[1] Determinism refers to the idea that everything in human life is caused inevitably by previous event or conditions. Extreme environmentalism refers to the idea that everything in human life is caused by the natural environment, and mild environmentalism refers to the idea that human life should be viewed as under the influence of the natural environment.

identifying man as a destructive agent. Eudoxus, a contemporary of Plato, however, developed the theory of climate based on increasing slope away from the Sun on a spherical surface, and emphasised the importance of climate in the life of man.

Aristotle (384–322 BC) is credited with the most speculative concept of environmentalism of his period. He markedly differed from Plato who was rather hesitant to accept the extreme form of determinism. Aristotle tried to conceptualise the varying habitability with differences of latitude. He contended that the parts of the Earth close to the equator, the torrid zone, were uninhabitable; that the parts of the Earth far away from the equator, the frigid zone, were constantly frozen and also were uninhabitable; and that the temperate zone in between constituted the habitable part of the Earth. *'Ekumene'*, the inhabited part of the Earth, was in the temperate zone, but much of it, said Aristotle, was not inhabited because of the ocean.

Emphasising the importance of climate in shaping the human skill and sustainability, Aristotle observes in his *Politics*: 'The inhabitants of the colder countries of Europe are brave, not deficient in thought and technical skill, and as a consequence of this they remain free longer than others, but are wanting in political organization and are unable to rule their neighbours. The people of Asia, on the contrary, are thoughtful and skilful but without spirit, whence their permanent condition is one of subjection and slavery' (quoted in Tatham, 1953, 128). Greeks, however, living in the intermediate region, he considered, combined the best qualities of both.

Eratosthenes (c. 234 BC) redefined Aristotle's zones of habitability, but he also emphasised the climatic determinism while describing the ekumene, the inhabited Earth. Posidonius, who lived shortly before the time of Christ, however, contradicted Aristotle's assertion that the equatorial part of the torrid zone was uninhabitable because of heat. The highest temperatures and the driest deserts, he said, were located in the temperate zone near the tropics and the temperatures near the equator were much less extreme. It is not clear whether he refuted the contemporary assertion of climatic control on human activities because his belief concerning the habitability of the equatorial region was overlooked.

Strabo (64 BC–20 AD) carried forward Aristotle's standpoint on habitability in his book *Geography*. Like Eratosthenes, he also redefined the habitable part of the Earth, the ekumene, but held the view of environmental control on human activities. He sought to explain how shape, relief, climate, and space relations of Italy affected the rise and strength of Rome.

Ptolemy (Claudius Ptolemaus), who lived in the second century after Christ, repeated Aristotle's heritage of equating habitability with latitude, and the effect of the position of the celestial bodies on human affairs.

With the death of Ptolemy came to an end the geographic horizons that had been developed and widened both physically and intellectually by the Greeks. But the concept of equating habitability with latitude developed by Aristotle did not die out and was carried forward by successive generations. However, the deterministic idea of the classical antiquity lacked empirical validation.

Determinism in the Middle Ages

The Middle Ages were a dark period for the development of science in Europe. At best, scholars made accurate, but sterile copies of the works of the ancients, rejecting anything which did not conform with the dogmas of the church. Ptolemy became the major authority in the medieval Christian world, and his works were translated from Arabic into Latin by Plato of Tivoli in 1138. The geographical ideas of Aristotle, particularly that of his deterministic concepts, were made available in Christain Europe by translation from the Arabic in the twelfth century.

The first medieval writer to make use of Aristotle was **Albertus Magnus,** whose book on the nature of places combined astrology with environmental determinism. The Greek theory of equating habitability with latitude became strongly implanted in medieval writings. Albertus even went beyond the Greeks. From them, he accepted the idea that people who lived close to the limits of the heritable Earth turned black; but then he insisted that if black-skinned people should move into the temperate latitudes they would gradually turn white.

One of the major controversies in the medieval Christian Europe was 'whether or not the torrid zone was habitable'. **Cardinal Pierre d'Ailly,** in his book published in the early fifteenth century attempted to explain climatic determinism, but he did not take any stand regarding the habitability of the torrid zone. Aeneas Silvius (Pope Pius II) in his book published in the mid-fifteenth century, while subscribing to the deterministic idea, insisted on the possibility that the torrid zone was habitable.

The **Arab scholars** during the medieval period greatly subscribed to the Greek idea of environmental determinism though some of them contradicted Aristotle's zone of habitability by new findings based on observation. One of the earliest Arab scholars **lbn-Hawqitl** insisted on the fact that considerable number of people lived in those latitudes that the Greeks thought to be uninhabitable. His assertion was based on his voyage along the African coast to a point some 20° south of the equator. Nevertheless, he supported the idea of environmental control on human affairs.

Al-Masudi, who died about 956 in Egypt, was more clear in his assertion on environmental control. He was quite emphatic in describing the effect of environment on the mode of life and attitudes of people. To him, 'the powers of the Earth vary in their natural vegetation and topography' (Husain, 1984). **Al-Maqdisi,** who in 985, prepared a new division of the world into 14 climatic regions, also described human activities, particularly in the torrid and temperate regions, emphasising the importance of climate on human actions.

Al-Biruni in his book on India (*Kitab-al-Hind*) in 1030 also made implicit references to the impact of monsoon on the Hindu culture, particularly while describing the cultural landscape and the socio-economic institutions of contemporary India. His views on the monsoon were based on his visit to India. **Ibn-Sfna or Avicenna,** who is credited with the concept that 'mountains were being constantly worn down by streams and that the highest peaks occurred where the rocks were predominantly resistant to "erosion", also referred to the natural laws bringing changes in the landscape. It was a deterministic view, highly idealistic in nature' (James and Martin, 1972, 51).

Ibn-Batuta, who travelled widely during the fourteenth century, confirmed what Ibn-Hawqal had implied earlier that the torrid zone in East Africa was not torrid and that it was occupied by numerous native populations as the environment there suited them. Nevertheless, he was an ardent supporter of the deterministic concept (Wright, 1925).

Ibn-Khaldun was the last Arab scholar to have contributed to and enriched the medieval muslim deterministic concept. His book *Muqaddimah*, in 1377, begins with a discussion of man's physical environment and its influence, and also with man's characteristics that are related to his culture or way of living rather than to the environment. He repeated the old idea of climatic determinism that the people turned black when they lived too close to the Sun and that when black people moved to the temperate zone they gradually turned white and produced white children. The physical environment impressed its characteristics on people in many subtle ways. Ibn-Khaldun insisted that the momadic culture was expressive of the desert environment. It may be said that he was *the first scholar to have turned his attention specially to man-environment relations.*

Renaissance and Determinism

Though the medieval Arab scholars strongly contradicted Aristotle's Ptolemic concept of habitable zones, but they strongly subscribed to the idea of determinism. However, the idea of environmental determinism was revived with full enthusiasm in western Europe during the renaissance. Explorations, discoveries, voyages and expeditions to different parts of the world from the early fifteenth century onwards substantiated the idea of determinism with new information. In 1490, almost 20 years after the first Portuguese ship crossed the equator into the southern hemisphere without burning up, an Italian writer published a compendium in which he described the torrid zone as sterile and uninhabitable.

Notwithstanding the revival of the idea of determinism during the renaissance, the German cosmographer Sebastian Munster hardly made any explicit reference to the idea of determinism. However, Cluverius's book on Italy, published posthumously in 1624, and his six-volume compendium of geography, which also appeared in the same year, did contain some references to the concept of the habitability zones. British scholar, Nathaniel Carpenter, who prepared a compendium in English, gave numerous examples to show that human character was determined by climate. He fully accepted the Aristotelian Ptolemaic concept that the habitability of a place was a function of its latitude.

The enlargement of geographic horizons during the Age of Exploration provided much speculation regarding the influence of the natural environment on human behaviour. **Jean Bodin,** in 1566, sought to describe the people of northern lands as brutal, cruel and enterprising; those of the south as vengeful, cunning, but gifted with the capacity for separating truth from falsehood. Inhabitants of temperate regions are more talented than those of the north, more energetic than those of the south and they alone possess the prudence necessary for command.

One of the most influential of the eighteenth-century determinists was the French political philosopher, **Montesquieu.** One of the major themes in his work on laws had to do with the influence of climate on politics. He sought to explain the determining effect of climate and soil on the character of the people as a guide to the law-giver. People in cold climates are stronger physically, more courageous, more frank, less suspicious and less cunning than those of the south who are like old men, timorous, weak in body, indolent and passive. Northerners who go to live in south quickly lose their vigour and acquire the passivity of those around them. Consequently, the hot climate is the cause of immutability of religion, manners, customs, and laws in the eastern countries. Legislators must take cognizance of these physical facts. Soil (i.e. 'the goodness of the land') is less potent than climate, but nevertheless has great influence on the form of government. Monarchies are more frequently found in fruitful countries, and republics in sterile ones. The barrenness of the Attica soil established there a democracy, and the fertility of Lacedaemonia an aristocratic institution.

However, Kriesel (1968) points out that a careful study of Montesquieu's works shows that he recognised the importance of other factors than climate alone—such factors as religion, the maxims of government, precedents, and customs. In any one country, as some of these factors act with stronger force, the others are weakened. Kriesel, therefore, describes him as a possibilist rather than an environmental determinist.

Nevertheless, Montesquieu was very persuasive in his discussion of the effect of differences of climate on behaviour. In fact, he drew up generalisations about climatic conditions by continents rather than by zones of latitude. Yet, he made the theory of climatic influence so plausible that these ideas persisted long after his time.

'The aim of the writers of the renaissance period, like that of their predecessors, was to understand the variation in the character of human types. They did not start from an interest in the earth; natural forces were mainly called in to supply reasons for human variation which otherwise were inexplicable.... Consequently, there was no systematic approach; each writer drew conclusions from his own experience and contradictory conclusions were common' (Tatham, 1952).

The eighteenth-century geographers, in fact, were not very much active in the framing of the concepts of determinism. One of the reasons was that contemporary geography was yet to become anthropocentric; moreover it was still dominated by the conviction that its function was purely descriptive. Human's relation to its environment was not looked upon as a proper subject for geographical research. Nevertheless, many of the current concepts were incorporated in the geographical descriptions of different parts of the Earth.

Immanuel Kant (1724–1804) was not a true determinist, rather he set forth a new concept of a chorological science, i.e. a subject which studies things which were mutually coordinated; not subordinated, in space. He believed that the geographical synthesis is distorted when nature is regarded as dominant and man as subsidiary. Nevertheless, he provided a short description, discussing the local geographical condition as having influenced the inhabitants on the coast of New Holland, and few more things like the impact of climate on the nature of man's sustainability.

August Zeune (1808), however, in his attempt to define natural regions by their unique interrelations of all physical and biological phenomena, was led almost inevitably to develop the thesis. 'Spaniards', he observed, 'must be lazy, languishing, sensual, inflammable with all the burning passions natural to a hot climate. In like manner the speech of peninsular Italy on account of the proximity of the sea has many sibilants and almost no gutturals, just as Low German, because the moist air of the north coastal lowlands is softer and more drawling than the harsher, quicker High German of the upland regions further south' (quoted in Plewe, 1938, 21–27).

Determinism in the Nineteenth Century

The nineteenth-century geographers were more involved in framing deterministic hypotheses and attempting at empirical validation of such hypotheses through field observations and experiences. Two of such geographers of the early part of the century—Carl Ritter and Alexander Humboldt—known for their 'positivist' approach based on empiricism, had attempted to give a new dimension to the idea of determinism and such hypotheses. Carl Ritter's anthropocentric viewpoint fastened their growth.

Ritter's view of science sprang from his firm belief in God as the planner of the universe. He regarded the Earth as an educational model for man, where nature had a God-given 'purpose' which was to show the way for man's development. Ritter did not regard the shape of continents as accidental but rather as determined by God, so that their form and location enabled them to play the role designed by God for the development of man. His *teleological viewpoint, therefore, seems to be a manifestation of an 'implied determinism'.* In fact, he sought to combine a basic teleological standpoint with a critical scientific precision, and offered an ambiguous explanation of determinism.

Many people said that Ritter was not a determinist because he was much cautious to indulge in facile generalisations. Furthermore, though he was interested in the effect of the Earth on man, the reciprocal action of man on Earth was to him equally significant. He believed that the unity of nature would be severely jeopardised if nature was regarded as dominant and man as subordinate to it; he felt that there was a mutual interaction between the two. Ritter, in fact, pursued 'idealism' in his explanation to this concept of unity of nature vis-a-vis the hypothesis of determinism.

Humboldt viewed the problem of determinism in a remarkably clear scientific manner. He realised that environment influenced man, but earlier was not convinced that evidence was available to permit the formulation of a hypothesis. Elaborating the influence of the configuration of the Mediterranean on the evolution of early civilisation, he writes, 'The influence of the sea was speedily manifested in the growing power of the Phoenicians and in the rapid extension of the sphere of general ideas' (1847, 489).

Humboldt included man and his works in the concept of nature and natural areas, but he did not consider man as a primary determinant—probably because he worked in an area in which nature was so overwhelmingly dominant. This he experienced during his long American travel during 1799–1804. Primarily, Humboldt was a determinist. Nevertheless, he believed in the 'areal associations of natural and organic phenomena' in which he emphasised, what Hartshorne had traced, a chorological viewpoint, i.e. mutual coordination and interaction between various kinds of phenomena in which nothing could be regarded as dominant or subsidiary. Humboldt provided an 'aesthetic' explanation to develop his concept of the unity of nature. However, he made a reductionist approach to the idea of determinism.

Frederic Le Play (1879), the renowned French sociologist of the century, provided a broad portrayal of the socio-geographic structure of human societies. He postulated that the development of the European people took place in three very different geographical environments, namely, the steppes, the maritime shores and the forested lands. The Asiatic steppes was the home of stable nomadic families under the control of patrirchs. On the maritime shores of Europe with their fishing resources, the boat and the habitation were patrimony of the family which was made up of parents, all unmarried children, and the eldest named son with his family. Forested land covering great areas of much variety, with grass openings, heath and varied soils, was the birth place of the unstable family that had also developed in the urban environments of Europe and had spread to America (Dickinson, 1969).

Demolins, who was the pupil of Le Play, provides an analysis of the social structures of the world's people on the basis of what he calls their 'geographical environment', their resultant type of work, and their resultant type of social organisation (based on the family unit). A historical interpretation runs through his work published in two volumes in 1901 and 1903. Groups develop or become markedly characterised in a particular geographical (physical) milieu and then move onwards, so that their social systems undergo changes in new and different environments. The environment rigidly controls the most extreme form of geographic (environmental) determinism.

The basic idea of Demolins is expressed in the preface to the first volume as follows: 'The primary and decisive cause of the diversity of peoples and races is the route which has been followed by the peoples. It is the route (the environment) which created race and social type.... It has not been an indifferent matter for a people which route they followed: that of Great Asiatic steppes, or of the Tundras of Siberia, or the American Savannas, or African forests. Unconsciously and fatally these routes fashioned either the Tartar Mongol type, Eskimo-Lapps, the Red-skin or the Negro. In Europe, the Scandinavian type, the Anglo-Saxon, the French, the German, the Italian, and the Spanish are also the result of the routes through which their ancestors passed before arrival at the present habitat' (Demolin, 1901–3).

Demolins carried forward Le Play's deterministic concept and stated that 'the presence of an exclusive grass cover determines a uniform mode of work—art of pastoralism' (ibid.). This means a complete dependence on animals, chief among which is the horse. The steppe is essentially adapted to the horse and it is the horse that adapts the steppe to man. The horse fills many roles; it provides food, it confers mobility. Mobility makes possible the work of the shepherd, preserves the link between families, maintains religious unity of the steppe and on occasion enables the great horders of Genghiz Khan or Tamurlaine to assemble and to conquer. Patriarchal family is characteristic of the steppe.

Darwinism and Determinism

In geography; Darwinism was interpreted primarily as evolution in the sense of a 'continuous process of change in a series of transformation'. **Charles Darwin** was primarily concerned with the *mechanism of the change* or, as *The Origin* was subtitled, *'the preservation of favoured races in struggle for life'.* Darwinian ideas, in fact, revolutionised the early nineteenth century hypothesis on determinism. They provided a mechanical explanation to such hypotheses as an alternative to the 'teleological explanation' of Ritter and the 'aesthetic explanation' of Humboldt to the hypothesis of determinism.

The contemporary hypothesis on determinism, particularly since the mid-nineteenth century, seemed to have been greatly influenced by the major Darwinian themes which include: (a) change through time or evolution, a general concept of gradual or even transition from lower to higher or more complicated forms; (b) association and organism man as part of a living ecological organism; and (c) idea of natural selection and struggle.

The first theme of the Darwinian heritage sought to identify the forces and processes which tend to constitute the natural laws which, in turn, tend to cause *changes within time-frames.* Or, in other words, biological evolutionary processes necessarily manifest the operation or action of natural laws, that the life had evolved from the amoeba through multitudinous forms to man under the selective action of natural forces. It was inevitable for the late nineteenth century geographers and other scientists to see in the differentiation of man the operation of natural laws.

The idea of *change through time, or evolution,* was taken up by the American geographer William Morris Davis in his famous cycle-of-erosion model of landform development in 1899. Even the subsidence theory of Darwin himself points to the fact of the operation/action of natural processes on the evolution of the coral reefs. The same standpoint seems to have been applied by the scientists in the latter part of the nineteenth century to account for the evolution of humankind.

Haeckel elaborated the idea of *'association and organism'* and outlined a new science 'Ecology' in 1869, which is the study of the mutual relations of all organisms living in one and the same place and of their adoption to their environment. This science attracted the attention of geographers and was saturated with Haeckel's materialistic philosophy. Man was only one of the organisms to be studied and was along with all other living things equally in the grip of surrounding forces. Perhaps Darwin's most significant contribution to ecological thinking was *to include man in the living world of nature.*

The idea of *'selection and struggle'* found its first legitimate expression in the view of British philosopher Herbert Spencer (1820–1903) who elaborated what is known as 'Social Darwinism'. He believed that human societies closely resembled animal organisms and that human societies must struggle in order to survive in particular environment much as plant and animal organism' do. *Social Darwinism* provided a new dimension to the contemporary thinking on determinism. In fact, the Darwinian philosophy of selection and struggle seemed to have been necessarily based on cause-effect relationship as it sought to clarify the environmental influence, selection and adaptation depending upon the requirement of the way of life. Most Darwinian writers on the effect/influence/control of environment were content to look for cause-effect relationships without enquiring too closely into the process.

Friedrich Ratzel (1844–1904) pioneered systematic human geography much on the basis of Darwinian philosophy and methodology, and he treated the contemporary hypothesis on determinism exclusively on a new explanatory dimension. He *developed a new paradigm of natural science in human geography,* in which he stressed the extent to which men lived under nature's laws. He regarded cultural forms as having been adapted and determined by natural conditions. Ratzel attempted to modify his environmental determinism in his later work. A different emphasis dominated the second volume of *Anthropogeographie* (1891) which discussed the concentration and distribution of population, settlement forms, migrations and diffusion of cultural characteristics. He did not merely explain phenomena in human geography in terms of natural conditions, but stressed the significance of the historical development and cultural background of populations.

The first volume (1882) treats the causes of human distributions, i.e. the dynamic aspect of geography, and the second, published 10 years later, deals with the facts of distribution, that is, the static aspect of geography. The first volume is an application of geography to history, and the second the geographical distribution of man' (Dickinson, 1969).

The Darwinian idea of 'selection and struggle' was explicitly used by Ratzel in his political geography, giving a new paradigm to the sub-branch. This idea, in fact, led Ratzel to develop what is called the 'increasing space consciousness and conception' on a national level. In 1896, he provided a *model for the spatial growth of states,* in which he sought to give a 'realist' expression of the idea of selection and struggle through the concept of '*lebensraum*'. 'Just as the struggle for existence in the plant and animal world always centres around a matter of space, so the conflicts of nations are in great part only struggles for territory' (Ratzel, 1897). It is clear that the organic analogy for Ratzel not only provided a simple and powerful model in analytical, but deterministic political geography, but also an apparently scientific justification for political behaviour. Through his new paradigm, Ratzel sought to justify the organic concept of Spencer, particularly in the light of idea of evolutionary struggle. The evolution and growth of states, therefore, are conditioned by the operation and action of natural laws under given time and space situations.

There is no doubt that Darwinism, necessarily, provided a new scientific impetus to the hypothesis of determinism, giving a new direction towards empirical validation of the hypothesis. About the same time or a little earlier than the publication of *The Origin* in 1859, the study of social statistics began to reveal an extraordinary regularity in social behaviour. Everything seemed to point to one and the same conclusion, that man was not so free as had been thought, and that his actions were largely controlled by natural or economic laws.

Environmentalism in history is well-documented in *History of Civilisation in England* (vol. I, 1857; vol. II, 1861) by **Buckle.** To him, civilisations in Africa and Asia were most powerfully influenced by the fertility of soil and those in Europe by climate. Climate, he said, tended to influence labour in many ways. Excessive heat enervates the labourer, a more moderate temperature invigorates him, while a short summer broken by a long winter in which low temperatures interrupt work encourages desultory habits. Less obvious, but equally closely connected is the relationship between climate and wages.

Buckle further pointed out that in a country like India, climate and fertility combined to produce dense population and low wages, an unequal distribution of wealth which in turn created inequality in the distribution of power and social influence. A civilisation, he added, in the New World reflected the power of climatic influences. He continued that in India, Mexico, Peru, Egypt where Nature overpowers man, religion is one of complete and unmitigated terror. Everywhere the hand of nature is upon us, and the history of human mind can only be understood by connecting it with the history and the aspects of the material universe. Buckle seeks to mention historical events as the outcome of man modifying Nature and Nature modifying man; yet it is Nature itself which determines when and how man shall be so active. In other words, natural powers are in every instance dominant; the determinism is complete.

The Darwinian environmentalism seemed to have *no impact on the contemporary Russian thinking,* simply because of the fact that in Russia itself greater studies of evolution were made. Those who conceived of the viewpoint of the role of man as passive, seemed to have developed independent standpoints, and they were mostly historians and social scientists, trained in geography. The famous Russian historian, Sergey Solovyev, pointed out that the nature of a country had important significance in history as the national character to a large extent depended on it. Vasily Klyuchevsky studied the effect of the forests, steppes and rivers on the history of Russian people. He said that each of these separately, by itself, took a lively and original part in guiding the genre de vie of the Russian people.

The Russian historians and other scientists seemed to have rejected the more extreme form of environmental determinism stemming from Spencer and Ratzel, and also the use of biological analogy to describe sequences of landforms as proposed by Morris Davis. However, some of the historians did support the ideas of climatic influence on national character or of the critical importance of the large Asian rivers in providing the setting for the development of early civilisations. But the geographers as a whole avoided such extremeness.

Environmentalism had its maximum development in the nineteenth century. Geographers, historians, social scientists and biologists made valuable contributions to its understanding. But the rise of Darwinism in the latter part of the century not only provided a new dimension to it, but also made room for some form of scientific validation of the hypothesis of determinism. The Darwinian legacy continued well beyond the nineteenth century, as more and more scientists and geographers of the twentieth century drew scientific bases for their empirical generalisations.

Determinism in the Twentieth Century

The Davisian concept of 'ontography' in 1902 was a mere manifestation of continuation of the Darwinian idea of 'natural selection and struggle'. *Ontography, therefore, is some kind of crude determinism,* which is necessarily concerned with the 'rational correlation of the items that fall under two categories: on the one hand, the items of inorganic conditions that constitute the physical environment of living forms, and on the other hand, the items of organic response made by living forms to their environment' (Davies, 1902, 245). The organic response of the living forms to their physical environment is expressive of adaptation to the environment, which necessarily involves 'selection' according to the suitability, and 'struggle' in the light of natural resistance. Ontography is an idealist concept expressive of the Darwinian determinism, which W. M. Davis conceptualised to account for his hypothesis on determinism within a geographical framework.

The most influential determinist of the early twentieth century, belonging to the Darwinian-Ratzelian heritage, was the American geographer **Miss Ellen Semple.** She wrote her book *Influence of Geographic Environment* (1911) to introduce Ratzel's ideas in *Anthropogeographie* to the English-speaking world. The opening para of the book makes the following statement: 'Man is a product of the Earth's surface. This means not merely that he is a child of the Earth, dust of her dust; but that Earth has mothered him, fed him, set him tasks, directed his thoughts, confronted him with difficulties that have strengthened his body and sharpened his wits'... she has entered into his bone and tissue, into his mind and soul' (Semple, 1911, 1–2).

On the influence of climate, Semple says: 'The influence of climate upon race and temperament has a direct and indirect effect cannot be doubted, despite an occasional exception. The northern people of Europe are energetic; provident, serious, thoughtful rather than emotional, cautious rather than impulsive. The Southerners of the sub-tropical Mediterranean basin are easy-going, improvident, except under pressing necessity, gay, emotional, imaginative, all qualities which among the Negroes of the equatorial belt degenerate into grave racial faults ... the blond Teutons of the north are a bleached out branch of the brunette Mediterranean race. This contrast in temperature is due to climate. A comparison of northern and southern people of the same race and within the same temperate zone reveals numerous small differences of nature and charecter, which can be traced back directly or indirectly to climatic differences, and which mount up to a considerable sum total' (Semple, 1913).

Commenting on the deterministic hypothesis of Miss Semple, Derek Gregory (1981, 1986) points out that she 'spoke of geographical factors and influences, shuns the determinant and speaks with extreme poison of geogaphic control'. But

Montefiore and Willian (1955) argued that 'claims of this sort failed to recognise that determinism was not a universal hypothesis capable of empirical validation, but rather a logical structure of cause-effect which required to be translated into the scientifically creditable vocabulary of "necessity" and "sufficient" conditions, and while many geographers had been drawn to "possibilism" as an alternative, others have in fact attempted to establish a more rigorous, even "scientific" determinism'.

E. G. Dexter (1868–1918) and Ellsworth Huntington (1876–1947) stressed the importance of climatic determination. Dexter in 1904 published an empirical study of weather influences which attracted wide interest and which significantly affected later workers. He compared the daily temperature, atmospheric temperature, humidity, wind, sunshine, and precipitation with the records in New York City and in Denver, Colorado, in respect to the behaviour conditions investigated, and found statistically significant correlations between weather and conduct, suicide and other crimes.

Huntington was often described as an imaginative thinker and interpreter of the effects of climate on human life. Correlating the periods of drought with historical dates, he developed the hypothesis that the great outpourings of nomadic people from Central Asia, which led to the Mongol conquests of India and China and the invasion of Eastern Europe in the thirteenth century, could be explained by the drying up of the pastures on which the nomads were dependent. He presented this view in his book *The Pulse of Asia* in 1907. In 1915, he wrote another book *Civilisation and Climate*, in which he developed the hypothesis that man's civilisations could only develop in regions of stimulating climate and that the monotonous heat of the tropics would forbid attainment of the higher levels of civilisation. He did much to establish the fact that there have been significant post-glacial changes of climate and that these played an important role. In his last book, '*Mainsprings of Civilisation*' in 1945, Huntington suggested that diet was as important as climate as an explanation of human energy.

One of the American geographers, **Albert Perry Brigham,** who was a contemporary of W. M. Davis had held a relatively mild approach to the hypothesis of determinism. His book *Geographic Influence in American History* appeared in 1903; the same year in which Allen Semple's book on the same subject was published. Brigham emphasised the origin of what he called 'geographic condition'.

Brigham in 1915 urged upon the geographers to proceed with caution and commonsense in asserting the existence of influences and that every possible test should be made to ascertain the validity of any general principles that were suggested. He was specially critical of generalisations concerning the influence of climate. He appeared critical of vague and unproved assertions of climatic influence on racial character, skin colour, or man's institutions. The infinitely variable factors of the total environment, he insisted, produce diverse results upon body and mind.

In America, particularly in the field of human geography, social Darwinism was under attack, and indeed most of the historians and other social scientists had already rejected it. Many geographers were also ready to follow Brigham in rejecting strict environmental determinism in avoiding simple cause and effect explanations for complex associations of things on the Earth's surface. But not all the geographers were aware of the validity of the criticisms of Davis's scheme of human response to physical controls. The persuasive teaching of Semple, the creative work of Huntington, and to a lesser extent the work of Whitbeck (1926) continued to gain support for some kind of environmental control of human behaviour. Long after the physical cause and human response paradigm had been dropped, some geographers continued to use the language of 'geographic factor' and 'environmental control'.

Sir Halford J. Mackinder, the lone British geographer of the early twentieth century who had carried forward the Darwinian paradigm of determinism in his early ideas, reached the conclusion that 'no rational political geography could exist which is not built upon and subsequent to physical geography'. He also opined that history without geography was mere narrative and that since every event occurred in a particular time at a particular place, history and geography, which deal respectively with time and place, should never be separated.

Mackinder applied the Darwinian idea of selection and struggle in his first work *The Geographical Pivot of History* in 1904, in which he argued: 'European-civilisation is the outcome of the secular struggle against Asiatic invasion ... through the steppe. There came from the unknown recesses of Asia ... in all the centuries from the fifth to the sixteenth, a remarkable succession of Turanian nomadic people ... against the settled peoples of Europe. A large part of modern history might be written as a contemporary upon the changes directly or indirectly ensuing from these raids' (Mackinder, 1904). The geographical pivot was inaccessible on all sides by natural barriers except to the Caspian Sea-Ural opening to the west which necessarily provided mobility for the Asiatic tribes to the European mainland. The horse and the camel were the main sources of mobility for the Asiatic nomads to carry on raids on the European natives. The idea of the horse-camel mobility of Mackinder resembled the view of Le Play of the nineteenth-century France, who also held that the horse was the only source of mobility in the steppe lands.

The famous dictum of Mackinder's 'heartland concept' in his subsequent work *Democratic Ideals and Reality* in 1919: 'who rules East Europe ... commands the world' sought to generalise what he had said in 1902 that no political geography could exist unless it was built on physical geography. Moreover, he attempted to generalise that East Europe was so positioned geographically vis-a-vis the Heartland that any attempt to command/rule the Heartland necessarily required the control of East Europe. The position of East Europe, according to him, was geographically so viable that the further warfare would be decided to a greater extent on the soil of East Europe. Mackinder's generalisations of environmental determinism were, thus, based on historical evidences and contemporary political events. His generalisation, particularly of the expectation and pressure of the Heartland state on the European-Asiatic coastal lands, remained valid for some decades in international politics.

Russian geography had been traditionally exercised by the environmentalist thesis which received an important impetus particularly through the efforts of L. Mechnikov, N. Baranskiy and G. Plekhanov. However, **Mechnikov** was rather ambiguous in his approach to the environmentalist hypothesis. He pointed out that though 'river as a synthesis of all physical geographical conditions was one of the factors that determined the development of human societies, but at the same time man had played a significant part in the formation of the geographical environmet'. He further stated that 'one should seek the principle of the rise and character of primitive institutions and their subsequent evolution not in the development itself but in the relations between the environment and the capacity of the people inhabiting a given environment for cooperation and solidarity' (Mechinikov, 1924).

N. Baranskiy in 1926 argued that 'the influence of natural conditions on man is taken into account in the Marxist scheme of social development to the extent in which these natural conditions form a natural basis for the material productive forces which determine the productive relations and through them the legal and political superstructure and, finally, forms of social conscience.' But this was attacked and finally devastated in 1938 when Stalin decreed that although the environment accelerates or retards the speed of development of society, 'it was not a determining influence'.

In the post-Stalin era, **Plekhanov** came out with an explicit statement which resembled Baranskiy's paradigm of 1926. Plekhanov asserted that the 'peculiarities of the geographical environment determined the development of the productive forces, the development of the productive forces determined the development of the economic forces and directly after them also all the other social relations' (Plekhanow, 1956). The Marxian philosophy, for Plekhanov, appeared to be the application to social development of the Darwinian paradigm of the adaptation of biological species to the conditions of the environment.

Before closing the discussion on environmental determinism, it is necessary to conclude that the hypothesis of determinism appears to be as old as the human history itself. Different periods of human history since the classical antiquity had different perceptions about it, necessarily based on pre-conceived and 'carried-over' notions over centuries. The Age of Exploration and Renaissance in the early medieval period provided some material grounds to the hypothesis of determinism, so that some form of generalisations could be made on it.

During the early part of the nineteenth century, geographers, social and natural scientists and others sought for the empirical validation of the hypothesis of determinism, citing examples of how people responded to their environments. But it was not until the rise of Darwinism in the mid-nineteenth century that the hypothesis received a new scientific dimension to further its universal accountability. It was made possible as a result of a 'paradigm shift' in the contemporary socio-biological thinking which tended to compensate weakness that remained in the hypothesis. The nineteenth-century Darwinian tradition seems to have well continued to the first half of twentieth century with much scientific precision, explanation and validation.

However, people in advanced societies, who were conscious of their material cultures and capacities vis-a-vis the natural control, sharply reacted to the extreme generalisations of the environmental determinism, and doubted their universal application/validation. To some people, massive innovation-diffusion made the deterministic-paradigm inadequate to account for man's creative genius, and creativity in shaping his environment. Therefore, an alternative paradigm was sought for. This led to the development of the paradigm of possibilism in which man was presented as an active rather than a passive agent. Glacken (1956, 1967), however, attempted to identify three·different modes of nature-society or man-environment relations which appear to have permeated the history of Western thought:

1. Humanity in harmonious relationship with nature;
2. Humanity as determined by nature; and
3. Humanity as modified by nature.

The man as modifier and conqueror of nature has dominated modern thought though the other two perspectives have by no means been completely absent. Wittfogel (1929) made a forceful attack on the tenets of environmental determinism in modern geography in the late 1920s of the past century. He said that human labour organised in different social forms tended to mould nature into different material bases for economic development of regional societies. This is what created the distinctive regional cultural traits, rather than, the environment *per se.* This man made himself: Societies are human creations rather than natural/environmental creations. Wittfogel, however, was not opposed to the idea of natural forms being a potent influence or factor in man's life on the Earth. To substantiate this view, he attempted to draw attention to the climatically determined need for irrigation which in the East (China and India) gave rise to a line of social development that was greatly at variance with the one followed in the rainfed agriculture being practised and pursued in the West, giving rise to entirely different kinds of civilisation in the two cultural realms (Willforgel, 1957).

RISE OF POSSIBILISM

The possibilist paradigm views that the physical environment tends to provide the opportunity for a range of possible human responses and that people have considerable discretion to choose between them through their creative genius and creativity.

Plato (428–348 B.C.), who was the master of deductive reasoning, said that the observable things on the Earth were only poor copies of ideas, or perfect predicates from which observable things had degenerated or were in the process of degeneration. Proceeding from general to the particular, Plato sought to infer that men tended to make changes in the land they occupied; and soil erosion and land destruction were parts of their material culture. It was he who formulated the idea of man as an agent of change on the Earth's surface. This deduction based on *a priori* theory became a legitimate basis for enquiries in the subsequent centuries. The view expressed by Plato about capabilities of man was very close to the philosophy of possibilism. It can, therefore, be said that a philosophy resembling possibilism had its origin in the classical antiquity, but Plato did not express it clearly.

The idea of Plato went into obscurity for thousands of years after him, but the contrasting idea of determinism continued to grow and flourish upto the eighteenth century when **Montesquieu** (1689–1755) developed a philosophy highly resembling the nineteenth/twentieth century paradigm of possibilism. It is interesting to note that Montesquieu, who had also developed the idea of climatic determinism, sought for an alternative explanation to recognise what he called the 'esprit general', or the general spirit of a nation. The interplay of 'physical' and 'moral' causes was extended to be the factor which established the character of a society or nation which guided the choice of the society or nation to operate freely in the environment.

In fact, Montesquieu held the view that man did possess free will and was able to make, consciously or subconsciously, choices from among a series of avenues of escape or adaptation. He also held that even though the physical environment remained relatively static, man was growing in ability and becoming more complex. 'Montesquieu's views of man's freedom vis-a-vis the physical environment were too crypto-possibilistic to allow easy assimilation by later readers. In these lights, it is understandable that Montesquieu could have been jogged out of a position in the lineage of possibilistic thought. However, his views formed a large part of the fabric of later French political and social thought' (Kriesel, 1968, p. 574).

Another eighteenth century French scholar who strongly and clearly believed in the creative power of man was **Comte de Buffon** (1707–1788). Buffon believed that man was commanded to conquer the Earth and transform it. He thought that man changed the face of the Earth in the process of developing a civilisation. He conceived of man as an agent of change and man did possess free will to make desired changes on the landscape. Buffon insisted that man could adjust to any climate on the Earth. 'Man was not compelled to react to any climate the way uncivilized native people would react. With proper clothing he could protect his skin colour' (James and Martin, 1972, 107).

Buffon also speculated on climatic changes mostly as a result of man-made changes in the landscape and suggested that the removal of forests and the drainage of marshes might lead to temperature increase. In fact, he insisted on the conservation of forests. Comparing Montesquieu with Buffon, it can be said that the former was certainly less of a geographer and conservationist than the latter, whose works considered the physical changes brought about by human agency in more detail than anyone prior to his time.

Montesquieu-Buffon's *crypto-possibilistic hypothesis* could not make much headway with the contemporary geographical scholars because the great majority of them strictly adhered to the philosophy of determinism and Darwinism. Darwinism provided a scientific basis to the idea of determinism, but the crypto-possibilistic philosophy necessarily lacked it. Immanuel Kant who believed in the 'organising freedom of (human being) bounded by the realities of the mechanisms of natural realm' was rather ambiguous in his approach to the creative power of man otherwise he would have been explicitly known as a possibilist. However, this view of Kant emerged in Germany in the closing decades of the nineteenth century which is known as 'neo-Kantian'—a philosophy attributed to the possibilism of French School of human geography.

Possibilism in Nineteenth Century

Though the great majority of scholars, including geographers, political scientists and other social scientists strictly subscribed to the paradigm of determinism and others were seeking for its empirical validation, a very small group of people, nevertheless, were attempting to develop a hypothesis contradictory to the paradigm of determinism. **George Perkins Marsh** (1864) of America was one of them. Of course, he never studied geography as such, but he developed a novel approach to the study of the relations of man to the land. He was conscious of man's power and action. He elaborated the view of man's effect on nature, to the modifications of the organic and inorganic parts of the habitat that resulted from human action.

Almost at the same time, in Germany, **Alfred Kirchoff** (1838–1907) was attempting an anti-Ratzelian approach to the study of human geography. Instead of describing the influence of the physical Earth on human affairs, he focused on man himself. Human societies were studied in relation to the physical features, but greater attention was given to the culture of human groups rather than to the physical Earth. This was the method adopted by Ratzel.

It was not until 1899 that a new dimension to the philosophy of possibilism was added by **Paul Vidal de La Blache** (1845–1918). As opposed to the 'environmental determinism' of the Darwinian-Ratzelian heritage, Blache set forth a conceptual framework of 'possibilism' which was later fully developed by a critical historian Lucien Febvre. However, it is from Ratzel's and Kirchoff's works that he is believed to have formulated the paradigm of possibilism.

'Nature', Blache insisted, 'set limits and offered possibilities for human settlement, but the way man reacts or adjusts to these given conditions depends on his own traditional way of living.' What Blache meant to say was that the physical environment provided a range of possibilities which man turned to his use according to his needs, wishes and capacities, in creating his habitat. He also said that in an area of human settlement, nature changed significantly because of the presence of man, and these changes were greatest where the level of material culture of the community was highest. He added further that through his occupation and imprint on the land, man created distinctive countries, be they states or minor unit areas.

Blache emphasised the concept of a way of living (genre de vie) in supporting and furthering his philosophy of possibilism. The 'genre de vie' refers to the inherited traits that members of a human community learn—what may be called a culture. Blache pointed out that the same environment has different meanings for people with different genre de vie: the genre de vie is a basic factor in determining which of the various possibilities offered by nature a particular human group will select. In fact, this was the idea which Ratzel had ultimately found to have some relevance. He made special note of cases where cultural differences were more important than differences in the physical character of the land. In a regional geography of Germany, in 1898, he pointed out the great contrast in the way people may use the land in two places that are physically very much alike. The two places are the low mountain regions on either side of the middle Rhine valley the Vosage mountains in France and the Scharzwald in Germany. The differences to be observed in these two regions are related to the contrast in 'genre de vie' between the French and the German communities.

The idea of possibilism, in the closing years of the nineteenth century, thus started on a sound philosophical and methodological foundation, with Blache seeking for an empirical validation of it. It, somehow, weakened the importance of contemporary Darwinian heritage of determinism. The philosophy of possibilism continued with Blache who carried it to the twentieth century with many more scholars subscribing to it.

Possibilism in Twentieth Century

Blache who provided a new dimension to the philosophy of possibilism in the closing years of the nineteenth century, and carried it to the next at the expense of the Darwinian paradigm of determinism, said in an article in the *Annals* in 1913: 'Geography seeks to measure and localise the great part that man plays in modifying the face of the Earth'.

Blache's book was published posthumously in 1921 and its English translation was published in 1926. He makes the following observations in support of the philosophy of possibilism: 'Nature provides man with materials which have their inherent requirements, their special aptitudes—also their limitations—which lead them to certain uses rather than to others. To this extent nature does make suggestions, and at times restrictions.... Nature is never more than an advisor.... Isolation and lack of fresh incentives seem to be a primary obstacle to such a conception of progress. Indeed, those human societies which have been isolated by geographical conditions either on islands or in mountain fastnesses, or in forest clearings, seem to be cursed with stagnation and incapacity to move' (Blanche, 1922–1926).

Blache (1899, 101) observes: 'Civilisation has appropriated the favourite crops. From the original plant countless varieties have been perfected to suit the requirements of different climates, with the result that its importance is often greater in regions where it has been acclimatized, than in those where it originated. For instance, wheat does not today have the largest yield in regions where it was first cultivated; the harvests of Mediterranean countries cannot be compared with those of the plains of Central Europe. The largest ears of corn are no longer grown on tropical plateaux, but in the United States, on the prairies of the middle West.... With regard to social groups, the types which become dominant in the march of progress and continue to develop are those which originally resulted from the collaboration of nature and man, and gradually became more and more emancipated from the direct influence of environment'.

Analysing the above statements of Blache, it can be said that he conceived of man not as a passive being, but as an active force, who is able to make consciously choices from among a series of avenues of escape or adaptation. He also sought to generalise his philosophy with examples through a regionalist historical approach.

Jean Brunhes not only carried forward Blache's philosophy of possibilism and spread it through France, but also transmitted it to other countries. He seems to have put the philosophy of possibilism into a sacred philosophical and methodological framework. In 1910, he published the first edition of his great work which was translated into English in 1920 by I. C. Le Compte and edited by Boman and Dodge.

Brunhes divided the essential facts of human geography into three categories: (i) the facts of the unproductive occupation of soil: houses and roads (including rural habitations, urban agglomerations and circulation pattern); (ii) the facts of plant and animal conquests: the cultivation of plants and the raising of animals; and (iii) the facts of destructive exploitation: plant and animal devastation, mineral exploitation. However, Brunhes's, interest seems to be *focused on the facts of human exploitation of the Earth,* irrespective of whether they show environmental influences or not. In the second part of his book, the link between the Earth and man is examined; it is not influences that are sought, but 'geographical relations betwen physical facts and human destinies'. His approach, therefore, seems to be more conducive to unbiased research.

On the *power of natural agents,* he said: 'The unrelenting power of natural agents begins in the physical world alone. Human geography is a field of compromise, nothing is absolute or definitive for the human species on the Earth except these general laws, and those fundamental conditions which determine the limits beyond which all life is excluded; and if men are not able to push back indefinitely all these limits in altitude, latitude, depth, etc. they are at least able to force or modify some few of them.... Although in a large number of cases we seem to dominate nature, she still keeps her right of pre-eminence, for at all points of the Earth, she imposes upon our activity restrictive conditions.'

On the *power of man,* Brunhes remarked: 'The power and means which man has at his disposal are limited and he meets in nature bounds which he cannot cross. Human activity can within certain limits vary its play and its movement, but it cannot do away with its environment; it can often modify it, but it can never suppress it and will always be conditioned by it.... The forces of physical nature are bound to each other in their consequences, in their relations, and in the consequences of these relations. Man does not escape the common law, his activity included in the network of terrestrial phenomena. But if human activity is thus circumscribed, it does not follow that it is fatally determined' (ibid., 103–04).

The *limits set by nature to man's action* vary from place to place on the Earth's surface and from one historical period to another. In marginal environments, such as the hot and cold deserts, and at low stages of culture, man's choice may be extremely restricted. In the more favourable areas of the warm and cool temperate zones, and in periods when man's techniques are highly developed, the possibilities are numerous. But, however many skills man acquires, he can never free himself entirely from nature's control. This is emphasised over and over again by the possibilities (Tatham, 1952).

It may be noted that Montesquieu's views on the fruits of labour were echoed by Brunhes who said: 'Through a form of work adapted to the natural 'conditions and through the collective training that results, societies of herdsmen or fishermen, groups of miners or planters, etc. are really modified and in the long run show definite tendencies.' It seems that to Brunhes, labour involved in the modification of some aspect of the physical environment was able to produce far-reaching consequences. However, at this point Brunhes's movement from specific to general, from labour as the mediator between man and nature, is a departure from Montesquieuan heritage. This departure does not involve the terminology that Montesquieu and the possibilists were using.

To Brunhes (1910), 'everything on the surface of the globe is for men a matter of habit, of sound understanding, of physical facts, and of skilful adaptations to these facts. Moreover, the adaptations take place promptly and at the right time, preceded, prepared for, and brought about by exact scientific investigations.... These investigations should also tend to moderate our ambitions and turn us away sometimes from undertakings that would mean such bold opposition to the forces of Nature that man would run the risk of seeing sooner or later his patient work annihilated at a single stroke. The more imposing and glorious man's conquest, the more cruel the revenge of the thwarted physical facts.' Brunhes explains it with the example of the drainage of coastal lowlands causing the land to sink and to be re-flooded by the sea, or the busting of dykes, etc.

The critical French historian **Lucien Febvre,** in his *Geographical Introduction to History,* published in 1922 and English edition in 1924, is credited with having first termed this approach/philosophy as 'possibilism'. Like Brunhes, he also endorsed the Vidalienne tradition of possibilism and carried forward the 'heritage' in more neo-Kantian way. To him, *man is not a passive being but an active force.* 'Through centuries and centuries, by his accumulated labours and the boldness and decision of his undertakings, he appears to us as one of the most powerful agents in the modification of terrestrial surface.... And this action of man on his environment is the part which man plays in geography.... There are no necessities, but everywhere possibilities; and man as master of these possibilities is the judge of their use. This by the reversal which it involves puts man in the first place, man and no longer the Earth, nor the influence of climate, nor the determinant conditions of localities' (Febvre, 1924).

Febvre, probably the most caustic of the possibilists, discussing the matter of uniform regions of physical and human geography within which everything had identical, or merely identical, characteristics enunciated the possibilist viewpoint: 'And this is a Ratzelian dogma. If the space is limited and not greatly differentiated, the physical and human types found there are monotonous. Our contention is quite otherwise. We admit regional frames in a general sense, but in the collection of physical features it represents we see only possibilities of action' (Tatham, 1952).

Having set down, in the context of regional geography, the fundamental difference between possibilism and environmental determinism, Febvre relaxed somewhat and made a modest concession. He stated that any of his contemporary determinists, who agree that all of the possibilities of establishing a human society which a region offered did not inevitably exert their influence at the same time or with the same force, were worthy of commendation only.

Febvre, speaking for the possibilists, stated their case that a homogeneous region did not necessarily produce, for all time, a homogeneous society. The inhabitants of any region were able to choose, from time to time and in quantity they desired, some of the benefits or chances for benefit which the region possessed. However, the number of choices or opportunities for choice was not infinite in each region; each region had an upper limit.

American geographer, **Isaiah Bowman** was a staunch admirer of the French possibilist paradigm. In his *Geography and the Social Sciences*, Bowman stated. 'As knowledge of the world spread, the association of event or condition with place widened, they became more complex, they had less or more significance with respect to mankind. The potato and maize plants were unknown to pre-Columbian Europe. Their discovery raised the questions, "Are they useful to the rest of humanity and where can they be grown?" The whole known world was in a sense resurveyed by the rough processes of trial and error and the result has been astounding. These two plants largely changed the economy of Europe. The soil had not changed man; man had gained a little more knowledge through a new plant. An element of one environment had been added to the elements, long fixed of many other environments.... The geographical elements of environment are fixed only in the narrow and special sense of the word. The moment we give them human association they are as changeful as human itself. That is why modern geography has so definitely steered away from determinism and towards a study of types of actually working regional combinations of human and environmental conditions' (Bowman, 1934).

Notwithstanding his firm belief in the possibilist paradigm, Bowman was emphatical in his assertion with regard to man's variable responses to the physical laws: 'While the physical laws to which mankind responds are variable in their application and in degree of effect, yet this is also true that all men everywhere are affected to some degree by physical conditions' (ibid.).

In a significant paragraph of his paper 'The Pioneer Fringe', Bowman referred to the part habits of thought used to play in precipitating crisis in civilisation: 'It may be shown that there has never been a civilisation that declined because it exhausted the possibilities of the land. No nation has ever fully developed its frontier. The Earth has never gone back on man, but man has found himself entangled in the unpredictable efforts of his own system. What really happens is that knowledge at the moment of strain, is not great enough to control the forces of nature and of systems of government combined' (Tatham, 1952, 42–43).

Carl O. Sauer (1924) was an activist of the possibilist movement in the United States during the inter-war period. He was, perhaps, more influenced by the German geographer Otto Schluter who focused on the study of the human habitat as the creation and work-place of human groups whose relations are conditioned by their group habitat and heritage. Sauer called for such field studies which aimed at exposing the areal expression of man's activities. Man, behaving in accordance with the norms of his culture, performs works on the physical and biotic features of his natural surroundings and transforms them into the cultural landscape. The design of the landscape includes the features of the natural area, and the forms superimposed on the physical landscape by the activities of man, the cultural landscape. To him, *Man is the latest agent in the fashioning of the landscape.*

Sauer's *landscape paradigm* 'appears to be an intrinsic part of the possibilist philosophy, the focus of which was the study of on-going processes leading to landscape change up to and including the present and the beginning at the pre-human stage of occupance. The human geographer must be obliged 'to make cultural procesess the base of his thinking and observation' (Sauer, 1927, 116). Physical environment should not be conceived as the principal source of explanation for the condition of the humanised cultural landscape, but rather as the setting and environmental patterning upon which agents sculpt their peculiar ideas in ways conditioned by their cultural make-up. He approahed the human use of the Earth ecologically and while he was critical of the paradigm of determinism represented in the United States by Ellen Semple, he himself was strongly influenced by ecological concepts of culture traceable to the same biological sources.

Voskanyan Whittlesey's concept of 'sequent occupance'—the view which establishes the genetics of each stage in terms of its predecessor—represents an antithesis of environmental determinism. In a sense, this concept manifests a 'realistic' approach to the French concept of genre de vie. 'Sequent occupance' tends to recognise the fact that any change in the genre de vie (way of living) of a region inevitably leads to the reappraisal of the significance of resource base. It studies the ways in which each material culture uses a region in its own ways. This is demonstrated in America where most regions experienced a sudden change from the Indian to European culture, and in many parts of Europe which proceeded from agrarian to industrial cultures. Sequent occupance emphasises the 'stages in the development of a region not through studies of local differentiation as a result of the long continued and largely undisturbed interplay of man and nature over centuries (as in the France of Blache), but by emphasising how easily shifts in regional character can take place (Wiltlesey, 1929).

Summing up

Geography in the *United States*, during the inter-war period, witnessed the publication of a large number of studies which sought to contradict the philosophy of environmental determinism, but attempted to make use of the paradigm of possibilism. 'Out of these studies of sequences of development of cultural landscape vis-a-vis settlement, certain principles began to emerge. One was the principle that same physical conditions of the land could have different meaning for people with different attitudes toward their environment, different objectives in making use of it and different levels of technological skills' (James and Martin, 1972).

British regionalist Marton I. Newbigin, in the early part of the twentieth century, sought to identify herself with the Blache-Brunhes heritage of possibilism. She started with a question, 'Why is it easier for men to make their living at some places than at others'. The answer she believed could be found by examining the relationship between the genre de vie and the productivity of land. The classification of region, she said, should be based on the kinds of relationships observed between human communities and their natural surroundings.

Another British regionalist, H. J. Fleure of Aberystwyth (1917, 1919) also carried forward this idea by defining seven kinds of global scale regions. He said that all human activities are primarily directed towards the accomplishment of three functions: nutrition; reproduction; and the increase of well-being. In seeking criteria for the delimitation of human regions, he eliminated the first two because without them 'a race would perish'. He then sought to classify his regions according to the measures of the Earth's response to man's efforts in the pursuit of well being.

In *Germany*, the reaction against determinism in fact started earlier than in other countries. We have seen how Ratzel modified his earlier deterministic approach. Alfred Hettner was the most significant contributor to further development of possibilism. He asserted (1927) that the geographical synthesis is distorted when nature is regarded as dominant and man as subsidiary. He believed that the unity of geography can be maintained through the Kantian concept of chorological science.

Russian geographers who sought to focus on the concept of 'indeterminism', in contradiction/ opposition to environmental determinism, appear to have built their conceptual framework on the philosophy of possibilism. Chernyshevski in the late nineteenth century sought to trace out the difference of character between peoples and their national character, and the general character of elements promoting progress without once raising the bogey of the geographical determination. A kind of reductionism is clearly manifested in his opposition to determinism.

Pokrovskiy, a true Marxist geographer, stated in 1931 that 'Man can master Nature and it is not Nature that is the foundation of his economic activity. Nature is the only material of this activity. The foundation of the economics of man is man's labour' (Pokrovskiy, 1931). *The philosophy of determinism was conceived of as being a scientific weapon of the bourgeois revolution.* Stalin (1952), who opposed and necessarily contradicted the philosophy of determinism, stated that the influence of environment was not a 'determining' factor in as much as the changes and development of society proceeded incomparably faster than the changes and development of the geographical environment.Geographical environment cannot be the chief cause of development of that which underwent fundamental changes in the course of a hundred years. Voskanyan, who raised doubt over the role of the environment, emphasised the activities of man in changing the natural features. He believed in the ultimate control of nature itself by man.

A Canadian geographer, Sebor, felt that a strong reaction to geographical determinism might lead the Marxist geographers along the road to possibilism. However, I. Gerasimov rejected the assertion of Sebor on the ground that 'possibilism' is simply a form of geographical determinism. Sebor's claim that Marxist Soviet geographers were possibilists was just as incompatible with Marxism as environmentalism, although he admitted that the idea that geographers should be interested in nature only from the point of view of the possibilities that it offers to the economy, is rather widely held among Soviet economists. He agreed that possibilism completely rejects determinism but, nevertheless, considered that it erred in regarding nature as only an object for exploitation. His explanation of the origin of possibilism was, to say the least, novel. After pointing out that relations between man and nature are not similar to the relations between the exploiter and the exploited, he went to say that in the capitalist world it frequently happened that the bourgeois scientists sought to apply the social relationships of their system, based on exploitation, to natural phenomena, hence the idea of geographical possibilism.

Critique of Possibilism

The philosophy of possibilism could still be legitimately regarded as a qualification rather than a negation of environmental determinism. Yet, its philosophy became so distorted that by the 1950s it could be seen as a threat to the scientific status of geography as an autonomous discipline.

First, the philosophy sought to hold science to depend on determinism and human geography required 'laws' similar in stringency to those of physical science. But this could be met by the recognition that the traditional emphasis on 'contingence' and 'probabilite' was consistent with modern physics.

Second, the distinctiveness of geography was held to depend on (and be defined by) the relation between society and nature, in which the physical foundation must in large part control the superstructure; 'logical end of possibilism is economic geography with a scatter of place-names'. But this was countered by a neo-determinism (probabilism) and then overtaken by the quantitative revolution and its redefinition of geography as spatial science.

More recently, this redefinition itself has been challenged by the supposed revivification of the Vidalian tradition through the rise of humanistic geography. But this often overlooks Vidal's materialism and relegates the physical environment to a secondary role in the society-nature relationship, rather than the equal role which possibilism originally envisaged.

STOP-AND-GO DETERMINISM

Striking a balance between extreme determinism and extreme possibilism, **Griffith Taylor** developed a new philosophy, called 'stop-and-go determinism' or *neo-determinism,* in the early 1940s. It may be, he stated, that the well-endowed parts of the world offer a number of different possibilities for making a living, but in some nine-tenths of the Earth's land area nature speaks out clearly: This land is too dry, or too cold or too wet, or too rugged' (Taylor, 1941, 440). Any settlers who fail to heed this nature-given limitation must face disaster.

Elaborating his philosophy of 'stop-and-go determinism' Taylor observes: 'Protagonists of the possibilist theory instance the carrying of fertilizer to the Canadian prairies, or the remarkable development of somewhat sterile northern Denmark as examples of human control which have determined the utilization of the regions concerned. I do not for a moment deny that man plays a very important part, but he does not take fertilizer to the "barren grounds"; nor would the Danes have developed their less attractive regions, if they had been free to choose among the good lands of the world. They have merely pushed ahead in Nature's "plan" for their terrain. Even when their example is followed in other similar parts of the world it will only indicate that man has advanced one more stage in his adjustment to the limits laid down by nature. *Man is not a free agent* (ibid., 443–45).

'The writer then is a determinist. He believes that the best economic programme for a country to follow has in large part been determined by Nature, and it is the geographer's duty to interpret this programme. Man is able to accelerate, slow or stop the progress of a country's development. But he should not, if he is wise, depart from the directions as indicated by the natural environment. He is like the traffic-controller in a large city, who alters the rate but not the direction of progress; and perhaps the phrase 'stop-and-go determinism' expresses succinctly the writer's geographical philosophy' (ibid.).

It will be seen that Nature has only 'in large part' determined the programme; man who plays an important part determines the rest. Moreover, man only follows nature's programme if 'he is wise'; presumably he can act foolishly, which admits the possibilist contention that within broad limits set by environment man can choose, at the very least Taylor concedes him the choice between wise and foolish action. Possibility of choice is also suggested by the name 'Stop-and-Go Determinism', since traffic lights are not usually placed on one-way streets which have no interconnections.

Commenting on the philosophy of 'stop-and-go determinism', Tatham remarks: 'Man, not environment, judges an action as wise or foolish by reference to some aim or goal he considers desirable. Until such a goal has been set up, wisdom and folly have no exact meaning. Taylor's definition suggests that the goal must be adjustment to Nature's plan, the carrying out of Nature's programme. From among the possibilities of wise and foolish action how man can recognize this plan? Obviously, as the proponents of possibilism admit, the opportunities offered by any environment are not all equal. Some demand little effort from man, others continual struggle; some yield large, others meagre returns. The ratio between effort and return can be looked upon as the price nature extracts from man for the particular choice he makes' (Tatham, 1952, 160)

Tatham (1952, 161) further adds: 'Once the possibility of alternative action is conceded, then it is difficult to see how "Stop-and-Go Determinism" can claim that man is not a free agent. That his liberty is curtailed all agree. In no environment are the possibilities limitless, and for every choice a price must be paid; proponents of possibilism admit this but within these limits freedom to choose exists. Man makes his choice, but man himself judges its relative wisdom or folly by reference to goals he himself has established. Limits to man's freedom beyond those generally recognised by possibilists are ... those imposed by man's conception of wisdom. There is nothing indeed that contradicts the assertion of Febvre that "there are no necessities but everywhere possibilities and man as master of these possibilities is the judge of their use"... Despite the extreme Determinism phraseology, closer examination then reveals "Stop-and-Go Determinism" to be very different from the old Determinism. It introduces the idea of choice, and since it conceives of choice being guided by consideration of a goal that is to be attained, it could just as *logically be named "Pragmatic Possibilism" or "Stop-and-Go Determinism"'.*

However, Taylor made it clear that he was not an old-fashioned determinist who sought to make extreme statements about the effect of climate on man, but who never sought to put their statements to scientific validation and examination. Here is what he had to say, with considerable satisfaction, about the outcome of his arguments concerning potential settlement in Australia: 'Thirty years ago I predicted the future settlement pattern in Australia. At Canberra (in 1948) it was very gratifying to be assured by the various members of the scientific research groups there, that my deductions (based purely on the environment) were completely justified. This aspect of geography is Scientific Determinism' (Taylor, 1952).

The Marxist geographer of the former Soviet Union, **V. Anuchin** (1960) had identical standpoint on the philosophy which Griffith Taylor had developed in the 1940s. Anuchin still insisted, against formidable opposition, that determinism (in the sense of indirect causalties or mediation) was 'one of the indispensable facets of dialectical thought'. Indeterminism 'sharply separates and counterpoises human society to the rest of nature' and so 'rejects geography as a science'. An argument like this was about more than scientific status, of course, and its political resonances were unmistakable. Anuchin's neo-determinism was charged with restoring 'the causal connection among phenomena and the unity of the material world', but at the same time it had to transcend the classical doctrine and understand the qualitative difference between various categories of the world. In short, 'monistic geography' had to reaffirm the essentially materialist dialectic between man and nature. While a recognition of this was at the heart of classical Marxism and was unequivocally accepted by earlier radical tradition in geography, it was suppressed in modern structural Marxism, and remained peripheral to much of modern Marxist and radical geography.

R. Peet observed that such a theoretical base was as yet 'conspicuously weak' and largely untouched. But the scars of the debate over environmental determinism are at last beginning to heal. The physical environment is no longer seen as the exclusive preserve of a physical geography estranged from the human and social sciences. In part, of course, this flows from (and feeds back into) the rise of political movements concerned about environmentalism and environmental issues. It is also in part the product of development within physical geography itself, and particularly the emergence of an applied physical geography, and in part the product of developments within the human and social science. It is now apparent that the emergence of an *structuration theory*[2] which has one of its concerns the constitutive character of society-nature relations has done much to clarify the significance of what Marx once described as the process whereby people act upon 'external nature' and change it and thereby simultaneously change their own nature.

Debate over environmental determinism and possibilism continued into the 1960s, 1970s and 1980s and was actively pursued in the United Kingdom in the first decade after the Second World War. O. H. K. Spate in 1957 proposed a middle ground with the philosophy of probabilism (the view that although the physical environment does not uniquely determine human actions, it does nevertheless make some responses more likely than others—'human action was represented as not so much a matter of an all-or-nothing choice or compulsion, but a balance of probabilities' (Spate, 1957, 10). This view was in fact perfectly compatible with the original Vidalian concept.

'Thus the lengthy discussion among geographers about whether man is a free agent in his use of the Earth or whether there is a "nature's plan" slowly dissolved as the antagonists realized the existence of merit in each case. Some geographers proceeded independently to study man-environment interactions outside the confines of these debates. But while environmental determinism was a view strongly held and widely preached by geographers about whether man is a free agent in his use of the Earth, respect for the discipline declined somewhat in the eyes of the academic community at large, who rejected the paradigm. As a consequence, geography's next paradigm, which had strong roots in environmental determinism, was very much an introspective and conservative one' (Johnston, 1979/1983).

CHANGING PERCEPTION OF HUMAN-ENVIRONMENT RELATIONS: ENVIRONMENTALISM

The old concept of the man-environment relationship and that of the concept of environmentalism (a synonym for environmental determinism) has changed with change in man's capacities and capabilities. To O'Riordam (1989, 79), '... human society has always recognised its capacity to destroy the environs as greater than its ability to restore the damage within a manageable period of adjustment'. The central objective of the current view on environment is 'to place humankind in the ecological setting, simply as one of the sentient species'. It has been largely supplanted by perspectives which see human agency as having a greater effect, albeit often malignant rather than benign'. One of these more recent approaches is a celebration and exploration of the close and enriching effective bond between people and the environments that they create, inhabit, manipulate, consume, visit or even imagine (Gold and Burgess, 1981).

The current debate on environmentalism revolves around two fundamental issues: (1) resource exploitation is inevitable for human survival and in this process it is inevitable that man shall take more than he returns; and (2) it revolves around the hope for a better future based on the faith that ultimately the moral fibre in human nature shall prevail, leading to greater concern for the survival of the species as against narrow personal gain.

The changing 'perception' on environmentalism, however, aims at human well-being, and is focused on the belief that this is possible only if Nature is accorded its rightful place as a friendly partner in the process of development and growth. It is

[2] *Structuration theory* refers to an approach to social theory concerned with interconnections between knowledgeable and capable human agents and the wider social systems and social structures in which they are necessarily implicated.

built upon a deep faith that survival of humankind is dependent on the survival of a healthy and ecologically balanced Earth environment. As O'Riordan writes: 'Environmentalism is a collage of values and views of the world, a general patterning of predispositions, being first and foremost a social movement, though with political overtones.... It is based on the philosophy that embraces Earth-centredness, a sense of ultimate communalism[3], non-violence and concept of time that is almost timeless' (1989, 80).

The current thinking on environmentalism also attempts a greater sensitivity and awareness about the 'environment' that can be linked to promoting and reorienting practical action 'towards' the environment. 'It is, indeed, one of the most trenchant criticisms of a rather different approach, which has clearly been preoccupied with the formulation of environmental policy and which consists of those versions of "systems analysis" which have been directed towards the control of the "ecosystem", is that its interests are essentially "technocratic" rather than "ecnocentric", concerned with means rather than ends' (O'Riordan, 1981); or to use the language of 'critical theory', technical rather than practical in the original sense of the word. Certainly, early investigations of interface between human and physical systems were mainly technical, involved with strategies of environmental management, but later investigations of 'control systems' have at least recognised the importance of ethical issues as well.

According to Bennet and Chorley (1978), the environmental imperatives require change in social values towards a more conservation-oriented system. The changing perception and/or the current thinking on environmentalism is very close to the concept of 'sustainable development' which was made popular through the 1987 *Report of the World Commission on Environment and Development*. According to the *World Commission Report*, 'sustainable development stands for development that meets the needs of the present without compromising the ability of the future generations to meet their own needs' (WCED, 1987, 8, 43).

O'Riordan (1989) attempts to distinguish between 'sustainability' and 'sustainable utilisation', and says that the former incorporates ethical norms in the Gaia[4] tradition, including emphasis on taking due consideration of rights of the future generations of all living species, but later, denotes a rate of resource uptake that equals the rate of renewal and replenishment.

[3] *Communalism,* here refers to a faith in the cooperative capabilities of societies to establish self-reliant communities based on renewable resource use and appropriate technologies.

[4] *Gaia* refers to a name of the Greek goddess—the nurturing mother figure from whom all sustenance on the Earth was derived. The current literature on environmentalism reflects a faith that is identified as *'Gaianism'*.

14

Recent Trends in Geography (Quantitative Revolution)

HIGHLIGHTS

- Paradigms in Geography
- The Quantitative Revolution
- Positivism

In Kuhn's terminology[1], geography was in the pre-paradigm phase until the time of Darwin. Kant did not set forth a paradigm for geography, and thus he was not the founder of a school of geography, but clarified the role of the subject and its position in relation to other contemporary sciences. Carl Ritter was probably the first geographer to have provided a clear description of his method, which appears to have approximated the Baconian Classical Model of how a scientist works.

The inductive Baconian route derives its generalisations out of observations: a pattern is observed and an explanation developed from and for it. This begins with an observer perceiving patterns in the world; he then formulates experiments, or some other kind of tests, to prove the veracity of the explanations which he has produced for those patterns. Only when his ideas have been tested successfully against data, other than those from which they have been derived, can a generalisation be produced.

Ritter used the inductive method as a framework for his presentation of data and as a means to arrive at some simple empirical generalisation. He offered a teleological explanation which sought to explain that everything that happened was directed to a purpose whose controlling conditions were laid down by God. Such a teleological philosophy cannot be tested empirically; and therefore does not qualify as a scientific explanation. It does, however, have the qualities of a paradigm as defined. Because of Ritter's teleological philosophy, it is difficult to identify the professionalism phase in geography. It is academically important to note that the Ritterian School of Geography, which was the first actively founded school, did not lead the subject into the first phase of a paradigm. It is somehow attributed to the growth of Darwinism.

PARADIGMS IN GEOGRAPHY

However, Darwinism did not completely discredit the major ideas of contemporary geography, and the study of development over time continued to be regarded as important. It was largely from the Darwinian tradition that Friedrich Ratzel led the subject into the first phase of professionalism, and then the *'deterministic school' founded by him represented the first paradigm phase in geography.*

An interesting aspect of the development of geography during the latter half of the nineteenth

[1] Kuhh provided a graphical interpretation of the development of science through distinct phases: pre-paradigm phase; professionalisation; paradigm phase I; crisis phase with revolution; paradigm phase II etc.

century was the way in which research workers attempted to develop the subject as a nomothetic science to a greater extent than might have been expected. It was especially after Darwin that scientists were looking for the governing laws of nature (and the materially conditioned social laws) and to a considerable extent adopted a nomothetic or law-making approach.

The inductive explanation was replaced by hypothetic-deductive method, which was especially a characteristic of the natural sciences. Research workers, starting from inductive arrangements of their observations or from intuitive insights tried to devise *a priori* models of the structure of reality. These were used to postulate a set of hypotheses which could be confirmed, corroborated or rejected by testing empirical data. A large number of confirmations led to a verification of the hypothesis, which was then established as a law. This law stands until its eventual rejection as a result of later research.

Scientific knowledge, obtained through the hypothetic-deductive explanation, is a kind of controlled speculation. An increasing number of human geographers sought to apply such a procedure during the 1950s. The method, known as *positivism,* was developed by a group of philosophers working in Vienna during the 1920s and 1930s (Harvey, 1969, 35). It is based on a conception of an objective world in which there is order, waiting to be discovered. That order—the spatial patterns of variation and covariation in the case of geography—exists and cannot be contaminated by the observer. A neutral observer, on the basis of either observation or his reading of the research of others, will derive a hypothesis about some aspect of reality and then test the hypothesis: verification of hypothesis leads to building of laws. A key concept of this philosophy is that *laws must be proved through objective procedures,* and not accepted simply because they seem plausible.

The hypothetic-deductive method has led human geography to develop into a model building theoretical science because it deals with quantifiable phenomena which seem to have known situation in time and space. However, Minshull (1970) observed that the hypothetic-deductive method was not applied in strict sense in the subject, that the determinists and geomorphologists stated the generalisations first and then gave a few highly selective examples as proof. Geographers could not test hypothesis· by verification procedures involving a number of repeated experiments, as could the physicists. Statistical tests that might play the same role as the experiments were not then sufficiently developed to cope with the complex geographical material.

The ruling deterministic paradigm of Ratzel-Semple-Huntington was challenged by a new *possibilist-regional paradigm,* developed by the French School of geography led by Paul Vidal de la Blache. He stressed that man has free will and participates in the development of each landscape in a unique historical process. The possibilist paradigm presented a model of man perceiving the range of alternative uses to which he could put an environment and selecting that which best fitted his cultural dispositions. Geographers were trained to concentrate on the study of the unique single region. This paradigm necessitated the adoption of the method called the *'participating observation'* that involved the field study of the region in order to identify the uniqueness of the man-environment relationship.

While possibilism and the regional geographical school established a new paradigm, but this paradigm did not immediately displace its predecessor. It was because of the strong position of geomorphology and physical geography that the deterministic explanatory model survived side by side with possibilisim. Debate over environmental determinism and possibilism continued into 1960s and was actively pursued in the United Kingdom in the first decade after the Second World War and Spate (1957), for example, proposed a middle ground with the concept of *'probabilism'*.

Perhaps, the major reaction against the possibilist paradigm came from the Australian geographer

Grifith Taylor, who felt that the possibilists built their arguments in temperate environments such as that of north-western Europe, which do indeed offer several viable alternative forms of human occupance. But temperate environments are rare; in most of the world (as in Australia) the environment is much more extreme and its control over man is much greater accordingly. He used the phrase *'Stop-and-Go' Determinism* to describe his point of view that man might attempt whatever he liked/wished with regard to his environment, but in the long run nature's plan would ensure that the environment won the battle and forced a compromise with its human occupant.

The *regional paradigm* which was dominant in the years before and just after the Second World War, was not only an attempt at generalisation, but at generalisation without structured explanation. It was thus of a very different type from the increasingly discredited law-making attempts of the previous paradigm. The main focus of the paradigm was on areal differentiation, on the very character of the Earth's surface, especially of the inhabited parts. Its picture of that variation was built on parallel topical studies of different aspects of the physical and human patterns observed.

The regional paradigm flourished in France, Great Britain and the United States. Richard Hartshorne's name is associated with the regional paradigm. By the 1950s, disillusionment with the empiricist philosophy was growing and slowly the topical specialisms came to greater dominance and gradually the regional paradigm lost much of its earlier acceptability.

Paradigm shift

It was *Schaefer who brought about a paradigm shift,* when he criticised the exceptionalists' claims made for the regional paradigm. He claimed that in geography the major regularities which are described, refer to spatial patterns and hence geography should be conceived as the science concerned with the formulation of the laws governing the spatial distribution of certain features on the surface of the Earth. It is these spatial arrangements of phenomena, and not the phenomena themselves, about which geographers should be seeking to make law-like statements (Schaefer, 1953). He assigned the nomothetic (law-producing) philosophy to geography as a spatial, social science, and recognised the problems of experimentation and of quantification, and favoured a methodology based on map correlations.

Schaefer, with his *spatial organisation paradigm,* initiated what may be called the quantitiative and theoretical revolutions in geography. However, the major advances towards a unifying methodological and philosophical basis of the quantitative school were made in the 1960s by Peter Haggett, Richard Chorley and David Harvey. Haggett's book *Locational Analysis in Human Geography* (1965) called for a debate within the subject on this paradigm shift.

In *Models in Geography* (1967), both Chorley and Haggett stated that they had looked at the traditional paradigm in geography and had found that it was only classificatory and under severe stress. They held the view that geography should adopt an alternative *model-based paradigm* and made it clear that the new development within the subject not only represented a wider range of methods, but demands a fundamental paradigm shift. Model building was set up as the aim of geographical investigation and inquiry, a task to be done with the help of quantitative methods. The geographers have the option to choose between the traditional and the new model-based paradigm (Holt-Jenson, 1981).

Webber (1977) identified an *entropy-maximizing paradigm* which focuses on location models (the probability of an individual being in a particular place at a particular time), interaction models (the probability of a particular trip occurring at a particular time), and on the location/interaction models. The entropy-maximising paradigm asserts that though the study of individual behaviour may be of interest, it is not necessary for the study of aggregate social relations. The patterns predicted by the models are functions of the constraints (which are the information provided at the meso-state), so that knowledge of these means that 'the entropy-maximizing paradigm' is capable of yielding meaningful answers to short-run operational problems and thus is of immense value for immediate planning purpose.

According to Weber, much of the economic system is variable, the constraints and the spatial form of the urban region may change. The research task facing the entropists is: (1) to identify the constraints which operate upon urban systems, which is partly economic problem; (2) to deduce some facts of the economic relations among the individuals within the system from the use of the formalisms; and (3) to construct a theory which explains the origin of constraints. Only when the third task has been attempted, may the paradigm be adequately judged.

Since early 1950s, the discipline of geography has been subject to frequent paradigm shifts as geographers have attempted to explore a multiplicity of philosophical avenues and research strategies to deal with a multitude of problems and observations. Berry (1978) suggests that pluralism and inter-paradigm conflict have been much more common than have periods of normal science. He observes: 'A diversity arising from the mosaic quality of modern geography, they have moved from one paradigm to another, and in the last decade [1960s] they have been extremely dynamic.... With multiple ideas and origins, modern geography could rightly be characterized as a mosaic within mosaic.' *Geography at present* is *in the throes of a paradigm crisis* because of larger accumulation of various kinds of problems which cannot be solved within the frameworks of the 'ruling paradigms'.

THE QUANTITATIVE REVOLUTION

Anglo-American geography underwent a radical transformation of spirit and purpose in the 1950s and 1960s, replacing an earlier 'idiographic concern' with areal differentiation by a 'nomothetic' search for models of 'spatial structure'. The quantitative revolution in geography was largely concerned with giving geography a scientific approach along with the application of statistical methodology to geographical research. The revolution involved the acceptance of those elements in positivism which had previously been disregarded, namely, the concept that there is one science and one methodology which extends from the natural into the human sciences. The quantitative schools set out to discover universals, to build models and establish theoretical structures into which geographical realities may be fitted (Burton, 1963). The quantitative methods and models have, to a greater extent, been developed because of their predictive values.

The movement which led the revolution in geography was started by physicists and mathematicians, and has expanded to transform first the physical and then the biological science. It has come largely as a result of the impact of work by non-geographers upon geography, a process shared by many other disciplines where an established order has been overthrown by a rapid conversion to a mathematical approach. A strong reaction to environmental determinism has served to delay the coming of the quantitative movement to geography and has postponed the establishment of a scientific basis for our discipline that the quantifiers hope to provide. It is not surprising, therefore, that the quantitative revolution was resisted most strongly by American geographers, for it was in the United States that the reaction to environmental determinism was strongest.

Characteristically, the source of strongest opposition is now the source of greatest support and the American geographers achieved notable distinction in quantitative techniques (Burton, 1963). The report of a National Academy of Sciences—National Research Council (1965) committee on 'The Science of Geography'—discussed geography's problem and method with the statement: 'Geographers believe that correlation of spatial distributions, considered both statistically and dynamically, may be the most ready keys to understanding of developing life systems, social systems or environmental changes. In the past, the progress was gradual, however, because geographers were few, the methods for analysing multivariate problems were rigorous and systems concepts were developed only recently.'

Space. The word 'space' appears to be closely associated with the quantitative revolution. It was only in the late 1940s and early 1950s that the conception of space became a recognisable tradition of inquiry in modern geography when Schaefer (1953) declared that spatial relations alone are the ones that matter in geography. *Space has now become the basic organizing concept of the geographer.* Blaut distinguishes between 'absolute' space and 'relative' space. He suggests that 'space' in absolute· conception 'is a distinct, physical and eminently real or empirical entity in itself'; and 'space' in relative conception 'is merely a relation between· events or an aspect of events, and thus bound to time and process'. This distinction attempts at the development of general theorems of spatial organisation, because if geography is to generalise, it must be able to replicate cases and it has to use relative space.

The shift to a relative spatial context is still in progress and is probably the most fundamental change in the history of geography as it opens an almost infinite number of new worlds to explore and map. Space becomes relative because it is related to the perception of an individual. In location theory, the builders of spatial models are also trying to use measures of relative rather than absolute location. Forer (1978) introduces the concept of *'plastic space'*—that is, space which is continuously changing its size and form as a result of socio-economic demands and technological progress, making it dynamic and truly relative.

The quantitative revolution has led to the identification of two major approaches to the study of geography:

1. **Spatial analysis:** It refers to the quantitative (mainly statistical) procedures and techniques applied in locational analytic work, although such procedures and techniques are not particular to the philosophy of positivism.
2. **Spatial science:** The presentation of human geography as that component of the social sciences which focuses on the role of space as a fundamental variable influencing the organisation and operation of society and the behaviour of individuals. It was formulated during the quantitative revolution and is usually closely related to the philosophy of positivism.

The following aspects of the conception of 'space' under the above two approaches appear to be the central features of geography:

(a) **Spatial interaction:** It refers to interdependence between geographic areas. This interdependence is complementary to the society-environment interdependence within a single area and, therefore, it is a major focus of quantitative inquiry.

(b) **Spatial structure:** It refers to the resulting arrangements of phenomena on the Earth's surface. It could be defined most sharply by interpreting structure as geometrical, from which it followed that the science of space (geography) finds the logic of space (geometry) a sharp tool. The revival of this classical geometric tradition appeared to be a central feature of the quantitative· revolution in geography and its reconstitution as spatial science.

William Bunge in his *Theoretical Geography* (1962) sought to extend the arguments of Schaefer to the effect that geography is the 'science of spatial relations and interrelations geometry is the mathematics of space; and so geometry is the language of geography'. The Hartshornian chorological paradigm, emphasising nature and interrelationships between specific places or regions, was rejected in favour of a geography (paradigm) based on spatial analysis which emphasised the geometric arrangement and the pattern of phenomena. The study of the whereabouts of things, their spatial distribution, which the regional-chorological paradigm considered as a deviation, formed the core of the geographic enterprise (Sack, 1974; Holt-Jenson, 1981).

The development and the philosophical discussion which followed the debate between Schaefer and Hartshorne during the 1950s led to an early acceptance of theory-building and modelling. The account which Edward A. Ackerman presented in 1958, regarding the theoretical discussion, controversy and revolution, attempted to encourage younger generations to concentrate attention on systematic branches of geography enabling the development of more refined theories and models.

The accleration of theoretical works was especially marked in institutions led by geographers who had studied the natural sciences, especially physics and statistics, and who had good contacts with developments in theoretical economic literature. The frontier between economics and geography became productive of new ideas and techniques during the 1950s at several American Universities. In the United States, there developed four schools of quantitative geography. Three schools were developed in the departments of geography in the Universities of Iowa, Wisconsin and Washington, with Washington as the most prominent centre of innovation. The fourth quantitative school, known as the 'Social Physical School', developed independently at Princeton University, under John Q. Stewart and William Warntz, who sought to draw inspiration from physics rather than economics.

'By far the largest volume of work in the spirit of Schaefer's and McCarty's proposals that was published during the 1950s came from the University of Washington, Seattle. The leader of the group of workers at Seattle was W. L. Garrison ...(he) was influenced by Schaefer's paper, although the dates of his earliest publications indicate that he was involved in applying the positivist method to systematic studies in human geography before 1953, and also Involved was E. L. Ullman, who moved to Seattle in 1951 and who had already done pioneering research into urban location patterns, and transport geography.' Garrison and his associates at Seattle were mainly interested in urban and economic geography, into which they introduced location theory based on concepts from economics with associated mathematical methods and statistical procedures.

Outside the United States, the department of geography at Lund University in Sweden, developed into a centre of quantitative and theoretical geography, attracting scholars from many countries. A logical academic contact developed between Lund and Seattle from the very beginning of the quantitative revolution. Hagerstrand, the Swedish geographer, taught at Seattle in 1959 and R. L Morill, one of Garrison's associates at Seattle, studied with him in Lund where his work on migration and growth of urban centres was presented.

Major Theoretical and Methodological Developments

Location Theory. It is a body of theories which seek to account for the location of economic activities. Location theory was formally admitted into the developing corpus of human geography mainly through the efforts of W. L. Garrison (1959) and his associates of Seattle, and on the other side of the Atlantic through the efforts of Peter Hagget (1965) who stressed on the 'fundamental role of locational concepts within human geography', and called for a reaffirmation of its geometric tradition. However, the classical location theories of Von Thunen on the use of areas in agriculture (1826) and Alfred Weber's study of industrial location (1909) were economic theories. Later on Ohlin, Hoover, Losch and Isard attempted to develop the concept of the areal and regional aspects of economic activity, which could be modelled into Neo-Ricardian economics, and which had some relevance in earlier works on theoretical geography.

Central Place Theory. It is a major theory within urban geography providing an account of the size and distribution of settlements within an urban system. Walter Christaller was the first geographer to make a major contribution to location with his famous thesis of 1933 which was translated by Baskin as 'Central Places in Southern

Germany' (1966). He dealt with urban settlements solely as centres for retailing goods and services. He used the concepts of *range* (the maximum distance a consumer will travel to purchase a good or service) and *threshold* (the minimum volume of business necessary for an establishment to be economically viable) and noted that different retailing functions varied in their ranges and thresholds. These were grouped into orders, or types of establishments with similar characteristics. Another approach in the Central Place Theory was developed by August Losch (1954) who attempted to incorporate manufacturing as well as retailing functions in his model of urban system. He identified a separate range, threshold and hexagonal hinterland for each establishment type. These two approaches were central to the new quantitative urban geography of the 1950s and 1960s (Johnston, 1986). However, in Kuhn's terminology, Christaller's model could not be accommodated within the reigning paradigm.

Social Physics. It is an approach to the study of human society which draws upon analogy with the physical world for analysis of human behaviour in the aggregate. J. Q. Stewart (1947) of Princeton University together with Willian Warntz (1959) developed social physics to create what they called *macro geography.* Relying on the postulate that interactions between places are directly proportional to the product of the masses, and inversely proportional to some function of the distance between them, Stewart pointed out the isomorphic (equal form or structure) relationship between this empirical generalisation and Newton's law of gravitation. This concept in human geography is known as the *gravity model.* He pointed out that 'human beings obey mathematical rules resembling in a general way some of the primitive laws of physics'. Warntz also applied analogy models from physics in his studies of population potentials.

Zipf (1949) was another influential authority on social physics after Second World War. He devised a *'principle of least effort'* according to which individuals organise their lives so as to minimise the amount of work which they undertake. Movement involves work, and so the minimisation of movement is a part of the general principle of least effort.

Diffusion of Innovation. It means the *spread of a phenomenon over space and time.* Though the concept of diffusion is not new in geography as it was developed in an organised way by Ratzel, but it was Hagerstrand of the University of Lund, Sweden) who quantified the concept, developed the diffusion model, and provided a theoretical structure to the concept. Hagerstrand became interested in the possibilities of investigating the process of innovation with the aid of mathematical and statistical procedures. He stressed the spatial processes, and his thesis of 1953, later translated by Pred as 'Innovation Diffusion as a Spatial Process' (1967), examined the diffusion of several innovations among the population of a part of central Sweden.

Hagerstrand's diffusion model represents an interaction matrix that suggests the contours of a generalised or mean information field, which structures the way in which information circulates through a regional system. These flows are moulded by both physical barriers and individual resistances which together check the transformation and so shape successive diffusion waves which break into the adoption surface. With the aid of the so-called 'Monte Carlo stimulation' which involves the use of random samples from a known probability distribution, he was able to construct a general stochastic model of the process of diffusion. The stochastic Hagerstrand model enabled the spread of innovation to be stimulated and later tested against empirical study. Much work has since been done, both on the process of diffusion and even more, on patterns of spatial spread which, it is assumed, result from diffusion processes.

Regional Science, Theoretical revolution in the contemporary American human geography led to the emergence of the regional science. It is a discipline linking economics, geography and planning and is concerned with theoretical and quantitative analysis of regional economics and problems. Regional science is the creation of Walter Isard (1956), an economist who developed spatial components into his models, in part to provide a stronger theoretical basis for urban and regional planning than had existed previously.

Regional science is economics with a spatial emphasis and it tends to focus on mathematical modelling and economic theorising. The focus of regional science has encompassed both quantitative empirical modelling and more purely theoretical analysis of location problems and regional economics. Major work in empirical modelling has included regional input-output models, industrial complex analysis and gravity model of spatial interaction and statistical work of central place theory. The development of regional science theory was strongly influenced by contemporary trends in economic theory. In human geography, it provided an *umbrella revolution.* Geographers such as B. J. L. Berry, M. Dacey, R. Morrill, W. Tobler and W. Warntz were associated with the Regional Science Association, and many of the classical papers on quantitative geography appeared in the *Journal of Regional Science*.

Locational Analysis. It is an approach to human geography which focuses on the spatial arrangement of phenomena on the Earth's surface; hence the approach is sometimes referred to as *spatial science.* Fundamental to the approach is the philosophy of positivism, giving emphasis on models and laws of spatial arrangements; hence it has a close link with quantitative revolution. Peter Haggett's *Locational Analysis in Human Geography* (1965) is a classic on this aspect. He wanted that the geometric tradition in geography must be revived in an approach that had an emphasis placed squarely on asking question about the order, location order, shown by the phenomena studied traditionally as human geography. He realises the need to adopt a systems approach that focuses on spatial patterns and linkages, to employ models as the stimuli to achieving understanding, and to use quantitative procedures as the means to precise statements (generalisations) about locational order. Haggett's major contribution lies in his depiction of pattern and order in spatial structures, being phrased within a decomposition of nodal regions into five geometrical elements (movement, networks, nodes, hierarchies and surfaces) and the sixth element, diffusion, was added in the second edition of the book (Haggett et al., 1977).

The impact of locational analysis with its focus on spatial arrangements, models and quantification on human geography was substantial. Location theory indicates the nature of the normal science of the new paradigm, with work on spatial patterns, the links and flows between places in such patterns, the dynamics of the patterns and the preparation of alternative patterns through model building exercises that identify efficient solutions.

Systems Approach and Systems Analysis

A system consists of a set of entities with specification of relationships between them and their environment. It usually contains a large number of entities and the relationships between them imply a high degree of interdependence. The study of systems, therefore, appears to be associated with the study of complex structures.

Systems Approach. Geographers have used forms of systems concepts since the dawn of the subject. Despite its venerability, a systematic approach has tended to remain a philosophical concept rather than providing guidelines for practical research. No methods and techniques had been developed to enable the analysis of complex systems in an accurate way before Second World War when systems concepts were invoked in descriptive contexts with particular reference to consideration of a balance in nature. The concept of systems is often associated with particular theorizing styles, i. e. positivism or functionalism. However, Williams (1983) observes the relationship between systems concepts and structuralism.

Ludwig Von Bertalanffy (1950) is credited with the development of the general systems theory. It aims to provide theoretical properties of different types of systems. It reflects an attempt to unify science via perspectivism instead of the more usual division of science through reductionism. *Its focus is on isomorphism, the common features among the systems studied in different disciplines.* Its subject matter is the formulation and derivation of these principles, which are common for systems in general. Many of the basic ideas have a long history, but in geography their formal incorporation into a 'metalanguage' occurred during the 1960s. This involved a set of recognisably scientific procedures which could be connected up to those of the quantitative revolution. This also involved a set of concepts which offered the prospect of a theoretical integration of physical and human geography.

R. J. Chorley is the first geographer to have introduced general systems theory in geography. His paper 'Geomorphology and General Systems Theory' (1962) was the first major devoted exclusively to a Systems approach within the framework of General Systems Theory. A major part of the paper is devoted to the application of the concept of open and closed systems to geomorphology. Many of the later works appear to have used the formal procedures and concepts to identify isomorphisms between different types of geographical systems. Ray and Others (1974) applied the concept of allometry to show that the growth rate of a component of an organism is proportional to the growth of the whole. Berry (1964) pointed out that cities are open systems in a steady state, as exemplified by the stability of their behaviour-describing equations. Woldenberg and Berry (1967) drew analogies between the hierarchical organisation of rivers and of central place systems.

The most explicit recognition of the general systems theory has come from macro-geography, where Warntz (1973) claimed that, properly understood, geography is general spatial systems theory. The advantages of the general systems theory to human geography lie in its inter-disciplinary approach, its high level of generalisation, and its concept of the steady-state of an open system. But it is realised that geography's strong empirical tradition means that it has more to contribute than to take from the general systems theory. However, to Chisholm (1967), the general systems theory seems to be an irrelevant distraction.

Systems Analysis. It is a methodological framework for investigating the structure and function of a system. As a method, systems analysis concerns abstraction rather than truth. The system must be seen as a useful abstraction or model which enables a particular form of analysis to be made. The keystone of the study of systems is connectivity. As Harvey (1969) points out, reality is infinitely complex in its links between variables but systems analysis provided a convenient abstraction of the complexity in a form which maintains the major connections. A system comprises of three components (Fig. 14.1):

(a) a set of elements;
(b) a set of links (relationships) between those elements; and
(c) a set of links between the system and its environment.

In a system, the elements have volumetric qualities and material flows along the links. As the system operates, so the various quantities may change. Every system has three basic aspects: structure, function and development. The *structure* is the sum of the elements and the connections between them. *Functions* concern the flows (exchange relationships) which occupy the connections. *Development* represents the changes in both structure and function which may take place over time (Johnston 1983; Holt-Jenson, 1981).

The structure of the systems can be treated in two separate frameworks—*closed systems* and *open systems.* Closed systems have definable sealed boundaries across which no input or output of energy occurs. Such systems are rare in geographical studies. Open systems, on the other hand, are those systems which have both inputs and outputs of energy to maintain the system (Fig. 14.2).

Fig. 14.1 Structure of a system. Elements, LS links between elements. LE links between the system and outside environment.

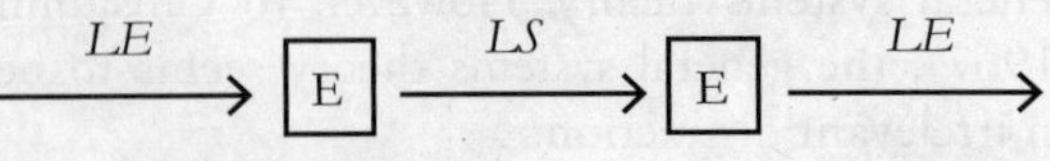

Fig. 14.2 Closed and open systems

Closed System

E → E

Open System

Energy inpur from outside → E → E → *Energy output*

Closed systems are extremely rare in reality, but are frequently created, either experimentally or, more usually in human geography, by imposing artificial boundaries, in order to isolate the salient features of a system. An open system interconnects with its surroundings. All real systems (such as landscapes) are open systems.

Chorley and Kennedy (1971) characterise systems in relation to their structures and flows within these structures. In fact, they identify *four types of systems.*

1. **Morphological systems**, which consist solely of the physical properties of their components and where the relationships between them are expressed through a web of statistical correlations. Much of the spatial analysis outlines such morphological systems.
2. **Cascading systems**, which consist of a chain of sub-systems linked by a cascading throughput such that the output of one sub-system forms the input for the next. Berry (1966) has linked two examples of cascading systems of Haggett's nodal regions and Isard's input matrix representation of an economy in his inter-regional input-output study of the Indian economy. However, factories can be portrayed as cascading systems, in that the output of one factory is in many cases the input for another.
3. **Process response systems**, which are formed by the interaction of morphological and cascading systems. The ecosystem is a process-response system concerned with the flows of energy through biological environments, most of which include, or are affected by, man.
4. **Control systems**, which are process-response systems and which are structured by the intervention of decision-making agencies at certain key points (values) so as to alter the disposition of the throughputs in the cascading system and hence change the equilibrium relationships in the morphological system. The ecosystem is also a control system in that the living components act as regulators of the energy flows. They further represent a major point at which human control system must interact with the natural world.

Langton (1972), on the other hand, *classifies systems into two types:* Simple action systems and feedback systems. *Simple action systems* are unidirectional in their nature; a stimulus in X produces a response in Y, which in turn may act as a stimulus to a further variable, Z. Such a causal chain is merely a reformulation of the characteristic cause and effect relation with which traditional science has dealt. In another language, it is a process law (Harvey, 1969). More important, and relatively novel to human geography, is the '*feedback system*', which is the property of a system or sub-system such that, 'when change is introduced via one of the system variables, its transmission through the structure leads the effect of the change back to the initial variable, to give a circulatory of action'.

The relationships between entities of the system and its environment can often be characterised as feedbacks. The feedbacks are of two types: positive and negative. *Positive feedbacks* are self-enhancing mechanisms which seek to enhance, accentuate and reinforce the changes in a system. Such feedbacks push the system farther and farther away from its initial condition. *Negative feedbacks,* on the other hand, are self-regulating mechanisms which counter -balance changes in a system and bring the system back to the state of equilibrium or to a situation which existed prior to the change.

With the positive feedbacks, the system is characterised as morphogenetic, changing its characteristics as the effect of B on C leads to further changes in B, via D. But with negative feedback the system is maintained in a steady state by a process of self-regulation known as homeostatic or morphostatic. Feedback may be either direct—'A' influences 'B' which in turn influences 'A'—or it may be indirect, with the impulse from 'A' returning to it via a chain of other variables. The concept of feedback with associated notions of homeostatics and morphogenesis, gave the nuclei of the systems theory of change. As a consequence, the nature of feedback should be the focus of geographical study.

Bennet (1975) has attempted to apply systems theory to a problem in human geography on the dynamics of location and growth in north-west England. Having represented the system—its elements, links and feedback relationships—Bennett estimated the influence of various external (i.e. national) events on the system's parameters, isolated the effects of government policy (industrial development certificates) on the system's structure and produced forecasts of the region's future spatio-temporal morphology .

The most comprehensive attempt to forge a systems approach to geographical study has been done by Bennett and Chorley in their book entitled *Environmental Systems: Philosophy, Analysis and Control* (1978) with the intention of providing a unified multi-disciplinary approach to the interfacing of 'man' with 'nature'. The book was prepared with three major aims: 'First, it was desired to explore the capacity of the systems approach to provide an inter-disciplinary focus on environmental structures and techniques. Secondly, to examine the manner in which a systems approach aids in developing the interfacing of social and economic theory, on the one hand, with the physical and biological theory, on the other. Third to explore the implications, of these inter-facings in relation to the response of man to his current environmental dilemmas. It is hoped to show that the systems approach provides a powerful vehicle for the statement of environmental situation of ever-growing temporal and spatial magnitudes, and for reducing the areas of uncertainty in our increasingly complex decision-making arenas.'

The use of systems analysis is based on the assumption (usually implicit) underlying much positivist work in human geography, that valid analogies can be drawn between human societies on the one hand, and both natural phenomena complexes and machines on the other. Individual elements in a system have pre-determined roles and can act and change in certain restricted ways only—depending on the structure of the system and its interrelationships with the environment. As a descriptive device, this analogy allows the structure and operation of society and its components to be portrayed and analysed, and it provides a source of ideas from which hypothesis can be generated. And once a system has been defined and modelled, systems can be used as a predictive tool, to indicate the nature of the element and links following certain environmental changes, such as the introduction of new elements and links.

The systems analysis has wide applications in both human and physical geography along with the interface where man and environment interact. Once the system has been successfully modelled, it can be manipulated using control theory which is a dynamic optimisation technique permitting optimal allocation along the time horizons, and shifts emphasis from mere model construction to model use (Haggett, 1980; Chorley and Bennett, 1981). Such a combination of models describing systems with a theory of systems control has a wide range of potential application in such fields as pollution control, catchment management, inter-area resource allocation and urban planning. It suggests a commonality of interest, focused on methods, between applied physical geography and applied human geography.

Gregory (1978) attempts to criticise both systems analysis and general systems theory on the ground that they are intrinsically associated with positivism. The concept of one systems theory which is relevant for all the sciences may be seen as a fruit of the positivist concept of one science, one method. He is further afraid that prominence given to control system may lead to instrumentalism.

Entropy and Information Theory

Weaver distinguishes systems of disorganised complexity, where the interrelationships are relatively weak, and those of organised complexity, where they are stronger. Traditional science has usually been concerned with simple systems with relatively small number of entities or variables. The 'entropy-maximizing methods' can be applied to systems of disorganised complexity, and the concepts and methods of bifurcation and catastrophe theory can be applied to systems of organised complexity.

The *entropy* of a system is an index of uncertainty. It is a measure of the amount of uncertainty in a probability distribution or a system subject to constraints. In the second law of thermodynamics, an increase in system entropy is an increase in system uncertainty. Thermodynamic entropy relates to the most probable configuration of the elements, within the constraints of the system's operations (Johnston, 1983, 118–19). The concept of entropy has been used in a wide variety of contexts, notably in information theory, and as the basis for the 'entropy-maximizing model' of spatial interaction.

The *information theory* is a mathematical approach in communication science, attempting to analyse the information or degree of organisation of a system. The theory and its associated methods have been used to describe settlement and population distribution in geographic space. In information theory, entropy refers to the distribution of the elements, across a set of possible states, and is an index of element dispersion. One can be completely certain about a distribution, in terms of predicting where one element will be, if all elements are in the same state. Conversely, one will be most uncertain when elements are equally distributed through all possible states, so that prediction of the location of any element is most difficult.

The information-theory measure of entropy is another descriptive index, and can be developed in a variety of ways: (1) as a series of indices of variation in population distribution; (2) as an index of redundancy in a landscape, where redundancy is defined as relating to a regular sequence so that it is possible to predict the landuse of a place as, for example, from knowledge of the landuses at neighbouring places; and (3) as a series of measuring of reactions to situation in states of uncertainty.

The usage of entropy as developed in statistical mechanics rather than in information theory was introduced to the geographical literature by Wilson in his book *Entropy in Urban and Regional Modelling* (1970). He points out: 'The application of the concept of entropy in urban and regional modelling, that is in hypothesis development, or theory building (Model and hypothesis are used synonymously, and a theory is a well tested hypothesis) … the *entropy maximizing procedure* enables us to handle extremely complex situations in a consistent way.' Wilson's initial example was a flow matrix. The number of trips originating in a series of residential areas, and the number ending in each of a series of workplace areas are known, but the entries in the cells of the matrix—which people move from which residential area to which workplace area—are unknown. What is the most likely flow pattern? Even with only a few areas and relatively small number of commuters, the number of alternations is very large.

He defines *three states of the system.* The first is the *'macro-state'*, comprising the number of commuters at each origin and the number of jobs at each destination. The second is the *'mesa-state'*, comprising a particular flow pattern: five people may go from zone A to zone X, for example, and three from the same origin to zone Y, but it is not known which five are in the first category and which three in the second. The third is the *'micro-state'* which is a particular example of a flow pattern—one of the many possible configurations of eight people moving from zone A, five to X and three to Y. Entropy-maximising procedures find the meso-state with the largest number of micro-states associated with it. In other words, it is the meso-state about which the analyst is least certain as to which configuration produces the pattern. The entropy-maximising approach uses entropy as the basis for finding the most likely macro-state of a system with specific constraints.

Weber has extended Wilson's argument that the purpose of entropy-maximising models is to draw conclusion from a data set which are 'natural' in that they are function of that data set alone and contain no interpreter bias. On the basis of Wilson's entropy-maximising model, Weber (1972) developed an *entropy-maximising paradigm* which focuses on locational models and on interaction models or both. The entropy-maximising model acts not only as a 'black box' (representation of an element) forecasting device, but also as a hypothesis. If the operation of the systems described is to be understood, the axioms—the constraints—must themselves be explained. Given the nature of the constrains (in Wilson's initial example, why people live where they do, why people work where they do and why they spend a certain amount of time, money and energy on transport), the task is a major one; entropy-maximising models aim to clarify it and indicate the most fruitful avenues for investigation.

Thus, the mathematical formulation of the information theory exhibits a close relationship with the mathematics of entropy in statistical thermodynamic and entropy-maximising models in geography and planning, and is central to quantitative revolution.

Models and Analogue Theory

'One of the most striking characteristics of geographical analysis which this subject has in common with the other natural and social sciences is the high degree of ambiguity presented by its subject matter.... Where such ambiguity exists, scholars commonly handle the associated information either by means of classifications or models. Much has been written on the subject of geographical classifications, which usually result from the accumulation of a back-log of information which is then dissected and categorised in some convenient manner: model building, which sometimes may even precede the collection of great deal of data, involves the association of supposedly significant aspects of reality into a system which seems to possess some special properties of intellectual stimulation. This is not to imply that classifications and models do not share common ground and, indeed, the genetic classifications of national economic stages and of shorelines form something of a link between them. However, in their extreme forms, classifications and models are sharply differentiated' (Chorley, 1964).

A **model** is defined as an idealised and structured representation of the real, or a simplified representation of reality which seeks to illuminate particular characteristics. The concept is a wide one, and a model is either a theory or a law or an hypothesis or a structured idea (Chorley and Haggett, 1967). A model or theory, when tested and validated, provides a miniature of reality and therefore is a key to many descriptions. It can offer insights towards understanding the real world. There is a single master key instead of the loaded key-ring. Model-building has a long history in many sciences, but its incorporation into geography is of comparatively recent origin (the 1950s and 1960s) and depends on so-called analogue theory.

An **analogue theory** is a formal theory of model-building which provides for the selective abstraction of elements from an empirical domain and their translation into a simplified and structured representation of a particular system. The theory has important connections with positivism, where models are typically used in tests which are intended to allow for the validation of hypothesis and thus the extension of general theories. These procedures were often formalised in statistical and mathematical terms, and as such played a vital role in search for general theorems of spatial organisation. Haggett et al. speak of essential links between model-building and the quantitative revolution, and causing a move from an idiographic to a nomothetic geography (Haggett, Cliff and Frey, 1977).

The use of analogy has long been recognised as a powerful tool both in reasoning process and in throwing new light on reality. Reasoning by analogy involves the assumption of a resemblance of relations or attributes between some phenomenon or aspect of the real world and an analogue or model. The basic assumption involved is that two analogues are more likely to have further properties in common than if no resemblance existed at all, and that additional knowledge concerning one consequently provides some basis for a prediction of the existence of similar properties in the other.

A model or analogue must, therefore, belong to a more familiar realm than the system to which it is applied. In other words, the problem is 'translated into more familiar or convenient terms such that a useful model involves a more simplified, accessible, observable, controllable, rapidly developing, or easily formulated phenomenon from which conclusion can be deduced which, in turn, can be reapplied to the original system or real world. In general, any two things, whether events, situations, creatures or objects, can be said to be analogues if they resemble each other to some extent in their properties, behaviour or mode of functioning. In practice, the term analogue is used rather loosely to cover a wide range of degrees of resemblance. This wide range of resemblance may lead the model builder to employ as varied analogues as a past situation or happening in the real world, a mathematical simplification or a construction built of string and wire' (Chorley, 1964).

There are *two basic functions of a model* as a representation of the real world, such as: (a) a scale model, a map or a series of equations, and some other analogue; and (b) an ideal map, a representation of the world under certain constrained conditions. Both are used in the positivist method to operationalise a theory, as a guide to the derivation of testable hypothesis. A model becomes a theory about the real world only when a segment of the real world has been successfully mapped into it. As a theory (model), it can be accepted or rejected on the basis of how well it works. As a model, it can only be right or wrong on logical grounds. A model must satisfy only internal criteria; a theory must satisfy external criteria as well. Reasoning by analogy is a step towards building of theories, such that a promising model is one with implications rich enough to suggest some hypothesis and speculations in the primary field of investigation.

Models need to be objective but 'tend to be subjective approximations, in that they do not include all associated observations or measurements, but as such they are valuable in obscuring incidental detail (noise) and in allowing fundamental aspects of reality to appear. This selectivity means that models have varying degrees of probability and a limited range of conditions over which they apply. The most successful models possess a high degree of probability application and a wide range of conditions in which they seem appropriate. Indeed, the value of a model is often directly related to its level of abstraction.

Many of the early models, e.g. the Von Thunen model of agricultural land use and the central place models of Christaller and Losch, were attempts to represent the abstract geometry of an idealised economic landscape through various translations of concepts of 'accessibility' and 'distance decay' (Haggett, Cliff and Frey, 1977). Further, since the itinerary of:

< abstraction— > generalization— > disclosure >

which their construction involved was supposed to provide for further extension, they represented necessary moments in the 'puzzle solving' activity required for the foundation of a new scientific paradigm. It was no accident, therefore, that Chorley and Haggett's early advocacy of a 'model based paradigm' for modern geography should have been phrased in such resolutely Kuhnian terms (Johnston, 1986).

Chorley (1964) provides a diagrammatic model for models. The diagram is simply composed of a series of 'steps' each of which contains some aspect of real world, model, observation, or conclusion. These are connected, sometimes in a very loose and varied manner, by 'transformations' whereby the reasoning process is advanced or checked upon. Each of these transformations introduces (in the language of information theorist) the possibility of 'noise' (incidental detail) in so far as the processes of simplification and translation involve, on the one hand, the discarding of information, and, on the other the possibility of the introduction of new information which is irrelevant. The most useful employment of analogues is that which succeeds in causing the least noise.

This sequence of steps represents the most difficult type of model building because, in developing a simplified but appropriate model for a given object, system or segment of the real world, huge amounts of available information are being discarded and therefore much noise (incidental detail) is being potentially introduced. The transformation of 'idealization' is especially difficult, and very much depends on the experience and genius of the model builder. It involves extracting from the mass of information about the real world those aspects which are held to be especially significant, in that they seem to fit together into some sort of pattern. One of the major sources of noise in this transformation is introduced by the phenomenon of 'feedback'. Idealisation largely involves reaching some conclusion about the relationships in the real world which really matter, so that irrelevant information could be snubbed to expose the significant relationships then susceptible to further examination.

A conceptual model is derived by such a means, which, having some basis of observed facts and regularities, contains a mental image of the significant 'web of reality'. The mental image may have come quite unambiguously from the simplification of previous empirical knowledge or may appear to have come out largely from imagination. Many attempts at model building depart from formality at this point and pass by direct reasoning to a hypothesis or some conclusion about the real world which, if successfully appraised against the real world, may form the basis of a theory (Chorley, 1964).

The conceptual model appears to be too complex to handle. It requires further simplification by removing/discarding still more extraneous information, and retaining only the simplest and most significant aspects of the basic matter. The transformation is less noisy than idealisation, because both ends of it are more completely realized. And in practice it may represent a whole series of simplification, the aim of which is to maintain the simplest and most significant aspects of the problem, while removing any irrelevant material which might obscure the fundamental relationships and present a satisfactory solution. The final product of abstraction is often a 'simplified model' which has been reduced to a condition where the fundamental symmetries and relationships be exposed for precise definition and further treatment.

In geography, most attempts at model building by abstraction have met with least success. It is notable that those which are currently judged as most successful are the conceptual models, rather than the simplified models in which excessive simplification has been avoided. The simplified economic models of von Thunen and Weber, with their assumptions of fixed markets and sources of raw materials, seem less attractive than the freer and more complex models of Christaller and Losch. However, most of the earlier models in geography (such as Ritter's notions regarding the development of peoples under differing geographical conditions, Mackinder's 'Heartland Theory', Malthus' population growth model, Frederick Jackson Turnier's 'Frontier Hypothesis of national development', and Suess' eustatic theory) belong to the group of such simplified models in which too much truth has been sacrificed for too much simplicity .

It is interesting to note that not all simplified models produce detrimental results. But some simplification may be significant or successful as it transforms a segment of the real world into a new dimension which is merely different and interesting. The derivation of some form of simplified model enables structures and relationships to appear which are capable of further exploitation, commonly in such a manner that prediction can be attempted. This exploitation can be achieved by means of mathematical, experimental, or natural models.

Mathematical Models. These models are the transformation of the ideas of the simplified models into the formal, symbolic logic of mathematics. They include those models which involve the adoption of a number of idealizations of the various details of the phenomena and in ascribing, to the various entities involved, some strictly defined properties. The essential features of the phenomena are then analogous to the relation-ship between certain abstract symbols, and the observed phenomena resemble closely something extremely simple, with very few attributes. The resemblance is so close that the equations are a kind of working model, from which one can predict features of the real thing which have never been observed. The construction of a mathematical model involves the language transformation from the words of the simplified model to the mathematical symbols (mathematization) so as to produce a mathematical system.

Mathematical system or models can be of two broad kinds, 'deterministic' and 'stochastic'.

(a) In *deterministic models,* if specified conditions exist at a particular moment of time, it is possible to specify what conditions existed at an earlier time or will exist at a later time. Such models are based on the classic mathematical notion of cause and effect, and consist of a set of mathematical assertions from which consequences can be derived by logical mathematical argument. Commonly, the reasoning exploits the assumed simple or multiple relationships between a number of interlocked factors which have been identified in simplified manner.

In physical geography, such mathematical deterministic models have been developed in large numbers so as to analyse/predict various aspects of the morphology of the Earth's surface. The most popular geomorphic application of deterministic models has been in the deduction of the forms accompanying slope recession, summarised by Scheidegger. Jeffreys deduced theoretically the form of resulting peneplane by developing a deterministic mathematical model. Miller and Zeigler applied it to predict the expected patterns of coastal sediment size and sorting. Most mathematical models of the atmospheric circulation are of the deterministic type, as are those involved in numerical forecasting.

Deterministic models have also been developed in human geography, but are not so common. However, Beckman applied the theory of the hydrodynamics equation continuity to investigate the cost of inter-local commodity flows. Lighthill and Whitham used the principle of kinematic waves to investigate the flow and concentration of traffic on crowded arterial roads and Richard applied the principle of the mathematics of fluid flow to investigate traffic flow.

Thus, in deterministic models the development of some system in time and space can be completely predicted, provided that a set of initial conditions and relationships is known (Holt-Jenson, 1981).

(b) *Stochastic models* specify sequences of events within a certain range of probability. Some authors distinguish between a 'probabilistic model' in which the outcome of individual traits is predicted and a stochastic model in which the development of a whole series of outcomes is modelled. A stochastic model may, therefore, include those situations in which the probabilities of an outcome of a particular trial are determined by the outcome of preceding trials. *Stochastic models have been increasingly developed in human geography rather than in physical geography.* Isard developed a spatial economic model of a statistical character. Hagerstrand's 'innovation wave' model is similarly based as is the mathematical theory of population clusters by Neyman and Scott which was inspired by the kinetic gas theory. Garrison suggested the employment of an electronic computer to develop a stochastic mathematical model for city growth.

Both the deterministic and stochastic mathematical models must be susceptible to logical mathematical argumentation, a transformation which is virtually noise free and involves the solution of the equations forming the basis of the mathematical system to provide logical mathematical interpretation into conclusion about the real world It is important to recognise that of themselves mathematical models do not provide explanation of the real world but merely allow conclusion to be drawn from the original mathematical assumptions (Chorley, 1964).

Experimental Models. It is through substantiation that a simplified model can be treated in order to examine certain phases of its operation or attempt to predict it. It is through this process that the concepts of the simplified model are reproduced as tangible structures by a transformation which is inherently very noisy. Experimental models can be further divided into the 'scale model' and the 'analogue model.'

(a) *Scale Models* are closely imitative of a segment of the real world with which they resemble in some respects (e.g. being composed mostly of the same type of materials) and the resemblance may sometimes be so close that the scale model be considered as merely a suitably controlled portion of the real world. The most significant characteristics of scale models are the high degree of control which can be achieved over the simplified experimental condition and the manner in which time can be compressed. Scale models have been *successfully developed in physical geography.* Friedkin developed scale model to investigate the phenomenon of meandering and Rouse applied its techniques in meteorological research. However, the fundamental problems attending their construction is that changes of scale effect the relationships between certain properties of the model and of the real world in different ways.

(b) *Analogue Models.* A more abstract refinement of scale model are the analogue models. They involve a radical change in the medium of which the model is constructed. They have a much more limited aim than scale models in that they are intended to reproduce only some aspects of the structure or web of relationships recognised in the simplified model of the real world segment. Analogue model is a great potential source of noise (incidental detail). Analogue models *abound in physical geography.* Though these models usually provide the basis for plausible hypothesis, but their use is valuable in physical geography.

Some of the analogue models so far developed in physical geography include the construction of hydraulic analogue model to simulate freezing and thawing of soil layers Lewis and Miller's use of a kaolin mixture to simulate some features of the deformation and crevassing of a valley glacier, Reiner's rheological models for simulation of different types of deformation, and Starr's dishpan analogue model to investigate the gross features of the atmospheric circulation in a hemisphere. It is, however, in the fields abutting *on* human geography that analogue experimental models have been most strikingly applied. The spatial economists have shown their interest in the use of analogue models.

Whether a scale or analogue experimental model is used, their construction is followed by experimentation leading to a set of experimental observations, which remain to be interpreted in terms of the real world in a manner similar to the theoretical interpretation of the conclusions from a mathematical model. Very occasionally these results from a mathematical model are transformed into experimental model (a very noisy process) as a further step in testing their agreement with the real world.

Natural Models. Natural models are those in which simplified models are further treated, exploited and used as a basis for further analysis. Prediction by their translation into some analogous natural circumstance is believed to be simpler and more readily observable. Two types of natural models can be conveniently differentiated, the historical natural model and the analogue natural model.

(a) *Historical natural model* implies a translation of the simplified model into a different time and/or place, on the assumption of historical repetition of events. As such it is an obvious source of noise or incidental detail. The nineteenth-century German and French geographers were attracted to the historical geographical analogue concepts. The most obvious geographical analogue which applied to a different time is presented by the analogue forecasting methods developed in meteorology.

(b) *Analogue natural models* involve the translation of the simplified model into a different natural medium—a process of the most noisy kind. Problems of social geography are being attacked on the basis of physical analogues. Garrison developed an ice-cap analogy of city growth. The social physicists have naturally been attracted to the analogue natural models. The model is also used in physical geography.

However, the construction of the historical natural models or the analogue natural models, is followed by or subject to appropriate observation and may yield observed natural results which, through reapplication, may lead to some testable conclusions about the real world situation.

Chorley (1964) observed that it is through a process of more or less 'direct reasoning' that some of the reasoning in geography, leading to appropriate conclusions about the real world situations, could be achieved. However, the value of this method is obviously determined by the creative genius of the operator. Successful appraisal implies the checking of conclusion derived from model building with real world, so that a hypothesis can be developed. And if successive checks of this kind lead to similar conclusions, then a 'theory' comes into being or may be developed.

The **value of model-building** in geographical 'studies' is critically high in importance as it helps in building up refined and systematic approach to understanding man-nature relationship in various forms and contextual frame of reference. Most of the branches or sub-fields of the discipline of geography are replete with models of varying order and character. Models in geography could be categorised in many ways, taking into account the very nature of geographic concern to understand the man-nature development syndrome and its various forms and stages. When one talks of models in geography then it is necessary to add that the *models and their conceptualization is intimately connected with quantification in geographical inquiry,* as the models tend to be systematically precise – and for precision a mathematical frame of reference is desirable, but it is not absolutely necessary (Sharma, 1990).

The rapid development of model building and the use of quantitative techniques could not have taken place without computers, but the computers did not determine the development of either model building or quantitative methods. Model building preceded the invention of the computer in many sciences, but in a discipline like geography which handles large quantities of data it would hardly have been possible to develop operational models worthy of the name without computer. The technological development had given the subject new possibilities which researchers had no hesitation in exploring.

Johnston (1986) observes that the claims and concepts regarding the accountability of the models, and their subsequent application in geographical inquiry, as developed by Chorley and Haggett, have been subjected to re-evaluation on three levels. (1) First, the models themselves have been subject to a (limited) reformulation. In the second edition of 'Locational Analysis', the authors admitted that the present stock of models may be 'unprepossessing enough'.[2] The limited advance was in part because the autonomy of location theory had been compromised and more inclusive models of the space economy had been constructed outside its traditional domain.[3] Those who continued to work within the orthodox framework were thus concerned more with developing statistical and mathematical models which could break open complicated data-structures (e.g. space-time forecasting, Q-analysis) rather than with seeking to establish in any direct way the theorems of a general spatial science. (2) Second, therefore, the original claim for model-building as the object of geographical enquiry was displaced and research efforts were directed towards methods seen as means rather than as ends. (3) Third, even this was attacked by some critics who insisted that there was such clear-cut connection between model-building and the epistemology of positivism that the whole enterprise 'ought to be rejected.

Reaction to the Quantification

Though the bulk of the Anglo-American geographers of 1950s and 1960s were crazy about the quantification of the discipline, but there were a number of people who opposed the scientific method being followed in geography. Two related issues were the main foci of contention: (1) 'whether quantification was sensible in geographical research', and (2) 'whether law making was possible in geography'.

Dudley Stamp (1966) was one of the principal opponents to the quantitative geography. He pointed out that there are many fields of enquiry in which *quantification may stultify rather than aid progress,* because there will be a temptation to discard information which cannot be punched on a card or fed into magnetic tape. There is also a danger that ethical and aesthetic values will be ignored.

Broek (1965) observed that since massive quantitative data on human behaviour are only available for the advanced countries and that only for at best a century, the theorists tend to construct their models from facts of the 'here and now', virtually ignoring former times and other cultures. The procedure becomes invidious when one projects the model derived from one's own surroundings over the whole world as a universal truth and measures different situations in other countries as a 'duration' from the 'ideal' construct. He, therefore, pointed out that search for general laws, at a high level of abstraction, goes against the gain of geography because it *removes place and time from our discipline.*

Minshull (1970) felt that the landscape was rapidly becoming a nuisance to some geographers, and that *many of the models will only apply to a flat, featureless surface.* He warned that there was a real danger that these ideal generalisations about the spatial relationships could be mistaken for statements about reality itself. He also suspected that there would be an overriding temptation not to test and destroy one's beautiful hypothesis or model but to prove it in a subjective way.

Lukermann (1958) reacted sharply at the attempts by the social physics school of the Princeton University to establish analogies with physics. He pointed out that *hypotheses derived by analogy could not be tested:* falsification was impossible. Statistical regularities and isomorphisms with other subject matter do not provide explanations, so that hypotheses derived from such models test only the models themselves.

Criticising Bunge's and Haggett's assertion that geography is a spatial science and that geometry is the language of geography, Sack (1972) pointed out that space, time and matter cannot be separated analytically in a science which is connected with providing explanations. *The geographical landscape*

[2] A comparison with the first edition (1965) of this book shows that the intervening period had seen remarkably few attempts of constructing new models, although some of the old ones had been reworked, e.g. the Von Thunen model was recast as the dual of the 'transportation problem' in 'linear programming', and the' 'gravity model' was shown to be an instance of a more general class of entropy-maximising models.

[3] For example, A. Scott detached the Von Thunen model from classical and neo-classical economics and extended it towards neo-Ricardian economics while the traditional core periphery models were replaced by ones more congruent with Marxian economics.

is continuously changing. The processes which have left historical relics and which are creating new inroads all the time must be taken into account as important explanatory factors·. The laws of geometry are, however, static—they have no reference to time. They are sufficient to explain and predict geometries, so that if geography aimed only at analysis of points and lines on maps, geometry could be sufficient as our language. But, *geometry alone cannot answer geographic questions.*

There is certainly a danger that models based on research within the western cultural experience may be elevated into general truths. But it is concluded that *a universal urban geography does not exist,* and urbanisation cannot be dealt with in a universal process. There are several fundamentally different processes that have developed out of differences in culture and time, and four such universes can be identified—North America and Australia; western Europe; the Third World; and the Socialist bloc—each with its own urban geography, which again would change through time (Berry, 1973). Owing to this fact, *generalisations on the basis of quantitative techniques may be misleading and found to be useless and inconsistent.*

Burton (1963) identified *five categories of critics* who were opposed to the quantification:

(a) those who felt geography was being led in the wrong direction;

(b) those who felt geographers should stick to their proven tool—the map;

(c) those who felt that quantification was suitable for certain tasks only;

(d) those who felt that means were being elevated over ends, and there was too much research on methods for methods' sake; and

(e) those who objected not to quantification, but to the quantifier's attitudes,

However, he believed that quantification had been proven to be more than a fad and that geography would develop out of a stage of testing relatively trivial hypotheses with its new tools so that *'the development of theoretical, model-building geography is likely to be the major consequence of the quantitative revolution'* (Burton, 1963). 'Whereas the adherents of the quantitative school could admit a certain lopsidedness in their approach by the late 1970s, their approach and argumentation had been far more orthodox at the end of the 1960s when it was thought that a definite choice between paradigms had to be made' (Holt-Jenson, 1981, 62).

Whatever may be the opposition to quantification in geography, the discipline may still be regarded as 'a science concerned with the rational development, and testing of theories that explain and predict the spatial distribution and location of various characteristics on the surface of the Earth'. The discipline is in the process of changing from an idiographic to a nomothetic science that requires the development and testing of theories and models through hypothetic-deductive methods in order to develop geographical laws (Yeates, 1968).

POSITIVISM

Scientific and philosophical discussion in the 1950s and 1960s had produced two chief categories of 'meta-theory' (superior theories or theories about theories): positivism and critical theory. 'Positivism' is connected with the naturalistic-pragmatic trend in modern thought and 'critical theory' with phenomenology and hermeneutics. Positivism has tended to dominate the English speaking world and Scandinavia, while critical theory prevailed in Germany, France and the Spanish-speaking countries.

Positivism as a philosophy was originally proposed by August Comte in the 1820s and 1830s, but it had drawn upon the earlier ideas of Saint-Simon. The primary purpose was to distinguish science from both metaphysics and religion. The concept of positivism began as a polemical weapon against the 'negative philosophy' prevalent before French Revolution. This was a romantic and speculative tradition which was more concerned with emotional than practical questions and which sought to change society by considering utopian alternatives to existing situations. The positivists regarded such speculation as 'negative' since it was neither constructive nor practical; it showed that philosophy was an 'immature' science. Philosophers, like other scientists, should not concern themselves with such speculative matters, but should study things they could get to grips with: material objects and given circumstances. This approach was to be recommended as the positive approach.

Comte himself wanted to direct the development of society, but stated that the nature of positivism is not to destroy, but to organise. An organised development should replace the disorder. created by the revolution. 'Free speculation, or systematic doubt … was identified by Comte as the metaphysical principle…. Metaphysics was later redefined as that which lies outside our sense perceptions or is independent of them. Comte regarded metaphysical questions as unscientific, and held that in a positive society scientific knowledge would replace free speculation or make it unnecessary' (Holt-Jensen, 1981, 77).

Comte's lectures on the philosophy of positivism were published as a book in 1829. His philosophy of positivism incorporated the following five basic percepts:

1. All scientific knowledge must be based on direct experience of reality, since direct observation is the surest guarantee that the knowledge acquired is scientific.
2. The direct experience of reality should be complemented by *'la certitude'*, that is, the unity of scientific method. This is implied that the different branches of knowledge were distinguished by their object of study (i.e., the subject matter) and not their method. In other words, science differs from one another in 'what' they study rather than 'how' they study.
3. The concept of unity of the scientific method required *'le precis'*, that is, a common scientific goal of formulating testable theories. This implied that there was no place for subjective value-judgements in scientific inquiry since, being based on ethical assertions, value judgements are not products of scientific observation, and are as such, not verifiable;
4. Positivist view of science incorporated the principle of *'l'utile'*, which means that all scientific knowledge must serve some useful purpose—it should be utilisable; it should be a means to an end, and a tool for social engineering.
5. The fifth precept was *'la relative'*, which means that scientific knowledge is essentially unfinished and relative, because knowledge keeps progressing by gradual unification of theories which in turn enhance man's awareness of social laws. Greater awareness demands more comprehensive theory.

Positivism, thus, in some connection called 'empiricism', has the central thesis that science can only concern itself with 'empirical questions' (those with a factual content) and not with 'normative questions' (questions about values and intentions). Empirical questions are questions about how things are in reality. In this context, 'reality' is defined as the world which can be sensed and experienced. This means that science is concerned with 'object' in the world. The subject or the subjects for which there is a world, or worlds, are excluded from the field of interest (see chapter on 'Scientific Method in Geography'). 'Positivism holds that … we cannot investigate such things as moral norms with our senses; we should keep away from normative questions; we cannot justify our tastes scientifically. Science can describe how things are, and experimentally or by some other measurement, discover the association of causes which explain why things are as they are…. Ideally, science is value-free, neutral, impartial and objective.'

In seventeenth-century England, acute political strife intermitterttly favoured independent research. Francis Bacon formulated the scientific method of interaction based on factual data derived from the evidence of the senses. John Locke formulated the basic positivist principle that all knowledge is derived from the evidence of the senses: what is not derived from the evidence of the senses is not knowledge. Reliable knowledge can only come from the basic observation of actual conditions. To be scientific is to be objective, truthful and neutral. Comte, who later defined positivism as a scientific ideal in line with Locke's principles, believed that alongside the natural sciences there should be a science of social relationships to be developed on the same principles. As natural sciences discovered the laws of natues, so scientific investigation of communities would discover the laws of society. He, however, admitted that social phenomena are more complex than natural phenomena, but believed strongly that the laws governing society would eventually be discovered and that subjective elements in research would be eradicated. This belief is central to Comte's proposition that social development takes place in three stages: (l) theological, when men explain everything in God's will; (2) metaphysical; and (3) positive, when causal connections are discovered between empirically observed phenomena.

The great stress laid by positivism on empirical data and replicable research methods enabled a marked development of science during the nineteenth century. Because metaphysical questions came to be regarded as unscientific, science developed its own objectives which were apparently free of belief and value postulates. Positivism tends to be anti-authoritarian in so far as it requires us not to believe in anything until there is empirical evidence for it and it can be investigated by controlled methods.

In the 1920s, a group of scientists known as the 'logical positivists' was founded in Vienna. The leading philosopher of this 'Vienna circle' was Rudolf Carnap. The group pursued a modern development of positivism, and extended its fundamental principle by arguing that formal logic and pure mathematics as well as evidence of the sense provide sure knowledge. The basis of **logical positivism** was a distinction between:

1. analytical statements, *a priori* prpositions whose truth was guaranteed by their internal definitions (i.e. tautologies). These constituted the domain of formal sciences, logic and mathematics, which were clearly vital in maintaining the coherence of:
2. synthetic statements, whose truth still had to be established empirically, through conventional hypothesis testing. These revisions provided empirical inquiry with a much more secure basis than the Comtean model, and the new geography was readily recommended within their framework (Harvey, 1969). However, the principle of verification was supposed to be the hallmark of the 'factual sciences', although it was subsequently challenged by Karl Popper's principle of falsification (see chapter on 'Scientific Method in Geography').

As a set of scientific principles, logical positivism is concerned with acquisition of knowledge in the form of general statements obtained through accepted scientific procedures of observation and analysis, that can be used in manipulating phenomena with a view to bringing about the desired results. This was the general concept of science that the positivists believed in. This concept incorporated a set of three related doctrines: (a) *'scientism',* i.e. the claim that the positivist method is the only true method for obtaining knowledge; (b) *'scientific politics'* which means .that positivism is the key to rational solution of all problems of society—it is the fundamental basis for social engineering;·and (c) *'value-freedom',* which means that scientific judgements are objective and independent of moral or political commitments.

The logical positivists were, in fact, opposed to everything that smacked of metaphysics and unverifiable phenomena. They, therefore, became bitter opponents of Nazism, which they saw as a mixture of irrational prejudice and ideological dogma. 'Positivist' became a term of abuse in Nazi Germany, and was applied to Alfred Hettner, among others. The leader of the Vienna Circle, Moritz Schlick, was murdered by-the Nazis and its other members driven abroad (Holt-Jensen, 1981).

Johnston et al. (1981, 1986) attempted to reveal how positivism determined the scientific states of its statements through:

1. Their grounding in a direct, immediate and empirically accessible experience of the world (*Phenomenalism*) which gave observation statements a peculiar privilege over theoretical ones, i.e. empiricism, and which guaranteed their generality through:
2. a unitary scientific method, accepted and routinely drawn upon by the entire scientific community; this depended on:
3. the formal construction of theories capable of empirical verification, their successful proof would serve to identify universal laws which had:
4. a strictly technical function, in that they revealed the effectivity or even the necessity (but not desirability) of specific conjunction of events — thus value judgements and ethical utterances were ruled out of the scientific world because they could not be empirically tested, and the statements which remained could be brought together by:
5. the progressive unification of scientific laws into a single and incontrovertible system.

The cumulative effect of these five claims was to move from the immediate through the unitary to the universal: to close the system around a particular version of the present and to refuse admission to alternative ways of being in and acting on the world. These alternatives were in fact extremely important to geography, whose founders often accepted and advocated conceptions of science which were denied by positivism. Both Kant and Humboldt, for example, rejected brute empiricism on which Comte's system was built and elaborated sophisticated philosophical systems in its place … yet, paradoxically, much of the subsequent history of the subject was dominated by the acceptance … admittedly … of some or all of these assumptions, so that when they were finally formalised during the quantitative revolution of the 1950s and 1960s the so-called 'New Geography' which resulted was 'less of a radical departure than a logical extension of ideas which were already accepted by many geographers. In so far as they were in any sense philosophically novel, they derived their originality from a commitment to logical positivism' (ibid., 358). Geography, since then, became 'positivist-led'.

Human geography was one of the social sciences to adopt positivist approach in a greater way. Geography accepted the doctrine of positivism in so far as it formulated a 'systematic' approach as opposed to regional approach, and a number of scholars introduced this new paradigm in the field of urban and economic geography and in various branches of physical geography. In this everything we witness in the phenomenal world perceived by 'sense percepton' can be verified.

The New Geography, during 1950s and 1960s, came to be known as 'Quantitative Geography' that aimed at analysing spatial data, development of spatial theory, construction and testing of mathematical models of spatial processes—reflective of a paradigm-shift from the earlier regional inductive approach to systematic and deductive nomological approach (Fotheringhaur, 2000). 'Schaefer's (1953) paper on "Exceptionalism", in fact, had opened the door to the premise of (logical) positivism; their admission was, however, slow and, as it were, through several side entrances rather than up the main steps' (Gregory, 1978, 32). Thus, as a principle, positivism or logical positivism was acceptable. The greatest contribution of positivism was that it served a very useful purpose of bringing a scientific approach in the discipline. The predictive aspects of physical and social phenomena were highlighted in modern contemporary geography, founded in the post-World War Second period, with a nomothetic base.

15

Modern Concepts in Human Geography (Critical Revolution)

HIGHLIGHTS

- Weakness in Positivist-Spatial Science Tradition
- Critical Revolution

We have seen in the previous chapter that the positivist/spatial science tradition has been subjected to bitter criticism by many researchers who have sought for alternative frameworks. The geometric approach to spatial problems using structures of point, line flow, zone and surface has immensely stimulated geographical research, but it has tended to divert attention from analysis of spatial process, and to emphasise inductivism.

As a result of dissatisfaction with this emphasis on spatial pattern and inductivism, many geographers of 1970s and 1980s sought alternative approaches to geographical problems. At the extreme, these alternatives have been based on an outright rejection of spatial science. For example, in human geography the spatial view has been characterised as a form of 'fetishism' which obscures the more fundamental social questions. It is claimed that there is no specifically spatial theory that is not a part of a more general social theory; in other words, human geography has no valid methodological base within the social sciences. For physical geography, the rejection of spatial science comes from a different direction, and physical geography witnessed the introduction of non-geographical criteria into it (Bennett, 1985, 212–13).

WEAKNESS IN POSITIVIST–SPATIAL SCIENCE TRADITION

Apart from methodological critiques, the *separate development of human geography and physical geography erodes the unity of the subject.* Where physical and human issues were frequently linked as part of population-environment concerns, now many geographers grasp the methods of social science which, in the most extreme Marxist and critical theorists cases, reject any substantial role for the influence of the physical environment. The physical geographers largely seek theoretical development divorced from social and human needs. It is now accepted by most geographers that the spatial geometric base of spatial science has little major utility, except in a few special situations.

Bennett (1985) feels the necessity of reappraising the nature and appropriateness of the methods of statistical inference within quantitative geography, and concludes that the quantitative methods should not be seen as only or even essentially those of statistical inference. Alternative methods are available, which are appropriate for the most common geographical situations of non-repeatable experiments, non-random samples which constitute the entire population, and where difficulties such as spatial auto-correlation and modifiable areal unit problem frequently arise. It may be noted that this problem arises only when aggregated data are available, for example, from census data tracts. As a result, the objects cannot be assessed directly. The problem is that any set of initial entities may be aggregated in a large number of ways to form regional or zonal data.

Quantitative geography is often characterised as the positivist of necessity. As such, it is alienating and distracting researches from the central question of social distribution (Harvey, 1973; Gregory, 1978; Bennett, 1985). (1) It creates a false sense of objectivity by attempting to remove the observer from the observed, which in turn facilitates the control and manipulation of society. (2) Its use of computers, machines and techniques subsumes people under a panoply of mathematics and machinery, reducing them to an atomistic level by eliminating consideration of social and humanistic concerns. (3) It is descriptive of existing behaviour and hence supports the status quo in society, especially the social distribution of well-being. (4) It allows no consideration of values and hence of the norms by which society should be organised. (5) It attempts to construct models and theories of universal generality by a mode of inductive logic moving from the particular to the general.

Some geographers[1] have ridiculed the positivist-spatial science tradition in geography, and even challenged the methodological base of geography as a whole. Thus the rejection of positivism also entails the rejection of space, and hence of geography as a whole.

Bennett (1985) contends that this *critique of quantification in geography as positivism is almost ill-founded.* Whilst it has had some useful benefits in redirecting attention to social questions and to applied areas, it has serious flaws. The flaws derive from two sources: first is an uncritical acceptance of the argument that statistical inference, its 'spatial science' relationship, and hence positivism, form the foundation of 'quantitative geography', and second is, ironically, an uncritical acceptance of the methodological arguments of other disciplines. Johnston (1983) has accepted that the Anglo-American geographic tradition was positivist; even Haggett, Cliff and Frey accept this assertion and also make it clear that their book is 'narrow and positivist in its philosophy'.

Bennett (1985, 222–23) draws four important conclusions:

1. Spatial geometry and spatial science were an inadequate base for geographical enquiry. Geography cannot be distinguished solely by its spatial approach. Although space is clearly a major aspect of geographical enquiry, space as such is merely a geometry or pattern and yields no process of understanding.
2. Statistical inference as developed in the 1960s within geography, and still largely dominant among quantitative workers, is almost entirely inadequate for geographical problems. This is due to non-repeatability of the experiments and the data; and also due to the facts that the sample frequently constitutes the population, that much prior knowledge is available and should be used, and that the spatial character of geographical data undermines most of the assumptions of inferential statistics (especially because of auto-correlation and modifiable areal unit problem).
3. Recent emphasis on applied and policy-related research has brought specialist sub-areas of the discipline to the forefront and opens the discipline to the danger of being embraced

[1] Peet and Slater (1980) claim that 'geography has no theoretical object that is specific to it, and ·its 'theory' is no more than a heterogeneous amalgam of spatial models developed by a variety of bourgeois scholars. Eliot Hurst (1980) argues for a 're-definition' of geography: 'Geography is revealed for most part as merely descriptive and scientifically bankrupt ... and is nothing more than a theoretical ideology based on technical practice alone'. Instead, a truly multi-disciplinary approach (dialectical and historical materialism) is required which can uncover the true relationships between economy and society in the totality of social existence. Geography is a 'fetishized domain', and even 'the co-opted Marxism of radical geography and "Marxist Geography" ... is superficial and doctrinnaire, incapable of transcending the orthodox problematic boundaries' (Eliot Hurst, 1980).

within the methodologies of disciplines outside the geographic core.

4. The critique of geography as positivism, though it has been useful in highlighting some of these weaknesses, and in asserting the importance of question of social distribution, has further weakened geography as a whole. It has established a falsely based and extreme methodological debate, induced divisiveness by disregarding physical geography and its relation with human geography, and accepted uncritically the methodology of social science, which has been developed without regard for geography's concern. Hence, geography cannot evolve through a pluralism of positivism. Quantitative geography is not the prerogative of any one method or philosophy, such as positivism; but is a tool developed for use within the value position or the research question in hand.

CRITICAL REVOLUTION

Geography, in the 1970s and afterward, underwent another revolution, and this time, *'the revolution was anti-positivism and critical'*. From the early 1970s, work by cultural and historical geographers has been presented as an attack on the positivism of spatial science. To replace the latter, a variety of *humanistic approaches* has been proposed, presenting a focus on the decision-maker and his perceived world and denying the existence of an objective world which can be studied by positivist methods. The intent is to reorient human geography towards a more humanistic stance, to resurrect its synthetic character, and to re-emphasise the importance of studying unique events rather than the spuriously general (Johnston, 1983).

However, from the mid-1960s there was a growing disillusionment with science and technology, and the popularity of the social sciences boomed. Within the latter, there was a shift in emphasis from the aggregate to the individual, an increase in the relative volume of research conducted at the micro-scale, and growing unease about the role of social scientists in planning mechanisms. Both the humanistic approach and behavioural (or behaviorist) geography reflect these trends.

Humanistic Geography

It is an approach in human geography, distinguished by the *central and active role it gives to human awareness and human agency, human consciousness and human creativity.* It is an attempt at understanding meaning, value and the human significance of life events. The revival of humanism in geography in the 1970s owed much to a deep dissatisfaction with the more mechanistic models developed during the quantitative/spatial science revolution. For this reason, its early steps were taken alongside behavioural geography (Gold and Goodey, 1984). But the two soon parted company and humanistic geography came to recognise the essential 'subjectivity' of both the investigator and the investigated in ways which departed from the formal structures of behaviourism.

Indeed, humanistic geography shared in the more general critique of positivism's claim to 'objectivity' (in which behavioural geography was itself inculcated). It came to be represented as a form of criticism through which geographers can be made more aware and cognizant of many of the hidden assumptions and implications of their methods and research, rather than as a coherent and robust methodology for the 'post-behavioural revolution' in geography. But *humanism was intended to be more than just a critical philosophy.*[2]

Anne Buttimer (1978) attempted to bring forward the tradition of Vidal de la Blache, and argued that historical and geographical studies go together. She emphasised the need to understand each region and its inhabitants from the 'inside'

[2] Critical philosophy refers to a tradition of social and political thought whose central concern is the historicity or social action, in particular, the connection between human agency and social structure that exist under capitalism and which can be recognised and restructured through a process of self-reflection.

(that is, on the basis of local perspective) and not from the perspective of the researching 'outsider'. There undoubtedly are affinities between the French School of la geographie humaine and humanistic geography.

Idealist Approach. Leonard Guelke (1981) advocated an idealist approach, and expressed his ideas to historical geographers, as a counter to the arguments that they should adopt the approaches and techniques of positivism. He points out, 'it is obvious that quantitative techniques will often be useful, but statistical methods put in harness with positivist philosophy is a dangerous combination. Historical geographers need to rethink not their techniques but their philosophy. This can be best achieved by moving from problem-solving contemporary applied geography, towards the idealist approach widely accepted by historians'.

A well-verified idealist explanation will be one in which a pattern of behaviour can be shown to be consistent with certain underlying ideas. The idealist philosophy combines two positions, according to Guelke: (a) *a metaphysical argument* that mental activity has a life of its own which is not controlled by material things and processes; and (b) *an epistemological argument* that the world can only be known indirectly through ideas, and all knowledge is ultimately based on an individual's subjective experience of the world, and comprises mental constructs and ideas. There is no 'real' world that can be known independently of the mind. Positivist spatial science is thus criticised because it believes in the existence of a real world, the nature of which it seeks to explain via general laws of behaviour.

Behavioural geography was equally criticised, not for its response to perceived images and subjective evaluation, but because of two assumptions within this field that: (a) identifiable environmental images exist which can be measured accurately; and that (b) there are strong relationships between revealed images and preferences and actual (real world) behaviour. Such an approach, according to Bunting and Guelke, puts human geography into a single-cause model, much like the earlier environmental determinism, and if one accepted this model, research in behavioural geography has failed to validate it. They argue for an idealist approach, which focuses on overt behaviour and its interpretation—in searching for the truth a scholar conducts a critical dialogue with his evidence and in due course he puts the results before his colleagues for their appraisal' (Bunting and Guelke, 1979, 453–58).

Guelke considered that geographers should attempt to discover what the decision-maker believed and not why he believed it. Thus, the human geographer does not need to develop theories, since the relevant theories which led to the action under study already existed in the minds of the decision-makers (Holt-Jenson, 1981, 70).

Hays (1979) points out the inherent weaknesses in the idealist approach that it ignores the possibility of either unconscious or sub-conscious behaviour. He further points out that the objective facts must influence behavioural outcomes, in addition to the thoughts of the actors. To him, idealism acts as a reductionist, but the world is more than a large number of independent decision-makers. Mabogunje also criticises the idealist approach of Guelke, and argues that such a retreat from objective theory formulations as a means of seeking explanation to certain events would exclude from our consideration the exploration of the consequences of social actions. Instead of retreating to a focus of particular cases, seeking special explanation for each situation in which a different value system can be shown to be operative, the geographers should attempt to build better theories encompassing these differences in value systems. Some scholars have expressed doubts as to how an idealist interpretation could be verified.

Hermeneutic Approach. Idealism implies one type of Hermeneutic Approach, which is the *theory of interpretation and clarification of meaning*. It took its place in the German tradition of 'Geisteswissenchaften', the human sciences.

Its arguments were then developed to embrace the reader (or interpreter) as well. The dialectic between the subject and the object leads on to what has been called 'double hermeneutics'. In geography, hermeneutics has been formally employed in a similarly general fashion to contest the epistemology of positivism and hence to reject the exclusive claims of spatial science. The critical impulse which hermeneutics entails has been generalised through claims for a presuppositional approach in human geography capable of giving direction to its emerging social conscience.

However, hermeneutics has much to offer to regional geography. On the one hand, it can provide the epistemology for a self-reflective, historical, geographical science of the region. On the other hand, it has specific contribution to make concerning the characteristics of the Earth, space and place, culture and especially, language, and thus it could be a means of reconstructing regional geography.

Phenomenological Approach. Phenomenology has attracted more attention among human geographers than idealism. The term was first used by Sauer as early as 1925, but it had only become widely known during 1970s. It was Relph who first attempted to introduce a phenomenological approach, when he pointed out that the basic aim of phenomenology is to present an alternative methodology to the hypothesis-testing and theory building of positivism, an alternative grounded in humankind's 'lived' world of experience. Phenomenologists argue that *there is no objective world independent of man's existence*—all knowledge proceeds from the world of experience and cannot be independent of that world: 'Idealism', on the other hand, accepts that there is a real world outside the individual's consciousness.

Kirk (1963) recognised two separate, but not independent environments: (a) *'phenomenal environment'*, which is the totality of the Earth's surface, and (b) *'behavioural environment'*, which is the perceived and interpreted portion of the phenomenal environment. Phenomenology in human geography is somewhat concerned with the phenomenal environment. The contents of that environment are unique to every individual, for each of its elements is the result of an act of intentionality; it is given meaning by the individual, without which it does not exist, but with which it influences behaviour. Phenomenology is the study of how such meanings are defined. It involves the researcher seeking to identify how the individual structures the environment in an entirely subjective way.

One of the well-known phenomenologists was **Yi-Fu Tuan** (1971), to whom geography is the mirror of man and reveals the essence of human existence and striving. To know the world is to know oneself, just as careful analysis of a house reveals much about both the designer and the occupant. The study of landscapes is the study of the essence of the societies which mould them, in just the same way as the study of literature and art reveals much of human life. Such study is clearly based in the humanities rather than in the social or physical sciences.

According to him, 'Humanistic geography achieves an understanding of the human world by studying people's relation with nature, their geographical behaviour as well as their feelings and ideas with regard to space and place. Scientific approaches to the study of man tend to minimise the role of human awareness and knowledge. Humanistic geography, in contrast, specifically tries to understand how geographical activities and phenomena reveal the quality of human awareness' (Tuan, 1976, 267).

There are *five basic themes* in *Tuan's humanistic geography:* (1) the nature of geographical knowledge and its role in human survival; (2) the role of territory in human behaviour and the creation of place identities; (3) the interrelationships between crowding and privacy, as mediated by culture; (4) the role of knowledge as an influence on livelihood; and (5) the influence of religion on human activity. Such concerns are best developed in historical and in regional geography; their value to human welfare is that they clarify the nature of the experience. 'The model for the regional geographers of humanistic leaning is ... the Victorian novelist who strives to achieve a synthesis of the subjective and the objective' (ibid., 270).

Buttimer (1976, 282) also favoured phenomenology, as a path to understanding, on which informed planning can be built. 'It could elicit a clearer grasp of value issues surrounding one's normal way of life, and an appreciation of the kinds of education and socialization which might be appropriate for persons whose lives may weave through several mileux.' **Berry** (1973) also supported the phenomenological orientation, calling for a view of the world from the vantage of process metageography. By metageography is meant that part of geographic speculation which deals with the principles lying behind perceptions of reality, and transcending them, including such concepts as essence, cause and identity.

Entrikin argues that 'the humanistic approach is best understood as a form of criticism'. Whereas the positivist/quantitative movement was characterised by a great numerical superiority of practitioners over preachers, the phenomenological movement (like the idealist) has been characterised by the converse—much preaching and little practice.

However, humanistic geography has moved far from the position defined by Entrikin. It has moved from its early attack on positivism to make an assault on structuralism, and at the same time, it has developed a much more incisive methodology for empirical investigation. Johnston (1986) distinguishes two basic streams of work. The first stream is characterised by a self-conscious drive to connect with that special body of knowledge, reflection and substance about human experience and human expression, about what it means to be a human being on this Earth, namely, the humanities. The second stream is, perhaps, more self-consciously theoretical—in fact one of its central concerns is to clarify the 'theoretical attitude' itself. It draws on a range of constraints derived from the human and social sciences, most usually from existentialism and phenomenology, and from ethnomethodology and symbolic interactionism.

Humanistic geography, therefore, appears to be concerned either with the study of individuals and their construction of phenomenal environments, or with the analysis of landscapes as repositories of human beings. As such, it may be considered separate from the subject matter of such human geography, notably behavioural geography, and its investigation of everyday activity within environments (most of them man-made). The phenomenological perspective has been adopted to such work, in the writing of Schultz on what he terms the 'taken for granted' world.

Much of everyday behaviour is unconsidered, in that it involves no original encounters. The behaviour is habitual, because all of the stimuli encouraged are processed as examples of particular types. Those types are not externally defined for the individual, but are created by him. The phenomenology of the 'taken for granted' world is the study of those individually-defined typifications of the unconsidered 'world of social reality' rather than 'a fictional non-existing world' constructed by the scientific observer. It may be that interactions within communities lead to common typifications. Quantitative methods may be used to identify the common elements, but as descriptive tools only. Quantification is not tied to positivism except when it is used to suggest laws and other generalisations.

Humanistic geography is based on a profound critique of positivist work, both that which makes major assumptions about the nature of decision-making and that which seeks inductive laws of human behaviour which can be scientifically verified. Its argument is for an understanding of man as a 'living, acting, thinking being' (Ley, 1980; Johnston, 1983). It is contended that the human condition can only be indicated by humanistic endeavour, for attitudes, impressions and subjective relation to places, and the sense of place, cannot be revealed by positivist research.

Behavioural Geography

In contrast to humanistic geography, Behavioural Geography may be seen as a developing criticism, from within the 'quantitative' movement, starting from disillusionment with theories based upon the concept of 'economic man' (Holt-Jenson, 1981, 71). It is an approach to human geography, and in particular to the processes responsible for human spatial behaviour, which especially draw on behaviouralism, or on cognition, as a key to understanding human spatial behaviour.

Behavioural geography *treats man as a responder to stimuli.* It seeks to identify how different individuals respond to particular stimuli (and also how the same individual responds to the same stimulus in different situations) to isolate the correlates of those varying responses to build models that can predict the probable impact of certain stimuli. The end-product is the input to processes aimed at either providing environments to which people respond in a preferred way or at changing the stimuli.

Behavioural geography has maintained strong ties with the positivist/spatial science tradition. Data are collected from individuals, but there almost all are concerned with the conscious elements in action; they are usually aggregated in order to allow statistically substantive and significant generalisations to be made about spatial behaviour, almost certainly in the context of the normative model of the spatial science school (Johnston, 1983, 157). Though it was once assumed to be a 'behavioural backlash' to a conventional spatial science, behavioural geography was really a 'logical outgrowth' of the commitment to positivism enshrined in the quantitative revolution.[3]

Kates (1962) was a major exponent of the behaviouristic approach. In his study of the flood-plain management, he said, 'the way men view the ranks and opportunities of their uncertain environments plays a significant role in their decisions as a resource management'. Kates developed a scheme which seemed to have relevance to a wide range of behaviours. His scheme was based on four assumptions:

1. *Men are rational while taking decision.* Such a decision may be either prescriptive—describing how men should behave—or the descriptive of actual behaviour. The latter appears to be the most fruitful, both for understanding past decisions and for predicting those yet to be made. Decisions are made on a rational basis, but in relation to the environment as it is perceived by the decision-maker, which may be quite different from either 'objective reality' or the world as seen by the researcher.
2. *Men make choices.* Many decisions are either trivial or are habitual so that they are accorded little or no thought immediately before they are made. Some major decisions regarding environment and its use may also be habitual, but such a behaviour usually develops only after a series of conscious choices has been made. This leads to a stereotyped response to similar situation in the future.
3. *Choices are made on the basis of knowledge.* Only very rarely can a decision-maker bring together all the information relevant to his task, and frequently he is unable to assimilate and use all that he has.
4. *Information is evaluated to pre-determined criteria.* In habitual choice, the criterion is what was done before, but in conscious choice the information must be weighed according to certain rules.

[3] The roots of behavioural geography are, however, much older. In Europe, the Finnish geographer Johannes Gabriel Grano, and his Estonian student Edgar Kant were attempting a behaviourist approach in the 1920s. Even in the United States, the behavioural and quantitative approaches were contemporary developments. It was at the University of Chicago that the first attempts by geographers to explore inductively were made by a series of investigations into human responses to environmental hazards, initially floods, guided by theories of decision-making, and influenced by methods used in psychology and sociology. It was Gilbert White, the director, whose thesis on human adjustments to floods was published in 1945. His associates at the University of Chicago developed a behaviourist approach for studying reactions to the natural hazards, basing this on Herbert Simon's theories of decision-making. White (1974) found it more important to map the personal 'perception' of the decision-maker than to describe the factual physical and economic conditions of the environment, since the decision-maker would act upon his own perceptions, not on the environmental factors themselves.

Kates' study aimed at understanding why people choose to live in areas which are prone to flooding. Their information was based on their knowledge and experience, and they could be scaled according to the certainty of their perceptions regarding further floods. In justifying their decisions, most were boundedly rational, and had made conscious choices in order to satisfy certain objectives (Johnston, 1983).

It was **Julian Wolpert** who introduced many human geographers to the behaviourist alternative to the normative approaches then quite popular. In his paper, entitled '*The decision process in spatial context*', published in the *Annals* of the Association of American Geographers (1964), he compared actual and potential labour productivity on farms in Central Sweden. He observed that the sample farm population did not achieve profit maximisation, nor were its goals solely directed to that objective. The farmers were 'spatial satisfiers' rather than 'economic men'.

Wolpert further continued this theme with studies of migration, aiming to model the decision-making which is behind the patterns of migration reported in census volumes and analysed by spatial scientists. He found the gravity model inadequate as a representation of such flow patterns (Wolpert, 1965). To him, such individual has an action space—the set of place utilities which the individual perceives and to which he responds—whose contents may deviate considerably from that portion of the 'real world' which it purports to represent. Once the first decision—to migrate—has been made, then the action space may be changed as the potential mover searches through it for potential satisfactory destination If necessary, he extends the space if no suitable solution to the research can be found (Wolpert, 1965).

Undoubtedly, Wolpert's papers heralded the development of behavioural geography. His behavioural approach has focused on topics related to decision-making in spatial contexts. Wolpert and his associates also prepared pioneering papers, related to political decision-making, applying the approach he set forth in his first work. Regarding the distribution of certain artefacts in the landscape, Wolpert pointed out that the location of a public facility, for example in an urban area, frequently is the product of policy compromise (Wolpert, 1970).

The ***aim in behavioural geography*** has been to derive alternative theories to those based on economic man. These theories are 'more concerned with understanding why certain activities take place rather than what patterns they produce in space'. Here the researcher uses the real world from a perspective of those individuals whose decisions affect locational or distributional pattern. He tries 'to derive sets of empirically and theoretically sound statements about individual, small group, or mass behaviour'. Man is considered an active decision-maker, not a reactor to particular institutionally-created stimuli (Golledge, *et al.* 1972).

It may be noted that the initial interest in resource management was followed by an extension from environmental perception and decision-making into aspects of attitudes and motivation. These were applied to studies of migration, the diffusion of innovations, political behaviour (especially voting), perception, choice behaviour, and spatial search and learning. 'By studying behavioural processes in these contexts, the aspiration was to increase geographer's understanding of how spatial patterns evolve, thereby complementing their existing ability to describe such pattern. Morphological laws and systems are insufficient of themselves for understanding; the amalgamation of concepts about decision-making taken from other social sciences with geography's spatial variable would allow development of process theories that could account for the morphologies observed' (Johnston, 1983, 130).

One aspect of behavioural geography has been the concept of *mental map.* This refers to the images of place, 'mentally stored by individuals and drawn upon as resources in their interpretation of spatial desirability, their organisation of spatial routines, and their decision-making transaction as satisficing agents. Mental maps are an amalgam of information and interpretation reflecting not only what an agent knows about places but also how he or she feels about them' (Johnston, 1986).

Mental maps are important to geographers not only as a means of examining an individual's areas of spatial preference, but also as an insight into the processes whereby decisions are made, opportunities perceived and goals determined and satisfied. 'If we grant that spatial behaviour is our concern, then the mental images that men hold of the space around them may provide a key to some of the structures, patterns and processes of man's work on the face of the Earth' (Gould, 1966, White and Gould, 1974, 76). Such maps are useful, it is believed, not only in the analysis of spatial behaviour, but also in the planning of the social investment.

According to Downs (1970), there may be two other approaches to the study of environmental images: (a) the *'structural approach'* which inquires into the nature of the spatial information stored in people's minds and which they use in their day-to-day affairs; and (b) the *'evaluative approach'* in which the question is, what factors do people consider important about their environment, and having estimated the relative importance of these factors, how do they employ them in their decision-making activities. With this evaluative approach, geographers moved into the wider field of cognitive mapping.

On the ***methods in behavioural geography,*** Johnston points out, 'The behaviouristic approach is an inductive one, with the aim being to build general statements out of observations of ongoing processes.... In terms of the accepted route to scientific explanation ... behavioural geography involved moving outside the accepted cycle procedure to input new sets of observations on which superior theories might be based. In doing this, the behaviourists did not really move far from the spatial-science ethos. Indeed many of their methods were those of their predecessors; Gould's mental map studies, for example, used the same technical apparatus as the factorial ecologies' (Johnston, 1983, 135).

However, **Pred** presented an alternative to theory building based on 'economic man' in his two-volume work *Behaviour and Location* (1969), and proposed the use of a *behavioural matrix* to provide a framework in which locational decision-making could be analysed.

Axes of behavioural matrix are quantity and quality of information available and the ability to use that information; economic man is located in the bottom right-hand corner. Because of the nature and importance of information flows, the position on the axis depends in part on the decision-maker's spatial location. His position would reflect aspiration levels, experience and norms of any groups to which he belonged. Different people in different positions in the matrix would vary in their decisions; therefore two at the same position may act on different bases and in different ways. Individuals do not stay at the same position in the matrix, and spatial patterns are not static.

In the second volume of the book, Pred introduced a dynamic element by shifting individuals through the matrix. As they shift, and change their decisions, so the environment changes for others. As people learn, they both acquire better information, and become more skilled in its use. They shift towards the bottom right-hand corner of the matrix, some of them in advance of others, who benefit from the activities of the 'decision leaders'. The unsuccessful are gradually eliminated, so that with time a concentration of good decision-makers close to the economic man position evolves. But changes in the external environment produce parametric shocks, which result in decision-makers becoming less informed and less certain; as a consequence they are shifted back towards the upper left-hand corner, and another learning cycle begins. As long as parametric shocks occur more frequently than the learning experience takes, an optimal location pattern will never emerge, except perhaps by chance.

Harvey (1969) criticised the model of the behavioural matrix, finding it ambiguous, unoperational and *an over-simplification of the complex nature of behaviour.* He also expressed scepticism about the viability of a behavioural location theory. Instead, Harvey suggested two alternatives to behavioural location theory—further development of normative theory and the construction of a stochastic location theory. To him, both these approaches offered more immediate pay-offs in terms of understanding spatial patterns than did behavioural efforts, because of the conceptual and measurement problems of the latter. Curry (1967) favoured the stochastic approach, while Weber (1972) attempted to model locational approaches.

The behavioural approach has not brought about a revolution away from the spatial science paradigm; in effect it has become an attachment to it. Behavioural geography is now widely accepted within the positivist orientation. It seeks to account for spatial pattern within the environment (both man-made and natural) by establishing generalisations about man-environment interrelationships, and then using these as a basis for change through environmental planning activities that 'modify the stimuli which affect the spatial behaviour of ourselves and others'. The behavioural approach, therefore, is *based on four major assumptions,* according to **Gold** (1980):

1. The environment in which an individual acts is that which he perceives, which may well differ markedly from the true nature of the real world.
2. Individuals interact with their environments, responding to them and reshaping them.
3. The focus of study is the individual, not the group.
4. Behavioural geography is multi-disciplinary.

The behaviourist approach appears to consist of two approaches: (a) The first is based on the study of overt behaviour using the traditional positivist formulation of dependent variables influenced by independent variables. This approach has involved the widespread application of statistical techniques. (b) The second approach is based on attempts to identify the mental constructs that lie behind overt behaviour. However, little has been achieved in the second approach, i.e. linking cognitive schemata to behaviour, and thus extending the predictive models of the first approach.

'In the early 1960s and 1970s, therefore, behavioural studies in geography typically examined the ways in which information is (selectively) abstracted, structured and stored in mental maps; fed into and channeled through decision-making systems, as individuals or corporations make diagonal moves through a notional behavioural matrix at movements in a recursive learning process; and disclosed in patterns of spatial behaviour through the analysis of revealed spatial preference and the reconstruction of individual-action-space' (Johnston, 1986).

Time-Space Geography

Related to the behaviouralist tradition, and somewhat also to the humanistic heritage, is Hagerstrand's Time-Space Geography. It may be seen as a critique, not so much of the 'quantitative movement' of the important aspects of social research in general. *It conceives of time and space as providers of 'room' (rum) for 'collateral processes'.* Time-space geography emphasises the importance of continuity and connectivity for sequences of events which necessarily take place in 'situations bounded in time and space and whose outcomes are mutually modified by their common localization' (Hagerstrand, 1976, 1984).

An important aspect of time-space geography is that *time and space are both regarded as resources which constrain activity.* Any behaviour which requires movement involves the individual or group traversing a path through space and time. Movements along the horizontal axis indicate spatial traverses while those along the vertical axis mark the passage of time. Any journey, termed a lifeline by Hagerstrand, thus involves movement along both axes simultaneously.

Time-space geography is based upon *four basic assumptions:*

1. Space and time are resources on which individuals have to draw in order to realise particular projects.
2. The realisation of any project is subject to three constraints: (a) 'Capability constraints', which restrict the activities of individuals through their own physical capabilities. They derive large measures from the individuals' livelihood position. (b) 'Coupling constraints', which operate to require certain individuals and groups to be in particular places at stated times. Such constraints define space-time boundaries. (c) 'Authority or steering constraints', which impose certain conditions of access to and modes of conduct within particular space-time domain. Together, these three define the time-space prism (Fig. 15.1) which contains all the available lifeline paths for an individual starting at a certain location and needing to return there at a given time.
3. These constraints are interactive rather than additive and together they delineate a series of possibility boundaries which mark out the paths available for individuals to fulfil particular projects. These boundaries correspond to an underlying and evolving logic or structure whose disclosure requires a way of dealing with projects in space-time terms of considerable conceptual precision.
4. Within these structural templates, competition between projects for free paths and open space-times is the central problem for analysis and is mediated by specific institutions which seek to maintain an essential space-time coherence (Hagerstrand. 1973).

Time-space geography is often described as a 'situational ecology' concerned to incorporate certain essential biotic and ecological predicates within human geography and social theory. It has the potential for throwing new light on

Figure 15.1 The time-space prism

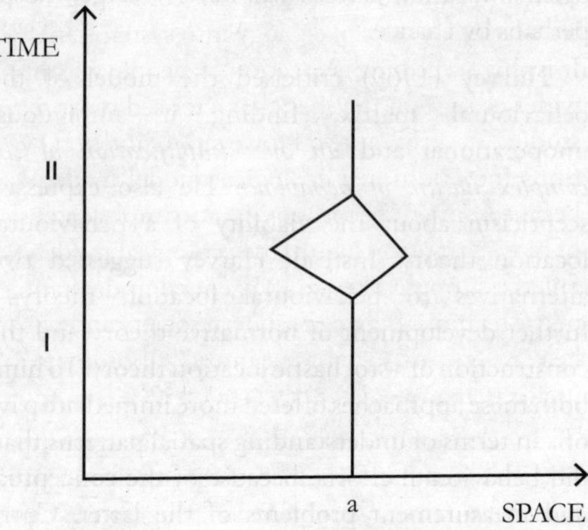

some of the very different kinds of questions customarily posed by 'old-fashioned' regional and historical geographers, as well as 'modern' human geographers.

Time-space geography is a challenge to turn to the 'choreography' of individual and collective existence—to reject the excesses of inter- and intra-disciplinary specialisation for a concern with collateral processes. It provides a method of mapping spatial behaviour and at the same time represents a reorientation of scale, away from aggregate data towards studies of individual behaviour. Hagerstrand's time-space geography appears to be essentially humanistic, aimed to provide insight into what is specially human in man's nature, and elucidates the specific human situation.

Human Ecology

One aspect of the critical revolution in the mid-1960s was the revival of Human Ecology, the *application of ecological concepts to the study of the relations between people and their physical and social environment.* It was a logical extension of the system of thought and the techniques of investigation developed in the study of man. Reacting to the contention of Eyre and Jones (1966) that human ecology was aimed to launch a deliberate assault on the quantitative revolution, Stoddard attempted to remove and/or reconcile the confusion by his translation of human ecology into systems analysis. It is this focus on the location of people within the wider ecosystem that has characterised most recent studies in geography. However, Chorley (1973) finds the traditional ecological model inadequate, because human geography is no simple extension of biogeography. To him, the ecological model may fail as a supposed key to the general understanding of the relations between modern society and nature, because it casts social man in too subordinate and ineffectual a role.

It is because of time-space geography of Hagerstrand that human ecology received explicit recognition in the contemporary geographic studies. He defined his theoretical system as a 'space-time ecology'. His so-called *'web model' of space-time interaction* should, in principle, be applicable to all aspects of biology from plants to animals to people. Its central subject matter, in fact, is to incorporate certain essential biotic and ecological predicates within human geography and to bridge the gap between biological and human ecology.

Holt-Jenson (1981) argues that those who describe geography as human ecology have often defined the concept too narrowly and have presented studies of man's relationships to his environment as if it only encompasses man's relationship with nature and not to his total physical and social environment. Kirk (1963) described the geographical environment in terms which may possibly provide a useful starting point for a discussion of systems in which both ecological and social science theories and concepts may be relevant.

Welfare Geography

One of the distinct consequences of the 'critical revolution' in the contemporary human geography was the emergence/development of Welfare Geography in the 1970s. *It is an approach to human geography that stresses on questions of inequality.* The welfare approach emerged from the radical reaction to the quantitative and model-building emphasis of the 1960s, which was thought to be insufficiently concerned with contemporary problems. It was developed as a possible reaction to the positivist/spatial science tradition. During the 1960s, a lot of research was reported, under the general title of factorial ecologies, but only in the 1970s was the factorial ecology set of procedures adopted to the task of mapping social welfare on any significant scale.

A movement towards the welfare approach in the contemporary human geography was in fact heralded by D. M. Smith and P. L. Knox. Smith's work, *The Geography of Social Well-being in the United States* (1973) was prepared in the light of American social indication movement and the growing belief there that gross national product (GNP) and national income are not necessarily direct measures of the quality of life in its broadest sense. His aim was to initiate the collection and dissemination of territorial social indication, to point out the extent of discrimination by place of residence which occurs in the United States. Knox (1975) stated that it was a fundamental objective for geography to map social and spatial variations in the quality of life, both as an input to planning procedures and as a means of monitoring policies aimed at improving welfare. *Smith and Knox set forth the tradition of the welfare approach in geography in an organised manner.* As Chisholm stated, their works represented the geographer as a 'delver' and 'dovetailer', as a provider of information on which more equitable social planning could be based.

A number of such works, suggesting spatial policies for social improvements, were done in the 1970s.[4] These works carried forward human geography towards 'welfare' issues such as poverty,

[4] Harries (1974) studied spatial variation in crime rates and the administration of justice, and argued that predictive models of criminal patterns could aid in the organisation of police activities. An interesting work on variations in the provision of health-care facilities was done by Shannon and Dever (1974) who also suggested for spatial planning for the improvement of medical services being

hunger, discrimination, crime, medical care, racial tension and access to social services. This corresponded to a major shift in societal concern, from narrow economic criteria of development to broader aspects of the quality of life.

The basis of the welfare approach is in *'who gets what, where and how'*. The present states of society, with respect to 'who gets what and where' may be represented by extension of the abstract formulation of welfare economics, and the practical objective is to give it empirical substance. The welfare approach in geography requires the development of social indication for the empirical identification of social inequality and injustice in territorial distribution. This may combine particular elements of social well-being in a composite measure. The method in welfare geography essentially appears to be descriptive, because descriptive research provides a basis for evaluation of social well-being. However, the early preoccupation with descriptive research in welfare geography has now given way to more process-oriented work on the question of how inequality arises.

Although originally proposed as an alternative framework for human geography, the welfare approach has now been merged with other lines of inquiry within geography directed towards the fundamental problems of inequality. Implicit in welfare approach is a recognition that the issues in question extend beyond the limits of single discipline, and in fact render-disciplinary boundaries increasingly irrelevant. The welfare approach logically requires a holistic social perspective (Johnston, 1986). Welfare geography attempts to make human geography more relevant to contemporary social problem.

Radical Geography

'Whereas welfare geography works in principle within the framework of existing economic and social system, radical geography which has been established more recently, *calls for both revolutionary theory and revolutionary practice'* (Holt-Jenson, 1981). Radical geography appears to be an outcome of a 'new critical revolution' in the contemporary human geography, which seemed to have occurred largely as a result of the critiques of positivist/spatial science tradition in human geography.

As a holistic, revolutionary science, *Marxism provides a film theoretical base for the radical movement in geography.* Marxism offers an opportunity to develop an integrated comprehension of reality as a whole. To Harvey (1973), Marxist theory provides the key to understanding capitalist production from the position of those not in control of the means of production—an enormous threat to the power structure of the capitalist structure. It not only provides an understanding of the origin of the present system, with its many-faceted inequalities, but also propounds alternative practices which would avoid such inequalities.

The Marxist theory renounced that the scientific laws of society are eternal. This view sharply contrasted with the claim of positivist science that scientific laws are universal in space and time. Economic laws are not eternal laws of nature, but historical laws appear and disappear. Society has inbuilt conflicts in it which tend to resolve themselves by change both in practice and in theory. Science and scientific development, according to the Marxist theory, must be understood

offered to the sick. Morrel and Wohlenberg (1971) investigated the spatial variations in poverty in the United States, providing both social policies and spatial policies. Bunge (1971) prepared a 'geobiography' for a part of the black ghetto of Detroit, which has a deep humanitarian orientation for the future of humankind, which he interprets as a need to ensure a healthy existence for children. Cox (1973) investigated the urban crisis in the United States, in the light of racial tension and riots, municipal bankruptcies and the role of government in the urban economy. In fact, he presented his analysis in terms of conflict over access to sources of power. Coates et al. (1977) attempted to present an understanding of spatial variations in well-being. They pointed out three sets of causes of such variations: the division of labour, accessibility to goods and facilities, and the political manipulation of territories.

in relation to social reality, To Harvey (1973), the *essential difference between positivism and Marxism is that positivism simply seeks to understand the world, whereas Marxism seeks to change it.* For Peet (1977), the Marxist science begins with a material analysis of society, proceeds through a critique of capitalist control of the material base of society and proposes solutions in terms of social ownersip of that material base.

The growing disillusionment in the American society, partly as a result of the Vietnam War, social inequalities and injustice, racial tension, and the unresponsive attitude of the authority to the needs of the under-privileged, on the one hand, and the Marxist theory, on the other hand, formed a broad symbiosis for radical geography. However, according to Peet (1977), radical geography developed largely as a negative reaction to the established disciplines, a reaction which was initially formulated within existing methods of positivism. By the early 1970s, the critique however, had developed a strong Marxist base, aiming to create a radical science, which seeks to explain not only what is happening, but also to prescribe revolutionary change. A journal of radical geography—*Antipode* was launched in 1969 at Clark University.

The first step in the development of a mature radical practice of geography was the emergence of a liberal viewpoint. This was an attempt to find more socially appropriate uses for the existing techniques and theories, and yet to maintain the basic ideas and values of the social system. When geographers began to question the ideology underlying the existing geographic models, they moved into the critique stage of development. Next came the extraction of elements–problems (Young and Sukhwal, 1984, 185).

Failure on the part of the liberal geographers to offer meaningful solutions to the major socio-economic crisis in the United States prompted Harvey (1973) to offer a revolutionary theory, thereby overthrowing the current paradigm. The immediate task, according to him, is nothing more than the *self-conscious and aware construction of a new paradigm* for social geographic thought through a deep and profound critique of our existing analytical constructs. The new paradigm would be built on a Marxist base, and it would achieve social reforms through the education process. The benefit of Marxist theory was that it can handle two crucial issues that positivist theory cannot—increasing injustice, and heightened economic and social instability.

However, Peet (1977) has argued that the early 'radical' work by geographers in the late 1960s was liberal in its attitude. David Harvey in his book *Social Justice and the City* (1973) made a major contribution to the case for a Marxist-inspired, materialist theory development within geography. Peet also moved to a Marxist position replacing his earlier paper on poverty by a Marxist interpretation, based on the assumption that inequality is inherent in the capitalist mode of production.

As a mode of analysis, *Marxism is a variant of the philosophy of structuralism which identifies three levels of analysis*: (a) the level of appearance or the *super-structure;* (b) the level of processes, or the *infrastructure;* and (c) the level of imperatives, or the *deep structure.* Marxism has a materialist base. It argues that the infrastructure comprises a set of economic processes and most Marxist work concentrates on the processes operating within the capitalist mode of production. Marxist analysis seeks to identify the processes operating in the infrastructure and to relate them to the pattern in the super-structure. In human geography, this means deriving general theories of historical materialism that will account for particular patterns. Marxist structuralism can also be presented as a variant of the realist philosophy of science which seeks to relate the contents of an empirical world to a set of infrastructural determinants—economic processes.

The Marxist/radical approach in geography has *four basic components*: (1) The first is the critique of positivist spatial science and behavioural geography, and of humanistic geography.

(2) The second is to provide general theoretical frameworks, within which empirical work can be set. (3) Thirdly, there is work that seeks to establish how individuals act within the structural imperative. (4) And finally, there is detailed empirical work that seeks to understand particular aspects of the subject matter of human geography within the structuralist framework.

The radical geographers, with their concern for social values and political action, have rejected the traditional concept of geography. Although the radicals generally remain interested in the human/environmental and spatial relationships, they spurn the systems of theory and methodology that are viewed as providing only partial accounts of reality and as serving the interests of a select social group or class. Most radical geographers accept geography as a legitimate field of study and feel that it has much to offer in finding solution to the world problems. The radical geographer's aim is the alteration of the operating societal processes by changing the relations of production. Since radical geography is purely an American enterprise, it had no academic recognition in the former Soviet Union where its desirability and viability was doubted for not having a good understanding of Marxist-Leninist thought.

Though critical revolution in geography occurred side by side with the quantitative revolution in the 1960s and the early 1970s, as a possible reaction to the positivist spatial science tradition, but recent studies in humanistic geography, behavioural geography, time space geography, human ecology, welfare geography and radical geography reveal that there are some attempts towards some *'revaluation' of the positivist approach.*

It is equally important to note that in geography, paradigms or rather schools of thought have not succeeded each other as Kuhn's model suggests, but, to a large extent, continue to exist in parallel, whilst the new schools slowly absorb the older ones leaving some former contradiction to linger on within the new structure. There is little evidence either of large-scale disciplinary consensus for any length of time about the merits of a particular approach or any of the revolutions that have been entirely consummated. Certainly the quantitative and theoretical developments followed by the critical and neo-critical revolutions have had a major impact, but there were many residuals of earlier paradigms. The failure to fit Kuhn's model to contemporary trends in human geography leads to the conclusion that the model is irrelevant to this social science, and perhaps to social science in general.

16

Concept of Landscape and Region

HIGHLIGHTS

- Landschaft and/or Landschaftskunde (Landscape) and Region
- Origin and Meaning of Concepts: Landschaft and Region
- Synthesis
- Typology of Region
- Approaches to the Study of Regions

Earth surface can be indefinitely sub-divided into segments of various sizes. When such a segment of Earth surface is set off by boundaries it is known as an 'area'. An area of the Earth's surface with a *distinctive appearance,* natural, man-made, or both is called a landscape. Any tract of the Earth's surface with either natural or man-made *distinctive characteristics* is called a 'region'. A region is identified by specified criteria and its boundaries are determined by these criteria. However, *in geography both the words 'landscape' and 'region' are often used synonymously.*

The concept of 'region' or 'regional' concept has for long been a major concern to geographers, particularly to those who believe in the 'geographical synthesis' of the earth-bound phenomena. It was Philippe Buache, the French geographer, who in the mid-eighteenth century came out with a concept of an Earth 'marked off into major basins bordered by continuous ranges of mountains. On the land these are drainage basins, and the mountains form the drainage divides between different river systems' (James Martin, 1972, 110). Perhaps his concept would not have gained such a wider support had it not been for German geographer, Johann Christoph Gattener, who identified the drainage basins as natural regions and used them as the frame of organisation for geographical texts.

'The followers of Gattener saw a first step toward a more scientific geography in replacing the traditional divisions into political units by the division into "natural regions" offered by the theory of the continuous network of mountains. When this theory proved untenable, they did not abandon the concept of "natural regions" but sought to define them in less simple terms. Thus Ritter furthered the establishment of the concept in general as a fundamental basis for regional geography' (Hartshorne, 1939/1976, 250).

In the second half of the nineteenth century, there was little concern over this concept. With the renewed interest in regional geography at the turn of the century, the concept of region as definite, concrete, if not natural, unit reappeared.

LANDSCHAFT AND/OR LANDSCHAFTSKUNDE (LANDSCAPE) AND REGION

The German word 'landschaft' literally and most generally means 'landscape', which is a concept more particularly associated with the continental European school of 'landschaft geography'. *A landschaft may either be a specific unit area or type of area.*

The word *'landschaft' has two distinct meanings*: (a) 'an extent of area/territory distinguished by a more or less uniform aspect', and (b) 'the aspect of the face of the Earth as seen in perspective but without connotation of areal extent'. Humboldt and Wimmer used landschaft to refer to the visual impression, the aesthetic appreciation of beauty of the observable phenomena over the surface of the Earth. However, the use of the word in the sense of an area of uniform aspect has never entirely disappeared.

The word *'landschaftskunde'* (landscape science) concerns both the study of small unique areas and the delimitation and classification of different types of regions. Hannerberg considered that 'landschaftskunde' might be taken over by the *systematic branch of geography,* while *'landerkunde'* remained as regional geography in its own right.

Otto Schluter, however, noted that accepting the landscape as the subject matter of geography would give the field a logical definition, like that of any other academic field except history. The non-material content of an area-such as political organisation, religious beliefs, economic situations or even the statistical averages of climate could not be considered primary objects of geographical study, although this could be introduced to explain the observable landscape.

Hettner and Schluter expressed their concern about the variations in the character of the face of the Earth, which became known later as *'areal differentiation'*. Both recognised that there were distinctly different kinds of areas on the Earth which were distinguished from their surroundings and which showed a certain degree of homogeneity within boundaries that could be defined. However, Hettner stressed the ways in which features of a region reflected the basic patterns of the physical Earth, but Schluter emphasised that the attention in the interrelation of these features tended to provide the region its distinctive appearance. He sought to distinguish between the landscape that predated or antedated the arrival of 'cultural' man, and the landscape created by the human culture, i.e. the cultural landscape.

Schluter's *'kulturlandschaft'* (cultural landscape) appears to have a greater resemblance with the French *'Pays'*—small homogeneous well-defined areas/regions manifesting intimate and reciprocal relationships between man and his immediate surroundings (*milieu*) through centuries. A *'pay'* is, therefore, an areal expression of the *genre de vie* of a particular human community whose occupation of the area/region over centuries seems to have created the landscape. Richthofen, earlier, realised that in addition to looking at the world as a whole, it was necessary to examine smaller and smaller segments of the Earth's surface.

Schluter defined the landscape as the total impact of an area on man's senses, including such invisible phenomena as wind or temperature. He specifically included man as a part of landscape. But some of his followers insisted on restricting the term only to those material objects that could be seen. Leo Waibel in 1933, sought to point out that the word came into common use in Germany just at the time when geographers were focusing their studies on smaller and smaller areas, for which the word *landschaft* (with its meaning as a small region) seemed more appropriate than the word for a larger region (*Gebiet*). However, the emphasis on Richthofen's concept of the interconnection among diverse things associated in area continued and the word *'landschaft'* seemed to carry the connotation of harmony of related parts.

'The uncertainty regarding the exact meaning of this word is responsible for a conceptual error that still persists. The sequence of ideas runs as follows: first, it is agreed that "Landschaft" is made up of concrete, observable features; second, it is agreed that the word "landschaft" is a synonym for a homogeneous area, or region. Combining these meanings it becomes possible to assert that *a region is a concrete reality and not just an intellectual construct*' (Hartshorne, 1939/1976, 263).

To most of the German geographers before World War Second, the concept of 'landschaft' referred to an area with a more-or less uniform appearance and its interpretation required the study of the material and the non-material phenomena in an area or region. Most German geographers, however, followed Schluter in identifying the study of landscape as the central theme of geography. 'Every landscape is a dynamic structure, a thing-area-time system of specified quality inside the whole geosphere' (Schmithusen, 1963, 13).

The German word *'landschaft' essentially appears to be analogous to Carl Sauer's 'cultural landscape'* which means a scientifically defined geographical region which is expressive of the human imprints across the territory. Sauer pointed out that geography is concerned with the study of things/ phenomena associated in area on the Earth's surface with differences from place to place, both physical and cultural. Man, behaving in accordance with the norms of his culture, performs work on the physical and biotic features of his natural surroundings and transforms them into the cultural landscape.

'The cultural landscape is fashioned from a natural landscape by a cultural group. Culture is the agent, the natural area is the medium. the cultural landscape is the result. Under the influence of a given culture, itself changing through time, the landscape undergoes development, passing through phases, and probably reaching ultimately the end of its cycle of development. With the introduction of a different—that is, an alien—culture a rejuvenation of the cultural landscape is superimposed on remnants of an older one' (Sauer, 1925/1927).

To understand the changes man has made on the face of the Earth, it is necessary to go back for enough time to establish the nature of processes. According to Sauer, 'the landscape is undergoing manifold changes. The contact of man with his changeful house, as expressed through the cultural landscape, is our field of work. We are concerned with the importance of site to man; and also with his transformation of the site.' (Sauer, 1925/1927, 53).

Despite some epistemological differences, the German 'Landschaft' and Carl Sauer's 'cultural landscape' seem to be analogous to each other. However, Schluter's 'kulturlandschaft' is more akin to Sauer's cultural landscape as both sought to recognise the importance of human agency in the evolution of the landscape. One of the best known phenomenologists, Yi-Fu Tuan, once remarked that 'geography is the mirror of man; to know the world is to know oneself. The study of landscape is the study of the essence of the societies which mould them. Such study is clearly based in the humanities, rather than in social and physical sciences' (Tuan, 1971, 1978). He preferred to use the term 'humanistic geography' for such studies which are regarded here as including both idealism and phenomenology.

There is much *controversy among the professional geographers with regard to the extent to which a landscape* is *analogous to a region.* Is a region different from a landscape? Is a landscape a concrete viable geographical entity or an arbitrarily chosen fragment of land? The word landscape here refers to the cultural landscape. A landscape may be closed-system if it exhibits a defined *'genre de vie'* with no input from the outside. A region in the Vidalean tradition is an area over which such an intimate relationship between man and nature has developed through the centuries which tends to constitute it.

In the evolution of a landscape, the role of the human agency is of paramount importance, but in the case of a region, if the Vidalean tradition is to be believed, it is the intimate nature of relationship between man and nature over centuries that concerns much in its evolution. A region may be an arbitrarily chosen fragment of land if its identification is not based on the indices of the intimate man/nature relations over centuries in the area. However, a landscape is not necessarily akin to a region. Nevertheless, they appear to be analogous to each other because of the role of the human agency *vis-a-vis* the man-nature intimate relations in this evolution. The landscape study and the interpretation of the region essentially include and involve existentialism and phenomenology.

Some have sought to divide landscape into (i) *natural landscape,* and (ii) *cultural landscape.* These two seem to indicate that they are separate components of the total 'landscape', with the former consisting of all the natural features of an area, the latter of all the man-made forms. Most American geographers who sought to use the term 'cultural landscape' meant simply the present landscape of any inhabited region. In the German literature, there is a distinction between 'natural' in the sense of nature excluding man, and 'actual' or 'real' in contrast to 'arbitrary' or 'artificial' and is used to indicate that a particular Landschaft is a region, is a real unit. Some German scholars attempted to contrast between 'natural' and 'actual' on a relative basis, i.e. on the extent of cultural development.

The original natural landscape, however, is not the same as the present theoretical natural landscape. Nature has not been static during thousands of years that man has lived on the Earth. Consequently, it can be said that the *natural landscape is a theoretical conception that not only does not exist in reality, but never did exist.* Something more or less like it did exist in the original primeval landscape which, in complete form, only the first man to arrive in any area could observe.

ORIGIN AND MEANING OF CONCEPTS: LANDSCHAFT AND REGION

The origin of the concept of the region can be found in the writings of the pre-classical geographers from whom it may be traced through Ritter and possibly Ratzel. Its greatest development in the present time appears to have been furthered first by the works of Schluter and Passarge, both of whom have presented it in opposition to the methodology of Hettner. Passarge in particular tended to emphasise the purpose of presenting the *total form* from a Landschaft in opposition to merely analytic method of Hettner. In fact, the German geographers are credited for having developed the concept of the Landschaft which is for all practical purposes analogous to the English word 'region'.

The German word Landschaft has long been used in common speech to indicate either: (a) the appearance of a land as is being perceived, or (b) simply a restricted piece of land. Both these concepts were introduced into German geography not later than the beginning of the nineteenth century. Hommeyer used the word to indicate an area of land intermediate in size between *'gegend'* and '*land'.* Humboldt used the term primarily in the sense of the aesthetic character of an area, though he might have occasionally used it in other sense also. Similarly, Oppel and Wimmer both regarded the landschaft primarily from the aesthetic point of view, but the latter also included social features as a part of the landschaft as a coherent whole.

The majority of German geographers of the early twentieth century tended to use the word Landschaft in either sense without distinguishing them. The corresponding French term 'paysage' apparently also permits of this double meaning. Even the writers who sought to use the term more or less synonymous with region are by no means in agreement on what the term included. Schluter repeatedly emphasised the difference between his use of the term and that of Passarge. The differences were much more fundamental than those involved in the attempts of the American geographers to define a 'region'. There were no differences in determining limits, but the differences were about the kinds of things which were included in the Landschaft, whatever its limits. The difficulty resulted from the fact that in using Landschaft to mean 'region', all these writers wish to carry over certain aspects of its meaning as 'landscape', but they were not in agreement as to which of the latter they wished to include. The American geographers; 'who sought to use the word "landscape", had given it, perhaps unconsciously, a special redefinition'.

To Waibel, the Landschaft is 'the section of the Earth's surface and the sky that lies in our field of vision as seen in perspective from a particular point.' The visible landscape forms the core of Grano's concept which includes human sensations of sound, smells and feelings of an area. However, Grano's concept is based on Hellpach's definition of the Landschaft as 'the total impression aroused in us by a piece of the Earth's surface and the corresponding section of the sky'. He seemed to have proceeded from the psychic sensation produced by an area to the objects in the area that caused the sensation. In order to conserve the strict meaning of Landschaft as the *sensational landscape,* Grano suggested other terms, but those who followed him simply transferred the word Landschaft or landscape to the objects in the area that are responsible for our sensations of landscape.

To Penck, the Landschaft included only 'what is perceptible in our field of vision'. This does not include man, but only his effect on the Earth's surface. Schluter, on the other hand, not only included men on the grounds that they were perceptible objects, but in order that they might appear as more than 'minute grains' in the landscape. In fact, he sought to make man a major Landschaft element. Passarge, in contrast, excluded not only man, but also animal life in general, because otherwise it was too difficult to separate one Landschaft (region) from another. He also sought to limit his Landschaft to natural, non-human elements, thereby including elements like natural vegetation that might no longer be in existence.

Each geographer's definition of the word varies according to what he thinks geography should study. 'Landschaft is therefore no more precise than 'geographic area' but has only the appearance of being more precise. To a geographer who is primarily a geomorphologist, a term covering the visible objects of an area seems sufficient. To one primarily interested in cultural geography, something must be done to make man, the active cultural agent, assume a stature proportional to his importance. For one who feels that almost all geographic features can be interpreted ultimately in terms of natural vegetation, climate and landforms, the Landschaft may be limited essentially to those features' (Waibet, 1933).

All of the above writers employ the word landschaft as limited to physically perceptible objects. Those who, however, wish to include the non-material phenomena as objects of direct study in a geography that studies the landschaft have a greater difficulty in defining their concept. Lautersach, perhaps, included non-material elements such as 'racial and linguistic conditions' in his concept. He did not define Landschaft, but his description of what was included in it appeared to represent all that any geographer would include in the study of an area. Maull (1925), however, considered the political state as an element of Landschaft, 'not so much on the grounds that the state tends to affect the visible landscape but simply because it is closely related in one way or another with other elements of Landschaft.'

The term is therefore nothing more or less than a synonym for 'region'. Landschaft may be used to indicate a unit political area as a whole, including its population, commonly one of smaller size than a land. To Krebs, the term Landschaft refers not so much to the particular individual area, but rather to certain aspects of its character that are considered as typical of many similar areas. To him the peculiar character of the Alps as a mountainous area might be called the 'Alpine Landschaft' which is repeated in other parts of the world, whereas the land or region of the Alps is unique. 'This is no fine distinction but a major one: the Landschaft of the Alps is limited to certain characteristics excluding relative locations and distinctive cultural characteristics, whereas the Alps as a "land" includes all significant characteristics' (Krebs, 1923).

The majority of the contemporary German scholars preferred to use Landschaft more or less in the sense that the Americans used region though each defined it in a different way. The word Landschaft was more preferred by the Germans, particularly when asked to choose between' 'Landerkunde' and' 'Land'. The Germans showed no interest in the use of the English word 'Region'. Waibel, however, observed that the contemporary drive in Germany was to see things as they were together to consider the totality of phenomena landscape.

To many American geographers, the German Landschaft meant 'Landscape', which is an area made up of a distinct association of forms. Sauer's (1925, 25) epoch-making essay on the 'Morphology of Landscape' provided a detailed exposition of the concept. Sauer's design of landscape includes': (1) the features of the natural area and (2) the forms superimposed on the physical landscape by the activities of man—cultural landscape. His concept of landscape requires to be defined either as the area minus its phenomena, or the area in so far as it is material. Sauer is believed to have considered that landscapes are individual units of area—'the geographic area is a corporeal thing which is approached by the characterisation of its forms recognised as to structure, and understood as to origin, growth and functions'. The contemporary *American geographers held the word 'landscape' as being synonymous with the word 'region'*. If they thought that the geographic region should be regarded as including non-material features, then the state should be included as an element of landscape, just as Maull included it as an element of the Landschaft.

Broek (1938), however, concluded that the word *Landscape had two quite distinct meanings.* It consists of: (1) 'the observable features of any arbitrary part of the Earth surface but in a wider sense than the common usage implies', and (2) 'a region that has a certain homogeneity in its morphology'. Broek held the view that the geographers should not limit themselves to 'directly observable phenomena but should also include invisible features'. The landscape for the geographer, therefore, is 'an abstract landscape free from time bounds and place bounds of the observers and supplemented by invisible, but nevertheless, significant data'. Broek did not attempt to come out with a precise statement on landscape; otherwise the statement would have been: 'the landscape consists of the observable features of any arbitrary part of the Earth's surface plus all non-observable features of significance.'

The use of the word landscape by the *contemporary English geographers* carried the same meaning as did their American counterparts. To Hartshorne, the landscape consists of 'the landforms, their plant cover, together with the fixed material features, in compact areal association, which are the result of the transformation of the land forms and plant cover through human occupance' (1959, 34). The 'landscape' is essentially synonymous with a piece of area having certain characteristics which set it off from other pieces of area. This concept represents the German Landschaft to which many German and American geographers adhered.

English-speaking geographers have long used the word 'landscape' to indicate an area with a certain homogeneity. Whatever difficulties have been experienced have not resulted from any confusion over the word 'region', but simply from the problems involved in determining what 'an area with a certain homogeneity' may be. The concept of a piece of area somehow distinct from neighbouring pieces of area is much more definitely suggested by region than by landscape. If it be claimed that a region sounds like a larger piece of area than the geographers may wish to select it as a unit, landscapes then suggest a very much smaller piece.

A large number of German and American writers in the early part of the twentieth century sought to use the word 'Landschaft ' or 'landscape' (if the words are identical in meaning) to indicate all the material facts in an area. They spoke of 'observable' facts, but any fact to be a fact must somehow be observed. Many, following Schluter, spoke of sensually perceptible features—all objects that are theoretically, directly observable as sights, sounds, smell and feelings. To many, this concept was suggested by the common meaning of landscape. Grano selected all the features of an area that were responsible for human sensation. To him, therefore, the word landschaft meant a concept of sensation. The word landscape does suggest some objective single reality outside our sensations. It is a concept that represents a certain distinct and real aspect of an area. In empirical science of geography, there is little need for any of the concepts of 'landscape' as sensations.

By defining landscape as more or less synonymous with area/region and at the same time retaining the connotation of its other meaning as a visible scene/sensually perceptible, one appears to have proved without much argument that an area/region is an objective unit. Thus Sauer, after making landscape appear identical with 'geographic area', sought to refer to either of these apparently interchangeably as 'a reality as a whole', a 'corporeal whole' that has 'form, structure and functions'. One of the basic purposes of geography is to discover the areal combination of phenomena and their order; it seems to be the 'phenomenology of landscape' that is to be studied.

If an areal section of the Earth surface, delimited in some particular way to constitute a 'region' can be demonstrated logically to be a corporeal thing, a concrete unit whole, it must be that by whatever name we call it. But if an area and region are regarded inadequate by some because geography does not intend to include everything in an area but only its material features, and there exists no term that expresses the area minus its non-material and/or immaterial features, it would be better to find out a term. Hartshorne suggests the word 'geographic region' which, according to him, 'could be precisely defined to mean that, since it has no established meaning in common thought.'

There is some *confusion with regard to the use of the word 'individual' to refer to any area, and/ or region.* Presumably all would agree that its particular combination of interrelated features makes it different from any other area. This is what the German geographers asserted with regard to the concept of region. Hettner spoke of the 'individuality' of any area or rather of any point on the Earth. 'It is only through misunderstanding of his statements that some perhaps believed that he regarded the area in question as an individual in the additional sense of a limited object or entity' (Hetlner, 1927). It is quite possible that the contemporary French geographers likewise were concerned simply with the unique character of area when they referred to them as 'individuals' though their interest in the personality of each area would have implied a more concrete concept. The confusion with regard to the use of the term may be avoided by using the term 'unique character ' for the former sense of 'individuality' and confining 'individual' to definitely limited objects.

The German word *'einheitlich'* presumably refers to 'unitary' or 'uniform' or 'homogeneous'. However, Grano used einheit in the sense of uniformity or homogeneity in order to contradict 'geographic individuals' which according to him referred to limited areas that might or might not be homogeneous. However, the German geographers were a little confused with regard to the use of words like 'unitary' and 'uniform' to refer to an area or region. The American geographers, on the other hand, preferred to use the word 'unitary' in the sense of an individual unit which might or might not be uniform or homogeneous.

Related to the concept of region are the concepts of the *'vertical ' and 'horizontal' aspects of unity.* However, the vertical concept of unity seems to be more applicable to any area, whether it be homogeneous or heterogeneous. Each of its parts is related in some way to others. In Germany, this vertical form of unity is called a Ganheit or Whole. In a Ganheit or a Whole, the parts are dependent not merely on each other but on the Whole and cannot be properly interpreted in terms of each other independently of the Whole. The Whole is something more than the sum of its parts, arid is relatively independent of outside elements.

Penck attempted to introduce a new term '*Gestall*' to express the unified form of larger areas, each consisting of a particular arrangement of regions, but others have adapted it for the geographic landschaft. To Bunges, a Gestall is a dynamic structure in which the parts reach into other functionally, and can be understood only in view of the whole. All living organisms, certain kinds of groups of living things, as in a family and to a certain degree in a people, and certain phenomena of experience may be examples of Wholes (Gestalten).

'To what extent may these various concepts be applied to particular sections of the Earth surface somehow defined as regions? Any area we select will have structure and will include forms that are in functional relation to each other. Even if we are not to consider the region as a (primary) Whole, something may still be gained if we can consider it as at least a loose unity, in the sense of a complex of related elements forming a relatively closed system—the unitary self-enclosed Landschaft.... The region has structure, form, and function and hence a position in a system. To have these attributes the region must obviously have fairly definite limits and since the region can only be defined in terms of area it must have fairly definite areal limit. The region thus determined is so constituted in the structure and function of its parts as to represent a relatively closed unit, in contrast to adjacent or distant similar units' (Hartshorne, 1939/1976, 267).

The *problem of drawing the limits of regions cannot be solved simply by substituting boundary zones in place of lines.* 'Even if it were possible to solve this problem of delimitation of areas in respect to a single element, the problem remains as to how can we decide which of various elements are to be considered as decisive' (ibid.). Passarge recognised that vegetation (or climate) and landforms could be considered as possible elements to solve the problem of delimitation of areas. Indeed, in most of his works he relied entirely on vegetation, though often this was confused with climate directly. Nevertheless, he recognised that even with but two factors it might be necessary to resort to arbitrary selection of lines of convenience. 'Thus, if forests overlap from mountains into the margin of a grass plain, the forested border may be included in the region of grass plain. If the mountains grade through foothills into the plain, but the forest grass boundary is sharp, then this line may conveniently be used as the Landschaft boundary. If, however, the transition area is of considerable extent it may be recognised as a landschaft' (Passarge, 1923). But Passarge did not realise that this simply introduces the double problem of delimiting the transition zone on either side.

To say that 'the elements of a region are geared together so that they are independent, each element reacting upon all the others and in turn being reacted upon by them, is to give the impression that if any one element varies in major degree the others will likewise vary in major degree and hence the complex of elements of a Whole would change in marked degree.... The different realms of nature and circles of phenomenon are closely knit together and consequently differences present in any one realm of nature and one circle of phenomenon will correspond to a certain degree with those of the others' (Hartshorne, 1939/1959, 250).

Most of the scholars who uphold the concept of region as actual units attempt to include at least the material features of culture as well as the natural features. Nevertheless, there is a lack of correspondence between natural and cultural features, and none of the scholars would claim that there is a complete correspondence.

'It is essential that the demarcation of the regions can be applied to the Earth's surface without effort and in an incontestable manner, it is necessary to demonstrate the existence of regions without mutual trespassing, and it is necessary that all of the facts of geography, physical and human, can be grouped in concordant with regional definitions.... The differentiation is not the same in each circle of phenomena.... Any particular large portion of the Earth surface is therefore homogeneous only in one relation, in others it is heterogeneous; and in the strictest sense only a particular spot on the Earth has complete individuality or character of its own' (ibid.).

Some geographers regard the *region as an organism that passes through a lifecycle and disappears with its death.* It can be said that Carl Ritter, in company with many of his predecessors, regarded the Earth as a whole an organism, and his description of the continents as 'individuals' or as 'organs' might be regarded as leading naturally to the concept of smaller areas as 'organism'. Likewise, Vidal de la Blache called the Earth's surface as 'the terrestrial organism'. For many geographers today, the term which Ritter used to describe the intricate interlacing of all Earth phenomena, animate and inanimate, material and immaterial, is accepted as a precise statement of the character of a region.

Carl Sauer (1925, 45), who sought to carry forward Vidal's concept of the unity of the 'terrestrial organism' in his 'landscape paradigm', wrote that 'we regard the geographic area as a corporeal thing.... We are to study the anatomy of that area ... the landscape is considered in a sense as having organic quality'. Sauer, perhaps, specifically emphasised the fictive character of the region as an organism. Fictive analogies, however, have a way of developing into direct statement of facts. Finch (1939, 12) once stated that 'a geographic region, or even an arbitrarily chosen portion of the Earth's surface, may be thought of having some qualities of a human being. It is a thing ... with physical and cultural elements so interwoven as to give individualism to the organism'. To Trewartha, the regions are 'functioning organisms' comparable with plants. Likewise, James made a similar shift from 'pseudo-organic' to 'organism'.

However, the organic concept of the region has been reviewed and criticised on several counts. 'To say that region or landschaft is an organism is to imply more than that it is in some particular unstated manner like an organism (Vallaux, 1906).' The region has the qualities that are inherent in organisms. 'Thus it is not sufficient to establish that a region includes animate and inanimate things in close interrelation. At the most, therefore, one would have not a closed individual organism but simply something organic part of an organism.... An organic combination of animate and inanimate matter, however, involves something more than a close interrelation of those elements; it involves a connection of the whole as the superstructure to which the physical elements are subordinate' (Hartshorne, 1976, 258). It is aptly believed that neither the Earth's surface nor any areal portion of it can possess some of the characteristics of organism which include power of adaptation, of cohesion, of reaction and of recreation. *'A region is acted upon; of itself it never reacts.'*

Bunge has sharply drawn the distinction between a region which he regards as a whole and an organism. In both cases, he observes a multiplicity of different parts combined to form a whole. In the organism, however, whole is capable of life because there is a differentiation between the parts such that each part has a particular function prescribed for it by its place in the whole organism. There exists, therefore, a general functional harmony that subjugates the individual part (organ) to the laws of the whole, and limits its separate existence. 'It follows from the description of an organism that its parts cannot be considered in themselves as organisms, but only as organs, members, or organic parts of an organism. *An organism* is *essentially indivisible, whereas any regional unit of the Earth surface can be divided into smaller units*' (Hartshorne, 1939, 258–59).

The analogy with an organism, Hartshorne asserts, is apt to be particularly misleading when we consider what is frequently called the growth of regions or development of the cultural landscape out of the natural landscape. 'In the organic growth, all the individual parts develop from a common origin ... are nourished from a common food supply and are controlled in their growth by some directive agencies. External elements introduced into a single part of the organism are either converted into materials that are spread through the whole, or are expelled, or are immediately recognised as 'foreign bodies' and isolated as in a cyst. What do we find comparable to this in the alterations of the area of the earth? The soil erosion of any single slope may be entirely independent of all conditions in other parts of the area; the growth of a single tree is dependent only on the immediately surrounding conditions; what takes place in all the rest of the area may be of no importance to it whatever. The rainfall conditions are largely the result of external forces quite independent of changes in the area itself' (Hartshorne, 1939/1976, 259).

The region as a whole does not undergo changes, but only the complex of different elements changes with changes in its elements. The development of a cultural landscape is therefore nothing but the development of the complex of its cultural elements. The changes may be largely the result of external influences. 'The analogy of an organism, if at all applicable, can be applied only to the whole Earth surface. The interpretation of changes in any single area of the world must be extended into neighbouring and even remote areas.' The region is not an organism and the landschaft as an area is not a concrete object nor an individual unitary whole but is only a more or less arbitrarily chosen fragment of land.

Generic and Specific Regions

Efforts to find a system of regional division of the world are no doubt as old as geography itself; indeed the very nature of the subject requires some basis of division of the Earth's surface. Although most of the early geographers were content to accept the division already provided by the political map, the efforts of early Greek geographers to divide the world into climatic belts represented an early attempt in a different direction. The first major effort in modern times was that of Ritter, which, however, was based on a teleologic philosophy much in contradiction to the modern sciences. Nevertheless, 'to Ritter remains the great merit that he recognised the necessity of a division of the Earth spaces based on a universal consideration of all features and made an attempt to put it into effect' (Ritter, 1833). Hettner followed Ritter in so far as the latter utilised not one element, but the combination of various elements and, further, had developed his system in part inductively from the character of individual small areas.

A different direction was given to the problem by the epoch-making, effort of Herbertson in 1905. Ritter, Hettner and many other German authors had associated individual regions together, as they are in fact, into larger realms and continents. Herbertson, on the other hand, considered his regions as *specimens of types,* thus associating widely separated areas in the same type. His system was, therefore, a classification of regions regardless of location and association in space, in contrast with an areal division of the world into major parts, each sub-divided into sub-divisions which were contiguous and together form an associated Whole. One system is a classification, according to 'internal character', the other 'an outline of the actual areal relations'. In the first case, the regions might be called *generic;* in the latter *specific.* Whereas most works dealing with the regional geography of particular continents employ a division into specific regions, most of the recent attempts towards the regional division of the world followed the generic or comparative system.

'The *generic method* reveals the inherent similatity of remotely separated regions; this similarity, however, is fundamentally incomplete, since the location of a region is one of its important inherent characteristics which may have marked effect on other inherent characteristics. Even though we find "Mediterranean" climate, vegetation and crops, combined with landforms similar to those of the Mediterranean region itself, in such areas as Southern California and Central Chile, these areas are not and cannot be "Mediterranean" in true sense of the term.' On the other hand, if we merely associate each of these areas with their neighbouring areas in a system of specific regions which might be called the actual system of regions, we lose the advantage of similarity of regions provided by the generic (comparative) method.

Almost every professional geographer has attempted to divide some larger area, such as continent, into *specific regions.* To divide any area into regions involves necessarily more than the recognition of the distinctive character of certain parts of it, ignoring those less distinct in character. An incomplete study may limit itself to the 'cores' or 'hearts' of regions, although this involves the questionable assumption that such areas are geographically more important than the areas of less distinctive character. A complete study must give attention to every portion of the area so that the geographers sooner or later must face the problem of definite, even though arbitrary, boundaries between the regions. The device of 'transition zones' does not eliminate the problem since these zones are themselves areas which must be delimited.

Hettner made a thorough study of the problem of dividing the world into specific regions. Recognising the manifold character of the world, he felt that a division based on all of reality was not possible on one principle, but could only be built up on a combination of several grounds. The problem, therefore, is to compare the different bases and measure their relative importance. The degree of importance is to be measured in terms of the effects on organic life, notably that of man.

'What is the order of importance of the multitude of factors involved? On one point there may perhaps be complete agreement. The presence of land in contrast to sea is of first importance not only for man but for most organic life.... If we confine our attention to the lands, we find no fixed order of importance of different elements that is applicable to all parts of the world: in one area, climatic differences may be of major importance, in another differences in relief, in still others, soil, mineral deposits or simply relative location. Consequently, it is necessary to use different grounds for division in different parts of the world, although within a single division ... On the other hand, different grounds may be combined in recognising a single region. Thus the Spanish Peninsula is distinguished as a subdivision, of major size, because of the separating effect of the sea on three sides and of mountains on the fourth, even though these factors of sea and a type of land are actually opposing elements' (Wittlesey, 1954).

A system of division based on a large number of elements whose relative importance varies greatly from place to place may appear complicated. These elements, however, are not completely independent of each other. Will not the study of their interrelations lead us to a somewhat simpler basis? To Hettner, 'every realistic system of division must be *genetic;* it must show causal relations present in the reality: It must search for the creative forces of the earth; it must seek to know the manner in which the phenomena of the Earth surface result from the composite influence of those forces; and it must likewise learn to reconstruct in mind the edifice of the Earth and thereby learn to understand the individual parts and spaces in their character and their significance' (1905).

It is clear that a system of regional division cannot be completely genetic as the genetic principle leads us into serious difficulties. These become clear when we consider the relation of human or cultural phenomena to the system of regional division. Though the works of man may be largely regarded as resultants in a genetic explanation, at the same time, human factors are causal factors in the development of many features. Though these causes may conceivably be explained as results of non-human causes, we are seldom in the position of being able to demonstrate that.

In a realistic system of world division, every region is recognised as unique and the regions are associated in their actual relationships. With such a system it is possible to take widely separated regions which show marked similarities in certain elements and study both their similarities and their differences. *In this sense, comparative regional geography is nothing new.* 'On the other hand a system of regional division which is based on certain major similarities of regions, regardless of where they are located, a system, which distinguishes regions not so much in terms of their unique reality but in terms of types into which they can be classified, essentially facilitates the study of comparisons' (Hartshorne, 1939/1959, 268). Such a system may be called a comparative system of division of the world into generic regions.

Natural Regions

There seems to be a controversy regarding the usage of the term 'natural region'. What actually constitutes a natural region? The term carries different meanings in different countries. In its broadest sense, nature obviously includes man, together with all his works. In this sense, it is nearly synonymous with the universe, or better, with the reality of the universe as distinct from our thoughts about it. Thus Kant, Humboldt and others have used *the word nature to mean the objective world outside of the observer's mind.*

Some geographers attempted to use *the term natural region to indicate 'something inherent and not arbitrarily imposed.'* The word 'region' itself carries the connotation of 'inherent character' in contrast to the word 'area'. If 'natural' is added in order to emphasise the concept of 'inherent regional character' as opposed to arbitrarily imposed divisions of area, it implies that there is in nature, i.e. in the real world, an unambiguous division of the Earth surface and the problem is simply to recognise it. No such division exists in reality and attempts to divide the world involves subjective judgement, not in the determination of the limits of individual factors, but in deciding which of the several factors is to be regarded as most important.

The 'inherent character' of an area is composed of a multitude of incommensurable elements which are interrelated in no complete sense. Temperature; rainfall, relief, slope of land, physical texture of soil, underground minerals and relative location are all in large degree independent of each other. In a world in which these factors all vary from place to place, the variations being in large degree independent, it is impossible to draw the division on the basis of all the factors simultaneously. One must, therefore, determine which variations are of greatest importance. The decision can only be subjective; the regions so constructed are in this sense arbitrarily imposed on reality.

Some scholars have attempted to use the term 'natural regions' merely to indicate that the basis of the regional division could be found in nature as a whole, including man, in contrast to a division based on a single element, as in the case of 'climatic regions', 'agricultural regions', etc. (In most cases, the division is actually based almost on one factor, namely, climate). On first thought, 'natural regions' may seem a legitimate abbreviation of 'regions on the basis of nature'. But we find that the shorter phrase claims much more than the longer one; as we shorten 'regions on a basis of real conditions' to 'real regions'. The conditions are reality, the regions are intellectual conceptions.

Hettner, however, sought to *distinguish between 'regions on a natural basis' and 'natural regions'.* He recognised that whatever the basis for division, its application required so many subjective decisions of judgement, that the resultant regions could not in any proper sense of the word be called 'natural'. He attempted to contrast his system of division with that which he termed artificial. His system then was 'natural' in the sense of realistic; more fully he sought for a logical basis founded solely on reality and on the whole of reality.

One might say that the use of the word 'natural' in its broadest sense tended to include all the phenomena observed outside the observer's mind—objective reality. During the nineteenth century, nature was used as a word to express the contrast between that part of reality that was independent of man and the part that was human. The fact that man alone was capable of contemplating such a contrast was sufficient justification for the terms with which to express it.

Man in his relation to 'non-human' factors reacts in an entirely different manner from all other living things; *his reactions are cultural reactions.* 'Since geography is a science developed by and for men—as distinct from all other biological beings—the relations between the world of man and the non-human world are of greatest concern in geography.... We are therefore justified in accepting the popular conception that in the universe the things that are cultural, i.e. of man are fundamentally different from everything else—which for want of a better term we call "natural"' (Fleure, 1919).

The term natural regions would presumably indicate regions considered in terms of their non-human elements. It is also possible that 'natural regions' in the limited sense might be used merely to indicate that the world has been divided on the basis of the natural, 'non-human' elements. However, this usage is misleading for two reasons. (1) Nature, as distinct from man, provides no basis for a division which would be significant or suitable at the same time for all natural features, not even for two or three natural features of great importance. (2) Neither does nature indicate which of its elements is of more significance than others. Needless to say, all such divisions by geographers have been made with reference to man's point of view—nature as man is concerned with it.

Some geographers preferred the term *'geographical regions'* to natural regions. For example, Unstead sought to recognise that the term geographical regions necessarily included all the significant features, both natural and human, but he observed that climate was the main criterion. His world map was, therefore, essentially 'climatic'.

A different direction was taken by Passarge and James. Inspite of the underlying consideration of climate in the systems of both these authors, the *individual regions were defined specifically in terms of natural vegetation.* The actual classification of vegetation regions was found to be very different in the two cases, and James put the high mountain areas into a separate category like that of Unstead who classified highland areas separately.

The climatic basis of regional classification of the world does not classify cultural regions as a classification of natural regions (as significant to man). Even in those areas where climatic differences are of first importance and where they are correlated fairly completely with vegetation and soils, these together cannot be credited with a monopolistic determination of regional character. A map of climatic areas is an indicator of variable reliability for many other elements, including cultural. To call it a map of natural or geographic regions is in either case to invite confusion.

Likewise, a regional system based on natural vegetation—even to the limited extent in which it is possible to construct one—does not seem to justify the claim expressed by such titles as 'natural' or 'geographic' divisions. A division of the world based on vegetation is simply a division of world vegetation and nothing more. Like a map of climatic division it suggests other things but tells us nothing.

A *regional classification based on either climate or natural vegetation* does not seem to be adequate for the purpose it is made. The question that arises is whether 'all the natural elements significant to man' should be considered for a system of world regional division instead of depending on one or two natural elements? However, it had not been done either by the German or the American geographers in the past. The synthesis of natural elements in areas represents a combination of several factors of importance to man. These several important factors are sufficiently independent so that they can be associated together in a great number of different combinations which may be very nearly as great as the total number of areal divisions, i.e. nearly every area is a type in itself. 'It may be suggested that this difficulty is to be avoided by making our types more generalised, so that each separate region could be considered as a sub-type of the more general type. But the combination of several factors, each of which varies independently, can be reduced to common types only if one assumes that some of the factors are of minor importance, no matter how much they vary' (Gilbert, 1960).

This proposition is important to the whole problem of comparative regional division on whatever basis it is made. It might be thought, for example, that classifications of climate that are based on actual climatic conditions rather than on genesis, represent the organisation of many semi-independent factors, like rainfall, temperature, winds, humidity, etc. The system of classifications of Koppen and Thornthwaite and others primarily depended on only two of these factors, namely, precipitation and temperature.

The number of actual combinations that must be recognised, however, need not be too large and each type may be represented by several areas.

Grano made use of seven independent symbols (six symbols for natural elements excluding climate and one for human elements) to identify a single type of area. To be sure, many of the elements can be regarded as factors of minor importance in any particular area, but none can be eliminated from the entire consideration since anyone may be, in some particular area, the most important factor. Consequently, the number of actual combinations which we would have to recognise as types would be little less than the total number of regions that would be recognised; we would have a system of regions but not of types.

The fact that the study of the world is organised on the basis of one particular feature inevitably leads the reader to think of that feature as the most significant feature in every area studied. If the discussion of a certain type of cultural development is largely confined to a chapter named in terms of a certain type of natural vegetation or climate, a relationship is thereby suggested, whether or not it has validity. The disadvantages inherent in any system based on a single natural feature are most apparent in one based on natural vegetation because of the marked discrepancy between the agricultural regions and areas of natural vegetation used as the outline.

The conclusion to be drawn from the consideration of regional systems based on natural elements is in agreement with that of Finch and Trewartha who said: *'It is deemed unwise to try to fit world culture patterns into the framework of the physical regions.'* In addition, however, Hartshorne concluded that 'it is hardly possible to fit the summations of natural elements which they call "physical regions" into a framework based on one or two natural elements' (Trewartha and Finch, 1952).

Cultural Regions

There are possibilities of constructing a regional division based on cultural elements. Some continental geographers retreated to such simple divisions as those based on political boundaries, states and provinces, or on racial or folk areas. A few attempts have been made to lead to a system for world division on a general cultural basis. A theoretical approach to the problem is therefore required.

What are the major aspects of culture which are of greatest importance to regional geography? If the question is referred to the fundamental concept of the nature of geography, it reads: What major aspects of man and his works are relatively homogeneous in limited areas, but in different areas differ notably in ways that are significantly related to the total character of the area? If areas could be found for which these conditions are true of all aspects of man and his works, these could be perfect cultural regions. Since the areal differentiation of various cultural aspects will not coincide, we must limit ourselves to major aspects. The distinction between major and minor, among phenomena that are mutually incommensurable, is not a question permitting objective answer, but on a subjective basis it is possible that there might be a fairly general agreement.

Presumably all will agree that the *'areal differentiation' in numbers of people, population density, is of major importance.* Hall and others, long back suggested that population density in different areas of the world might be considered as a possible criterion for cultural regional division, as it could be easily measured. He even emphasised the need for such a study of the whole world. How effective such a basis would be for world regional division can be judged better when we have such detailed maps on a common basis of at least some areas of the world of notably different culture.

It, however, seems probable that the same form of distribution may be found in areas of different culture and may represent in many ways entirely different regional character. It can be concluded that population density, even in its most detailed and complicated form, provides us with one major key which alone is not sufficient. Aside from their numbers, what are the ways in which people differ in different parts of the world? All those differences which, however important as between members of any local group, are not significant between people of different areas and therefore can be eliminated for geographic purposes. Further, in comparing regions we can disregard certain differnces that are of major importance within a region—including the whole complex of differences between 'urban' and 'rural' areas. On this basis, we can eliminate a whole host of cultural phenomena of man, since in so many ways people of the world are alike, do the same things and do them in more or less the same way.

There are a great multitude of cultural phenomena which differ significantly from region to region. For this very reason, geographers should not study any one or two phenomena without first considering their importance both in themselves and in relation to other cultural phenomena that are geographically significant. 'If one were to attempt a full list of such phenomena it would include a large number of material objects which men make, and make differently, in different parts of the world.... These would include: the physical characteristics of the population; the manner and the substance of thought, speech and writing; the way in which people eat, dance, walk, or ride; the character of their clothing, shelter—both for man and beast—and the grouping of these shelters in settlements; the way in which the people work and play and the tools and implements used in doing each; the domestic animals which they use in different ways; and implements, shelter, fuel and power, and finally various alterations to which they subject the Earth's surface, including alteration of vegetation, soils, land-forms, bed-rock and even underlying formations' (Broek and Webb, 1973).

Any one of all these phenomena may have its own geographic regional expression and might furnish something of a key to the geography of cultural phenomena. Serious attention should be given to those criteria which are in themselves of greatest importance in determining regional character. Theoretically, if there are such criteria, we can use only those which we are able to observe, classify and to some degree measure quantitatively. On the basis of both the conditions, the physical characteristics of the population may be eliminated. Nevertheless, the *racial map of the world* might be of even greater importance than that of mere number of the people.

Since most of the cultural phenomena are in part products of the way in which people think, *it is suggested by geographers that cultural regions are required to be determined by finding areas in which 'people think alike'*. There is some degree of uniformity in thinking by people who speak the same language in contrast to those who speak a different language. *Linguistic regions* may, therefore, show similarities in a large number of the cultural phenomena. But many of the human phenomena are hardly affected by linguistic differences.

Geographers are faced with the problem of determining the comparative geographic importance of different individual cultural phenomena. If they combine the areal spread of a phenomenon with its relative importance to man, they might get the major factors for a theoretical basis of measurement. Without attempting a complete analysis, one might measure the importance of a phenomenon to each individual in terms of the part it bears in his total activity.

For most of the people of the world, a major part of their activity is concerned with ways and means of making a living, i.e. *economic activities.* By far the most important of the economic activities is the use of the land to secure plant and animal products. *The intensity and the manner in which the land is used may be a basis for regional division of the world* that will be significant for the largest possible number of cultural phenomena important to man and to the Earth surface as well.

In terms of differences in the intensity and manner of *agricultural landuse,* the land areas of the world can be divided into: (1) those which are not used at all—the uninhabited areas; (2) the areas in which man utilises the wild vegetative and animal life; and (3) the areas where the surface is dominated by man's cultivation, i.e. where man has removed the wild vegetative cover and replaced it with plants of his own choice. While the largest part of the land area of the world is included in the second class, the most important areas are, of course, those of the third.

Within the cleared and cultivated lands, the differences in manner of use which are geographically most significant are the differences in the plants and animals produced. The intensity of agricultural landuse in these cultivated areas is measured in the first instance in the extent to which they are cleared and cultivated rather than left in forest or wild grass. Differences in the intensity and manner in which the land is used across the Earth's surface may be added to the system of criteria for the world regional division in terms of cultural phenomena.

However, a common basis for dividing each of the major classes of lands, namely, in terms of the actual vegetation (and animals) can be established. These are real rather than theoretical features which we can observe and measure by various methods, and can therefore classify and map. Such a system would provide at least a major part of a valid basis for establishing culture—*geographic regions.* A world division on this basis will show significant associations with the areal differentiation of natural features.

The regional division of the world in terms of cultural phenomena provides no method for measuring the relative importance of fundamental cultural features not associated with production from the surface of the land such as mining and manufacturing. Nor does it provide a sufficiently inclusive classification of characteristics of rural areas. Thus, different areas may have nearly the same association of crops and animals. But the methods, tools and equipment of production may be sufficiently different in the several areas as to require recognition in the regional classification.

Not many attempts have been made to construct regional systems based on cultural features. Those which have been done are confined to a few major aspects of cultural geography with perhaps no consideration of their relation to the general problem of culture–geographic division. In many studies in cultural geography, the attempt is made to proceed directly from individual elements—once they are classified and their distribution determined—to regional determination. This appears in the *attempt to divide an area into 'agricultural regions' on the basis of single crop.*

'While Engelbrecht and others in Germany, as well as many in America, attempted to determine agricultural regions on the basis of single crops, Hahn (1892) struck at a more basic element in considering the manner of cultivation—in particular the contrast between hoe culture and plough culture. Waibel (1933) studied the distribution of forms of landuse by the fallow system, the three-field system and various forms of rotation. Kniffen, however, emphasised that in the study of cultural geography in general a necessary step is the recognition of areal associations of elements, or what be called *element-complexes.*'

An early attempt towards agricultural 'element complexes' was represented by the map of Agricultural Regions of the United States, prepared by O. E. Baker and others, and published by the US Department of Agriculture in 1915. The first attempt to classify types of agricultural landuse for the world in term of 'element-complexes' and to divide the world map on that basis was the system developed over a long period of time by Wellington Jones and Derwent Whittlesey. In 1936, Whittlesey revised the whole map and replaced it by a new map with a write-up: 'Major Agricultural Regions of the Earth'.

Valuable as these studies are, they cannot be combined into a single uniform system, in part because each of the authors constructed an independent system. Furthermore, in some of these, the regional determination is based on the individual elements. Finally, no clear distinction is maintained between 'specific' regions, each unique in itself, and the 'genetic' regions of a comparative system of division.

Each of the systems—one prepared by Jones and Whittlesey and the other by Hartshorne and Samuel in 1935–36—sought for a comparative' regional division of the world which appeared to be significant for the greatest possible number of features associated with landuse. Busch presented a model which sought to determine agricultural types and areas in which they were found. It was, however, based on particular combination of crops and livestock. The *criteria of crop-animal complexes* was found to be more easily applicable and provided better results in areas limited to a single major cultural type. Hartshorne and Dicken found it most nearly successful when they applied it to areas in both Europe and North America. But the problem of determining which factors in the crop-animal associations are most effective in determining types that will reflect the largest number of significant farm features is not one that admits of clear and unquestioned answers.

Hartshorne and Dicken sought to base their *division on the criteria of the landscape as being observed.* The present landscape cover consists of vegetation, fallow fields, quarry pits, buildings, etc. Whether the vegetation be natural, wild, or cultural is for the moment immaterial. In contrast to the concept of the natural landscape, or natural vegetation, this concept of the present landscape cover is a reality; it can be observed and analysed. Further, its various elements can be arranged in an order of importance which does not vary. Furthermore, the landscape cover of the world is not a hopeless kaleidoscope of varying features, but in major aspects shows marked similarity within limited areas with greater contrasts between those areas. This results from two separate forces both of which produce real synthesis in the landscape. Nature produces the synthesis of the plant associations, theoretically the 'natural vegetation' (a) where man has affected these without destroying them, and (b) where man has more or less disrupted the synthesis of the association of his own choosing by substituting new plant associations. In the process, various human species over million years altered the biological character of nature. *This conversion of the 'natural landscape' into the 'cultural landscape'* over time appears to be 'perhaps the greatest act of man, a geographical and historical event of the highest importance'.

From this point of view, the distinction of importance in present landscape is not that between those areas where 'natural landscapes' may actually be found, i.e. the glacial regions and the inhabited lands affected positively or negatively by human occupance. The contrast is expressed in a multitude of different aspects between areas whose landscape cover is largely under the control of man and those where any effect of man's occupance is minor in comparison to the forces of nature. Within these *wild landscapes* one could distinguish, theoretically between those that are exclusively natural and those which have been negatively affected by man. The areas of wild landscapes may be sub-divided on the basis of appearance, distinguishing areas of ice, bare Earth, tundra, grass, savanna and forest. Within anyone of these major sub-divisions one could distinguish between areas of continuous wild landscape with cultivated clearings.

In the *cultivated landscapes,* the landscape cover is represented not merely by the cleared and cultivated fields whose character changes so markedly through the seasons, but also by various features including fences, farm-building, villages, roads, town, etc. These features are functionally associated with the farm features, and therefore they may be considered as parts of a complete areal complex. Any division of the cultivated landscapes consistent with the general system, however, would presumably be used on the major features of the landscape cover, namely, the cultivated fields.

It is apparent that a great variety of cultural features with greatest geographical significance are represented, directly or indirectly, in the present landscape. A regional division based on that seems to be significant for a great many features.

The *alternative to the system based on the present landscape cover is one based on the synthesis of all features involved in the productive use of the land surface.* The individual cultural features are not independent in their variations; they are in fact synthesised by man in his individual organised units of landuse. The basic units are represented by the individual farm, plantation, or ranch. The farm, as an organised unit, includes not merely the land and the plants and buildings on it, but also the livestock, tools, methods and intensity of production, and the use of the products. The farm, therefore, represents not merely an element-complex, but is a primary 'Whole'. A particular type of landuse units have areal expression. The individual farms are very much alike over considerable areas and differ notably from units in other areas.

The land-use units are associated in areas to form larger element-complex of various orders of magnitude and coherence. In most of the countries, the farms of a particular community in which all the families live in a compact village form, an element complex that may well be considered a unit, though hardly a 'Whole'. The association of a much larger area of farm-land with neighbouring agricultural town or city, likewise, forms an element-complex of a relatively loose form, in which highways and railways constitute connecting elements. Though this larger areal complex does not appear to be a Whole, it does show a definite character that can be considered in the recognition of 'generic' types of regions. The areal distribution of various types of landuse is significantly related to other geographic features, both cultural and natural, and the landuse complex is the most important of all element-complexes in human geography.

If we are to have the *culturo-geographic regions on the areal distributions of landuse units,* considered as primary Wholes, then those units should be studied in fact as actual Wholes. In order to comprehend either element-complexes of the more highly organised and closed units, called the Wholes, we must study individual cases. The study of a unit as a whole requires analysis as well as synthesis. But the elements analysed are to be studied not over the whole world, nor over a whole region, or even a country, but within the unit farm. It is essential first to establish the relation of each of these elements to the total farm (landuse) as a unit Whole.

SYNTHESIS

Throughout the discussion of regional systems of world division based upon cultural complexes, the question which has been frequently raised requires to be answered. 'Can such a system—whether based exclusively on cultural features or on actual features of the landscape cover, both cultural and natural—provide a sound framework for a field that is concerned with all features, natural as well as cultural? Will we not face difficulties in comparing such a system of essentially cultural complexes with the complex of natural elements—whether the present, theoretical natural element or the actual original fundament?

This comparison, however, is not to be stated as 'the relation of cultural regions to natural regions'. 'The determination of regions in either case requires the measuring rod of significance to man. Though this presents no difficulty in the recognition of 'cultural regions', it eliminates entirely the concept of 'natural regions' in any literal sense of the term. What is the situation when we compare the culturo-geographic regions with the regions of the individual natural elements? Either system of culturo-geographic regions is based on a large number of cultural features, no one of which is dependent on a single natural element. In fact, each is dependent, in part, on a number of natural elements and the manner and degree of dependence is different for each cultural feature' (Broek and Webb, 1973). In consequence, the sum total of cultural features as represented by the cultural regions will differ markedly from the regional classification of any one of the natural elements or of any combination of a few of them. 'If one takes the maps of cultural regions of either Whittlesey or of Hartshorne and Dicken and compares them with the world maps of climate, relief, soil, etc., it is obvious that the explanation of any single type of cultural area in the cultivated lands will require consideration of all the natural elements, and further, that it cannot be completely explained by that consideration alone' (ibid.). The racial difference between regions in China and those in North America will at once require the consideration of cultural factors.

The insistence that the study of the cultural features of an area must start with the cultural features does not mean that in the general study of the geography of any region one must start with cultural features. Assuming that one may beg the question of delimiting the region, the only scientific requirement of course is that one start with available facts. The available facts, as Finch strongly put forward, are the present, actual features of the area, regardless of whether they are cultural or natural. In other words, *he recommended the procedure of starting with present features of the landscape.* 'The interpretation of the natural features would depend exclusively on non-human factors, so that the study of an area little affected by man might quite reasonably begin with the consideration of natural features' (Herbertson, 1905, 303). But the consideration of the cultural features even within such an area must, to be sound, proceed from the observations of the cultural features.

Such a system should be based on a single element-complex or synthesis whose varying forms have areal expression. Since neither nature nor man synthesises all geographic features in this way, *a single system of regional division including all geographic features is not possible.* As many regional systems as seem desirable can be established on the basis of individual elements or on individual element-complexes and can be compared with each other. If all these maps are superimposed on one map, the total would represent the actual geography of the world as it is. But it would not establish definite regions, nor would it establish definite types of areas, but rather would show all the differences actually existing in specific regions. For some purposes, the world division into type areas based on actual landscape cover may prove the most useful. The largest number of different features, however, is represented by a world division into type areas based on the great number of cultural features synthesised by man in his productive use of land.

A cultural region thus refers to an area which is well-defined in terms of boundary and that has fundamental unity in the composition, arrangement and integration of significant traits which distinguishes it from other cultural regions. The Earth surface therefore consists of a mosaic of cultural regions with distinct sets of 'genre de vie'.

TYPOLOGY OF REGION

Geographers tend to draw distinction between regions on various grounds. The features which distinguish regions may be singular or plural. The region of covered bridges (a covered bridge region is an area in which the distinguishing element and/or feature is a part of the rural landscape) is a *'single-feature region'* which maintains a well defined boundary. A *'multiple-feature region'*, on the other hand, is one in which three different sets of features—core, domain and sphere—tend to overlap and intervene, so as to make the region plural.

Since culture is a multi-faceted concept, most cultural regions tend to have multiple rather than single distinguishing feature. Geographers use the term *'core areas'* to define such regional heartlands which have a special significance in the study of the regional mosaic of the Earth. A *'domain'* is defined as an area where the culture is dominant, but with less intensity than in the core and where local differences in social organisation are evident. A *sphere* is an outer zone of influence and peripheral contact in which the members of the culture become significant local minorities. The boundary of a multiple feature region does not seem to be well defined because of the gradual weakening of the domain. Whittlesey sought to identify a variety of different kinds of regions.

Uniform region. The *'uniform region'* is a discrete distribution that is defined in terms of specified criteria and homogeneous throughout in terms of these criteria. Uniform regions are defined in terms of single feature or association of several features. Examples of such uniform regions are the unit areas on Finch's map of Montfort or the Cotton Belt/Com Belt as defined in 1927 by O. E. Baker. Whittlesey (1956) defined the 'Compage' as a uniform region in which all the features of the physical, biotic and social environment are functionally associated with the human occupance of the Earth. As a revival of an ancient word meaning 'a way of joining and connecting matter', the term 'Compage' was introduced into geography by Whittlesey to lend greater precision to several aspects of regional geography. It implies a highly diversified though unitary complex of elements.

'Nodal' or 'functional' region. The *'nodal'* or *'functional' region* is an area that is tied functionally to a node or to several nodes. The definition of such a region involves the measurement of movements or spatial interaction, e.g., the area functionally connected with a central place; the service area of a market; the area reached by a newspaper; or the territory that is tied to a centre of government by the lines of political authority. The nodal/functional region is bounded by lines that mark the disappearance or weakening of the tie to its own focus in favour of some focus. The lines that mark the extension of a nodal/functional region are similar to isopleths in that they do not separate different kinds of things. Furthermore, the lines of connection to the node usually run at right angles to the boundary of the region. In essence, a nodal or functional region is a multiple feature region.

Formal regions. Many of the early studies in regional geography had been concerned with *formal' regions,* identified on the basis of the presence or absence of particular distinguishing features. Hartshorne (1939) had prefigured the significance of the alternative concept of a functional region. R. E. Dickenson (1947) acknowledged the force of 'regional homogeneity' which underpinned the identification of formal regions, but argued that the fullest measure could be discovered only through analysis of the character, intensity, extent and interrelations of regional associations and the ways in which they are interlocked and separated from each other in space. Dickenson's formulation was its break from physical geography and physical environment and following from this its provision of a 'central regionalising principle that lies behind the spatial structure of society'. This was the *city region:* 'an area of interrelated activities, hindered interests and common organisation, brought into being through the medium of the routes which bind it to the urban centres'.

Dickenson's central concern with the human organisation of space was reinforced ten years later by A. K. Philbrick (1957), for whom 'the task of any intellectually viable human geography was to analyse the areal structure of human occupance independent of the natural environment, through an explication of the functional organisation of human occupance in area'. He fashioned a set of theoretical constructs—locality, localisation, interconnection and spatial discontinuity—which were supposed to provide a new framework for geographical inquiry. 'Both Dickenson and Philbrick had been greatly impressed by Christaller's Central Place Theory, and together these three men formed a spring board for the eventual translation of functional regions into spatial systems' (Gregory, 1986, 394).

'The emergence of several avowedly non-functionalists has led, in the 1980s, to a renewed interest in regions and in the reconstruction of a theoretically informed regional geography. Within Time-geography, for example, Hagerstrand's (1953/1967) emphasis on sequences of events which necessarily unfold in situations bounded in time and space and whose outcomes are thereby mutually modified through their common "localisation", clearly presages a return to Kant's "physical classification" as a distinctive and a regional basis for geographical inquiry. Building on Hagerstrand's foundations, Thrift (1983) described the region as meeting place of Human Agency and social structure. For him, a region is made up of a number of different but connected settings for interaction—a particular intersection of different Locales—which help to structure the intricacies of interaction, the specificity of particular times and spaces, the sense of living as meeting' (Gregory, 1986).

To Giddens (1984), regionalisation 'is a notion that should be seen as having a major role in social theory. Regionalisation is best understood not as a wholly spatial concept but as one expressing the clustering of contexts in time and space. As such it is a phenomenon of quite decisive significance to social theory, on both a theoretical and empirical level. No single concept helps more to redress the misleading divisions between micro and macro sociological research; no concept helps more to counter the assumption that a society is always a clear-cut unity with precisely defined boundaries to it'.

Pred (1984) provided a similar highly schematic model of the *region as a 'process'*. In drawing attention to the 'contingency' of regional configurations, however, he also paralleled some of the substantive theses of Marxist Geography concerning the regional geography of the division of labour, layers of investment and uneven development.

Across a broad and rapidly advancing front, therefore, the region is coming to be recognised as an indispensable focus of any truly contextual theory, though not its single pivot, and as being constantly constructed, destroyed and reconstructed in changed form as a result of the interaction of local and non-local processes of social development.

APPROACHES TO THE STUDY OF REGIONS

Fochler-Hauke (1959) distinguishes five different approaches to Landschaftskunde which include: landscape morphology; landscape ecology; landscape chronology; regionalisation; and landscape classification (or systematisation). However, the first three of these branches demonstrate three different approaches to the study of regions.

1. **Landscape Morphology.** It is a form of regional geography which was particularly well developed in Germany during the inter-war periods. Otto Schluter, who played a central role in its development, stated as early as 1906 that geographers should consider the form and spatial structure created by visible phenomena on the surface of the Earth as their unifying theme. All human distributions of non-material character such as social, economic, racial, psychological and political conditions are excluded from this study as ends in themselves.

The French scholar Jean Brunhes (1910) said: 'We should analyse the landscape and the characteristic interplay of observable phenomena there. The visible landscape is a result both of natural conditions and forces and a manifestation of the work of man. *The landscape itself creates a synthesis,* there is no longer a gap between physical geography and geography of man; both have the same object—the variable landscape—and also there is close contact in terms of methods.'

Alfred Hettner opposed Schluter's limitation of geographical study to the visible landscape. He was concerned with the 'uniqueness of area', whether this uniqueness was evident in the visible landscape or not. He, however, recognised the focal interest of landscape, but refused to recognise the limits set by it on the study of the human facts in space.

Despite this disagreement, landscape morphology can be seen as an attempt to study region on the basis of Hettner's principles. He had provided the German geographers with a philosophical basis for geography as a distinct discipline, but no clear-cut practical methods of enquiry. Schluter provided a practical approach which proved to be useful in field studies everywhere. The landscape morphologists agreed with Hettner on the main philosophical issue, that geography is a chorological science which should focus on regional synthesis and only delve into historical developments to the extent necessary *to* explain the contemporary situation.

2. **Landscape Ecology.** It is less familiar, but it represents *functional approach* in regional geography. The functional approach has grown up to study the relationships between centres

and their umland, i.e. the well-developed theme of functional or centred regions. Landscape ecology is also concerned with the relationships which occur within a region in terms of transport, contact fields, etc. Clearly a functional region does not need to restrict itself to the study of centred regions, even if this has come to be a most important theme in recent geography. It may also study the total interplay between the factors in an ecosystem whether dominated by man or by nature.

3. **Landscape chronology.** It manifests a major concern for the development of regions over time. The same category includes a type of regional geography which was developed in North America between the wars, described by Whittelesey in 1929, as 'sequent occupance'. *Sequent occupance* studies the ways in which each culture uses a region in its own way. This is demonstrated in America where most regions experienced a sudden change from the Indian to the European culture, and in many parts of Europe which progressed from agrarian to industrial cultures. Sequent occupance stresses the stages in the development of a region, not through studies of local differentiation as a result of the long interplay of man and nature over the centuries, but by emphasising how easily shifts in regional character can take place.

4. **Regionalisation.** The landscape consists of the observable features of any arbitrary part of the Earth's surface plus all non-observable features of significance. It consists of the landforms and their plant cover together with the fixed material features, which are the result of the transformation of the landforms and plant cover through human occupance. The German term 'Landschaft' which is literally analogous to the English word landscape also means a scientifically defined geographical region in German. Geographers have come to use landscape as being synonymous with 'region'. The region or landscape (Landschaft) is said to constitute a definite individual unit that has form and structure and is, therefore, a concrete object so related to others like it that the face of the Earth may be thought of as made up of a mosaic of individual landscapes or regions. The region is an organic object comparable to biological organism.

The regional concept refers to the mental image of an Earth's surface differentiated by an exceedingly complex fabric of interwoven strands and produced by diverse but interrelated processes. The region is thought of as a device for selecting and studying areal groupings of the complex phenomena found on the Earth. Any segment or portion of the Earth surface is a region if it is homogeneous in terms of such an areal grouping. Its homogeneity is determined by criteria formulated for the purpose of sorting from the whole range of Earth phenomena the items required to express or illuminate a particular grouping, areally cohesive. So defined, a region is not an object, either self -determined or nature-given. It is an intellectual concept, an entity for the purpose of thought created by the selection of certain features that are relevant to an interest or problem and by the disregard of all features that are considered to be irrelevant.

Nevertheless, the region has a form, structure and function; and region, being identical with landscape, it intrinsically refers to itself as 'a reality, as a whole, an objective unit and a corporeal thing, as well as a concrete unitary object' (Holt-Jenoon, 1981, 24). The region is unique in respect of its tool combination of major characteristics. *The distinctiveness and/or the uniqueness of the region* can be expressed in three basic concepts: (a) the interrelation of different kinds of phenomena that are directly or indirectly tied to the earth; (b) the differential character of these phenomena and the complexes they form in different areas of the Earth; and (c) the areal expression of the phenomena of complexes. On account of its individuality, personality and uniqueness, the region is suggestive of a primary whole, to which the German word

Gestalt is used. Gestalt is expressive of the unified form of larger areas, each consisting of a particular arrangement of regions. It is a dynamic structure in which the parts reach into each other functionally and can be understood only in view of the whole.

5. **Landscape Classification.** We can distinguish between *'natural landscape'* and *'cultural landscape'*: the former consisting of all the natural features of an area, the latter of all man-made features. The cultural landscape means the present landscape of any inhabited region. The natural landscape, however, would be found in an area never disturbed and invaded by man. In the inhabited areas of the world, the concept of natural landscape is purely theoretical; throughout the civilised world at least the natural landscape has been seen by no living person. The Germans distinguished between 'Urlandschaft' (the landscape that existed before major changes were introduced through the activities of man) and 'Kulturlandschaft' (landscape created by human culture).

There are two principal types of systems of regional division, both of utility in different ways. A realistic system of 'specific' regions is based on consideration of all factors involved, including relative location in reference to land and sea. A comparative system of 'generic' regions considers only the internal characteristics of the areas without respect to relative location; strictly speaking, it does not establish regions, but simply areas of particular types. Natural regions are actually regions based on a combination of certain natural factors, according to their importance to men of particular culture and technology. It is the fundamental unity in man's ethno-cultural traits that leads to areal assodation of cultural element-complexes in the space that tends to produce cultural regions. Cultural regions are required to be determined by finding the areas in which people think alike and maintain a common *genre de vie.*

There can be the uniform or formal region and the nodal or functional region. The former is defined in terms of specified criteria and homogeneity, whilst the latter is an area that is tied functionally to a node or to several nodes. A functional region intrinsically constitutes a *Whole* or a *gestalt.*

Nevertheless, the regional entities which are being constructed on the above criteria and principles, are in the full sense mental constructions. They are entities only in our thoughts, even though we find them to be constructions that provide some sort *of* intelligent basis for organising our knowledge of reality.

17

Geography: A New Synthesis

HIGHLIGHTS

- Empiricist Philosophy of Regional Geography
- Scientific Explanations/Analysis
- Trends Towards 'A New Synthesis'
- Geographical Knowledge Under Globalisation (Post-Modern World)
- Cosmopolitanism and Its Geography
- Sites for the Production of Geographical Knowledge
- Structure of Geographical Knowledge
- Geography among the Disciplines
- Political Projects

Geography is concerned with the development of knowledge of the Earth as the home of humankind, focusing on the physical environment, on the interaction between physical environment and human society, and on the spatial organisation of that society. This definition of geography tends to focus on the idea of synthesis. The view that geography is a synthesising discipline has always been basic to its philosophy. It is the raison d'etre of geography. A *belief in synthesis has been the teleology of geography,* the purpose that justified the activities of geographers (Holt-Jenson, 1981, 110).

The discussion on the philosophy and methodology in this chapter will tend to focus on the idea on a new synthesis. The western geographers were much concerned with the multitude of philosophies and methodologies, and at last some form of unanimity was achieved in the twentieth century.

EMPIRICIST PHILOSOPHY OF REGIONAL GEOGRAPHY

Prior to the 1950s, geographers adopted an empiricist philosophy, which places emphasis on depiction of the observed world. It was a discipline that provided descriptions of places. Further, each separate place was presented as singular, as an assemblage of phenomena into a whole (an idea of synthesis implied) which was not repeated elsewhere. The task of geography was to identify those separate assemblages, describe them and account for them. The debate over this philosophy focused on the singularity of regions. However, regions are not singular, but they are unique. A unique phenomenon results from a particular interaction of general principles and so can be accounted for by understanding what those principles are and how they have been combined in that particular instance. A singular phenomenon is one that cannot be accounted for by any general principles.

It was argued by the British geographers in the 1950s that it was possible to develop general laws regarding the elements of the physical environment, which could be combined to provide accounts of the environmental complex making up a region. *Thus, systematic geography and regional geography were to be seen as complementary,* with the former providing the general understanding that could be synthesised in the latter when it referred to specific unique places.

In the United States, the protagonists opposing the exceptionalist view offered two sets of arguments: that it was possible to identify laws of spatial form, i.e. arrangement of phenomena on the surface of the Earth; and that it was possible to identify general laws of human behaviour which produce those spatial forms. The wide acceptance of these two sets of arguments meant that regional geography led to the development of systematic geographies as the origin of its inputs. Regional geographers are those who synthesised the findings of geomorphologists, climatologists, urban geographers, and so on into an explanation of the nature of a place.

By the 1950s, disillusionment with the empiricist philosophy of regional geography was growing. Slowly the systematic specialism aroused the interests of the new generation of researchers, and the regional synthesis was ignored. The philosophy adopted in change-over or shift was positivism. Associated with the philosophy of positivism is a particular methodology, i.e. the scientific method. The positivist methodology is a continuous looping sequence of thinking, speculation and testing.

Much geographical works in the 1960s focused on the testing, on the empirical validation of hypotheses. The emphasis was on measurement of the phenomena under consideration and of their empirical coincidences, and on the use of statistical inference to establish the veracity of the hypotheses. Thus, the development of positivism was widely known as the quantitative revolution. Geography came to be known as *a law-seeking discipline,* using the procedure of the natural sciences and the technology of quantification. The common method is the hypothetic-deductive method and the model discipline is physics. The positivist-inclined geography does not seem to have abandoned the idea of synthesis. Instead, it puts emphasis on the 'unity of phenomena'.

SCIENTIFIC EXPLANATIONS/ ANALYSIS

From the positivist-led philosophical, methodological and empirical studies, Harvey (1969) attempts to identify six recognisible forms of scientific explanations in geography in the form of general and empirical laws. They are cognitive explanation, morphometric explanation, cause-effect explanation, temporal modes of explanation, functionalism and functional modes of explanation, and systems analysis.

1. Cognitive explanation. It is the simple description of what is known, resulting from a more or less successful ordering and classification of the data which have been collected. No theory is involved explicitly but, because the classification usually follows some pre-determined ideas about its structure, this involves an element of theory.

2. Morphometric explanation. It is a special form of cognitive description where systematisation and classification form a geometric, spatial, coordinate system. Morphometric analysis can lead to certain types of predictive and stimulation models. In this explanation, the emphasis is on measurement, whereas studies of landscape morphology usually take the form of cognitive descriptions.

3. Cause and effect explanation/analysis. It develops from the assumption that previous causes can explain observed phenomena. We look for causal relationships which are, in their simplest form, of the type 'cause A leads to effect B'. This implies that 'cause B cannot lead to result A'. Causal laws may be discovered by the hypothetic-deductive method, or more simply, by comparing data from different phenomena in a region. Cause-effect explanation is useful for the analysis of geographical problems.

It can be used to analyse simple phenomena, to regulate associations, to construct theories, to formulate laws, and so on. But causal explanation has its limitations. It may be applied with a high degree of success in a large number of situations but, at the same time, we must be cautious that the causal explanation is not the only basis for scientific analysis and explanation. Causal analyses provide only one of the several possible research tools, and cause-effect laws provide only one form of explanation of reality.

4. Temporal modes of explanation. They provide scientific procedures for describing or explaining phenomena in relation to their development over time. Temporal explanation tends to form a distinct type of causal analysis. History can be seen as a causal series which began at the vaguely defined dawn of history and ends today. It is better termed as a historical analysis. In practice, it will never be possible to understand such a comprehensive causal series; the analysis must be restricted to some determined period of time.

The assumption in the explanation is that a particular set of circumstances may be explained by examining the origin and subsequent development of phenomena by the operation of process laws (Harvey, 1969). Temporal modes of explanation are very common in studies of geomorphology and historical geography. Holt-Jenson (1981, 111–12) points out that 'although indepth studies by geomorphologists in recent years have brought about a fairly comprehensive understanding of the processes which govern the formation of landscapes, human geographers have made little progress in the formation of laws which govern processes.'

5. Functionalism and functional explanation. Functionalism is a perspective which views the world as a set of differentiated and independent systems. Their collective actions and interaction are instances of repeatable and predictable regularities in which form and function can be assumed to be related. They explain these form-function relations in terms of their role in maintaining the continutity of systems (Bennett and Chorley, 1978). While much of empirical work in geography could well be construed as functionalism in form, its precepts tended to remain implicit rather than overt in (traditional) geographical thinking (Harvey, 1969).

The positivist-led geography in the 1960s stressed the need for functional analysis as an alternative to the cause-and-effect explanation. Functional explanatory models are very common in geography. It is often assumed that teleological and functional forms of explanations are closely related to each other. What is interesting is the close association between the traditional view of geography as a discipline seeking synthesis and the functional-teleological analyses and forms of explanation. It has been regarded as logically sound to make the synthesis (or wholeness) our goal, so that each phenomenon should be analysed with a view to explaining its purpose or function within the whole. However, a teleological explanation relates to the purpose of a phenomenon and a functional explanation refers to the functions, and functional analysis and explanation is contained within teleology (Holt-Jenson, 1981).

Functional analysis/explanation attempts to analyse phenomena in terms of the role they play within a particular organisation. In most cases, functional analysis only demonstrates the necessary location conditions for a town or factory to flourish. Harvey (1969) takes a central place or market which served the needs of an economy for the efficient exchange of goods and services. Functional analysis can be useful in enabling us to discern some of the conditions necessary for the phenomenon to function within a given system.

Functional mode of explanation may also be a useful point of departure for the formulation of individual theories and, perhaps more significantly, for research into alternative methods for the study of complicated systems and the structures of organisations. It is also useful to know how central places function in an economy because this question raises a range of other questions and draws our attention to the complexity of systems within which researchers work. Useful derivation from functional analysis depends on the extent to which we can specify a system clearly. This is why functional analysis has been so ardently discussed by geographers in recent years (Holt-Jenson, 1981).

In the 1970s, as human geography was once again brought much closer to the other social sciences, functionalism advanced on a broader front. Particularly important was the rapid incorporation of ideas both from classical and structural Marxism, and from somewhat different structuralism. However, the 1980s, witnessed a series of debates on the relation between functionalism and Marxism. It is now widely held that the functionalist model requires a new level of sophistication if it is to survive the more general critique of simple functionalism in social theory.

6. Systems analysis. It is a methodological framework for investigating the structure and function of a system. Harvey (1969) maintained that systems analysis could be distinguished from any metaphysics based on general systems theory because it is a methodology pure and simple. But in practice there must be a reciprocal relation between the concepts made available by systems theory and their translation into the methods of system analysis. Systems analysis provides a convenient calculus for examining geographical problems, particularly since these are frequently of a multivariate nature. But its use cannot be independent of the concepts that allow us to find an interpretation for that calculus.

Much of the geographical works around 1970 made use of general systems theory, and the application of systems analysis has served to highlight some of the impending methodological difficulties and problems that the geographers, adhering to positivism, have faced. Systems analysis has been used as a methodological tool to build a new form of 'geographical synthesis'.

TRENDS TOWARDS 'A NEW SYNTHESIS'

Haggett (1972, 1979) attempted to develop 'a new form of synthesis', which proceeds from the traditional division of the subject. To him, it would be more useful to divide the discipline in relation to the way in which it attempts to analyse its problems. He divided geography into three main groups of analysis:

1. Spatial analysis deals with the variation in the localisation and distribution of a significant phenomenon or group of phenomena. It is a type of systematic geography which seems to have been developed as a result of the positivist/spatial science tradition.

2. Ecological analysis concerns itself with the study of connection between human and environmental variables. It is based on the concept of homogeneous region, that each phenomenon within it is similar, but differs from equivalent phenomenon of other regions. Ecological analysis appears to have close affinity with chorology and regional geography which seeks to understand the region as an entity.

3. Regional complex analysis attempts to combine the results of spatial and ecological analysis. Appropriate regional units are identified by real differences. Connecting lines and flows between the individual region may then be observed. Complex regional analysis concerns itself chiefly with functional regions.

'Modern geography', points out Holt-Jenson (1981), 'attempts to focus its attention on spatial analysis, a systematic geography constructed with newer, expanded models; ecological analysis, a regional geography based on homogeneous region, and a regional geography based on functional regions. This simple division may have some relevance for the further development of geography towards a science of synthesis.'

Soviet geographer Anuchin (1973) also stresses the need to have synthetic geographical studies. To him, the existing main branches in geography, inspite of their relevance, are completely unsatisfactory when one is concerned with the evolution of the variables involved in regional complexes of the government which control the possibilities of the development of production. The results of synthetic geographical studies would provide practical forecasts of consequences of interference with natural processes, which are taking place inevitably. What is interesting is the justification in asserting that geography can only add something useful to research in the future if it can make the geographic synthesis work.

While geographers in the 1960s and 1970s explored the positivist philosophy and associated scientific method, others extended the search to other philosophies, other definitions of knowledge and other scientific methods. Quite a large number of geographers inclined towards critical theory, discarded what they called 'neo-classical economics', and sought for a new society through the development of a radical geography based on historical-dialectical methods and Marxist theory.

Other critics of the positivist-quantitative geography attempted to establish humanistic geography which focuses on the experience of place and environment and its methods involve means of identifying and reporting these meanings. It is thus opposed to the type of positivist work which emphasises regularity and order, and which seeks to explain any event or piece of behaviour in terms of class events. Such work denies the individuality and humanity of people, ignores emotions and meanings, and treats humans in the same way as machines (Ley and Samuels, 1978).

Geographers committed to critical theory also stress on the unity of science. Marx also wrote about a unified science which consisted of nature, society and human psychology. But while positivists maintained that a unified science should be based on the methods of the natural science, Marx considered the philosophy of the social science to be potentially far superior to that of the natural sciences. The eventual fusion/amalgamation of the two fields of study would come about through the socialisation of the natural sciences (Harvey, 1973). William Bunge (1973), known for his commitment to critical theory, observed that 'geography is the integrating science so we call upon coworkers in geology, sociology and so forth to discuss planning for a region or even the lesser labour of just understanding a region with no ambitions humanly to improve it.'

Positivist-led quantitative geography or critical geography has not in fact given up the idea of synthesis. Instead, many of its leading advocates have been specially concerned in a search for synthesis through new forms of analysis. Modern geography can, however, develop methods which can be used in the formation of regional synthesis. The synthesis needed, especially in planning, would provide a basis for the development of the discipline. Geographers should try to understand the synthesis, revealing the wholeness (Holt-Jenson, 1981).

It is the dichotomies in geography which tend to weaken the very raison d'etre of synthetic geography, or attempt to weaken the geographical synthesis. James (1972) pointed out that acceptance of the many dichotomies is a semantic trap and that the attribution of fixed meanings or interpretations to word symbols may result in unreal conflicts between them. He suggested that *the following dichotomies have done particular damage* to the raison d'etre of geographical synthesis and geographical thinking:

1. that geography must either be idiographic or nomothetic, but not both;
2. that physical and human geography are clearly differentiated branches of the discipline with separate concepts and methods;
3. that geography must be either systematic or regional;
4. that geographical methods must either be inductive or deductive;
5. that geographical philosophies and methodologies must either be positivist or critical theory;
6. that geography must be classified as either a science or an art (the quantitative-qualitative argument).

For Holt-Jenson (1981), geography straddles all these dichotomies, and tends towards synthesising the discipline. Geography, like any other discipline, must be both idiographic and nomothetic. There is a continuous transfusion from one to the other, and most research consists something of both. An integration of physical and human geography is one of the dominant functions of geography justifying the existence of the synthetic character of the discipline. The systematic and regional geography are to be seen as complementary, with the former providing the general understanding that could be synthesised in the latter when it referred to specific and unique places. Many geographers have regarded regional geography as the core of the subject, viewing systematic geography as the area in which laws are formulated and regional geography as the field in which they are tested empirically. The culmination of regional geography there becomes the verification of geographical laws and the presentation of a synthesis of the physical and human phenomenon within an area or region. Both induction and deduction are generally used in the same scientific analysis.

The new critical geography shows some of the same characteristics as typical of scientific revolution in general and the positivist quantitative revolution in particular. Dichotomies between positivism and critical theory do not necessarily seem to split the very synthesis of the discipline. Geography attempts to synthesise between science and arts, for it is neither a purely natural science nor a purely social science; it lies at the median of the two. If any of the traditions is taken out, geography will then cease to exist.

Geography today represents a well-established tradition of research and scholarship within the academic discipline. It is one in which the legacy of past ideas remains embedded, but current thinking is still being worked out.

GEOGRAPHICAL KNOWLEDGE UNDER GLOBALISATION (POST-MODERN WORLD)

The all-encompassing political-economic process which has come in recent years to be called 'globalisation', has depended heavily upon the accumulation of certain kinds of geographical knowledge. The further development of this political-economic system will undoubtedly influence geography as a distinctive discipline and as a distinctive way of knowing what permeates social thought and political practice. Reciprocally, geographical understanding may affect future paths of political-economic development (through, for example, the recognition of environmental constraints, identification of new resources and commercial opportunities or the pursuit of juster forms of uneven geographical development). A critical geography might go so far as to challenge contemporary forms of political-economic power, marked by hyper development, spiralling social inequalities and multiple signs of serious environmental degradation.

The aforesaid dialectical relationship between political-economic and socio-economic change on the one hand, and geographical knowledge on the other, however, requires to be looked into with the following three basic observations:

First, the history of this dialectical relationship is a fascinating area of inquiry (for example, in the whole relationship between geographical knowledges, state formation, colonisation, military operations, geopolitics and the perpetual seeking out of commercial and economic advantages). The past legacy, however, weighs heavily upon contemporary geographical knowledges and any broad-based attempt to transform the latter must, at some point, confront the particularities of past achievement.

Secondly, it is dangerous to presume that there is some settled way of understanding or a unified field of knowledge called 'geography' even within the academy. A discipline that ranges from palace-ecology and desert morphologies to post-modernist and queer geography obviously has an identity problem. The presumption that there is some yet to be discovered 'essentialist' definition of geography's subject-matter, its methods and its 'point of new', has to be challenged, though it is a long time since any one dared write so confidently a book as Hartshorne's *Nature of Geography*.

Thirdly, the strategic point which is even more important: there is a significant difference between geographical knowledge in different institutional settings (e.g. state apparatus, the World Bank, IMF, the media, the public at large, NGOs, the tourist industry, multinational corporations, transnational organisations, financial institutions, etc.) and the geography taught and studied within departments that operate under that name. The tension between geography as a distinctive discipline and geography as a way of assembling, using and understanding information of a certain sort in a variety of institutional settings is important. Geographical knowledge of the latter sort are widely dispersed throughout society. They deserve to be understood in their own right (e.g. how the tourist industry or cable television has created and promoted a certain geographical sense in society). Different institutions, furthermore, create a demand for different kinds of geographical knowledge (the tourist industry is not interested in highlighting the geography of social distress). If academic geography does not or cannot meet these various demands then someone else will surely do.

From these preliminary remarks following immediate conclusions can be drawn:

1. Need for general studies in comparative historical and geographical settings to better understand how the dialectical relationship between forms of geographical knowledge and socio-economic and ecological development occurs.
2. Need for careful studies of how geography as a mode of understanding is formulated, used and applied in different institutional settings (e.g. the military, green peace, the state apparatus, multinational corporations, etc.)
3. Need to better understand the links between geographical discourses that emanate from particular institutions and the way geographical knowledge is created and taught both within and without the specific discipline of geography.
4. Need to think through the principles that might govern the 'proper' application of 'sound' geographical knowledge in specific settings. Here geography has a potential role of considerable importance, as both arbiter and judge of appropriate uses of properly formulated geographical knowledge.

COSMOPOLITANISM AND ITS GEOGRAPHY

It is now claimed that global governance, management and regulatory activities are now being mobilised through ideals of 'cosmopolitanism'. Writers like David Held have argued eloquently that such a cosmopolitan perspective is essential to the development of democratic institutions of global governance to regulate neo-liberalism. But what kind of geographical knowledge is presupposed in such an argument?

In order to conduct any adequate global dialogue, Nussbaum says, 'We need knowledge not only of geography and ecology of other nations—something that would already entail much revision of our curricula—but also a great deal about their people, so that in talking with them we may be capable of respecting their traditions and commitments. Cosmopolitan education would supply the background necessary for this deliberation.' Cosmopolitanism without a 'sound' and 'proper' understanding of geography and anthropology is an empty ideal.

David Harvey (2002), however, says, 'The assertion of Nussbaum follows no less a figure than Kant whose founding arguments on a cosmopolitan ethic are frequently appealed to in the general literature. Kant recognized both geographical and anthropological understandings as 'necessary preconditions' for the discovery and application of all other forms of knowledge including that of a cosmopolitan ethic.' However, a study of Kant's geogrpahy reveals a serious problem. For not only is Kant's account unsystematic and incoherent but it is also prejudicial in the extreme. 'Humanity', Kant says, 'achieves its greatest perfection with the whole race'. Geographical knowledge of this sort appears deeply inconsistent with Kant's universal ethics and cosmopolitan principles. It immediately poses the problem: What happens when universal ethics get inserted as principles of global governance in a world in which some people are considered inferior and others are thought indolent, smelly, or just plain untrustworthy? It inevitably leads to 'prejudices' of unspecified dimensions causes, 'humanised space' disaggregated and discriminated against. Biased or 'empty' geographical knowledege, deliberately constructed and maintained, provide a licence to pursue narrow interests in the name of universal goodness and reason.

A hefty dose of geographical enlightenment is, therefore, a necessary pre-condition for any kind of reasoned global governance. But what kind of geographical knowledge might be implied here? Geographers tend to be suspicious of cosmopolitan ideas (in part for good reason). But geography, uninspired by any cosmopolitan vision, either becomes a matter of mere description or a passive tool of existing powers (military, administrative, economic). Liberating the dialectic between cosmopolitanism and geography seems a critical pre-condition for the achievement of any juster and saner socio-ecological order for the present century. How can geographical knowledge be reconstituted to meet the needs of democratic global governance inspired by a consmopolitan ethics, for example, justice, fairness and reason? These are big questions, but essential to contemplate not only from the narrow standpoint of geography as a discipline, but more importantly from the standpoint of geographical knowledge (no matter where produced) in affecting future trajectory of the global socio-ecological order and its associated patterns of political-economic power.

SITES FOR THE PRODUCTION OF GEOGRAPHICAL KNOWLEDGE

Professional geographers, like economists, sociologists and political scientists, do from time to time generate their own data sets and produce novel information to fuel their enquiries. But much of their work rests on the analysis of data, information and perspectives developed elsewhere. There is very little formal recognition within geography of how geographical knowledge assembled in different institutional settings vary according to distinctive institutional requirements, culture and norms. If geography as a discipline aspires to be judge and arbiter of the proper application of sound geographical knowledges, then a first step down that path is to provide principles to evaluate the production of geographical knowledges in different institutional settings. Many geographers attach themselves to external institutions. But this is often viewed as a private or personal matter. Rarely do we sit back and reflect upon the consequences of such attachment for the discipline as a whole. Consider, for example, some of the primary sites for the production of geographical knowledge and how the qualities of such knowledges vary from site to site.

1. The State Apparatus, with its interests in governmentality, administration, taxation, planning and social control, has steadily built up from the eighteenth century onwards as a primary site for the collection and analysis of geographical information. The process of state formation was and still is dependent on the creation of certain kinds of geographical understanding (every thing from mapping of boundaries to the cultivation of some sense of national identity within those boundaries). For the last two centuries, the state has been perhaps the primary site for the production of geographical knowledge necessary for the creation, maintenance and enhancement of its power. However, the interests of particular states lead to particular kinds of geographical knowledge related, interestingly, to geographical and geopolitical conditions. The 'hidden geography' of geographical knowledge has rarely been expressed except elliptically and occasionally.

2. Military Power deserves to be categorised separately because it is in this arena that correction between privileged geographical knowledge and the pursuit of power becomes most obvious. Geographical knowledge is here often held in secret. Access to it is a matter of national security. The connection between geographical knowledge and the military has always been extremely strong.

3. **Supra-national Institutions** have increasingly become major sources of new geographical knowledge, particularly since 1950. The World Bank, the UN Development Program, the ILO, the WHO, the WTO, UNESCO, FAO, and the like form a huge and rapidly growing domain for the production of variety of geographical knowledges (often of a specialised sort on topics such as world health, agriculture, labour, the environment and the like). Other supra-national institutions, like the European Union, the OECD, NAFTA, APEC, and other such trans-continental institutions take less of a global perspective but, nevertheless, also operate as key sites for the production of particular geographical knowledge.

4. **Non-governmental Organisations** have proliferated in recent decades, making the production of geographical knowledges throughout civil society at large a much more complicated affair in part because the objectives of such organisations vary greatly. Such organisations incorporate vast amount of geographical knowledge as do human rights groups, like Amnesty International, environmental groups (like the World Wildlife Fund or Green Peace) and the vast array of organisations dealing with specific issues (violence, the situation of women and children, education, poverty, health, refugees, etc.)

5. **Corparate and Commercial Interests** have their own way of assembling and analysing geographical knowledge for their own particular purposes. The vast business of consultancy today operates with particular force as corporate and commercial interests seek out expert opinion on marketing possibilities, locational preferences, resource availability (both human and natural), environmental constraints, security of investment, business climate, amenities for personnel, and the like.

6. **Media, Entertainment and Tourist Industries** are a prolific source of geographical knowledge. In this instance, however, we are largely concerned with the projection of images and representations upon a public at large and the predominant effects of those images and representations upon the population subjected to them. The impact is primarily aesthetic and emotive rather than 'objective'. The selectivity entailed in the choice of images is often problematic. But the variation in images and representation within the media, entertainment and tourist industries is enormous and it forms a highly problematic, but influential field within which geographical knowledge get shaped and reshaped.

7. **Education and Research Institutions** generate a lot of disciplinary-specific geographical information. Economists, sociologists, anthropologists and political scientists all produce and modify information that has geographical content and often re-shape that knowledge to their own disciplinary purposes. Geographical knowledge is found throughout the whole educational and research system. It is quite proper that such knowledge become widely diffused rather than circumscribed within one unified disciplinary frame. This can be perceived within the discipline of geography either as a threat or as a marvelous opportunity to engage in constructive dialogue about the proper use of sound geographical knowledge in many distinctive spheres of endeavor.

Institutionalised geographical knowledges of the sort, as mentioned above, are particularly important to geography as an academic discipline. But there are far wider and more general kinds of geographical knowledge embedded in language, local ways of life, the local symbiosis achieved between nature, economy and culture, local mythologies and diverse cultural practices and forms, commonsense prescriptions and dynamic socio-linguistic traditions. Specialised geographical knowledge, however, abound. Geographers along with anthropologists have paid close attention to localised 'structures of feeling' and ways of life and in doing so have helped frequently to highlight the conflict between institutionalised knowledges directed towards governmentality and localised knowledge that guides effective loyalties and socio-environmental identities. They have traditionally provided and continue to provide the backbone of argument for an authentically independent discipline of (human) geography.

It is years now since Foucault taught us that knowledge/power/institutions lock together in particular modes of governmentality, yet few have cared to turn that spotlight upon the discipline of geography. They have been unmindful of Foucault's other key observations on the importance of discipline, surveillance and punishment to the functioning of all institutions. Nevertheless, it requires to be recognised that geography as a discipline is situated at the confluence of a vast array of geographical discourses, constructed at quite different institutional sites with often seemingly incomparable and incommunicable rules of operation. We should view the confluence of these divergent discourses within the discipline of geography as an opportunity and an advantage rather than as source of mystification and confusion.

The extraordinarily diffused presence of geographical knowledge across the different disciplines and their dispersal throughout many major institutions provide a ready-made network for the diffusion of 'strong' geographical ideas, constructed within the discipline. Instead of geography weakly retracting institutionalised discourses, it is possible to imagine geography as a discipline sending strong innovative impulses throughout the academy and across multiple institutional sites based upon the collective work which geographers produce. However, geography will not survive unless it develops 'strong' geographical ideas expressive of some of the 'unities' that we come to identify among the highly differentiated discourses that converge within our disciplinary frame. 'Strong' ideas will be listened to and command respect elsewhere. And those strong ideas must be born out of experience gained through the specific personality of our discipline as a convergent point of multiple geographical knowledges.

STRUCTURE OF GEOGRAPHICAL KNOWLEDGES

The structural components of geographical knowledge necessarily identify the unities that underlie highly diverse geographical knowledge and suggest foci around which 'strong' ideas may cluster. Four structural elements stand out.

1. Cartographic Identification. Map making and cartography have been central to the history of geography. Maps have always been and continue to be created and used in an extraordinarily wide range of institutional and disciplinary settings and for a variety of purposes. Cartography is, plainly, a major structural pillar of all forms of geographical knowledge. Given its fundamental role in geography as well as in other institutional settings, it provides one thematic point of convergence from which 'strong' ideas about the role of geographical knowledges may be derived.

Cartography is about locating, identifying and bounding phenomena and thereby situating events, processes and things within a coherent spatial frame. It imposes spatial order on phenomena. In the contemporary manifestation, it depends heavily upon Cartesian logic in which *res extensa* are presumed to be quite separate from the realms of mind and thought and capable of full depiction within some set of coordinates (a grid or graticule). The innovation of thematic, synoptic and iconic maps extended the range of what could be represented cartographically in important respects (synoptic charts in meteorology and climatology becoming basic tools for analysis, for example). Cartographic operations can be found right throughout the academy at the same time as they are fundamental to the work of many institutions (the state, the military, the law, etc.). Information is now often stored digitally and in GIS there exists a powerful tool for automated storage, analysis and instantaneous presentation of data and information in an ordered spatial form.

Cartography has opened up the possibility for the 'rational' organisation of space for capital accumulation, the partition of space for purposes of efficient administration or for the pursuit of improvements in the health and welfare of population. The history of cartography is now also being written from a broad-based comparative perspective, revealing much about cultural and temporal differences in understanding of human positionality in the world. The evaluation and historiography of cartographic forms is well under way by geographers, historians, anthropologists and a wide range of scholars from other disciplines.

The literature on 'cartographic consciousness' on 'mental' and 'cognitive' maps is now growing rapidly, suggesting an emergent field of enquiry that links thematics in geography with much of cultural and literary theory. How urban life is experienced and practised, has much to do with how we form and reform mental maps of the city, for example. Cartography, thus, provides one central structural support to all forms of geographical knowledge and is made up of many intertwining threads. It is a central pillar of all forms of geographical knowledge and deserves thorough consideration as such.

2. The Measure of Space-Time. Maps have traditionally taken the form of two-dimensional spatial representation. They rest, therefore, upon a certain conception of space and an ability to order and locate position, things and events in that space through precise measurements. Ways of representing, understanding and shaping space appear as an element common to all forms of geographical knowledge. Here, too, we encounter a commonality, a unity, within which there exists a whole world of difference.

Many geographers· now claim that 'space' is the central, privileged and even defining concept of their discipline. However, this claim appears to be rather far-fetched and potentially misleading. Most of the physical sciences (physics and cosmology in particular) and engineering have a long history of dealing with the concept of space (and space-time) and it has likewise been the object of extended reflection in philosophy, literature, anthropology and many of the social sciences. So, while the concept of space may be central to the discipline of geography, it is in part received, like cartography, as vector of multiple discourses about space, many of which emanate from elsewhere even as they converge within the discipline. To put it this way is not to imply that there is nothing new to be discovered or thought about space (or space-time) within geography or that there is no indigenous tradition to which we can appeal. Indeed, the recent convergence of multiple discourses about space and space-time within geography makes this a key point from which 'strong' ideas have emerged.

To the degree that time, space and matter (or process) are fundamental ontological categories in our understanding of the world, geography internalises within itself the same problematic as the other disciplines. Questions of the absolute, relative and relational conception of space (and time) are posed as is the issue of whether or not time can meaningfully be separated from space. 'Space-time' or 'spatio-temporality' is, therefore, the relevant category. This quite properly implies that all geography is historical geography, no matter where it is to be found.

The importance of this dynamic conception of spatial ordering and spatial form will shortly become apparent. Without it, geographical knowledge tends to become dead and immovable structures of thought and understanding when their most exciting manifestation invariably comes from observing them (or even setting them) in motion. All the elements of spatial structure become integral to our understanding of how phenomena are distributed and how processes work through and across space over time. It is just important to see the spatial frame itself as malleable and variable (relative and relational), as an actively produced field of spatial ordering that changes sometimes quickly and sometimes glacially over time. Space must be understood as dynamic and in motion, an active moment rather than a passive frame, in the constitution of physical, ecological, social and political-economic life.

Space, like cartography, is as much a mental as a material construct. This is so not only in the sense that the measurement systems and the mathematical constructs (geometries and calculus) that are used to represent spatiality are products of human thought. The spatial and temporal imaginary, the construction of alternative possible words and the senses of space and time that course through consciousness and which present themselves in works of art, poetry, novels, films and multi-media form—all of these provide a vast array of metaphorical meanings with which it is possible to explore hidden connectivities and analogies. So-called 'mental' or 'imaginary' space and time are rich terrains through which to work in order to understand personal and political subjectivities and their consequences when materialised as human action in space and time.

3. Place/Region/Territory. The 'region' is possibly the most entrenched of all geographical concepts. Within the discipline it has proved the least flexible, mainly because of its central role in those essentialist definitions of the subject which rest exclusively on the study of chorology or regional differentiation. Terms like 'locality', 'territory' and 'place' have often been substituted for 'region' in geographical discourses, both within and outside the discipline. Extensive literatures on the 'local and the global', on 'deterritorialisation and re-deterritorialisation', and on the changing significance of 'place' under conditions of hyper-mobility across space, testify to the vibrancy of the topic and the diversity of conceptual apparatus with which it is approached.

The central idea is that there is some contiguous space that has the character of an 'entity' of some sort defined by some special qualities. Sometimes the boundaries are clearly demarcated (as with administrative territories), but in other instances they are left ambiguous or even unconsidered (many ideas of 'place' fail to deal at all with the activity of bounding where a place begins and ends). Sometimes the region is defined in terms of homogeneous qualities (e.g. of landuse, soils, geological forms) and sometimes in terms of coherent relations between diverse elements (e.g. urban functional regions). Sometimes region is defined in purely materialist terms (physical qualities of terrain, climatological regime, built environments, tangible boundaries), but in others it depends on ideas, loyalties, a sense of belonging, structures of feeling, ways of life, memories and history, imagined community and the like. In either instance, it is important to recognise that regions are 'made' or 'constructed' as much in imagination as in material form. The approaches to place/regionality/territory are wondrously diverse, no matter where they are found.

The scale problem also enters in, with a hierarchy of labels often deployed that begin with neighbourhood, locality and place, and proceed to the broader scale of region, territory, nation-state, and globe. Region then becomes territorialisation at a certain geographical scale. Scaling is not a problem unique to the social side of matters. Whatever the procedure or methodology, once continuous space gets carved out into distinctive regions of whatever sort, the pictures we form of and the operations we are enabled to conduct upon geographical information multiply enormously. Comparative studies of geographical differentiation and uneven geographical development become much more feasible.

Furthermore, as human populations frequently organise themselves territorially, regionality becomes as central to consciousness and identity-formation and to political subjectivity as does the cartographic imagination and the sense of space-time. Beyond the obvious cases of nation-state formation and movements for regional autonomy (much more prominent in recent years despite or perhaps because of the forces of globalisation), the general processes of political articulation resting on everything from community boosterism to 'not in my backyard' politics transforms the world into complex regional differentiations, inter-regional relations and rivalries. Geopolitical struggles between territories and regions have, therefore, been of considerable importance in geographical understandings. By force of historical circumstance, all national liberation movements must also define themselves geopolitically if they are to succeed, turning the geography of liberation into geopolitical struggle.

But it is not only the interaction between geographical entities that need to be treated in a dynamic way. The processes of region formation are perpetually in flux as social and natural processes re-configure the Earth's surface and its spatially-distributed qualities. New urban regions form rapidly as urban growth accelerates, climate change generates shift in biotic conditions, water regions and the like. Populations shift their perceptions and allegiances, re-invent traditions and declare new regional formations or racially transform the qualitative attributes of the old. Like space-time and the cartographic imagination, the dynamics of the process are by far the most interesting. Regionality, the dynamics of place and space, the relationship between the local and the global, are all in flux, making the uneven geographical development of the physical, biotic, social, cultural and political-economic conditions of the globe a key pillar to all forms of geographical knowledge.

4. Environmental Qualities and the Relation to Nature. All societies develop means to evalutate, appreciate, represent and live within their surrounding environments. Local knowledges concerning the uses of various processes and things, the appreciation of the qualities of local fauna and flora (indigenous resource knowledges), of changing meteorological and climatic conditions, of soil types, of natural hazards, the construction of symbolic meanings and the development of capacities to represent and 'read' the landscape and its signs effectively—these sorts of knowledge have been fundamental to human survival since time immemorial. The nature of such knowledge varies greatly, depending upon technologies, social forms, beliefs and cultural practices all of which instantiate a certain view of the relationship of human life to life and nature in general.

The question of how people do and should understand the relationship to environment and nature, forms the fourth pillar to all forms of geographical knowledge. But, as with the other structural supports, the issue is not unique to geography, but has a wide-ranging presence across all manner of other institutional sites. Thinking about it within geography has been strongly influenced by these external institutional needs.

The world was consequently understood as a spatially diversified bundle of 'natural' resources waiting to be discovered, exploited and transformed into systems of production of various sorts. Commercial geography reflected this trend. The accurate description of physical and biotic environments, of climate, soil and water regimes, of resource complexes and possibilities, largely for utilitarian purposes, has remained central to geographical endeavours ever since. This kind of geography was always profoundly materialist but often crassly and a-historically so. Close observation of geographical variations in ways of life, forms of economy and social reproduction has also been integral to the geographer's practice ever since merchant capitalism came to regard such knowledge as essential to its practices. Geographical knowledge was deeply affected by imperial and colonial practices coupled with the exploration of commercial opportunities and markets. The objectification and exploitation of Nature under capitalism went hand in hand with the objectification and exploitation of peoples. Many forms of geographical knowledge were complicitous with that politics.

The dialectics of socio-environmental change can, however, take many twists and turns and be understood from a variety of perspectives. The long history of 'environmental determinism', a doctrine that periodically returns in a variety of guises, provides one angle of thought which runs counter to the triumphalist humanism that underlies so-called 'possibilist' doctrines of economic development and change. The resurrection of environmentalist discourses, even within the confines of a major institution like the World Bank (where the issue of 'Is Geography Destiny?' has been seriously debated in recent years) poses interesting challenges since this is a style of things that has long been suppressed or abandoned within geography as a discipline. It would not take much to resurrect the argument and invoke geography's historical experience with it as evidence in the debate.

The more favoured posture within geography concerns anthropogenic influences in 'changing the face of the Earth', recognising the extensive role played by human settlement and action upon everything from the morphology of landscape, habitat transformation to climate change. Instead of seeing humanity as a mere 'object' of evolutionary forms, the trend has been to see ourselves as 'subjects' actively transforming the environments in which we live with all manner of intended and unintended consequences (for ourselves as well as for biotic and physical environments).

We should be prepared to think about this issue in a much more dialectical mode, treating the subject-object distinction as arbitrary, and understanding that in changing the world we change ourselves and that we cannot change ourselves and our society without changing our environmental conditions. Social and political projects are always ecological and environmental projects. The fluidity of that idea is constrained, however, by the ways in which relatively permanent features crystallise out to act as barriers to further change. Environmental transformation (whether arrived at through human action or occurring by virtue of the dynamic forces always at work within the environment in general) limit socio-economic transformations (e.g. nuclear power stations, once constructed, required a certain kind of science and organisation in order to be managed over the time horizon of their existence as environmental hazards require massive organisational forms if their destructive consequences are to be avoided).

Environmental issues (like that of space and region) get much more interesting when it is recognised as a dynamic process and when it is treated as a dialectical rather than pure analytic problem. Geographers have already contributed much on an issue that pervades thinking and practices in a wide array of other institutional settings.

GEOGRAPHY AMONG THE DISCIPLINES

The four structural elements to be found within all forms of geographical knowledge collectively form structural supports for a unified methodological field of activity to be called 'Geography'. A number of points can be made about the positionality of the field among the disciplines.

Work within this field is not confined to the discipline of geography. A scholar in literary theory studying, say, the works of Wordsworth, might examine his poetry against a cartographic background of the city-country, might pay minute attention to the conceptualisations of space and time that symbolise a distinctive way of life, thought and personal subjectivity, might pay close attention to environmental qualities and the portrayal of the relation to nature, and, finally, might examine the way in which the poetry helped to produce the idea of 'The Lakes' as a distinctive region eliding into the creation of a tourist industry (based on Wordworth's 'Tour Guide' writings) which in turn helped produce a distinctive regionality on the ground.

It is possible to imagine palaeoecologists, geomorphologists, sedimentologists, economic geographers, cultural historians and rural sociologists all taking somewhat similar steps in their research design. 'Thinking like a geographer' then entails an understanding in each one of these operations of how the four structural pillars of geographical knowledge can be worked and woven together in specific instances and settings to produce profounder insights into socio-ecological conditions and processes of change. There are some deep commonalities and unities in how seemingly disparate geographical knowledges are structured and it is surely worthwhile examining more carefully how such structure works.

The problem for geography as a discipline has been its search for an 'essence' and for an exclusively defined 'nature' which sets it clearly apart from all other disciplines within the social and natural sciences. Taking essentialist definitions of other subject matters like biology and economics as given, the best that geography can do is to claim some 'hybrid' status, to hold itself up as some model of higher order 'synthesis' or to set itself apart by indulging in 'exceptioinalist' claims. The latter can be based on the peculiarities of thinking that derive from the deep contemplation of region and space relations, paying particular attention to the seeming recalcitrance of geographical information in the face of general theory (ergo the idea that general laws and universal statements are impossible in geography).

But there is an entirely different mode of thought that avoids essentialist definitions and meanings and which seems far more appropriate to our actually existing circumstances. Analogical reasoning seeks connections and inter-relations, pushes forward metaphors and underlying unities within seemingly disparate phenomena, seeks analysis to illuminate phenomena in one area by examination of another. Above all, it seeks translation between different modes of thoughts. It is profoundly open and avoids all the turf-wars and exclusions that typify a world dominated by essentialist and purist categories. The moment in the history of geography that was particularly fertile in this regard was that led by the collaboration of Chorley and Haggett to produce collective works like *Models in Geography*. At the heart of that enterprise lay analogical reasons opposed to the essentialist definition earlier sought in, say, Hartshorne's *Nature of Geography*. What is so impressive about the current situation is the widespread occurrence of analogical reason. Spatial themes, for example, permeate literary and social theory. Some of the spatial and cartographic metaphors deployed in literary theory today are wildly inappropriate. Part of our scholarly job is to place such transfers of thought and feeling on reasonably solid ground.

But now seems the moment when geographers are superbly placed to be a central guiding force within the networks of knowledge being created by widespread appeal to analogical reason throughout all spheres of academic activity. But for geographers to take advantage of this positionality, it is necessary to abandon essentialist attitudes (the negative effects of which are all too plain to see in other spheres of knowledge like multi-culturalism, nationalism, or gender studies). There is no 'nature' of geography to be found. The search for an essence is profoundly misplaced if not counter-productive. But 'thinking like a geographer' is everywhere. Learning to think 'soundly' and 'properly' as a geographer is a profoundly important attribute in today's world. This is where the unified methodological field of geography is to be found at work. Harvey, however, says: 'As the example of Kant's cosmopolitanism and its murky tradition all too easily shows, not knowing "how to think properly like a geographer", how to weave together the four structural pillars of geographical knowledges into a system of geographic wisdom, has long-lasting negative effects upon the collective prospects for emanicipatory socio-ecological change.

POLITICAL PROJECTS

Even the most objectivist and neutral-sounding scientist will acknowledge that broad context of scientific activity and learning has a great deal to do with human emancipation from want and need, that the improvement of human understanding is a necessary condition for the betterment of society (whether it be in material or non-material ways). The claim of objectivity and neutrality is always a circumscribed claim (pertaining to certain limited and carefully defined aspects of the overall learning enterprise).

The supposed neutrality of geographical knowledge has at best proved to be a beguiling fiction and at worst a dowright fraud. Geographical knowledge in bourgeois era, for example, have always internalised strong ideological content. In this scientific (and, predominantly positivist) forms, natural and social phenomena are represented objectively as things, subject to manipulation, management and exploitation by dominant forms of capital and state power. In their more scientific, humanist and aesthetic incarnation, geographical knowledge project and articulate individual and collective hopes and fears while purporting to depict material conditions and social relations with the historical veracity they deserve.

Although it aspires to universal understanding of the diversity of life on Earth, geography has often cultivated parochialist and ethnocentric perspectives on that diversity. It has often been, and all too often still is, an active vehicle for the transmission of doctrines of racial, cultural, sexual or national superiority. Ideas of geographical or 'manifest' destiny of 'natural' geographical rights, of the 'white man's burden' and the civilising mission of bourgeoise or of the global reach of an American definition of democracy (thoroughly corrupted by money power as it may be), are liberally scattered in geographical thinking and are deeply embedded in popular geographical lore. Cold War rhetoric, fears of 'orientalism' or of some demonic 'other' that threatens the existing order have likewise been pervasive. The 'facts' of geography presented often as 'facts of nature' have often been used to justify imperialism, neo-colonial domination, expansionism and geopolitical plans for dominance. Geographical information can also be presented in such a way as to prey upon fears and feed hostility (the abuse of cartography is of particular note in this regard).

Many forms of geographical knowledge have been tainted by virtue of their connection with the instrumental ends for which they were designed and institutional frameworks to which they were beholden. But this is not to say that they are useless, irrelevant or too contaminated to be touched (any more than we might dismiss the uses of specific technologies because they are invented for purposes of military domination and destruction). The problem, as much within as without, is to take these varied forms of knowledge, appreciate the circumstances of their origin, evaluate them for what they are and, if possible, transform them or translate them (with the aid of analogical reason) into different codes where they might perform quite different functions.

There are abundant examples of attempts to mobilise geographical knowledges to humanistic ends. Concern for the unwise use of natural and human resources, environmental degradations and insufficient or unjust spatial distributions (of population, industry, transport facilities, ecological complexes, etc.) have led many geographers to consider the question of 'rational' configuration of both. This aspect of geographical practice, which emerged with the early geological, soil and landuse surveys, has increased markedly in the last 60 years as the state has been forced to intervene more actively in human affairs. Positive knowledge of actual distribution (the collection, coding and presentation of information) and normative theories of location and optimisation have proved useful in environmental management and urban and regional planning. These techniques entailed acceptance of a distinctively capitalist definition of rationality, connected to the accumulation of capital and social control. But such a mode of thought also opened up the possibility for planning the efficient utilization of environments and space according to alternative and multiple definitions of rationality.

Harvey (2000) writes: 'Geographical knowledges have the largely unrealised potentiality to express hopes and aspirations as well as fears, to seek universal understandings based on mutual respect and concern, and to articulate firmer bases for human cooperation in a world marked by strong geographical differences. The construction of geographical knowledges in the spirit of liberty and respect for others, opens up the possibility for the creation of alternative forms of geographical practice, tied to principles of mutual respect and advantage rather than to the politics of exploitation. Geographical knowledges can become vehicles to express utopian visions and practical plans for the creation of alternative geographies. They can openly confront and articulate the legitimate and frequently conflictual aspirations of diverse populations. They can infuse cosmopolitan projects, founded on ideas of justice, tolerance and reason, with geographical understandings that do not automatically negate their universal claims. They can become embedded in alternative politics to good effect. They can provide effective means to mobilize knowledge of the world for those emancipatory ends to which all learning and all science has traditionally aspired.'

18

Applied Geography

HIGHLIGHTS

- Pure and Applied Geography
- Whether 'Pure' Geography or 'Applied' Geography
- Positivist Approach and Applied Geography

James and Martin (1972, 10) have identified five different kinds of questions of geographic character:

'(1) There are "generic" questions that have to do with the content of Earth space but that cannot be effectively answered without a framework of concepts to guide the separation of the relevant from the vast complexity of the irrelevant.

(2) There are "genetic" questions that have to do with the sequences of events leading from past situations through geographic changes to present condition; these are studied by the methods of historical geography.

(3) There are "theoretical" questions that deal with the formulation of empirical generalizations or general laws, perhaps even with basic theory, and with the methods of drawing logical deductions.

(4) There are "remedial" questions that have to do with the application of geographic concepts and skills to the study of practical economic, social, or political problems.

(5) There are "methodological" questions that have to do with experiments in new methods of study, new techniques of observation and analysis or new cartographic methods.'

The fourth 'remedial' questions of geographic character are of special significance as they necessarily point out to the application of geographical knowledge and skills to the study of practical problems of government or business. It was as early as in 1896 that Hugh Robert Mill, one of the contemporaries of Sir Halford J. Mackinder, drew up a plan for using the sheets of the Ordnance Survey (1 inch to the mile) as basis on which to plot categories of land quality and land use for all of the British Islands. In 1900, he provided sample studies of two sheets to demonstrate the utility of such detailed field-mapping. Mill's attempt may be thought of as being the first such attempt to apply geographical skill to land use vis-a-vis land quality mapping, but his proposal was ignored by the British geographers of his period. However, the idea that Mill developed finally echoed in Dudley Stamp's celebrated work during the 1930s.

It needs to be mentioned here that Hugh Robert Mill's heritage with regard to the mapping of potential land quality based on the 1 inch to the mile sheet of Ordnance Survey, was given a concrete expansion in regional studies in Great Britain by sociologist Sir Patrick Geddes (1854–1932). Although not a geographer himself, Geddes had a major influence on British geography, especially in the field of regional survey, regionalisation and applied geography. Field study—observation and recording in the field—was basic to his teaching. This led on to what Geddes called 'regional survey', embracing place, work and folk; it is alternatively described as geography, economics and anthropology, or as environment, function and organism (Dickinson, 1969, 204). Geddes saw an important application of regional survey in regional planning. He was rather impatient with geographers who defined their subject as a 'descriptive science', which tells us whether Geography should be an 'applied science' which tells us what ought to be (Stevenson, 1978, 57).

In this way, visionary planners Ebenezer Howard with his model of suburban landuse and functional structure attempted to plan garden cities, and this had significant influence on British geographical thought before World War I, and finally laid the foundations for studies in applied geography which have continued ever since. Geddes influenced the study of regionalisation. A more direct and explicit reflection of Geddes' idea was found in the work of Charles B. Fawcett (1883–1952). In his book, *The Provinces of England* (1919), Fawcett incorporated somewhat 'obscure' ideas of Geddes into workable form, and made one of the first identifications of 'functional regions'. However, the main translator of Geddesian thought into geography was Herbert J. Fleure (1877–1969) who rendered Geddes' ideas accessible and acceptable to geography. Even Andrew J. Herbertson's (1865–1915) scheme for a division of the world into natural regions reflected Geddes's influence (Holt-Jenson, 1981, 34).

One concept derived from the work of Geddes was that of the regional survey of potential land quality and landuse as a basic input plans for economic development. This was the concept that Hugh Robert Mill was believed to have developed in 1896 and 1900, and Geddes made it more expressive and applicable. L. Dudley Stamp in 1930s translated Mill-Geddes' ideas into his workable paradigm for the regional survey of potential land quality and landuse. He organised and directed the first British land utilisation survey during 1930s, employing some 22,000 school children in the mapping of landuse on a scale of 1/2500 in their home district under the supervision of school and university teacher (ibid., 35).

One of the most important geographic thoughts that developed during the interwar period was the application of geographic concepts to politics. This has given the German name Geopolitik (geopolitics). According to Karl Haushofer, who pioneered Geopolitik in Germany, it was the art of using geographical knowledge to give support and direction to the policy of a state. Geopolitik was later conceived of as 'applied' political geography. The transformation of political geography into Geopolitik (geopolitics) in Germany in 1920s was a real manifestation of the remedial character of geography for Germany that sought for the application of geographical skill to the study of the political problems of interwar Germany with regard to its geographical space. The very concept of 'lebensraum', in Germany during that period, was necessarily aimed at making the application of geographical principles more operational in the pursuit of national interest concerning the acquisition of more territories. Instead of calling it applied political geography, the German geographers of contemporary Germany preferred the words 'Geopolitik' to other comparable words.

At the Paris Peace Conference in 1919, geographical concepts and principles were applied in re-drawing the map of the post-World War I Europe. In fact, it was the American geographers who played the key role under Isaiah Bowman and Mark Jefferson. The former was given the title of Chief Territorial Specialist of the American Commission to Negotiate Peace and the latter was appointed as Chief Cartographer. Several maps were prepared on scale of 1/1,000,000 or 1/3,000,000. These were the general maps of Europe as a whole or the somewhat more detailed maps of the Balkans. But there were also a great number of very large-scale maps, such as the map of Alsace–Loraine on a scale of 1/250,000. All these maps were used at the Paris Peace Conference for the study of various boundary proposals. In fact, the boundaries of Europe were re-drawn on the basis of these maps prepared for the purpose. At the Paris Conference, the map became everything. A major part of the work was in the field of cartography in which a map-making programme of unprecedented size and detail was thus undertaken.

Thus, there are two instances in Europe in which geographic principles were applied to resolve political problems: application of geography to sustain the state policy of territorial aggrandisement, called Geopolitik in Germany; and application of geographic vis-à-vis cartographic principles for re-drawing the boundaries of Europe at the Paris Peace Conference. However, it cannot be an exaggeration if it is said that both referred to the reasonability of 'applied political geography'. But at the same time, one of them caused the development of the other, i.e. the re-drawing of boundaries of Europe at the Paris Peace Conference led on to the development of German Geopolitik because the realignment of boundaries had placed Germany at great 'spatial' disadvantage.

To quote James and Martin (1972, 343), 'As the number of professionally trained geographers increased, a certain proportion of the younger generation—then as now—expressed impatience with theoretical studies and demanded that geographic investigations be clearly relevant to the practical problems involving public or private policy. In the 1920s and 1930s, there were some geographers who were disenchanted with the experiments with field methods applied to small areas, and there was a demand for some tangible connection with the "overriding" economic, social, or political problems of the day. There was no clear answer to the question about how one might demonstrate a tangible connection, so each scholar who decided to use geographical studies for practical ends had to formulate his own justification. The result was the appearance of a wide variety of studies in what might be called "applied geography", in the sense that the purpose was to provide the basis for planning remedial action.'

The first large breakthrough in the use of professionally trained geographers to study practical problems came during World War I and its aftermath. During the 1920s and 1930s, not only were certain wartime projects continued and completed, but new kinds of applied research were undertaken. A large number of geographers were called into both military and civilian services during World War II, and since then the application of geographical knowledge and skills to the solution of practical problems has grown rapidly.

Though there is a general belief that the root of the development of applied geography lay with the work of Robert Hugh Mill of Britain in 1896, but there was a section of geographers who strongly felt that it was in Prussia that the usefulness of geography for practical purposes was first realised. Geography was potentially useful for both commercial and military reasons in a world of imperial rivalry. In France, this usefulness was expressed in the creation of many commercial geographical societies in the decade following the defeat of Germany. Geography's usefulnesss was known in Germany after the Franco-Prussian war in 1871. The Prussian State education bureaucrats must have thought there was some value to a geographic perspective in Germany in 1874. The government in unified Germany imposed geography on universities with the objective that the subject would be useful for its particular purposes at that time. A section of geographers strongly believed that, 'neither Alexander Von Humboldt nor Ritter was in any direct sense the creator of modern geography as a university discipline. Their ideas were used by the creators so that they could become symbolic founding fathers but their impact was limited' (Taylor, 1985, 97).

The question that was raised was as to who were the creators of modern geography? The answer was, the creators of modern geography were to be found in the bureaucracy of the Prussian State in the period immediately following the creation of the German Empire. In 1874, Kaiser Wilhelm-I decreed that all royal universities should create chairs of geography. This political decision led to the immediate imposition of geography throughout the German universities. Hence, by the end of the nineteenth century, geography was an established discipline in the German university system. It was the practical use of geography that the state bureaucracy in Prussian State realised in their quest for fair share of colonies. In Germany, the established higher education system proved to be the chosen vehicle for the advance of geography' (Capel, 1981, 41).

These developments in Germany, imposing geography on universities, not with the purpose of enriching the intellectual and academic quality and acceptance of the discipline, but with the purpose of the receptivity of the discipline for reasons other than intellectual/academic excellence. While the 'committed' German geographers (inclined to Humboldt-Ritter geography), sought to achieve methodological and philosophical distinctiveness for their discipline so as to equalise the emerging pace of other social sciences, the state bureaucracy sought to use the discipline for the state's requirements. This mutual exclusiveness that developed within the discipline finally led to 'methodological controversies' to the extent as to have stood in the way of the discipline to be qualified as a subject for real and profound research. This controversy necessarily led to the dichotomy of 'pure' and 'applied' geography.

PURE AND APPLIED GEOGRAPHY

Grano (1981) identifies two types of outside influences on the development of the discipline. Within, academic geographers had to be given an intellectual foundation to satisfy intellectual peers, and outside in the wider world, geography had to be justified as a useful activity on which to spend public money. These two pressures produce 'pure' and 'applied' geographies. Success in both activities is necessary for the long-term survival of the discipline.

The dichotomy of 'pure' and 'applied' in any discipline is a myth based on a theory of knowledge that claims a separation of knowledge from society. A 'pure' science is supposed to be an autonomous product of scientists' researches that generates theoretical knowledge. It is characterised by having no direct use for solving any particular problems. In contrast, 'applied' research is explicitly problem-oriented although it will claim to be based on the theoretical foundation set by 'pure' scientists. This is particularly clear in fields of social enquiry where the emphasis has always been more on the teaching than the research function in universities. 'Pure' research represents a set of intellectual concepts brought together to provide the theoretical foundation of a discipline. Geography is shown to be a 'natural' segment of the world of ideas. 'Pure' geographers define geography's place in the civilisation of knowledge and promote the status of the discipline within academics. They produce the necessary core of ideas to justify the existence of geography. 'Applied' geographers accept these ideas and may use them to solve problems (Taylor, 1985, 100).

All disciplines tend to emphasise their problem-solving capacity, and applied geography is found to be in the ascendancy within geographical studies. In contrast, in a period of expanding and emerging economies and social optimism, outside pressures will diminish and academics can be expected to render less external pressure. Geographers will be able to contemplate their discipline and feel much less guilty about this activity. The 'burst' of 'pure' geography would occur when pressures from academics would seem more threatening than pressures from the outside. This suggests that since the creation of geography as a university discipline, and one should expect the balance between 'pure' and 'applied' geography to approximate the long economic waves of the world economy. Therefore, three 'applied geographical' tendencies can be assumed to have occurred: (1) in the late nineteenth century, (2) between the world wars, and (3) in early twentieth century and the post-1945 economic boom.

Geography was created as an applied discipline by the Prussian State at the beginning of the late nineteenth century depression. Though the standard discussions of this period had emphasised the methodological debates, the discipline was very clearly balanced towards its applied aspects. In the age of imperialism, geography was a tool of nationalism and imperialism, both political and economic. During the methodological debates, however, the seeds of 'pure' geography had been germinating, but culminated as a result of the widespread acceptance of region and regional synthesis as the core of a 'new' geography in the early twentieth century. The first 'pure' geography was expressed differently in different national schools—French 'pays', German 'landschaft', British 'natural regions', Russian 'zones' (Grano, 1981)—but there was only one core geographical philosophy based on the holism of the founding fathers: Humboldt and Ritter. In fact, the justification for geography was taken back as far as the late eighteenth century by reference to Immanuel Kant's division of knowledge into systematic, historical and geographical parts.

The first 'pure' geography remained the theoretical basis of geography throughout the first half of the twentieth century. But in the inter-war depression such 'pure' concerns were put aside as geography was required to show its usefulness. The most important, rather, notorious example was the transformation of (political) geography into geopolitics in Germany. In Britain, Dudley Stamp was carrying out the massive Land Utilisation Survey and generally geographers were contributing to the growth of the planning movement in many countries. The emphasis here is on technical knowledge, and the practical communicative aspects of the first 'pure' geography are becoming lost. It is at this point that Hartshorne (1939) in his *Nature of Geography* attempts to attack on those who seek to reform the nature of geography by making the discipline more problem-oriented, more scientific and more in tune with national interests. His great work stands, therefore, as a defence of the first 'pure' geography in a period of emphasis on applied geographies.

Geography entered the post-War era of economic prosperity and social optimism as an academic anachronism. With the social sciences fully established as specialised disciplines of social enquiry, geography's holistic pretensions were becoming intellectual liability. Woolridge and East (1951, 25–26) and David (1957) attached regional synthesis as empty intellectual rhetoric. Where were the grand syntheses that were the goal of geography? They sought for the reduction in geography's place in universities. Such internal pressures led to the development of a second 'pure' geography that rejected the theory of the predecessors. It was at this stage that developed the famous Hartshorne-Schaefer debate. Regional geography was discarded/dismissed as idiographic and geographers were instructed to begin the nomothetic task of finding morphological laws. Schaefer developed the spatial school that involved a new theoretical structure accompanied by the acceptance of statistical techniques in the 'quantitative revolution'. In Britain, Haggett and Chorley (1967) proclaimed a 'new model-based geography' just as Mackinder (1887) had done two generations earlier: a second 'pure' geography had been created. The 'spatial specialist' Schaefer had succeeded where the 'physical specialist' Geeland had failed and geography was dragged screaming into the twentieth century. The process of specialisation, partially held in check by the old holistic philosophy, now overwhelmed the discipline. However, geography began to splinter apart in the manner predicted many times before by the older generations. Geographers in the Anglo-American world came to have their academic activity dominated by organised groups of specialist researchers—'study groups' and 'specialty groups' respectively. There was no particular direction before the geographers to follow as the discipline became a collection of diverse and loosely related groups of researchers.

In one sense, the recession has saved geography. Instead of having to face up to the fundamental problems of legacy of the second 'pure' geography, outside pressures have diverted attention back to applied geography (Taylor, 1985). In the USA, 'applied geography' has even reached the status of being the new paradigm (Frazier, 1978). There had been much discussion on the role of geography in government, in the wider job market, and of its need to consider national priorities. Geography's usefulness was proved in public and private fields and in individual entrepreneurship. It was found that the technical baggage associated with the spatial school constitutes a very useful set of equipment in showing how geography could contribute to solving problems today. Computer models, modern cartography and information systems all, now make geographers as spatial analysts employable in the current session.

The second 'pure' geography had been found, after the event, to have been positivist. The result was the division of human geography into three broad groups: those that largely maintained the second 'pure' geography, (2) those that developed a humanistic critique, and (3) those of a more radical disposition (Johnston, 1983). The first group continued to dominate as more and more applied geography was spawned. With rare exceptions they had not entered into the methodological debate, but concentrated on getting on 'doing' geography. The two other groups, however, seriously undermined the existing 'pure' geography. The humanistic pointed out to its mechanistic view of the individual and the resulting neglect of the truly human aspects. The radicals condemned the failure to come to terms with power relations in capitalist society. The result was that 'pure' geography on which the current growth of 'applied' geography was based, however, got discredited, but with only minor overall effects.

The second 'pure' geography continued to underlay most activities of modern geographers who ignored the methodological controversies going on around them. It seemed that there ·could be parallel with the late nineteenth century's methodological debates. From this previous period of applied geographies, the political, commercial and colonial geographies are now largely forgotten, but the methodological debates are remembered (Taylor, 1985). It is now firmly established that the development of 'pure' geography takes place when there is no problem of economic recession and economic prosperity is at hand. The late twentieth century and the early twenty-first century have witnessed economic growth, and this growth has sustained high order specialised research in different social sciences, including geography. The time was found to be conducive for the emergence of 'pure' geography in earlier years of present century. In this way, 'pure' research is generally held in higher esteem than applied research.

However, the pressures for more applied work built up in the late 1970s and through the 1980s, as a reaction to the economic recession that afflicted the countries considered here during that period. In addition, government policies towards higher education, especially in the United Kingdom, placed greater emphasis on applied work, both in terms of providing vocational and professional training for students and in the pressure to cover institutional costs through 'earnings' from research activity. Thus, the debate over applied geography concerned pragmatic issues of disciplinary survival as well as concerned with regard to relevance (Johnston, 1996, 319). Now, it seems 'certain' that for the development of 'pure' geography, a sound economic prosperity with political stability is a necessity. But when social and economic problems, such as recession, develop, the need becomes to find out remedial measures of the problems, and in a process, whatever studies are carried out, they are all applied in nature. But, it is not that all applied geographies are problem-biased. There are many instances where both 'pure' and 'applied' geography have developed side by side and in some cases they have been found to be complimentary to each other, particularly in methodological areas.

WHETHER 'PURE' GEOGRAPHY OR 'APPLIED' GEOGRAPHY

In view of growing economic recession, societal and political problems, a section of geographers concerned with these human problems expressed their disenchantment and disillusion to 'pure' and/or 'academic' geography, and sought for a new problem-solving geography. In the words of Kasperson (1971, 13): 'the shift in the objects of study in geography from supermarkets and highways to poverty and racism has already begun, and we can expect to continue, for the goals of geography are changing. The men see the objective of geography as the same as that for medicine—to postpone death and reduce suffering.'

A similar but somewhat 'different' statement was made by W. Zelinsky (1970, 499), reflecting a growing disillusion within geographical circles with past achievements and a wondering about future directions. Zelinsky developed three basic arguments:

1. That people are inducing for themselves a state of acute frustration and a crisis of survival;

2. That these conditions originate, and can only be solved, in the advanced nations; and

3. That the current 'growth syndrome' has profound geographical implications.

Material accumulation can no longer be considered progress, Zelinsky said, because it is unsustainable; efforts are currently misallocated on a massive scale, and there is a major geographical task involved in its sensible reallocation.

Zelinsky (1970) identified three roles which geographers could play in facing the perceived oncoming disasters:

1. The 'geographer as diagnostician', applying the geographic stethoscope to a successful demography, mapping what he calls geodemographic load, environmental contamination, crowding and stress;

2. The 'geographer as prophet', projecting and forecasting likely futures;

3. The 'geographer as architect of utopia', educating with regard to problems and solutions and providing support for unknown leaders who have the political will to guide society through the coming 'Great Transition'.

Zelinsky further said: 'how woefully deficient we are in terms of practitioners, in terms of both quantity and quality, how we are still lacking in relevant techniques, but most of all that we are totally at sea in terms of ideology, theory and proper institutional arrangements' (ibid.).

These criticisms are not only directed against the geographers, but also against the scientists. Science, being a twentieth-century religion, has failed to avert the oncoming crisis. Zelinsky identified five essential axioms as the foundation of science:

1. That the principle of causality is valid for studying all phenomena;
2. That all problems are soluble;
3. That there is final state of perfect knowledge;
4. That findings have universal validity;
5. That total scientific objectivity is possible.

The social sciences have failed to lineup to these for several reasons: their immaturity; their use as a refuge for mediocre personnel; the difficulty of their subject matter concerning interpersonal relationships; their problems of observation and experimentation; and the political and other problems involved in applying their proposed solutions. Further, and the major cause of their failure, the natural science model is irrelevant to the study of society. The natural science approach adopted by positivist spatial scientists helps to describe the world, but not to understand it (Zelinsky, 1970, 141).

There are two major components to Zelinsky's case: geographical research should be relevant to the solution of major societal problems; and positivist-based spatial methodology may be inappropriate for such a task. House (1973), for example, reviewed the tradition of involvement in public policy by British geographers, and Stoddart (1975) identified the late nineteenth-century views of Recluse and Kroptkin as 'the origin of a socially relevant geography'—the latter was discovered by the 'radicals' (Peet, 1978).

It was claimed that geographical works should be more relevant to major social problems raised queries about the nature of that relevance, and soon became apparent that there was no consensus on what should be done and why traditionally, geographers have advised governments in the role of information-gatherers and 'masterful synthesisers' and they had not involved in the final stages of policy-making; they had been delvers and dovetailers, but not deciders. To Chisholm (1971), the challenge to human geography was to define such norms that could make the geographers 'deciders'. He further said, 'the danger with empirical science is the absence of guidance at the normative level as to which various options one should take.' Chisholm had worked for different governments in the role of 'decider' which was neither delver nor dovetailer. As a member of independent commissions (the Local Government Boundary Commission in the 1970s and the Local Government Commission in the 1990s), he was involved in advising the British Government on the best way to redraw important components of England's administrative map. It was suggested that the focus of research should be some of the social and spatial inequalities in the society, and the first challenge should be the study of the distribution of power in the society, thereby identifying the mechanism for allocating scarce resources. Research could then isolate the disadvantages of relative powerlessness and provide the basis for policy aimed at redistributing resources.

By 1971, many geographers were frustrated by a sense of failure to deal with major social issues, but although some members were taking notice of the sufferings of the outside world, Smith, in his 1971 report, mentioned that there was a deliberate ignorance by American geographers on conditions of human welfare and social justice, and therefore, American geography was about to undergo another revolution to counter a situation in which geography was once pre-occupied with issues other than societal issues. To him: 'the conditions which have helped to spawn radical geography in the United States include the existence of large oppressed radical minorities, inequalities between rich and poor with respect to social justice, a power structure and value system largely unresponsive to the needs of the unprivileged, and an unpopular war with sapping national, economic and moral strength. These conditions do not exist in Britain or exist in a less severe form, and the stimulus for social activism in geography is, thus, considerably less than in America' (Smith, 1971, 156–157). In 1994, Smith urged to place social justice at the heart of human geography in his book *Geography and Social Justice.* It is claimed that the British geographers had long been concerned with 'relevant' issues, than their American counterparts, with respect to the Third World.

It was felt long back that 'pure' or 'academic' geographers should provide a knowledge-base on which policy could be built, which implied close involvement with the education of future generations of policy-makers. To Blowers (1972, 291), 'the issue is not how can we cooperate with policy-makers, but whether and in what sense we should do it. It is a question of values'. He argued that the sort of activities that Berry (1972) proposed that 'just a new fact involving new entrants to the field seeking their turf ... the majority of the new revolutionaries ... are essentially liberals—quick to lament the supposed ills of the society and to wear their bleeding hearts like emblems or old scholastics—and quicker to avoid the hard work that diagnosis and action demand', would be strongly supportive of the status quo, and unlikely to produce fundamental social reforms. Research in geography should highlight particular problems, and teaching should place emphases on 'a man in harmony with nature rather than master of it, on social health rather than efficiency, and on the quality of life rather than on the quantity of goods' (Berry, 1972).

There was a long debate at the 1974 annual conference of the Institute of British Geographers (IBG) on how should geographers contribute to the solution of societal problems. Coppock in his presidential address (1974) urged upon the geographers to identify that their contributions could make to encourage research relevant to those contributions, and to enter a dialogue with those who advise on and implement public policy.

Analysing the reasons for geography's inability to be relevant to the domain of public policy-making, Hare (1977, 64) observed: 'a major reason for a lack of geographical contributions to public policy may be the poverty of their training: in recent years, we have swept geography departments into the social science divisions of faculties of arts and sciences where from playing second fiddle to geologists or literary critics, we learned to play second fiddle to environment and sociologists.' Geographers would have to rebuild their discipline based on the centrality of society-environment interactions, with a new brand of physical geography that leans heavily on biological ideas and concepts. Thus, we must reassert the old, essential truth that geography is the study of the Earth as the habitat of man, and not some small sub-set of that gigantic theme. Regional synthesis should be stressed too.

To Hall (1974, 49) '… geography, most clearly of all social sciences, has neither an explicit nor an implicit normative base ... spatial efficiency ... is rather a description of what men seek to do in actuality ... not ... an objective to be achieved or to be objective function to be maximized'. Policy–makers must seek their norms elsewhere; geographers meanwhile must develop a new political geography which will aid them in understanding the essential role of political decisions in structuring spatial systems.

However, at the meeting of the Institute of British Geographers (IBG) in 1974, it was claimed that governments, as pay-masters, already constrained what geographers could do research on, and as a result geographers were being used; their only alternative was political action.

Harvey (1974) in his contribution, entitled, 'What kind of geography for what kind of public policy?' argued that individuals wishing to become involved in policy-making were stimulated by motives such as personal ambition, disciplinary imperialism, social necessity and moral obligation: at the level of the whole discipline, on the other hand, geography had been co-opted, through universities, by the growing corporate state, and the geographers had been given some illusion of power within a decision-making process, designed to maintain the status quo. Harvey portrayed the corporate state as a 'proto-fascist', a transitional step on the path to the barbarism of Orwell's 1984. The function of academics, according to him, was to counter such trends to expunge the fascism, ethnocentrism and condescending paternalism from within their own discipline and to build a humanistic subject, thereby assisting all human beings 'to control and enhance the conditions of our own existence' (23–24).

The current mode of analysis in geography offered little for the solution of pressing societal concerns and problems, said Harvey (1973): 'There is an ecological problem, an urban problem, an international trade problem, and yet we seem incapable of saying anything of depth or profundity about any of them. When we do say anything, it appears trite and rather ludicrous .... It is the emerging objective social conditions and our patent inability to cope with them which essentially explain the necessity for a revolution in geographic thought.'

POSITIVIST APPROACH AND APPLIED GEOGRAPHY

Most of the 'applied' works in geography in the 1970s and 1980s of the last century followed empiricist-positivist approach. Even the works undertaken before and after World War II, also revealed some positivist trends. Applied geography as information-gathering and synthesising has a substantial record of empiricist-positivist bias, starting with Stamp's Land Utilization Survey of Britain in the 1930s and his subsequent involvement in the preparation for post-War planning. Other aspects of landuse planning were of interest to geographers in the 1930s and 1940s, and were discussed several times at the Royal Geographical Society. Similar geographical involvement occurred in the United States.

After World War II—during which geographers made many contributions as information synthesiser and gathers, the latter including the development of air-photo interpretation—landuse planning was established on a large scale, and trained geographers provided a large proportion of its personnel. Many academic geographers remained active in applied works. The developments in GIS during the 1980s and 1990s have stimulated very substantial increases in mapping and enumerations. Smith (1985) observed: 'the pre-dominance of remote sensing (and associated digital mapping) reflects a view of geography as a technologically sophisticated means of gathering and displaying information, in the tradition of the geographers as map-maker linked to the contemporary preoccupation with information technology … that appears to conflate the needs of the discipline with a conception of the needs of the society in which the emphasis is much more economic than social'. Alongside this, various policies have been evaluated, such as those aimed at changing the distribution of industrial activity, and much of the entropy-maximising systems modeling was intended to provide procedures for the joint activities of landuse and transport planning.

Most of this work used as empiricist and (usually implicit) positivist framework. Geographers are perceived as having valuable skills in the collection and ordering of data, as in landuse survey. The presentation of such data frequently assumed the existence and desirability of maintaining certain causal relationships; planning agricultural landuse, for example, often assumed a clear causal link between the physical environment and agricultural productivity and that of industrial location assumed the need for efficiency via the minimisation of total travel costs. As positivist work on the allocation of landuses and traffic flow increased, so the potential for geographic inputs to spatial planning was promoted (and many people initially trained as geographers became professional planners). Most of this pragmatic application of technical skills, though there were some attempts to evaluate policy impacts and to develop a theory of decision-making in this context. Geographers' empiricist-positivist role continued to be advanced, in response to economic and social crisis, and the perceived need for valid data.

Alongside those who argued for a greater commitment to applied geography in the empiricist/positive mould, and therefore for an implicit acceptance of a particular ideology, others challenged this as the best way to respond to pressing societal problems. They argued for a reappraisal of how geographers could assist in understanding the genesis of those problems rather than in the suggestion of solutions which rarely tackled the root causes (Johnston, 1996).

Since applied works need to be self-explanatory in nature and predictive in character, they cannot be successful without using empiricist/positivist approach. Applied works involve the collection and reporting of information to produce particular form of explanations that necessitates the use of positivist-led methodology because the results ought to be 'law-specific'.

A lot of empiricist/positivist-led research work was reported in the 1960s and 1970s, under the general title of factorial ecologies, application of multi-variate statistical procedures to large area matrix as a means of representing spatial variations in population characteristics. Smith (1973) and Knox (1975) had adapted factoral ecology procedures to the task of mapping social welfare in the 1970s. Knox (1975) promoted the mapping of social and spatial variations in the quality of life as a fundamental objective for geography, to provide both an input to planning procedures and a means of monitoring policies aimed at improving welfare. He used statistical procedures to produce accurate portrayals of spatial variations in meeting these needs. With the resultant maps, geographers must decide whether they are playing ·a sufficient role in awakening human awareness of the extent of the disparities or are under an obligation to help society improve the situation.

Smith (1973) identified territorial social indicators in the American context, to measure the level of social well-being. In fact, he initiated the collation and dissemination of territorial social indicators, to illustrate the extent of discrimination by place of residence that occurred in the United States; multi-variate statistical procedures generated maps on inter-state, inter-city and inter-urban scales. His goal was to present the basis for a better understanding of the origins of inequality as a geographical condition and of the difficulties in the way of plans to promote greater equality in human life chances.

These two works represented the geographer as delver and dovetailer, producing information on which more equitable social planning could be based (Chisholm, 1971). Other studies performed similar roles, and also suggested spatial policies which could lead to improvements. Harries (1974) studied spatial variations in crime rates and the administration of justice, and argued that empiricist/positivist models of criminal pattern could assist the organisation of police work. Shanon and Dever (1974) and Phillips and Joseph (1984) investigated variations in the provision of healthcare facilities and argued for spatial planning which would improve the services offered to the sick; Morril and Wohlenberg (1971) studied the geography of poverty in the United States, preparing both social policies—higher minimum wages, guaranteed incomes, guaranteed jobs and stronger anti-discrimination laws—and spatial policies (such as an extensive programme of economic decentralisation to a network of regional growth centres) which would alleviate this major social problem.

Bunge (1971) also prepared a 'geobiography' of his home area in Detroit's black ghetto, through mapping variations in human welfare. Cox (1973) studied the urban crisis in the United States—the racial tensions and riots, municipal bankruptcies and the role of the government in the urban economy—presenting his analysis in terms of conflict over access to sources of power. One of the major dimensions of his analysis was related to policies, designed to involve spatial reorganisation to achieve the desired equity (including metropolitan government integration), community control, population redistribution and temporal improvements.

Society-Environment Interface and Applied Geography

Geographers have long realised the necessity of maintaining a balance in the society environment interface, so that the fragile structure of our biosphere can be kept safe and secure. For example, the nineteenth-century British scholar Mary Somerville, in her *'Physical Geography'*, published in 1848, expressed concern for man's consistent exploitation of the resources, causing changes in the resource balance. George Perkins Marsh of the United States found Mary Somerville's observations about man's destructive use of the Earth very stimulating, and he made frequent references to her work, in his book *'Man and Nature'*, published in 1864. In Russia, Alexander Ivansvich Voeikov in 1901, expressed concern about the destruction of the grasslands owing to overgrazing and about the destruction by supposed modifications of climate that resulted from changes in the cover of vegetation. In America, it was not until 1905, that Nathaniel Southgate Shaler of the Harvard University and his student Louis Agassiz returned to the theme of Mary Somerville and George Perkins Marsh, concerning man's destructive effect on Earth's resource vis-a-vis society-environment inter-face. And many more such works were conceptually developed in the past, for example, the work of Warren Thornthwaite in the 1930s, the work of Carl Sauer particularly, at the symposium of 'Man's Role in Changing the Face of the Earth', beside a similar but pioneering work by Cumberland in 1947.

Ehrlich and Commoner also made similar comments. The former argued for the primacy of population increase, and popularised the concept of zero population growth, while the latter claimed that technological advances and consequent rapid depletion of resources plus deposition of pollutants created the major problem. Both arguments have clear geographical components and it was stressed that geographers have a strong record of activity in resource conservation.

Nevertheless, it was said that 'developments in geography have been such that several phases of national pre-occupation with environmental problems have not produced a general awareness of our interests and skills' (Mikesell, 1974).

According to Johnston (1996, 339), there were two types of work done by the geographers in the society-environment interface. First kind is a type involved in the traditional geographical concern with description and analysis that necessarily concerned with the problems of the physical environment of urban areas. The second type of work focused on issues of environmental management with particular emphasis on its economic aspects and on societal response to environmental hazards. A topic of special interest was the study of leisure, of the growing demand for recreational facilities, and the impact of recreational and tourism activities on the environment. Nonetheless, Mikesell (1974) concluded that geographical contribution to environmentalism had not been clear up to 1974.

Mikesell (1974) observed that Meadows' et al. (1972) 'The Limits to Growth' attracted remarkably little attention from geographers. He said that many of the social· and philosophical issues debated during the environmental crusade have not been paid adequate attention to by geographers. This conclusion of Mikesell was supported by analysis of the contents of contemporary geographical journals. Stoddart (1987) made a powerful argument not only for more work on the society-environment interface, but also for its centrality to the whole of geographical activity. He felt geography had become diffused and lacked a central focus, and that should be: 'Earth's diversity, its resources, man's survival on the planet. This called for a unified discipline, human and physical, in which the task is to identify geographical problems, issues of man-environment within regions—problems not of geomorphology or history or economics or sociology, but geographical problems, and to use our skills to work to alleviate them, perhaps to solve them. Focusing on the big questions, about man, land, resources, human potential, would involve geographers reclaiming the high ground.

Similar calls have been made by other geographers in response to the relevance of much that Stoddart would disregard in contemporary social science has been promoted as necessary to an appreciation of society-environment relations. Stoddart's argument was clearly set in the cultural ecology mould. A similar observation was made by Kates (1987, 526), who regretted the dominance of spatial science within geography, and claimed that when environmental issues became important on the public and political agenda in the early 1970s, 'No discipline was better suited than was geography to provide intellectual and scientific leadership. The natural science for the environmental revolution should have been the science of the human environment. Instead, intellectual leadership was split among biology, economics and engineering ... but none of these offered a truly integrated view .... The theory of human environment, then, was the theory of plant or animal ecosystem, or of pervasive externalities, or of technological and managerial fixes.'

One of the major concepts that emerged from the increased concern over environmental problems in the late 1980s was sustainable development, a considerably ambiguous term which was generally taken to imply continual increases in material living standards without any diminution in environmental capacity to meet the needs of the future generations. This was a focal concern of the 1992 Earth Summit at Rio de Janeiro, but the difficulties of implementing global environmental policies have been over-stated and could readily be addressed, within a satisfactory time-scale, using instruments of economic policy that tended to promote continued economic development.

The tradition of C. F. White that he established in 1945, with regard to flood control and policy imperatives, continued to direct some applied works in the field, bypassing local and regional bodies. In 1965, C. F. White prepared a paper for the Bureau of the Budget on national policy options for dealing with floods and similar hazards. The result was the establishment of the National Flood Insurance Programme, after which White used to organise annual seminars in Boulder to bring 'public-sector' officials and academics together 'to discuss disaster response and mitigation policies. He and many others who had studied environmental policy, later contributed to its implementation through what may be called the 'citizen -scholar' model, which allowed the scholar 'to choose his and her research concerns on the basis of perceived social need, but attempt to conduct the inquiry face of influence from outside the inquiry itself. The scholar is committed to social and political action as follow-up to the research, based on findings and the quality of enquiry.

White (1972) developed an approach which, according to Wescoat (1992), resembled the philosophy of pragmatism, because of the four 'comparable' components:

1. recognition of the precariousness of existence,
2. a pragmatic conception of the nature of inquiry—problems arise and are tackled within situations, for these are no absolute truths; hence
3. a tradition of learning from experience; and
4. a belief in public discourse and democracy.

Geographers have expressed concern about other pertinent questions, within contemporary society, such as nuclear weapons and nuclear power. They have criticised attitudes to civil defence policies and likely deaths from nuclear blasts and fall out, for example, with others setting that concern within a developing geographical contribution to peace studies, and have addressed issues relating to the siting of nuclear power stations and the transport of nuclear wastes. To Johnston and Taylor (1989), humankind is capable of very little except the destruction of the planet. The obvious devastation caused by much of that human activity which for half-a-century had been given the small collective label of development together with such associated events as the Chernobyl nuclear disaster of 1986, had had effect of reinforcing the unease about the general direction in which humankind was found to be moving. To be successful, humanity must act within the constraints of the environment and be aware both of its limitations and of its possibilities. Since the 1970s, the ideas of humankind as master, capable of transforming nature, has been increasingly replaced by that of humankind as a participant in an environmental totality, and as such having to treat nature with care and respect.

Environmental degradation, much of it clearly attributable to human activity, has increased a consciousness of the essential unity of the planet and the dangers of what Bunge (1989) referred to collectively as the destruction of human space. Bunge concluded that 'this planet is not too small for peace, but it is too small for war'.

O'Riordan (1981), however, identified four conclusions with regard to liberal humanitarian tradition concerning environmentalism:

1. Modern environmentalism challenges many aspects of western capitalism;
2. It points out paradoxes rather than clear solutions;
3. It involves a conviction that better modes of existence are possible; and
4. It is a politicising and reformist movement, based on a realisation of the need for action in the face of impending scarcity and a lack of faith in the western democracies.

A new social, environmental order is required to be set forth with three possibilities: centralised, authoritarian, and anarchist, but there could be liberal option, to which O'Riordn (1981) opined, 'we must individually and collectively seize the opportunities of the present situation to end the era of exploitation and enter a new age of humanitarian concern and cooperative endeavour with a driving desire to reestablish the old values of comfortable frugality and cheerful sharing.'

Policy and Applied Research

Most of the studies referred to here have concerned with identifying problems and suggesting solutions. Underlying their varied approaches has been the basic thesis that geographers should be much more involved in the creation and monitoring policies. Berry (1973) has categorised planning policies into four types:

1. Ameliorative problem-solving: involves identifying problems and proposing immediate solutions, as with the removal of

a traffic bottleneck. Such solutions are likely to stimulate further problems in the future, since it is only the proximate cause that is tackled (the features of bottleneck) rather than the real cause (the growth of traffic).

2. Allocative trend-modifying: planning towards the future involves identifying trends, evaluating what is likely to be the best outcome of the several which they imply and then allocating resources to steer system being planned towards the end.
3. Exploitative opportunity-seeking: planning with the future identifies trends and then seeks to gain the maximum benefit from them, irrespective of the possible long-term consequences. Compared to the previous category, this one has a dominantly short-term focus.
4. Normative goal-oriented: planning for the future begins with a statement of goals, a vision of the future, and then prepares a strategy which will ensure that they are achieved.

'Relatively little policy-making is of the fourth type; most involves elements of other three, with general statements about goals, but no clear strategy regarding a foreseeable future. Some critics interpret this as meaning the geographers who become involved in policy-making and evaluation are likely to be uncritically accepting the dominant forces in society, which leads to arguments that their claimed scientific objectivity and neutrality are on (often unrealised) cloak for ideological political judgments about the nature of society. Whatever their individual motives, such geographers are acting on behalf of interest groups, private and public, whose sustenance depends on maintaining an unjust and unequal structure to society' (Johnston, 1996)

Clark (1982), however, argued with the principle of policy analysis and the involvement of academics in policy-making. He suggested that academic contributions to policy analysis should be guided by the following four propositions:

1. Academics must acknowledge their own values and beliefs in presenting alternatives and impact assessments,
2. Policy analysts must be advocates for particular causes rather than supposedly independent and objective adjudicators of knowledge,
3. Policy science should be critical of the status quo, and
4. Sponsoring institution must encourage/advocate briefs and make those briefs accessible to the public.

Accepting these, academics who become involved in policy analysis should be considered part of the political process whose role should be to ensure that choices would be brought squarely into the open and be dependent upon the political as opposed to expert, process. Any rigorous social scientific enquiry must be based on nationally defended value judgements; objectivity, however, does not require objectivity.

Pacione (1990), on the role of urban geographers in applied work, said that practitioners have paid little attention to conceptual issues underlying what they do. He identified the following eight principles or guidelines to remedy those failings:

1. The notion of value-free research is an illusion,
2. Towns, as examples of places, are meaningful entities on which to work,
3. A spatial perspective is of substantial value,
4. The main emphasis of applied urban geography is on problem-solving,
5. A realist position provides the context for such works,
6. Analysis must integrate various spatial scales,
7. A wide methodological tool kit of quantitative and qualitative procedures must be employed,
8. Geography must integrate funds of many disciplines.

Johnston, reacting to Pacioue's eight principles or guidelines, argued that, unless Pacioue is prepared to tackle the fundamental issues of what problems are and how they can be tackled, he is unlikely to help those of us who want both an end to the problems that we currently perceived and a society which would no longer produce such problems. Few problems are solvable, which points to the need for resolution between opposing points of view, none of which have any claim to absolute truth.

According to Mitchell and Draper (1982), 'when functioning as an advocate, a consultant, the geographer must consciously decide how to resolve a conflict which may arise regarding the promotion of one perspective versus critical assessment and balanced judgement about all viewpoints. There are also issues of ethics, when functioning as a pure researcher, the geographers must balance a concern for obtaining necessary information against a concern for respecting the dignity and integrity of those people or things being studied which also applies to relevant research. They claimed that geographers have largely ignored these ethical issues, and their professional bodies unlike those of other disciplines have promulgated no codes of conduct. Conflict is frequently likely between striving to "discover truth", and respecting the rights of those being studied, however, and they advocated individual, institutional and external controls.'

Most of the critiques of the arguments for greater involvement by geographers in policy analysis are entirely sympathetic. They make a case for sensitive geographical involvement: others question the grounds for such involvement and instead focus on the development of revolutionary theory. The latter focuses almost entirely on applied work in the tradition of positivist, empirical science, in which the goal is perceived to improve well-being through contributing to one or more of:

1. the preparation of public (i.e. state) policies,
2. the development of commercial (i.e. profit-making) strategies, and
3. the attack on environmental problems.

The nature of applied geography has been restricted by some work at the society environment interface, but most applied geography in the English-speaking world accepts, and seeks to serve, the ruling ideas.

While launching the journal *Applied Geography* in 1981, its founding editor following Stamp, defined it as addressing some of the great world problems, the increasing pressure of population on space, the development of the under-developed areas, or the attempt to improve living conditions. Two decades later, these problems remain. Indeed they are growing: environmental problems such as pollution, damage to wild-life, destruction of habitat, soil erosion and resource depletion; the problems of human deprivation and inequality. Briggs (1981) identified the fundamental basis of applied geography as use of resource: 'The exploitation of scarce resources represents a dominating theme to human existence. It is from the pursuit of these resources and from the attempt to decide between alternative policies of exploitation that not only environmental damage, but also the greater part of political, social and economic problems emerge, they can be seen as expression of Man's inability to organize himself and his world to his best long-term advantage. To overcome that inability, the applied geographer must be brave.'

Most of the applied works undertaken by academic geographers have been commissioned by one or more arms of the public sector, and most geographers who have applied their skills in non-academic careers have probably entered the public sector too. Given the importance of the welfare state in the first four post-World War II decades, this has not been surprising. Nevertheless, much works have already been done for private sector too, in economic geography—geographers were involved in work for supermarket, and similar companies in the United States during 1950s, others building on their expertise in spatial data handling, moved into information measurement, including some who worked in the prestigious university business schools.

The pressure on university academics to obtain more external support for research activities in the 1980s and 1990s has stimulated much more work for the private sector—a trend accentuated by the attempts to reduce the size of the provision of public sector and to increase competitiveness and efficiency in the provision of public goods. Some of this work has involved spatial analysis in various forms. Much more of it has used geographers' skill in spatial data handling (adding value to geocoded data) in the growing field of geodemographics, whereby marketing and other campaigns are spatially targeted to people living in areas where demand for particular goods and services is most likely to be generated. Other work still is in the areas of resource and environmental management. In most of the developing democracies of the Third World, funds are more available to academic geographers to carry on 'applied' work on various social and environment-related work, including transport problems. Both, the private and public agencies have been appointing geographers because of their spatial data handling capacity for enumerative work in the areas of the planning.

Applied Geography in Some Selected Countries

Geography in **Britain** has included some notable contributions towards the solution of practical problems. Dickinson (1969) has traced out the influence of the French sociologist Fredric Le Play on the Scottish regional planner, Patrick Geddes. Geddes developed the concept of the regional survey of potential land quality and landuse as the basis on which to draw up a plan for economic development. As mentioned earlier in the chapter, that in 1896 Hugh Robert Mill suggested for using the sheets of the Ordinance Survey (1 inch to the mile) and four years later provided an example of how such mapping of relevant information could be done and how it could be used. His suggestion was discussed, but not operationalised. The British geographer who put these earlier ideas into practice was L. Dudley Stamp. After being convinced that Mill and Geddes were right about the need for a survey of Britain on basis for planning, he began searching the ways to carry out such a survey by plotting categories of land quality and landuse on the ordinance map at 6 inches to the mile. Stamp recognised that quite aside from the practical importance, a survey would be a basis for planning; the work of making it would have an excellent educational experience for children.

With great energy and patience, Stamp undertook to organise and direct what became known as the British Land Utilization Survey. He took the help of some 22,000 school children in his pioneering work. Geography students and staff at various universities cooperated in providing the professional personnel. The work was started in the summer of 1931, and by the end of 1935, the mapping was completed. The importance of these maps was realised when the Second World War began in 1939. Under the directive of the Ministry of Agriculture, all the maps were published between 1939 and 1945. The survey maps were used in a variety of ways and purpose in addition to the planning of emergency crop expansion during the war. It would scarcely have been possible to increase the production of wheat so rapidly had it not been for the existence of these field studies.

L. D. Stamp in his *Applied Geography* (1960) makes the observations, '... the unique contribution of the geographer on the holistic approach in which he sees the relation between man and his environment, with its attendant problems, as a whole, the relationship is discerned by survey in the field, and the gathering of facts systematically and objectively, but the twin goals of survey and analysis are achieved fully only when studied cartographically.' These survey and analysis procedures were extremely relevant to some of the most pressing world problems—population pressure on space, economic development and the improvement of living conditions. Many years later, in 1965, Stamp became Sir Dudley Stamp in recognition of his contribution to the survival of his country.

In 1946, the Town and Country Planning Act established a new survey to be done at a scale of 1/25,000 and to be kept continuously up-to-date. It was at the instance of Dudley Stamp that the legislation was passed. In Lisbon, at the International Geographical Congress, 1949, Stamp's proposal that a world landuse survey be established, was adopted, and an international commission was set up to supervise the work. The maps were to be published at 1/1,000,000.

British geographers studied public policy, welfare, and the social good. A growing concern with relevance, with the inequalities within society, environmental degradation, and the very structure of the capitalist system began to emerge. This point of view was, however, attacked, and a new set of political values were postulated and for the first time a political value-land (applied) geography was postulated, albeit by a minority.

The Institute of British Geographers (IBG) took initiative to work on various societal problems, and also sought to contribute to the solution of such problems. The debate on how geographers should contribute to the public policy was a major issue at 1974 conference of IBG. Since 1996, both the Royal Geographical Society and Institute of British Geographers started working together towards contributing in a greater way to the solution of practical societal problems.

The British geographers, however, have claimed that the applied geography in Great Britain started with the agricultural and landuse mapping under the supervision of Sir Dudley L. Stamp in 1930s, but a look at the history of the British geography has revealed the development of another kind of applied geography, involving the delimitation of political boundaries. Among the distinguished boundary specialists, the name of Thomas H. Holdich stands out. He became involved in boundary surveys in 1884 when he was appointed chief of the survey team to demarcate the boundary between Russia and Afghanistan. This was an applied work of very high relevance in contemporary central Asian political geography. And from that point of view, Thomas H. Holdich was 12 years ahead of Hugh Robert Mill, as his applied work on the boundary delimitation was done in 1884, while Hugh Robert Mill's suggestion for land-quality field-mapping using the sheets of the Ordinance Survey (1 inch to the mile) came in 1896.

Thomas H. Holdich was superintendent of Frontier Surveys in India from 1842 to 1848. At the beginning of the twentieth century, he was selected to establish the boundary between Argentina and Chile in Patagonia. He successfully used geographical methods in the solution of a particular set of problems, arising out of territorial disputes between Russia and Afghanistan; Argentina and Chile. Even the delimitation of the McMahon Line on the map (8 inch to the mile) by Great Britain and Tibet, delimiting the Tibet-Assam boundary in 1914 at Simla was another example of the application of geographical principles and skill to solve the vexed problem of the boundary which had remained ill-defined for centuries. It was an example of 'applied' work, because it was designed to resolve the problem of boundary delimitation in the Eastern Himalayan region. The British geographers and scholars who were in India during their colonial period had done extensive works of 'applied' nature.

In Great Britain, 'applied' geography was as developed as 'pure' geography, and in some cases, the development of 'applied' geography moved ahead of 'pure' geography, probably because of social and political necessities.

In **Germany** and **France**, 'applied' geography could not develop to the extent as it was developed in Great Britain. In fact, these two European countries experienced the development of 'pure' geography, designed to generate theoretical knowledge, or to identify the place of geography in the division of knowledge, intended to promote the status of the discipline within academics. Nevertheless, between the inter-war period, particularly in Germany, the transformation of political geography into geopolitik or the application of geographical principles and skill to political-territorial problems that Germany was made to suffer on account of the demilitarisation of the Rhineland, could be cited as an example of German geographers' attempt towards a problem-oriented applied study, with regard to the state's spatial requirements. In France, however, the involvement of geographers in comprehensive planning processes started since 1960s. After much debate in the 1960s and early 1970s over the question of applied geography (la geographie appliquée) as a contradiction of traditional values in the profession, the younger generation embraced it, both as a theoretical and as a significant source of employment (James and Martin, 1972, 198).

Geographical works of 'applied' nature, in **Russia**, began to pour into, between 1880 and 1914. In this regard, the contributions of the followers of Petrovich Semenov—A. I. Voeikov and V. V. Dokuchaiev—were of great significance and relevance. They were primarily research workers whose innovative studies of climate and soil provided model problems and solutions to those involved in problem-oriented studies in contemporary Russia. A. I. Voeikov was one of the contemporary late nineteenth century Russian geographers to have recognised the destructive effect of man's use of land. His work was comparable to the works of Mary Sommerville of Great Britain who worked much earlier and George Perkins Marsh of the United States, whose work echoed Sommerville's view-point with regard to man's destructive use of land.

V. V. Dokuchaiev's contribution to his innovative studies of soil was a milestone, in the sense that it attempted to identify and highlight the major soil forming processes and of the natural zonation of soils, according to climate, and then attempted to apply the concept of the method developed in the study, to make problem-oriented analysis of the soil studies. His views were enthusiastically received in the United States. Dokuchaiev trained a number of students at St. Petersburg who continued to develop his ideas, particularly, N. M. Sibirtsev, who died in 1900 at the age of 40 only, developed the concept of 'zonal' soils as distinct from 'azonal' soils—the former reflecting climatic patterns, the latter more closely related to underlying parent materials or such local conditions as poor drainage. K. D. Glinka, in 1908, brought Dokuchaiev's work outside Russia. Similarly, based on Dokuchaiev's work, L. I. Praslov compiled and prepared a soil map of Russia on a scale of 1/1,000,000—a project which was later carried forward by I. P. Gerasimov.

The development of 'applied' geography, in true sense of the term, took place during the Soviet period in Russia, when in 1918, the Commission for the Study of Natural Productive Forces was established in the Academy of Sciences, and one of its sub-divisions was the Department of Industrial Geography. The first assigned task was to make an inventory of Russia's natural resources. In 1920, in setting up studies on which the construction of a network of electric power stations and transmission lines was to be based, Lenin who was a strong supporter of economic regionalisation, specified that he wanted a 'geography of each of the sections of the electrification' and that he expected to see a map of the main power stations. Lenin insisted, '... the maps should clearly show the regions to be served by the central power station, the type of industry to be included and everything associated with these regional stations' (Saushkin, 1966, 9).

The State Planning Commission (GOSPLAN) was set up in 1921, but a special commission on regionalisation was appointed and given authority to recommend a national plan for the division of the national territory into functional units. The commission formulated the concept of economic region in 1922, which would be 'economically as integrated as possible ... would represent one of the links in the entire chain of the national economy' (Saushkin, 1966, 12).

GOSPLAN divided the Soviet Union into 21 such regions and then proceeded to a detailed study of each of them. It was in 1920, that L. L. Nikitin made his first investigation of the writings of the pre-revolutionary geographers. The older data were combined with the latest information regarding natural resources, physical conditions, population and types of economy in each of the regions. Never before in Russian geography had such a vast, diversified body of material been generated on a regional basis.

Russian geographers of younger generation got themselves involved in regional development, and many of them worked on problems of industrial location and resource development. One of the major policy decisions was the plan for the organisation of inter-regional combines that helped to develop regional plan in the early Soviet Union. N. N. Kolosovskiy was the author of the plans for the great Ural-Kuznatsk Industrial Combine. A theoretical model for an integrated industrial region, combining basic resources with steel production and with industries using steel, had been drawn up in 1927 for the Dniper Basin; but Kolosovskiy's plan called for the movement of raw materials and finished goods between regions.

There were two groups of geographers in the early Soviet period: the geographers who worked on practical problems of economic planning advanced professionally more rapidly than those confined themselves in the university teaching. To those involved in practical problems of economic building, the methodological and philosophical issues and discussion in geography appeared to be irrelevant, rather they were seeking for answers to 'real' problems. N. N. Baranskiy (1960) at the Research Institute in Moscow had consistently fought for regional approach because this seemed to him the only way that 'geography could make a useful contribution to practical question of economic development. However, in the 1930s, Stalin's policy of strong centralised control of the economy made regular approach less relevant and negated the whole idea of an 'economic' region as a 'major territorial production complex with a specialisation of natural significance'. Geographers had lost their 'practical' relevance during the Stalin-era. Soviet geography, during the period, or the whole of the 1950s, suffered from inherent inner contradictions, resulting from dichotomoies, and inconsistencies with regard to methodological and philosophical foundation. There was considerable debate between I. P. Gerasimov and V. A. Anuchin, and out of the debate developed what was then known as 'Constructive Geography' in the (erstwhile) Soviet Union.

Constructive geography was geography applied to the practical purposes of building the socialist economy. Geographic concepts and methods in those days of the 1960s had meaning in terms of what they contributed to economic development planning. In place of regional science, the Soviet geographers preferred the term 'landscape science'. Isachenko made use of new terms with old meaning to focus attention on applied geography as landscape science, which he related to the Dokuchaiev's 'law of zonality'. A classification of landscapes with a hierarchy of scales or degrees of generalisation was worked out for the whole of the Soviet Union to be mapped on a scale of 1/4,000,000. The main purpose of doing this was not just to provide maps for teaching purposes, but as a basis for identifying regions useful for planning processes. A landscape is a combination of interrelated environmental components (local climate, landforms, soils, plants, and animals) occupying a discrete territory. It exists objectively in the natural environment. But the study of landscape just for the purpose of describing them, said the Soviet geographer Gerasimov, was not enough: constructive geography must use this knowledge for the effective transformation of nature. He offered four examples of constructive geography:

1. The study of the geographies of natural and cultural landscapes of the wooded steppe in the central Chernozem zone. The purpose was to investigate the heat and water budget on the Earth's surface in virgin land and in cultivated land and then to experiment with various technical devices for controlling natural processes and to increase agricultural productivity.
2. The study of the irrigated lands of Central Asia. The purpose was to find ways to control salt accumulation, to use water more efficiently, and to increase the irrigated

area. Questions investigated include the fate of the Aral Sea and what the drying up of this body of water would mean for the total economy of the region.

3. The study of means of reclaiming the swamps of the Ob Valley by using properly placed dams and diversion canals. The study also includes the hydroelectric potentials.
4. Studies of the water in Lake Baikal for the specific purposes of reducing pollution, regulating the flow through the Angara River and finding new ways of putting this natural resource for better use.

Most of the urban studies undertaken during the 1960s and 1970s by the Soviet geographers, were meant for the practical purpose of providing the background for planning projects, and they were necessarily 'applied' in nature. Since 1970s, and particularly after the publication of V. A. Anuchin's *Fundamentals of the Management of Natural Resources* in 1978, a major shift was noticed among the Soviet geographers to devote more time to environmental and regional problems, and to demonstrate the practical value of an integrated geographical approach to their resolution. The Russian geographers in the 'defederated' Soviet Union have continued with this approach. They are now more concerned with growing social inequalities in the federation, and are seeking to address those problems through sustained efforts. The Russian geographers are now increasingly putting emphasis on welfare and environmental issues, and in this regard the achievements of the Geography Faculty of Moscow University and the Russian Academy of Geographers deserve special mention for their pioneering works in the field of applied geography.

The development of applied geography in **China** and **Japan** mostly occurred in the 1950s and has continued till date. Wen-Kiang Ting, a geologist trained in Scotland and Co-Ching Chu, a climatologist trained at Harvard, were the two earlier twentieth-century geographers in China who were credited with having given an 'applied' nature of area and region-based geographical works, before China was taken over by the communists and became the Peoples' Republic of China in 1949. Ting and Chu brought the British and American models of applied geography to China, but their geography was later abandoned by the young Chinese geographers on the ground of being 'imperialist' in nature—an instrument for aggression.

Nevertheless, the Chinese geographers identified geography as a fundamental science, because every kind of economic construction needed and required fundamental knowledge in geography. Geography was, however, given a prominent place in the twelve-year plan for the development of science and technology. Chinese geographers followed the Soviet model of 'constructive geography'. As a result, they gave a new impetus to the idea of resource inventory. Several of the physical geographers who had studied in Britain reported on L. Dudley Stamp's survey of land and landuse. In China, they started a similar inventory in the 1950s in spite of renunciation of the British model, and a large number of college and school teachers, students and other individuals were assigned to the work. In 1958, soil surveys in Szechwan involved 140,000 people and in Kwangtung more than 170,000.

In addition to these surveys of relatively large areas, the Chinese geographers also carried out detailed studies of small and critical areas. They gave much attention to the identification of microclimates, where measurements of heat and water balance were correlated with crop yields. In mountainous areas of the south, many enclaves of tropical microclimates have been discovered in which tropical crops of coffee, cocoa and rubber could be produced. Defining such microclimates was not enough; rather, the means of changing them for increased productivity were also sought and searched out. The river basins were studied in detail, and geographers were at work on preliminary studies of a huge new dam in the Yangtze gorge. A project was surveyed to bring water from the Yangtze system to the Yellow River basin. In the development of geography and the application of geographical principles and skills to practical problems, the role of the Institute of Geography of the Academy of Sciences in Peking needs to be highlighted, because it coordinated, and supervised all such research projects that were carried out with the objective of solution to major problems—physical and social. Chinese geographers were never involved in methodological and philosophical debates, rather they attempted for an integrated and geographic approach to demonstrate the practical value of the discipline to resolve practical problems.

Although **Japan** has a rich tradition of 'pure' geography, having been developed in the early years of the twentieth century, but the development of applied geography actually took place after the end of the Second World War when the Geographical Survey Bureau (in 1970 Geographical Survey Institute) was established in late 1945, to carry on a varied programme of mapping at different scales. The Tokyo Geographical Society was founded in 1879 and the Association of Japanese Geographers was established in 1925.

The Geographical Survey Bureau, founded in 1945, not only produced topographic maps, but also many thematic maps of such things as land classification and landuse, transportation, population and various kinds of economic data. The application of geographical methods to the study of practical problems of urban and regional planning was a notable development—notable because before the war very few professional geographers were in any way involved in planning operations. But in the contemporary post-War period, geographers were employed in considerable number in almost every city and perfectural planning commission and in the planning work of the larger metropolitan areas. A private agency, the Japan Centre for Area Development Research, has been undertaking since 1980 studies relevant to the policy problems of Japan as a whole.

Applied work has been done by Japanese geographers in the areas of social and regional construction, and regional planning, agricultural productivity, economic reconstruction, landuse and land classification, environmental and resource management, disaster management, agricultural regionalisation, trade and commerce, Pacific Ocean and defence, etc.

In the **United States**, the first major breakthrough in the use of professionally trained geographers to study practical problems came during World War I and its aftermath, although some works of this nature were done in the late nineteenth century and early twentieth century also. However, during the 1920s and 1930s not only were certain wartime projects continued and completed, but new kinds of applied research were undertaken. A large number of geographers were called into both military and civilian service during World War II, and since then opportunities for employment in various branches of government and also in private business firms have grown rapidly. Many studies involving the geographical analysis of location or of areal spread were still made by non-geographers, but in the contemporary period there was growing appreciation of the value of the professional training that the geographers underwent.

One of the first uses of the knowledge and skills of the trained geographers was in connection with certain commodities studies. The work had to be done under the patronage of the Division of Planning and Statistics of the US Shipping Board. It was a wartime need. However, the board was asked to classify all imports into one of three categories: those commodities so necessary to the war efforts that all available supplies must be imported; those commodities essential to the war effort, but not in such a short supply that all available production had to be imported; and those commodities not required for military or civilian needs. A number of geographers, under Walter S. Tower who was made the head of the Commodities Section of the Shipping Board, were appointed to prepare the report on the commodity supplies which was highly appreciated. The discovery that geographers had useful skills for commodities studies led to the setting up of another Division of Planning and Statistics as the War Trade Board. The geographer selected to head the research unit was Harlan H. Borrows, who recruited few more geographers of his choice to complete the project/study.

The US President Wilson had set up an organisation under Col. E. M. House that would gather the most complete collection of information possible and prepare it for use at the coming Paris Peace Conference. The group which was formed, called the Inquiry, consisting of some 150 specialists, including a good number of geographers. A major part of the work that the geographers were assigned was in the field of cartography in which a map-making programme was made, showing pre-War political boundaries, the complete drainage system, the roads and rail roads, and the cities and towns. Some of the maps were prepared on a scale of 1/1,000,000 or 1/3,000,000, and a few maps were prepared on a scale of 1/250,000. Isaiah Bowman, who was the director of the American Geographical Society (AAG) since 1915, supervised the geographical studies. He wrote a book, based on his experiences at the Paris Peace Conference in 1919, entitled: *The New World,* and it was perhaps one of the most authoritative studies in political geography. It provided informative detail of practical problems of particular regions, with adequate description of the local setting and the historical background so that the readers could understand what was going on in the post-War world. For many years after the War, Isaiah Bowman and the American Geographical Society were involved in boundary studies and programme of mapping. Bowman not only settled the European boundary problems at the Paris Peace Conference, but also offered solution to the Guatemala-Honduras boundary problem.

Under Bowman's direction, the American Geographical Society had undertaken a major research programme leading to the compilation of a map of Hispanic America on a scale of 1/1,000,000, conforming to the standards and format of the international map of the world. Although the Hispanic America project sponsored and supported publication of a series of research studies, but the most important is the study of the Central Andes by A. G. Ogilvie, the Scottish geographer. It was based on the compilation from the drawings of the Millionth Map and from Bowman's copious notes. The map was used to help adjudicate disputes between Chile and Peru in 1925, Bolivia and Paraguay in 1929, Columbia and Peru in 1932, and Columbia and Venezuela in 1933. The Hispanic America Map, completed in 1946 in 107 sheets (over 300 square feet in extent) was a contribution to the millionth map of the world.

Another quite different application of geography to the solution of practical problems had to do with studies of land quality and landuse. It was long recognised that detailed information concerning land resources was needed if plans for better resource use were to be properly guided. In fact, efforts to classify land in terms of its potential use began in the early days of the independent United States and were carried to new levels of utility in the Great Surveys of the West and in the work of the US Geological Survey in the later part of the nineteenth century. It was in that perspective and programme that the Michigan Land Economic Survey was planned in the 1920s. Carl O. Sauer and Wellington D. Jones provided the leadership to the Michigan land economic survey. Later they were joined by Arbor Sauer and Parish S. Lovejoy—the forester and naturalist. By the mid-1930s, detailed mapping was done with land quality and capacity of the Michigan Area which was later defined as being a distinctive contribution to the expanding field of land classification and landuse planning.

Similarly, Harlan H. Borrows had made three major contributions to water-planning studies. He played an important role in the formulation of policy regarding multipurpose river development projects that were developed during the 1930s. Besides, he had designed the procedure for integrated regional studies as essential basis for policy planning. Between 1935 and 1938, he drew up a plan which was accepted by the states of Colorado, New Mexico and Texas for the allocation of the water of the upper Rio Grande.

He also offered cooperative solutions for the water of the Pecos River, the Red River of the North and the Columbian Basin.

A good number of geographers under the direction of G. Donald Hudson were employed in the Department of Regional Planning Studies, with the objective of preparing a detailed survey of the Tennessee River Basin. The basin was, however, surveyed by the unit area method to provide an inventory of the nature and extent of land resources. Between 1941 and 1943, with the help of the so-called area analysis method, many small depressed areas in different parts of the United States were studied, and the approach was found to be useful in the guidance of efforts to rebuild the depressed economies. The so-called area analysis method was based on an outline to be filled in by field observation and was summarised under four main headings: (a) the employment pattern; (b) the condition affecting employment and income (natural resources, economic activities and institutions), (c) direction of desirable adjustments and (d) the proposed programme of remedial action.

Since Word War II, the whole approach to studies of this kind has been revolutionised through the use of computer programs, with data supplied from new remote sensing devices.

The American geographers, particularly G. Donald Hudson and Clarence F. Jones had assisted Latin American geographers in the field of land classification studies. Some of the Latin American geographers were trained in the United States, for example, Rafael Pico (who was trained in the Clark University, and had obtained his PhD degree from the university), was chairman of the Puerto Rico Planning Board. The Puerto Rico Rural Land Classification Survey was carried out between June 1949 and August 1951. The survey, however, demonstrated the value of this kind of inventory. The survey information was used to replan the landuse pattern, and also to plan the routes of new roads and to locate the numerous small manufacturing plants in relation to population and accessibility.

Since that time, many somewhat similar surveys were undertaken in Latin America, some by the Organization of American States, some by the Agency for International Development, and some by Latin American government agencies, as in Brazil. Some surveys as in Chile were done by private agencies of the United States on contract.

Isaiah Bowman was allotted fund by the Social Research Council to make an in-depth study of pioneer settlement. In fact, in 1925, he turned his attention to problems relating to thinly populated areas on the margin of settlements, and for that, had submitted a proposal to the National Research Council to obtain funds for studies of pioneer area. His proposal was finally endorsed and approved by the NRC. He also got funds from the Council of American Geographical Society.

Pioneer belt studies were not solely concerned with the possibilities of new settlement, but such studies were also concerned with the need to withdraw from less favourable places. Bowman proposed to study pioneer movements all around the world and to identify certain general conditions: not only the kinds of physical institutions considered favourable, but the attitudes and objectives that led people to become pioneers, and the economic, social and political institutions that could best support pioneers. Bowman also proposed to investigate the particular and unique conditions in specific pioneer areas, knowledge of which would be essential to the transformation of policy. His proposal ranged widely over all the fields of the social sciences and was essentially interdisciplinary in concept. His book *The Pioneer Fringe* in 1931 dealt with the nature of the problems and offered examples from the western United States, Canada, Australia, southern Africa, Siberia, Mongolia, Manchuria and south America. A volume containing 27 cooperative studies of particular pioneer regions came out in 1932. Finally, Bowman in collaboration with Karl Pelzer, summarised the results of the whole undertaking in a report on the world's potential pioneer areas in 1937.

The demand for the services of geographers in World War II far exceeded the supply of experienced and properly trained professionals. Geographers were required in all kinds of work performed during World War I and also in many research studies. They worked as commissioned officers and non-commissioned drafters assigned to intelligence agencies. Some of the geographers helped to prepare the compilations of information about countries or parts of countries either as a basis for planning military operations or as a guide to military government after the war. A large number of geographers in the Office of the Strategic Services were assigned to cartographic work, while others worked on various countries where the Joint Army-Navy Intelligence Service (JANIS) handbooks were needed. Many geographers worked on special problems and prepared background reports for the guidance of those who were responsible for decisions. Geographers prepared specific atlases which provided information of environment of the areas regions, with regard to the variety of equipment necessary to carry out field operations. This work was continued and expanded after the war. Geographers were required to make detailed studies of the beaches and the terrain behind beaches of the enemy-occupied territories for the landing of the allied troops.

One group of geographers received special training in the study of transportation facilities, but the very important function that could be performed only by an experienced regional specialist was the interpretation of capabilities and intentions of foreign countries. The group of geographers, dealing with the study of transportation facilities, undertook to provide descriptions, photographs and maps showing the condition of transportation facilities immediately after the armies started their advance eastward. The materials gathered proved to have great practical value and utility for those commanders in charge of logistics.

However, those geographers who had called themselves regional specialists had focused on parts of the United States or Latin America. The number of geographers who had specialised in European, Asian or African countries was very small. As a result, the work of foreign area interpretation was done by language specialists, historians or others who happened to have a familiarity with areas in question. The further result was that many persons—geographers and others—who were assigned to such positions proved inept and unreliable. To E. A. Ackerman (1945): 'In the three years from 1941 to 1944 American geographers dealt almost constantly with a series of difficult professional problems... Nevertheless, geography unquestionably has wider recognition than ever before in this country.... War-time experience has highlighted a number of flaws in theoretical approach and in the past methods of training men for the profession.... The geographer perfectly or even adequately trained for the specialty into which he was thrown has generally been exception. The unfamiliarity of most young American geographers with foreign geographic literature: their almost ignorance of foreign languages, their bibliographic ineptness, and the lack of systematic specialties are but a few points which may be cited in proof.' This situation of ignorance and lack of professional training in the systematic geography was attributed to a widespread belief in the essential quality of the subject.

However, it is equally important to mention here that despite a limited failure, American geographers had played a very decisive role in conceptualising the US foreign policies and relations in the war-torn world, and for that Isaiah Bowman's role is required to be highlighted. He was made a member of the Stettinius Mission to London (1944), at the Dumbarton Oaks Conference (1944) and the San Francisco Conference (1945), by the US President, and at these Missions and Conferences, he (Bowman) showed the relevance of geography and its application to resource inventory. At all stages, leading to the creation of the United Nations Charter, Bowman put forward the relevance of the geographical point of view and its value.

The application of geography to practical problems took place in many other sectors before the decade of the 1950s. One of these was in marketing research for private business firms. William Applebaum had prepared a thesis on secondary commercial centre of Cincinnati which had a significant applied value. Since 1931, and especially since World War II, most business firms engaged in selling to the public added market research departments, and the demand for persons with geographic training to work in the field of market research increased considerably.

In 1961, a whole issue of *Economic Geography* was devoted to a series of papers detailing examples of this kind of applied research. Economic geographers had been making studies in the location of economic activity ranging from iron and steel plants to floor millings.

Geographers such as Gilbert White and Joseph A. Russell made studies of natural hazards, including major river floods, and of military matters from the geographical point of view. A number of American geographers have worked with the agencies of the United Nations, and have made environmental studies in the wake of ever-increasing pollution. They have studied the effects of weather and climate upon humans (physiological climatology). M. I. Glassner had advised the government of Nepal in negotiating a transit treaty with India and had functioned as consultant to the UN Development Programme for the land-locked countries of Asia. There are almost endless applications of geographical point of view to the problems of the real world (James and Martin, 1972, 362–64).

In **India,** geography is yet to receive its 'legitimate official recognition' as a problem-solving discipline like that of 'economics' which is regarded as the only possible problem-solving discipline in social science. Economics is, thus, much ahead of geography in terms of application of its principles and skills to the problem. It is a fact that applied geography is developed in those countries which are in the category of 'developed' nations, and where geography is given due recognition. Here, in India, the decision-making elites do not consider geography as a potential' discipline, in terms of its social and political, and economic relevance. Geographers in India are mainly involved in the theoretical aspects of the discipline, or in other words, they are more concerned with the development of 'pure' geography, rather than 'applied ' geography. Perhaps, the Indian geographers have failed to justify the importance of the application of geographic principles and skills to the practical problems, and this could be one of the reasons that both the public and private sectors do not require geographers' involvement in their market researches. Even in the field of planning, the role of geographers in India is very limited. Very few geographers have really shown their worth in the various kinds of planning, including the spatial planning.

K. V. Sundaram was one of those few Indian geographers to have been appointed at the Town and Country Planning Organization, Delhi in 1958, where he served for 15 years. For the next 14 years, he rose from the Joint Director to Joint Advisor at the Planning Commission. In the decade of the 1990s, Sundaram worked as a consultant under the banner of United Nations and other international organisations like United Nations Centre for Regional Development, UNESCO, Food and Agricultural Organization (FAO) and Centre on Integrated Rural Development for Asia and Pacific. His geography was more 'applied' than 'pure' though he attempted to bring together both 'pure' and 'applied' geography into one unified geography. His work is mostly in the field of the various types of spatial planning, including regional and area-based planning.

L. S. Bhatt was attached with the Planning Commission and was member of Economic Advisory Council, Government of Uttar Pradesh, beside being the Technical Adviser to the Regional Development Plan for Western Ghats. He was also a member of the State Planning Board, Kerala. He has a number of applied research projects to his credit whose operational values in the field of planning were recognised.

The late V. L. S. Prakasa Rao, the late Moonis Raza, C. R. Pathak and R. P. Mishra have developed the theoretical aspects of applied geography, particularly those related to regional planning. They have suggested different ways and means regarding the application of geographic skills and principles to practical problems-development.

Though the government agencies like the University Grants Commission and the Indian Council of Social Science Research, both in New Delhi, have been providing funds to geographers and other social scientists to carry out research projects pertaining to social and physical problems, but the reports produced on the problems and once submitted to the agencies, hardly come out. Nevertheless, numerous reports are gathering dust in these agencies; they could have provided remedies to the problems analysed in the reports, had they been published.

Nonetheless, some departments of geography in India have done commendable work in the field of national concern in which the relevance of geography to practical problems was realised and understood. The Planning Commission in 1959 sanctioned a project, named the Damodar Valley Diagnostic Survey to three departments: Calcutta, Patna and Kharagpur. Similarly, after the Sino-Indian War of 1962, the Terrain Analysis Cell of the Ministry of Defence approached geography departments to undertake projects of terrain evaluation in areas of specific needs. The Terrain Evaluation study in Bikaner and Jaisalmer Districts (1967–69) by the department of geography, University of Delhi, and Terrain Evaluation of Punjab Plains, Punjab Himalayas, and Shipki-tract, Rajasthan border, 1965–68, by the department of geography, Panjab University, Chandigarh, exemplified efforts to fulfill defence needs of the country (Kapoor, 2002, 19). These works had immense strategic and practical values, and recommendations that these works had made were executed in full spirit by the concerned authorities.

However, it needs to be mentioned here that no such responsibility was attached to any departments of geography in India, after the submission of the terrain evolution report by the Ministry of Defence, Government of India. The various Ministries of the Central Government and the State Governments, and other public and private agencies have never assigned any such responsibilities to individual geographers, or group of geographers, or department of geography in the past decades to carry on studies from geographical point of view, designed to offer remedial measures to practical problems. There are various areas where geographers can do better than other disciplines of social science, because of their spatial outlook and cartographic/map-making capacity. There is a demand for greater involvement of geographers in the nation-building process in India, largely as a response to the challenges in social content, but for the government apathy towards geography, it has not yet become possible.

Applied Geography—an Overview

Following Grano (1981), Taylor (1985) identified two external influences on the disciplinary developments of geography: within academic geographers had to be given an intellectual foundation to satisfy intellectual peers, and outside in the wider world geography had to be justified as a useful activity on which to spend public money. These produce 'pure' and 'applied' geography respectively. Both are necessary to a discipline's future, but the relative emphasis placed on each will vary over time and space: 'Outside pressures will be particularly acute in periods of economic recession when all public expenditure has to prove its worth. All disciplines will tend to emphasise this problem-solving capacity and we can expect applied geography to be in the ascendancy.... In contrast, in a period of expanding economies and social optimism, outside pressures will diminish and academics can be expected to be under less external pressure. Geographers will thus be able to contemplate their discipline and feel much less guilty about this activity. We can expect bursts of pure geography to occur in these periods'.

There is a kind of cause-effect relationship between the cycles of 'pure' and 'applied' geography and cycles of economic prosperity and recession and to quote Hagerstrand (1977, 329): 'when the world is stable and/or unhampered liberalism prevails, then there is probably not much to do for geographers except surviving in academic departments trying to keep up competence and train school teachers in how wisely arranged the world is'.

The period since late 1970s witnessed applied geography in the ascendancy, an expectation generally supported by experience of those years. Pressure to justify the discipline in terms of utility to economic goals was strong and the search for research contracts of all types became much more determined. Attempts were made to promote the discipline that emphasised its utilitarian aspects. That pressure for applied geography has not been released, largely because the attitude within the state apparatus, particularly but not exclusively in the western countries, did not favour a return to the type of funding of higher education, which would sustain a further period of pure geography. Instead, academic departments were continually required to canvass for research funds from outside sponsors.

The desirability of a continued period within which applied geography is on the rise has been the forms of further debate. Some clearly accepted and argued for that ethos, but their main concern was either with geography using its perceived particular skills, what may be called the application of status quo theory, or with exploring how geographers could make a committed contribution towards the achievement of change (Clark, 1982).

According to Bennett (1985, 212), 'The role of geography to response to the increasing stimuli provided by "applied", social, political and policy issues which are tending to shift the emphasis of the subject towards frame disciplines and away from a central geographical core.' He argues that underpinning the shift away from both spatial science and statistical inference as central methodologies for geography has been an increasing emphasis on its applied or 'frame' disciplines. By such a shift, says Bennett, geography naturally becomes meshed with the methodologies of other disciplines. The more geography becomes enmeshed in its applied areas, the more important this subjective analysis becomes, deriving from and used in conjunction with the techniques and methods of quantitative analysis.

There could be a tendency to fragmentation of the discipline as a result of its interlinkage with the methodologies of other disciplines. And, there is a danger to geography because other disciplines have become more 'effective' at academic 'imperialism' than geography, and the overall trend of academic value has tended towards sub-specialisms and away from the broader, all embracing approach which was essential to the traditional nature of geography, and must still be retained as the essence of geographic method. Thus, although it is essential for geography to become involved in applied research, this is not without its danger, in the present intellectual climate, but it can be resolved only by re-asserting the geographical core—that it combines physical and human geography, and defines the role of geography as an environmental and social study, not as mere social science.

Nevertheless, one of the driving forces for the development of applied studies has been the need for geomorphology to re-examine its position at the interface of Earth, environment, social and economic sciences. The need to be more relevant to geography, one of its parent disciplines, has prompted many geomorphologists to work in areas where they can contribute to the interests of human geographers, planners, and environmentalists. It is obvious that geomorphological applications are in harmony with the growing awareness by human geographers and with public and political interests in the environmental management problems. Geomorphologists are now required to have an understanding of the socio-economic and technical planning of their surveys so that they are fully compatible with the needs of the social science disciplines and those of urban planners.

Jones (1983) has successfully argued that there is clear area, the man-environment arena, where the geosciences and socio-economic-political sciences overlap and there is an urgent need for research on: (a) natural hazard impact, (b) environmental auditing, (c) resource assessment, (d) impact assessment and (e) ex-post audit or ex-post assessment. Such work, all of an applied nature, clearly states one possible future for geomorphology and indeed Earth sciences as a whole in which it strives to re-establish its links with the 'human' sciences.

19

Scientific Method in Geography

HIGHLIGHTS

- Scientific Method
- Key Elements in Scientific Approach
- Theory and Fact
- Law
- Logic
- Reductionism
- Hypothesis
- Key Elements in Scientific Practice
- Routes to Scientific Explanation: Induction and Deduction
- Geographical Application of Scientific Method: Some Problems
- Relevance of Scientific Method In Geography

Ever since geography underwent a positivist-led revolution in the 1950s of the last century, followed by a paradigm-shift that necessarily replaced an earlier 'idiographic' concern with 'areal differentiation' by 'nomothetic' search for models of 'spatial structure', the discipline appeared to have grown into a 'science' with certain characteristics. To most of the young geographers of the Anglo-American realm, who were trained in mathematics and statistics, and to some of them who drifted and/or shifted to geography, from the natural sciences, geography appeared to be a true 'science'—a law-seeking discipline rather than an idiographic science. Geography, according to them, sought to strive for an increasingly nomothetic component. Its fundamental research, according to Ackerman (1958, 17), however, 'need not necessarily be law-giving.... Much fundamental research in geography has not been law-giving in strict sense but it has been concerned with a high level of generalization, and it has given meaning to other research efforts which succeeded it. In this sense it has a block-building characteristics... and is likely to rest on quantification.... Accurate study depends on quantification and should furnish a theoretical framework with capacity to illuminate actually observed distribution pattern and space relations.'

The positivist-led geography then called for the development of theory, for the application of quantitative methods, and for a focus on laws and generalisations to form the building-block for further nomothetic research. As a 'science', geography with its positivist character, necessarily concerned itself with empirical questions (those with a factual content) and not with 'normative questions' (questions about values and intentions). Empirical questions are questions about how things are in reality. In this context, 'reality' is defined as the world which can be sensed. This means that science is concerned with objects in the world. Science can describe how things are, and discovers the association of causes which explain why things are as they are.

Contemporary positivist-led geography, therefore, sought to investigate two pertinent questions of geographic character: (i) 'theoretical' questions that deal with the formulation of empirical generalisations, explanations or general laws, perhaps even with basic theory, and with the methods of drawing logical deductions, and (ii) 'methodological' questions that have to do with experiments in new methods of study new techniques of observation and analysis, or new cartographic methods. And researchers moved towards that direction to explore more possibilities within the scientific framework. It was the positivist-led modern geography with a nomothetic search of models for spatial structure that necessitated the application of scientific methods in geography, for 'geography could be regarded as a science concerned with the rational development, and testing of theories that explain and predict the spatial distribution and location of various characteristics on the surface of the Earth' (Yeates, 1968, 1).

SCIENTIFIC METHOD

Science is popularly defined as an accumulation of systematic knowledge. It is a method of approach to the entire empirical world, i.e. to the world which is susceptible of experience by man. It is, furthermore, an approach which does not aim at persuasion, at the finding of ultimate truth, or at conversion. It is merely a mode of analysis that permits the scientist to state propositions in the form of 'if – , then –'. Thus, no matter how systematic any body of knowledge may be, it is not science if it merely begins with axioms, or 'self-evident' propositions, and ends with deductions from those axioms. The sole purpose of science is to understand the world in which man lives.

The term 'scientific method' denotes the logical structure of the process by which the search for trustworthy knowledge advances. The primary task of scientific method is to explain empirical phenomena. There is no need to argue that geography ought to be a science. Geography simply is a science by virtue of the fact that it is a truth-seeking discipline whose raw materials consist of empirical observations. There is no suggestion that geography should undergo any sort of epistemological restructuring (Marshal, 1985, 113).

Hay (1985, 129) identifies four categories or groups of geographers with regard to the appropriateness or inappropriateness of scientific method in geography. The first group consists mainly of physical geographers who believe that their discipline is a field of a natural science and, therefore, do not doubt that scientific method is appropriate. In the second group are those human geographers, who see scientific methods as being appropriate to their discipline as social science, although they may also recognise that such an application poses certain problems not encountered in classical natural sciences like physics and chemistry. The third group consists of those who believe that the subject matter of geography makes scientific or quasi-scientific methodology inappropriate. Most recently, a fourth group has emerged which seeks to apply Marxist methods in geography and believes that such methods are scientific not in the mould of classical natural sciences.

In order to understand these differing views, it is necessary to identify the key elements of scientific thinking and practice, to outline some of the philosophical problems involved in scientific method not always evident to scientists themselves, as well as to examine some additional issues which arise when scientific method is applied to geography and also to similar disciplines.

KEY ELEMENTS IN SCIENTIFIC APPROACH

The scientific method is often characterised by five elements: theory (and fact), law, logic and reduction, which necessarily sustain the scientific thinking and/or approach. However, there is one more element, i.e. 'hypothesis', that also sustains and provides the required input to scientific practice and process in geography, and also in related sciences.

THEORY AND FACT

Basic to modern science is an intricate relation between theory and fact. Popular opinion generally conceives of these as direct opposites. Theory is compared with speculation and thus a theory remains speculation until it is proved. When this proof is made, theory becomes fact. Facts are thought to be definite, certain, without question and their meaning to be self-evident. Theory (speculation) is supposed to be the realm of philosophers. Scientific theory, therefore, is thought to be merely the summation of facts which have been accumulated upon a given discipline and/or subject. It is, indeed, a fact that (1) theory and fact are not diametrically opposed, but inextricably intertwined; (2) theory is not speculation; and (3) scientists are very much concerned with both theory and fact.

A fact is regarded as an empirically variable observation. It could never have produced modern science had it not been gathered. Theory refers to the relationship between facts, or to the ordering them in some meaningful way. It also refers to an organised and coherent body of assumptions and arguments. It may be directed to the explanation of a unique phenomenon (Wegner's theory of continental drift had only one world to consider) or a whole class of phenomena (the theory of air masses). Theories may be used to account for different phenomena. Without theory, science could yield no prediction. Without prediction there would be no control over the material world.

It can, therefore, be said that the facts of science are the product of observations that are not random but meaningful, i.e. theoretically relevant. Thus, facts and theory are interrelated/intertwined in many complex ways. The development of science can be considered as a constant interplay between theory and fact.

Theory is a tool of science in five ways: (1) it defines the major orientation of a science, by defining the kinds of data which are to be abstracted; (2) it offers a conceptual scheme by which the relevant phenomena are systematised, classified and interrelated; it summarises facts into (a) empirical generalisations and (b) systems of generalisations; (4) it predicts facts; and (5) it points to gaps in our knowledge. However, facts are also productive of theory in five ways: (1) They help to initiate theories; (2) they lead to the reformulation of existing theory; (3) they cause the rejection of theories which do not fit the facts; (4) they change the focus and orientation of theory; and (5) they classify and redefine theory (Goode & Hatt, 1952, 8).

The Role of Theory

1. **Theory as orientation.** A major function of a theoretical system is that it *narrows the range of facts* to be studied. Any phenomenon and object may be studied in many different ways. Each science and each specialisation within a broader field abstracts from reality, keeping its attention upon a few aspects of given phenomena rather than upon all aspects. Only thus can the work of science be managed. The broad orientation of each field, then, focuses upon a limited range of things, while ignoring or making assumptions about others. It is in the light of these considerations that much of the nineteenth-century geography may be understood, for a major task of great geographers of the century was to define the task and object of the study for the future science. Theory, then, helps to define which kinds of facts are relevant.
2. **Theory as conceptualisation and classification.** Every science is also organised by *a structure of concepts,* which refer to the major processes and objects to be studied. It is the relationships between these concepts which are stated in the facts of science. Such terms make up the specialised vocabulary that the scientist uses. They change as the science develops, for different phenomena come to be of major importance. However, it is clear that if knowledge is to be organised, there must be

some system imposed upon the facts which are observable. As a consequence, a major task in any science is the development of systems of classification, a structure of concepts and an increasingly precise set of definitions for these terms. Much of geography witnessed the development of elaborate conceptual schemata. These pointed to certain phenomena as the most important to be studied and thus helped to organise the facts of man-land relations and social relations as well.

3. **Another task of theory: Summarising.** A further task which theory performs is to summarise precisely what is already known about the object of study. These summarises may be divided into two simple categories: (1) empirical generalisations, and (2) systems of relationships between propositions.

Although the scientist may think of his field as a complex structure of relationships, most of his daily work is concerned with a prior task: the simple addition of data, expressed in empirical generalisations. Geographers may be studying the 'impacts' of the environment upon man or vice-versa, in order to summarise the observation in a set of descriptions. The demographer may tabulate births and deaths during a given period of time in order to ascertain the crude rate of reproduction. These facts are useful, and are summarised in simple or complex theoretical relationships. As a body of such summarising statements develops, it is possible to see relationships between statements. A number of summarising propositions on various social actions and activities within the environmental perspective can be made, but they can also be seen as related to one another: ways by which a group gives a different status to an individual, pattern of asserting group control, ceremonial expression of group unity, etc. Theorising on a still longer scale, some may attempt to integrate major empirical generalisations of an era. From time to time in any science, there will be changes in this structure of relationships between propositions. However, major shifts occurred in the mid-nineteenth century when a section of geographers, particularly German geographers, moved from older systems of theory towards a more acceptable system.

It is through systems of propositions that many of our common statements must be interpreted. Facts are seen within a framework rather than in an isolated fashion. Let us look at a few examples: 'Nature is merely an adviser'; 'Migration is a reflection of cultural diffusion'; 'The crime rate is higher in slum areas than in middle class areas'. If we study such apparently simple statements more closely, it will be clear that behind each of them is a complex series of observations, a set of assumptions about the effect of social factors upon behaviour, and a system of propositions about the way in which groups act. There is an implicit or explicit fact-chain or theory which gives such simple statements their full meaning. Theoretical clarity demands that the scientist must be more conscious of the thought system being employed than is the average man.

4. **Theory predicts facts.** If theory summarises facts and states a general uniformity beyond the immediate observations, it also becomes a prediction of facts. This prediction has several facts. The most obvious is the extrapolation from the known to the unknown. For example, we may find that educated women have relatively lower fertility rate as compared to the illiterate women and as a result the crude birth rate among the educated women declines. Thus, if education is spread or introduced into a native rural culture one can find this process taking place, i.e. of falling birth rate with increasing literacy and education. We can, thus, predict that in a region where education has already been introduced, the region experiences declining/falling birth rate.

We expect the same patterns simply because: (1) we believe we know which factors cause these patterns, and (2) we believe that these factors will be found in the new situation. This is a common sense way of saying that *behind our empirical generalisations is a body of theory.* A given theory may be incorrect, but it does make prediction about observation of phenomena. It is a set of directions, stating how certain operations, observations and calculations are to be made with a prediction about the outcome.

5. **Theory points to gaps in our knowledge.** Since theory summarises the known facts and predicts the facts which have not yet been observed, it must also point to areas which have not yet been explored. As noted above, the simple fact of prediction suggests where to test the knowledge. If a theory states a general relationship, such as inverse correlation between income and fertility, we can see immediately where further facts might be sought. We can break our income classes into smaller groups to see whether fertility might be higher (instead of lower) at the extreme upper income groups; we can ascertain whether this pattern is to be found in rural areas or in urban areas, or in other countries; or we can study the historical relationship between income and fertility. These are only examples. However, theory also points to gaps of a more basic kind. While these gaps are being filled, changes in the conceptual scheme usually occur. It might be noted, in addition, that 'seeing the gap' is very easy once it has been seen.

An example may be taken from man-environment relationship. Although a substantial body of knowledge was built upon concerning the impacts of environment on human activities and its control over the activities of human groups with relatively inferior 'genre de vie', but no attention was paid to the responses expressed by human groups with superior 'genre de vie' possibilists. This was a major gap in the deterministic theory. Such a gap would not be visible if our facts were systematised and organised. As a consequence, it can be said that theory does suggest where our knowledge is deficient. However, it becomes clear to us why one research problem is productive and another sterile.

Scientific Theory

'A scientific theory may be considered as *a set of sentences expressed in terms of a specific vocabulary.* The vocabulary may contain primitive terms which cannot be defined and "defend" terms which may be formed from the primitive terms. The sentences may similarly be divided into primitive sentences—axiomatic statements—and derivative sentences—theorems… terms such as "point", "line", "place" form the primitive terms collected together in an initial set of axiomatic statements…. In addition to the primitive term and the axiomatic statements, scientific theories also possess certain rules which govern the formation of the derivative sentences…. But a theory is useful in empirical science only if it is given some interpretation with reference to empirical phenomena…. The text of the theory provides a translation from the completely abstract theoretical language to the language of empirical observation…. The text of a theory not only identifies the empirical subject-matter which the theory refers to. It also identifies the domain of the theory… (that) may be regarded the section or sections of reality which the theory adequately covers' (Harvey, 1969, 88–89).

Harvey (1969, 90–91) identifies the following *advantages of a formal statement of a theory:*

1. 'The formal statement of a theory requires the elimination of inexactness and, as a consequence of this, ensures complete certainty as to the logical validity of the conclusion. The empirical success of a theory relies entirely upon the success of the text in linking the abstract symbols of the theory to real world events'.
2. 'The elaboration of formal theory, provided the basic postulates are good ones, can help suggest new ideas, prove unsuspected conclusions, and indicate new empirical laws.'
3. 'The formal statement of a theory requires turning a spatial or temporal sequence into a completely non-spatial and non-temporal set of relations. Even in theories which explicitly include time or space as variables the treatment is abstracted…. It is "therefore" characteristic of formal theory to state all propositions—whether primitive or derived—

as if they were universal propositions. Again, it is the text which has to perform the difficult task of linking these universal propositions to empirical events which have a location in space and time.'

It appears from the above points that the key problems in the application of formal theory in the empirical sciences is the provision of an adequate text. The text associated with a formal theory necessarily performs two essential tasks: (i) it identifies an abstract symbol with a particular class of real-world phenomena, and (ii) it may place the abstract symbols within a particular context which may include specific mention of location in space and time.

The text associated with a formal theory should not only link abstract symbols with abstract concepts. It should also specify how abstract concepts may be reduced to factual statements. This, therefore, raises the whole problem of the nature of concept-formation in the sciences and the operational problem of giving idealisation and theoretical concepts some adequate definition. There are innumerable idealisations to be found in both the natural and social sciences. Indeed, explanation would be impracticable without such idealisations. Some idealisations can be theoretically defined, and they are characteristic of the natural sciences where physics, in particular, has achieved a high degree of unification in its theoretical structure. However, many idealisations in both the natural and social sciences cannot be referred to any well-established theoretical structure, either because the idealisation is itself inappropriate, or because the requisite general theory is yet to be established. Idealisations in social sciences are fundamentally different from the theoretical concepts of natural sciences. The reasons: the text is deficient and/or weak, and the domain not clearly defined. The failure to achieve significant explanatory power in social sciences is the result of the paucity in such disciplines of the requisite general theory.

The universality of the theoretical statements may be transferred to any situation in space and time since it appears that the statements are universal in fact. In the social sciences this is not so, and therefore one of the important functions of the text of a theory is to identify the domain of objects and events to which such theories can be applied—this domain may simply be defined by a set of spatial and temporal coordinates. *A theory without a text and a well-defined domain is useless for prediction.* However, the greater success of the physical sciences, relative to the social sciences, in providing a text for theoretical structures accounts for the greater predictive success of the natural, as opposed to the social, sciences.

A scientific theory, however, needs to be tested or assessed not only for its own internal consistency, but also for its consistency with the world as observed

The Role of Fact

Theory and fact are in constant interaction. Developments in one may lead to developments in the other. Theory, implicit or explicit, is basic to knowledge and even perception. Theory is not merely a passive element. It plays an active role in the uncovering of facts. Similarly, fact has an equally significant part to play in the development of theory. *Science actually depends upon a continuous stimulation of fact by theory and of theory by fact.*

1. **Fact initiates theory.** Many of the human-interest stories in the history of science describe how a striking fact, sometimes stumbled upon, led to important new theories. This is what the public thinks of as a 'discovery'. Many of the stories take an added drama in the retelling, but they express a fundamental fact in the growth of science, that an apparently simple observation may lead to significant theory. Merton (1949) has called this kind of observation 'the unanticipated, anamolous and strategic datum'. Attempting to account for the anomalous datum, however, appeared to have led to an interesting development of theory.

Almost every 'discoverer' was preceded by others who saw his discovery first and thought no further about it. This was the case with most of those who attempted for the unanticipated, anomalous, and strategic datum for the development of theory. The fact initiates theory only if the scientist/researcher is alert to the possible interplay between the two.

2. **Facts lead to the rejection and reformulation of existing theory.** Facts do not completely determine theory, since many possible theories can be developed to take account of a specific set of observations. Nevertheless, facts are the stubborn of the two. Any theory must adjust to the facts and is rejected or reformulated if they cannot be fitted into its structure. Since research is a continuing activity, rejection and reformulation are likely to be going on simultaneously. Observations are gradually accumulated which seem to cast doubt upon existing theory. While new tests are being planned, new formulations of theory are developed which might fit these new facts.

The relation between fact and theory may be expressed in syllogistic terms. A theory predicts that certain facts will be observable: 'If X conditions exist, then Y is observable; if Y is not observable, then X condition does not obtain.' However, if X condition does exist, and Y is not observable, then the original proposition is denied. However, such a syllogistic pattern of logic does not guarantee that the original theory is correct when the facts are predicted and conformity merely guarantees that certain other theoretical propositions are not correct.

3. **Facts redefine and classify theory.** New facts that fit the theory will always redefine theory, for they state in detail what the theory states in very general terms. Facts clarify that theory throws further light upon its concepts. Finally, they may actually present new theoretical problems: that is, the redefinition may be far more specific than the theory. An example is the general hypothesis that when individuals from a rural population, particularly the tribal, enter the urban environment they experience a considerable amount of personal disorganisation. This process has been studied in most detail for immigrant groups and for children of such immigrants. It is normally held that many changes in habit pattern will occur in this adjustment process. One of these is a decline in fertility. As a consequence of these notions, we could predict that when the rural, particularly tribal, people settle in the urban areas and large cities their birth rate will drop. Actually, the net reproduction rate of urban tribal people is much lower than that of the rural tribal people and the fact is, therefore, in accordance with the theoretical prediction. The theory, however, is a general expectation, while the demographic facts are specific. The theory does not state how much the difference will be. In actuality, the fertility of urban tribal people is even lower than that of the non-tribal urban people. We are thus left with a redefinition of the theory towards greater specificity, and the older theory simply does not account for these new facts.

The facts do not reject the older theory: they are simply more complex and definite than the prediction of the original theory, and they call for further research. Indeed, it is one of the major experiences of research that actually testing any existing theory is likely to redefine it. The concepts that have been accepted as simple and obvious, turn out to be elusive, vague, and ill-defined when we fit them to the facts. It is not that the facts do not fit. It is rather that they are much richer, more precise and definite than concept or theory. Further, many such redefinitions and classifications may in turn lead to the discovery of new hypotheses. For so long as our theories use general terms and make rough predictions, it is difficult to disprove them.

However, facts become a stimulus to the redefinition and classification of theory even when they are in conformity with it. This process leads, in turn, to reformulation of theory and the discovery of new facts.

The growth of science is seen in new facts and new theory. Facts take their ultimate meaning from the theories which summarise them, classify them, predict them, point them out, and define them. However, theory may direct the scientific process; facts in turn play a significant role in the development of theory. New and anomalous facts may initiate new theories. New observations lead to the rejection and reformulation of existing theory, or may demand that older theories be redefined. Concepts which appeared definite in meaning are clarified by the specific facts relating to them. The geographer must accept the responsibilities of the scientist, who *must see fact in theory and theory in fact.*

LAW

The second key element in scientific thinking is 'law'. 'Any fully developed scientific theory contains, embedded within it, certain statements about unvarying relationships. These laws may be evident at the level of everyday experience or only at the level of scientific investigation, for example, by controlled experiment or microscopic investigation. As with theories there is a predisposition among scientists to seek laws which cover broad categories of phenomena.... There is also a preference within science for deterministic laws: that "wherever" A and B are present C "will" result. But it is recognized that some laws have a probabilistic form even if they represent a transient stage in the development of the discipline and will give way to deterministic laws as the discipline develops' (Hay, 1985, 130).

The credit to put the relevance of 'law' in geography goes to Schaefer who said that *geographers should seek to make law-like statements.* 'A science', according to him, 'is characterized by its explanations, and explanations require laws.... To explain the phenomena, one has described means always to recognize them as instances of laws.... In geography... the major regularities which are described refer to spatial patterns.... Hence geography has to be conceived as the science concerned with the formulation of the laws governing the spatial distribution of certain features on the surface of the Earth' (Schaefer, 1953, 227). Geographical procedures would then not differ from those employed in the other sciences, both natural and social: observation would lead to a hypothesis—about the interrelationship between two spatial patterns, for example, and this would be tested against large number of cases, to provide the material for a law if it were thereby verified.

A law should be unrestricted in its application over space and time. It is thus a '*universal statement of unrestricted range.* This suggests at least one important criterion for distinguishing a law. 'The *universality criterion* requires that laws should not make specific or tacit reference to proper names. Consider the proposition that towns of similar size and function are found at similar distances apart. The term "town" can be defined only with reference to human social organization and it carries with it ... an implicit reference to the proper name 'Earth'. Within such a context the statement may be true, but the universality criterion has undoubtedly been offended. To get round this difficulty, we may attempt to define "town" in terms of a set of properties which we claim are possessed by towns and only towns. In an infinite universe, however, there may well be some phenomenon which possesses all the properties listed without being a town. Again, we are not justified in regarding the statement being a proper law' (Harvey, 1969, 102).

However, strict laws are only found in physics, and to an extent in chemistry. These are truly universal in nature. It makes the development of laws in biology, zoology, geology, physical geography, etc. redundant, except in so far as such disciplines can reduce their statements to those of physics. The social sciences and human geography are even more seriously affected.

Harvey (1969, 102–03) contradicts interpreting universality in such a strict manner, as does Smart (1959). To quote Harvey (1969), 'There are two ways in which we may justify some relaxation of it. With purely empirical proposition it may prove useful to draw a *distinction between philosophical and methodological universality.* Philosophical universality involves the belief that universally true statements can be made. Such a belief may be supported by reference to some set of metaphysical propositions … or else it depends upon showing that a statement is in fact universally true. The latter course is essentially an inductive step and, therefore, a degree of uncertainty is involved. A proposition can never be shown empirically to be universally true. This applies as much to the strict laws of physics as it does to the "mere generalizations" of biology and economics. *Philosophical universality implies methodological universality*, but the reverse relationship does not hold. We may regard statements as if they were universally true without necessarily believing that they are or even assuming that they will ultimately be shown to be so…. In such a case it becomes a matter of deciding whether it is useful and reasonable to regard a statement as if it were universally true, and hence, law-like.'

A substantial part of Braithwaite's analysis of scientific explanation is concerned with establishing how laws are related to a surrounding structure of theory. It is impossible to determine whether a statement is or is not a law simply by referring to the truth or falsity of the generalisation it contains. A major criterion in determining whether a statement is or is not a law is the relationship of that statement to the system of statements that constitutes a theory. If this criterion is accepted, then the ideas are required to be adjusted regarding the verification procedures necessary to transform a scientific hypothesis into a scientific law. A generalisation may be set up or established as true or false simply by direct reference to empirical subject matter. The truth of an empirical law has to be established by this method too, but in addition, it requires support from other empirical laws, theoretical laws (that cannot be given any direct test), and also from other lower level empirical laws that helps it to predict.

A key concept in this respect is that *laws must be proven through objective procedures* and not accepted simply because they seem plausible. As Bunge (1962, 3) puts it, 'The plausibility or intuitive reality of a theory is not a valid basis for judging a theory. A valid law must predict certain patterns in the world, so that having developed an idea about those patterns, the researcher must formulate them into a testable hypothesis—"a proposition whose truth or falsity is capable of being asserted". An experiment is then designed to test the hypothesis, data are collected, and the validity of the prediction evaluated…. One successful test will not turn it into a law: replication on other data sets will be needed since a law is supposed to be universal' (Johnston, 1983, 72).

After sufficient (undefined) successful tests, therefore, a hypothesis may be accorded law-like status and is fed into a body of theory which comprises a series of related laws. There are *two types of statements within a full theory*: 'the axioms or givens', which are statements taken to be true, such as laws; and the deductions, or "theorems" from those initial conditions, which are derived consequences from agreed facts—the next round of hypotheses.

However, Jones (1956) pointed out the impossibility of discovering universal laws about human behaviour and indicated the existence of two types of law in physics: the 'determinate' laws of classic physics, which apply microscopically; and the 'probabilistic quantum laws' which refer to the behaviour of individual particles. Golledge and Amedeo (1968) attempted to indicate that science recognises several types of law, and also that the veracity of a law-like statement can never be finally proven, since it cannot be tested against all instances, at all times and in all places. They indicated *four types of laws which have relevance for human geographer*: (1) Cross-sectional laws, which describe functional relationships, but show no causal connection, although they may suggest one; (2) Equilibrium laws, which state what will be observed if certain criteria are met; (3) Dynamic laws, which incorporate notions of change, with the alteration in one variable being followed by (perhaps causing) an alteration in another. Dynamic laws may be historical, showing that B preceded by A and followed by C, or developmental, in which B would be followed by C, D, E, etc.; and (4) 'statistical laws' which are probability statements of B happening given that A exists. All laws of the other three categories may be either deterministic or statistical with the latter almost certainly the case with phenomena studied by geographers.

However, according to Sack (1972), space, *time and matter cannot be separated analytically* in an empirical science which is concerned to provide explanation. He attempted to show that geometry is not an acceptable language for such a science, i.e. geography. Nevertheless, geography is closely allied with geometry in its emphasis on the spatial aspects of events (the instances of law), but geometry alone is insufficient as a basis for explanation and prediction since no processes are involved in the derivation of geometries. Bunge (1973), however, responded to this statement, claiming that spatial prediction was quite possible with reference to the geometry alone, as instanced by central place theory and Thunian analysis. Sack (1973) responded by saying that the static laws espoused by Bunge are only special cases of dynamic laws having antecedent and consequent conditions and that: 'Although the laws of geometry are unequivocally static, purely spatial, non-deducible from dynamic laws, and explain and predict physical geometric properties of events, they do not answer the questions about the geometric properties of events that geographers raise and they do not make statements about process' (Sack 1973, 569).

Geography, according to Sack, is concerned to explain events and it requires substantive laws: such laws may contain geometric terms, such as 'the frictions of crossing a certain substance', but these terms of themselves are insufficient to provide explanations. He identified two types of laws relevant to geographical work: (l) *Congruent substance laws* which are independent of location: statements of 'if A then B' are universals which require no spatial referent; (2) *Overlapping substance laws* which involve spatial terms: 'if A then B' in such cases contains some specific reference to location. Both types are relevant and necessary in providing the answers to the geographical questions, so case may be made for a necessary 'spatialness' to the substance laws of human geography.

Thus, positivist-led geography has wider application of laws for successful and fruitful analysis of geographical phenomena together with spatial patterns. The concept of law has a much wider significance in such geography which is being conceived of as a science with law-seeking episteme because it postulates a three-fold hierarchy of scientific statements from factual statements or systematised descriptions through a middle-tier of 'empirical generalisations or laws', to general or theoretical laws.

LOGIC

But laws are not only type of connecting statements used in scientific theory, indeed a theory which consists entirely of laws based on experience and experiment is viewed with disfavour; it is seen as more satisfactory if most of the laws and other links in the theory can be shown to be logically derived from a much smaller number of fundamental assumptions and laws. Scientists have tended to use mathematics (algebra and geometry) as the language for expressing and developing this logic, but other abstract languages are also used (for example, chemical equations and bonding diagram).

Logical validation is one of the most commonly used methods of validation and certainly one of the most difficult to apply. It refers to either theoretical or 'common-sense' analysis which concludes simply that the items being what they are, the nature of the continuum cannot be other than stated to be. Logical validation, or 'face validity' as it is sometimes called, is always used because it automatically springs from the careful definition of the continuum and the selection of the items.

REDUCTIONISM

It is the final key in much scientific thought—*the idea that the laws and theories of a discipline can be re-expressed as special case of the outworking of the laws of a more fundamental discipline.* Reductionism is usually taken to apply to any doctrine that seeks to explain a higher-order phenomenon in terms of a lower-order phenomenon. Such a doctrine can be held in various forms, and applied in many different areas of intellectual endeavour. One form of reductionism is the notion that laws of all other sciences can be in principle reduced to, or expressed in terms of, the laws of micro physics; another is the thesis that all mental faculties can be expressed as events in or states of the brain. *Reductionism is defined more formally as concepts or statements redefined in terms which are more elementary or basic.* In many cases, the communication of knowledge thus raised cannot be expressed in ordinary verbal language, and two other methods must be added. The first is that of the use of words with specialised meanings, and the second is the language of pure symbols associated with mathematics and symbolic logic.

A geographical explanation may be said to be reductionist if it attempts to account for a range of phenomena in terms of a single determining factor. In human geography, the most common form of reductionism is probably that which asserts that all terms which refer to groups or collectivities can be, in principle, expressed as descriptions of the behaviour of individual actor. This view, however, has come to be known as methodological individualism. Some Marxist theories are said to be reductionist because they attempt to explain the diversity of sound behaviour by reference simply to the economy.

'Studies in physical geography have, in general, no option but to be carried out using the methods of contemporary science, whether these be reductionist or attempts at holism-like system analysis. We seek an understanding of landslides or glacial motion in terms of mechanical principles, and apply chemistry in the study of weathering processes and soil formation. The behavior of larger, more complex systems is routinely analysed with elaborate computer models, as in drainage basin hydrology or weather forecasting. But, at this point we ought to ask whether at the next level of complexity, that of human societies together with their environment, both living and non-living, reductionist approaches are appropriate or applicable, and with what degree of success.... Should geography try to be like physics? Should it be possible to express everything in laws ... which are 100 percent applicable in the sense of temporal prediction (... the laws of physics do not exhibit 100 percent probability but they are not far short for practical purpose).... The techniques of the mathematician and statistician have been paramount, for example, in the elaboration of models of spatial interaction, the use of idea from catastrophe theory and Q-analysis, and the widespread employment of bivariate and multivariate linear models for the analysis of data' (Simmons and Cox, 1985, 46–47).

To quote Hagget and Chorley (1967), '... those subjects which have modelled their forms on mathematics or physics ... have climbed considerably more rapidly than those which have attempted to build internal or idiographic structure'. There have been remarkable developments and advances in the development and adaptation of modelling and analytical techniques, but whether empirical studies which have used these have provided any more generalisations and accurate predictions than in non-reductionist approaches is arguable. The great emphasis on space as the central element in geography makes the discipline more like physics.

People unsympathetic to quantification, for its association with 'hard data', have often criticised the view for its apparent lack of humanity; its cold objectivity (if it should exist) does not appeal to all. However, reductionism of a different variety may not be accepted that which views all human patterns in terms of a single-factor explanation such as class struggle at the heart of classical Marxist theory, that appears to be too simple to explain the great variety of society-environment relationships which have been observed on the face of the Earth.

It is rather difficult to comment on the success of reductionist methods in geography, given the fact that geography has yet to achieve a great success in terms of laws. Some regularities have been pointed out, such as rank size rules of cities; the spatial patterning of towns as service centres, however, the bulky literature concerning them contains evidence of numerous exceptions and assertions of the culture-bound nature of the findings. 'Geography appears to have carried forward a more just world, no more and no less than any other divisions of learning with which it shares a reluctance to be committed to single element solutions, especially those of an ideological character' (Ackerman, 1958, 1963; Ackerman et al. 1965).

However, on reduction, Harvey (1969, 94–95) points out, '… The problem of finding adequate empirical definition of theoretical concepts can be solved by the provision of an adequate general theory. The development of powerful basic axiomatic statements will make possible precise definition of the idealizations on which current theory rests. The procedure may lead to the reduction of the large number of idealizations and concepts in social science to special cases of more axiomatic statements.… Many of the concepts and idealizations used in the natural sciences may be ultimately defined by reference to the basic concepts of physics. The unification of disparate theoretical structures into one system of statements involves the reduction of disparate idealizations to special cases of a few basic postulates. This phenomenon of reduction may also be found in the social sciences, and the development of general theory in the social sciences may well depend on such reduction. The postulates of economics may be reducible to a particular subset of postulates in psychology.… The degree to which reduction can take place, however, is a controversial issue, and even if it is conceded that total reduction is ultimately possible, this is far from being practicable at the present time. On the other hand, it cannot be denied that there is considerable benefit to be had from the integration of diverse concepts and statements into some more general theoretical framework.… The development of general theory in the social sciences—and the reduction of some concepts which this implies—may enable more precise definition of certain idealizations and hence facilitate the statement of an appropriate text for some of the theories developed in the social sciences.'

HYPOTHESIS

Theory, law, logic and reduction—these four elements are the key parts of scientific thinking, but there is however a fifth element—the research hypothesis—which provides a link to the area of scientific practice. In a well-developed natural science, a research hypothesis predicts the outcome of an experiment or observation if the theory is correct. In this way, a theory or its extensions can be tested in contexts other than those for which it was originally devised.

The formulation of deduction constitutes a hypothesis; if verified, it becomes a part of a future theoretical construction. In practice, a theory is an elaborate hypothesis which deals with more types of facts than does the simple hypothesis. A theory states a logical relationship between facts. From this theory, other propositions can be deduced that should be true if the first holds. These *deduced propositions are hypotheses.* Hypotheses do not necessarily have to be true, however. The truth of many hypotheses, the researchers formulate is most often unknown. Hypotheses, therefore, are tentative statements about things that the researcher wishes to support or refute. An hypothesis is a provisional statement that guides empirical work in several scientific epistemologies. An hypothesis, therefore, is a structured speculation that must be tested empirically. If it proves to be valid, then a positive addition is made to the stock of theory; knowledge has been increased. It it proves invalid, knowledge has also been increased, albeit in a negative sense (Johnston, 1985, 15).

Types of Hypotheses

There are several different kinds of hypotheses used in social and/or geographical analysis, studies and research. However, the primary types of hypotheses are : (1) research hypotheses, (2) null hypotheses, (3) scientific hypotheses, and (4) statistical hypotheses.

1. **Research Hypotheses.** Hypotheses derived from the researcher's theory about some social and/or geographical phenomena are called research hypotheses or 'working' hypotheses.

The social investigator usually believes that his/her research hypotheses are true or they are accurate statements about the condition of things he/she is investigating. The investigator believes that these hypotheses are true to the extent that the theory from which they were derived is adequate. Theories are, in one sense, suppositions about the true nature of things, and thus regarded as tentative statements about reality. Until they have been verified to the scientist's satisfaction, the hypotheses derived from theories must also be regarded as tentative suppositions about things until they have been tested. Testing hypothesis means to subject it to confirmation or disconfirmation (Black and Champion, 1976, 127)

2. **Null hypotheses.** Null hypotheses are, in a sense, the reverse of research hypotheses. They are also statements about the reality of things, except that they serve to refute or deny what is explicitly indicated in a given research hypothesis. *Null hypotheses are hypothetical models used to test research hypotheses.* The question that arises as why does the social investigator want to bother with so-called null hypotheses? Why doesn't the investigator test the hypothesis directly and let it go at that? These questions have been asked time and again by every researcher confronting null hypotheses for the first time. There are at least four explanations why null hypothesis models are used, none of which, however, may answer this question satisfactorily. They are:

   1. Trying to show the truthfulness of research hypotheses would imply to some, at least, a definite bias towards trying to confirm one's suppositions and possibly ignoring those things that would tend to refute our belief.
   2. There are those who would argue that it is easier to find fault with something, i.e. an idea, belief, or hypothesis than to look for those things that would support it.
   3. It may be summed up in one word: convention. It is conventional in social research to use null hypotheses. Null hypotheses, however, also perform specific functions in relation to probability theory and tests of research hypotheses.
   4. Under a probability theoretical model, hypotheses have a likelihood of being either true or false. Null hypotheses are particularly useful in such theoretical models. The null hypotheses is an expression of one alternative outcome of a social/physical observation. The probability model specifies that the null hypotheses may be either true or false but not both simultaneously. Neither the research hypotheses nor the null hypotheses is absolutely true or absolutely false under any given test of it. Both probabilities (being either true or false) co-exist for each type of hypothesis always.

3. **Scientific Hypotheses.** In scientific investigation, however, the term hypotheses is often given a somewhat more restricted meaning. To Braithwaite (1960, 2): *A scientific hypothesis is a general proposition about all the things of a certain sort.* It is an empirical proposition in the sense that it is testable by experience; experience is relevant to the question as to whether or not the hypothesis is true, i.e. as to whether or not it is a scientific law.'

A scientific hypothesis, in Braithwaite's tradition, is a particular kind of proposition which, if true, will be accorded the status of a scientific law. The testability of a hypothesis is crucial, but there are many hypotheses within a theoretical system which cannot be directly tested against sense perception data. Thus: 'The empirical testing of the deductive system is effected by testing the lowest level hypotheses in the system. The confirmation or refutation of these is the criterion by which the truth of all the hypotheses in the system is tested' (Braithwaite, 1960, 13).

Since scientific hypothesis is often regarded as being a proposition where truth or falsity is capable of being asserted, the truth and falsity of it (hypothesis) can be determined only with respect to the domain of some theory.

4. **Statistical Hypotheses.** These are statements about statistical population that, on the basis of information obtained from observed data, one seeks to support or refute (Winter, 1962). The statistical population may refer to either people or things. It is generally the case in the test of statistical hypotheses that *observations about people or things are reduced in some way to numerical quantities,* and decisions are made about these quantities. To subject these hypotheses to empirical test, what is required is to reduce the variables used in them to measurable quantities. Research hypothesis and corresponding null hypotheses can be transferred into a statistical hypotheses that may be evaluated by numerical means.

Statistical hypotheses are usually established to delineate: (1) differences between two or more groups regarding some trait or collection of characteristics that they possess, association between two or more variables within one group or between several groups, and (2) point estimates of sample or population characteristics (Johnston, 1983, 71–74; Harvey, 1969, 100–06).

Levels of Hypothesis

Apart from the aforesaid four types of hypotheses, three broad levels of hypotheses may be distinguished on the basis of the level of abstraction, which are as follows:

1. Some hypotheses state the existence of empirical uniformities. These hypotheses frequently, though not always, represent the scientific examination of common-sense propositions. They usually represent, also, a problem about which some 'commonsense' observation already exists. There are many types of such empirical uniformities which are common in social science and/or geographical research.

However, these investigations do not involve the testing of hypothesis at all, but are merely adding up the facts. These are not useful hypotheses for they merely represent what everyone already knows.

2. Some hypotheses are concerned with complex ideal types. These hypotheses aim at testing the existence of logically devised relationships between empirical uniformities. One such hypothesis was Ernest W. Burgess's statement on the concentric growth circles that characterise the city (Goode and Hatt, 1952, 61). This hypothesis was then tested against a variety of variables in a number of cities. That this ideal type does represent the actual patterns of city growth is not accepted by all ecologists, however, and so this formulation remains a hypothesis until a more crucial test of it is made.

Another hypothesis, concerning an ideal type also, results from these same ecological empirical uniformities. This was the notion that areas tend to represent certain characteristics in a series of predictable patterns. This was called the hypothesis of the 'natural area'. Much research has been done on this hypothesis, and the results, although they have modified the original statement somewhat, have generally supported it. With the growth of supporting evidence, notions about natural area have become a part of geographical theory rather than remaining hypotheses.

It is important to see that this level of hypothesising moves beyond the expectation of simple empirical uniformity, by creating a complex referent in society. The function of such hypothesis is to create tools and problems for further research in otherwise very complex area of investigation.

3. Some hypotheses are concerned with the relation of analytic variable. These hypotheses occur at a level of abstraction beyond that of ideal types. The hypotheses of empirical uniformities lead to the observation of simple differences, and those dealing with ideal types lead to specific coincidences of observations.

The study of analytical variables requires the

formulation of a relationship between changes in one property and changes in another.

On the basis of the above discussion, three major points can be identified: (1) that a hypothesis is a necessary condition for successful research; (2) that formulation of the hypothesis must be given considerable attention, to clarify its relation to theory, remove vague or value judgemental terms, and specify the test to be applied, and (3) that hypotheses may be formulated on different levels of abstraction.

Functions of Hypotheses

Theories are relatively elaborate tools used to explain and predict events. The social scientist develops a theory to account for some social phenomena, and then he devises a means whereby the theory can be tested or subjected to verification or refutation. Seldom does the researcher test theory directly. Most of the time he/she conducts tests of hypotheses that have been generated and derived from that theory. If the hypotheses 'test out' as the researcher has specified, or if his empirical observations are in accordance with what has been stated in the hypotheses, we say that his/her theory is supported in part. It usually takes many tests of different hypotheses from the same theory to demonstrate its predictive value and its adequacy as a tool of explanation for some event or sequence of events.

A major function of hypotheses is to make it possible to *test theories.* In this regard, an alternative definition of an hypothesis is that it is a statement of theory in testable forms. All statements of theory in testable form are called hypotheses.

Some hypotheses are not associated with any particular theory. It could be that as a result of some hypothesis, a theory will be eventually constructed. Consequently, another function of hypotheses is to *suggest theories that may account for some event.* Although it is more often the case that research proceeds from theories to hypotheses, occasionally the reverse is true. The social investigator may have some idea about why a given phenomenon occurs and he/she hypothesises a number of things that relate to it. He/she judges that some hypotheses have greater potential than others for explaining the event, and as a result, he/she may construct a logical system of propositions, assumptions and definitions linking his explanation to the events. In other words, the researcher devises a theory. Working from the hypothesis back to the theory is not necessarily poor methodology. Eventually, the investigator is going to have to subject the resulting theory to empirical test to determine its adequacy. The predictive value of the theory can be assessed at that time.

Hypotheses also perform a *descriptive function.* Each time a hypothesis is tested empirically, that tells something about the phenomenon it is associated with. If the hypothesis is supported, then the information about the phenomenon increases. Even if the hypothesis is refuted, the test tells something about the phenomenon that is not known before. The accumulation of information as a result of hypothesis testing reduces the amount of ignorance we may have about why a social event occurs in a given way.

Hypotheses also have some important *secondary functions.* As a result of testing certain hypotheses, social policy may be formulated in communities, penal institutions may be redesigned and revamped, teaching methods may be altered or improved solutions to various kinds of social problems may be suggested and implemented, and supervisory practices may be changed in factories and business. Testing hypotheses refute certain 'common sense' notions about human behaviour, raises questions about explanations we presently use to account for things, and most generally alters our orientation towards our environment to one degree or another. All hypotheses have to do with our knowledge of things, and as this knowledge changes, we change also (Black and Champion, 1976, 137–38).

Testing Hypotheses

Testing hypotheses means *'subjecting them to some sort of empirical scrutiny* to determine if they are supported or refuted by what the researcher observes'. Testing hypotheses means that the researcher will need to do a number of things. Following are the two prerequisites to hypotheses testing:

1. A real social situation is needed that will suffice as a reasonable testing ground for the hypothesis. If the hypothesis concerns managerial behaviour, it will be necessary for the investigator to study some real organisation or organisations where managerial behaviour can be taken into empirically. This particular prerequisite is frequently spoken of as 'getting access to data that will enable the investigator to verify or refute his/her hypotheses'. Once a given social setting is selected, the relevant data in that situation must be obtained to make the hypothesis test a valid one.
2. The investigator should make sure that his hypotheses are testable. This means that he/she should limit his/her investigations to empirical phenomena or events that can be taken into through the senses. The variables used in the hypotheses tested should be amenable to measurement of some kind. If they are not subject to measurement, the resulting test of the hypothesis will be relatively meaningless. Testing hypotheses must be a part of the empirical world. This is a fundamental requirement wherever the scientific method is employed in studying what is and why.

Terms that cannot be taken into empirically, render the hypothesis irrefutable and untestable. How can a scientist reject a hypothesis containing variables that he cannot experience in some empirical form? For example, if a researcher were to hypothesise that 'evil spirit causes delinquency', he/she can neither support nor refute this statement by using conventional scientific methods. He/she obviously has empirical tools to determine the incidence of delinquent or non-delinquent behaviour, but by what empirical means is he/she able to assess meaningfully the influence or impact of 'evil spirits' on delinquent behaviour? Unless there are empirical means of evaluating the impact of non-empirical phenomena on particular variables, the researcher cannot validly subject the hypothesis to true scientific test. However, it is possible that terms that are presently indefinable empirically, might at some later date become amenable to the senses through the discovery of new means of measuring such phenomena. This always exists as a possibility (Black and Champion, 1976, 142–43).

KEY ELEMENTS IN SCIENTIFIC PRACTICE

It is difficult to identify the best term for the first such element because all words commonly used carry with them specific and contentious overtones—but the word *'test'* seems to be used most often. A scientific theory requires to be tested or assessed not only for its own internal consistency, but also for its consistency with the world as observed. Such tests may be focused upon assumptions (are the specified assumptions of the theory realistic?), to test laws (are the specified relationships observable?), and most often to test hypotheses (is the outcome postulated by the hypothesis observable?). In the popular conception of science these tests are most commonly carried out by laboratory experiments in which sealed vessels and pure materials ensure that effects are not suppressed or spuriously created by undetected contamination.

One of the most dramatic tests of a theory occurs when it successfully predicts the outcome of an experiment or observation never before conducted. This provides the basis for the common idea that a second element—*successful 'predictio'*—is the hallmark of science (Gardner, 1982). Prediction is an effort to foretell future occurrence on the basis of past information. When that information takes the form of a scientific law, confirming a universal pattern of regularity in the relations among variables, predictions can be highly accurate. But it has two other aspects : first there is a logical sense that if there is a valid explanation for C—'because A and B, then C'—it can also be used to predict 'if A and B, then C'; second, successful prediction can be a useful practical tool either in unconditional form (The Sun will set at 18.30 hours) or in conditional form (if you apply Y tonnes of fertiliser D per hectare then yield will increase by X per cent).

Scientific prediction must be fashioned within the context of established theoretical premises. These, in turn, are encased in carefully etched out descriptions. It would be presumptuous, indeed, to expect that prediction could develop independently of progress made in these related areas. No amount of rhetoric is going to alter the fact that sound prediction rests upon appropriate descriptive and explanatory knowledge. As a result, most of what follows is designed deliberately to enhance these latter scientific objectives under the assumption that such practices will invariably lead to greater predictive accuracy.

Such tests often involve a third element in scientific practice—*'measurement'*—'which refers to the procedures by which empirical observations are made in order to represent symbolically the phenomena and the conceptualizations that are to be explained' (Di Ranzo, 1966). Measurement is the assignment of number to nominal social and/or psychological properties of individuals and/or groups according to rules, and the correlating of these numbers with these properties symbolically. However, measurement may be defined again as the assignment of numbers to, and the symbolic correlation of the numbers with, nominal social and/or psychological properties of individuals and/or groups. It is, indeed, a reality that important features of natural and human worlds cannot be handled well by qualitative categories, but the problem is even more acute in testing because some laws and theories yield hypotheses which are quantitative in form. A considerable amount of scientific effort has, therefore, been devoted to developing methods of measuring the characteristics of phenomena—an effort which has been parallel in geographical efforts to measure evapotranspiration, soil fertility or the level of economic development.

The development of measurement is closely linked with question of 'statistical inference'. It was recognised at an early stage that even experimental results include errors arising from the observer, from instrument calibrations and from small disturbances or contaminations of the experiment. Much early statistical works were related to inferences in the presence of errors. The use of statistics to infer population characteristics from a sample of observations (in which geographers have shown much interest) also become important.

Underlying these elements in scientific thought and practice there is often an implicit attitude known as '**scientism**': *the belief that scientific methods are the only methods of acquiring real knowledge.* The consequences of this belief are dangerous in two ways. On the one hand, a devotee of scientism may attempt to apply scientific methods in appropriate contexts and end up by redefining concepts (e.g. the beauty of a landscape) in terms of much more limited scientific measurement. The other scientific response is to exclude such concepts from all discussions simply because they are increasingly less tenable for two reasons: (1) investigations of the history and philosophy of science have identified many gaps and inconsistencies of scientific method, and (2) as other disciplines have attempted to follow scientific method, new dilemmas have become apparent (Hay, 1985, 132–33).

ROUTES TO SCIENTIFIC EXPLANATION: INDUCTION AND DEDUCTION

There are two alternative routes to explanation, or which are followed in establishing a scientific law, according to Harvey (1969, 32–36). The first by 'induction'—proceeding from numerous particular instances to universal statements—and the second that of 'deduction'—proceeding from some *a priori* universal premises to statements about particular sets of events.

1. **Route (1).** It is also known as the Baconian Route or 'Inductive Route to Scientific Explanation. According to Harvey (1969, 32–33), 'Sense-perception data provide us with the lowest level information for fashioning scientific understanding. This information, when transformed into some language, forms a mass of poorly ordered statements which we sometimes refer to as 'factual'. It is partly ordered by the use of words and symbols to

describe it. Then, by the process of definition, measurement, and classification, we may place such partially ordered facts into groups and categories and therefore impose some degree of seemingly rational order upon the data. In the early stages of scientific development, such ordering and classification of data may be the main activity of science, and the classification so developed may have a weak explanatory function.... The status of empirical laws established by such a route is a matter of some controversy. It should be noted that each step along this route so far involves inductive inference. Thus laws established by this route are alone sometimes called inductive laws. Some maintain that inductive laws cannot be accorded the status of scientific law.'

This route to scientific explanation, however, does not describe how the scientist should proceed, but it does describe one of the ways in which a scientist might describe his/her action so as to meet with the approval of other scientists. This route involves a dangerous form of generalising from the particular case, as the acceptance of the interpretations depends too much on the charisma of the scholar involved. Churchman (1961) observed that 'facts, measurements and theories are methodologically the same'. Applying an *a priori* classification system to a set of data may thus be regarded as an activity similar in kind to postulating an *a priori* theory (Harvey, 1969, 33).

2. **Route (2).** 'The second route whereby we may justify scientific conclusions clearly recognises the *a priori* nature of much scientific knowledge. It firmly rests upon intuitive speculation regarding the nature of the reality we seek to know.... This involves some kind of intuitive picturing of how that reality is structured. Such *a priori* pictures ... later identify as *a priori* models. With the aid of such pictures we may postulate a theory. That theory should have a logical structure which ensures consistency and a set of statements which connect the abstract notions contained in the theory to sense-perception data. The theory will enable us to deduce sets of hypotheses which, when given an empirical interpretation, may be tested against sense-perception data. The more hypotheses we can check in this fashion, the more confident we may feel in the validity of the theory provided, of course, that the tests prove positive.'

'In the process of elaborating or seeking to test a theory, we may resort to another kind of model—an *a posteriori model*—which expresses the notion contained in the theory in a different form say, in mathematical notation. In some circumstances, model building may here amount to developing an experimental design procedure, and a primary function of this procedure is to lay down the rules whereby we may define, classify, and measure the variables which are relevant for testing the theory. By using such experimental designs we may amass evidence to confirm the hypotheses contained in the theory' (ibid. 35).

In the nutshell, this route begins with an observer perceiving patterns in the world; he/she then formulates experiments, or some other kind of test, to prove the veracity of the explanations which he/she produced for those patterns. Only when his/her ideas have been tested successfully against data other than from which they are derived can a generalisation be produced. Scientific knowledge, obtained via the second route, is 'a kind of controlled speculation. The control really amounts to ensuring that statements are logically consistent and insisting that at least some of the statements may be successfully related to sense-perception data' (Harvey, 1969, 35). It is such a procedure that an increasing number of human geographers sought to apply during the 1950s and 1960s.

The method, known as positivism, was developed by a group of philosophers working in Vienna during the 1920s and 1930s (Guelko, 1978). It based on a conception of an objective world in which there is order waiting to be discovered. Because that order—the spatial patterns of variation and covariation in the case of geography—exists, it cannot be contaminated by the observer. A neutral observer, on the basis of his observations or his reading of the research of others, will derive a hypothesis (a speculative law) about some aspect of reality and then test that hypothesis: verification of his hypothesis is translates the speculative law into an accepted one (Johnston, 1979/1983, 71).

Deduction occurs when facts are gathered to confirm or disapprove hypothesised relationships among variables that have been deduced from propositions. Whether there were facts that precipitated the propositions does not really matter. What matters is that research is essentially a hypothesis-testing venture in which the hypotheses rest on logically (if not factually) deduced relational statements. Burgess and Akers (1966) have attempted to reveal how particular hypothesis can be generated from many general and inclusive assertions. Deduction is a type of reasoning in which the conclusion follows necessarily from the given premises, but it does not increase content, although one may require intellectual abilities of a higher order in order to trace all the steps that lead from the premises of a deductive argument to its conclusions (Marshal, 1985, 115).

Most writers on the scientific explanation have argued that the appropriate logic is that of deduction. Thus, the view that scientific explanation must always be reduced in the form of logical deduction has had wide acceptance. Braithwate (1960, 12) has also sought for the systematic organisation of scientific knowledge as a 'hypothetic-deductive' system. He pointed out: 'A scientific system consists of a set of hypotheses which form a deductive system, that is, which is arranged in such a way that from some of the hypotheses as premises all other hypotheses logically follow. The proposition in a deductive system may be considered as being arranged in an order of levels, the hypotheses at the highest level being those which occur as premises in the system, and those at the lowest level being those which occur as conclusions of the system, and those at intermediate levels being those which occur as conclusions of deductions from higher level hypotheses and which serve as premises for deductions to lower level hypotheses.'

With regard to the advantage of deduction, Harvey (1969, 36) suggests that: '... if the premises are true then the conclusions are necessarily true. If ... we have certain degree of confidence in a set of premises we may possess the same level of confidence with respect to any logically deduced consequence. This property has led to the use of reduction wherever possible. Theories are thus invariably stated as deductive systems of statements.... The application of such theories to the actual explanation of events is rendered as logical deduction.'

The form of explanation, which Hempel calls 'deductive homological' (covering law explanation) consists of: (a) one or more laws of nature, and (b) a list of specific initial conditions or circumstances which, when taken together, show that an event must necessarily have occurred or that describe the setting of the event to be explained. From these premises, the occurrence of the event in question can be inferred by a strictly deductive chain of reasoning. In this form of explanation, prediction and explanation are symmetrical and deduction ensures the logical certainty of the conclusion. It is assumed that the final outcome of this process will be the discovery and verification of a set of natural laws from which all events can be rigorously deduced. When that point is reached, science will have achieved its ultimate goals (Marshal, 1985, 118).

Induction involves moving from particular instances of relations among variables to the formulation of hypotheses and from these to the development of propositions. Many scientists had claimed or appeared to have claimed that laws and theories were derived from the observation of repeated regularities. This method is often referred to as induction. In one or another of its forms, induction is the way most social scientists go about business of expanding knowledge.

Theodorsen and Theodorsen (1969, 199–200) have attempted to distinguish between two basic types of induction—enumerative and analytic. *Enumerative induction* is the most common form of induction used in social science research today. Most often enumerative induction involves generalisation from samples with varying degrees of representativeness. Usually, but not invariably, these generalisations are derived through the application of statistical procedures to the data. Accompanying these studies are usually statements pertaining to 'probability' of generalisations to larger and more inclusive populations based on findings from the sample methods. *Analytic induction* is a procedure whereby there is a case-by-case analysis of specific features to determine which conditions are always present prior to the occurrence of certain types of conduct.

Induction is important to our concern with scientific method because of the role it plays in the formulation of empirical generalisations. These generalisations may conveniently be divided into: (1) summative, and (2) extended. A summative *generalisation* describes a property which has been confirmed by the actual observation of all the relevant cases, for example, the statement, 'All the men in this room are old'. Since the men in question have actually been examined, the generalisation does not go beyond the evidence and therefore does not involve inductive reasoning. However, the statement, 'All swans are white' is an example of an *extended empirical generalisation*—as one goes beyond the evidence on which it is based. We may have examined thousands upon thousands of swans and found them all to be white, yet we cannot guarantee the truth of the general statement because it remains possible that a non-white swan may some day be found (as in fact did happen with the discovery of black swans in Australia). In fact, we cannot guarantee the truth of an empirical generalisation unless, as in the summative case, we can point to all the singular instances upon which it is based. 'There is no logical justification for extending belief in the premises to belief in the conclusions. The failure of logicians and philosophers to find (or agree upon) such logical justification has led many to reject its use entirely in the presentation of scientific knowledge' (Harvey, 1969, 37).

Let us consider the role that the inductive reasoning plays in the *fallibilist* conception of the quest for knowledge.[1] The philosophy of fallibilism tends to reject the traditional positivist view that science begins with the collection of observational data and proceeds inductively to the establishment of general laws. The starting point of the fallibilist conception lies in the realisation that there is an asymmetrical relationship between logical states of a general law and that of its negation. It is increasingly believed by the fallibilist that a general law is never conclusively verifiable but always conclusively falsifiable. *Falsification is possible but verification is not.* To be scientific is to recognise explicitly that knowledge is approached by means of a process of conjectures and refutations (Popper, 1962).

Induction is involved in the development of conjectures (theories, hypotheses), but not in the search for refutation. Induction is also involved in theory-building by virtue of the fact that theories incorporate (often implicitly) laws or law-like statements that extend beyond the evidence upon which they are based. For the fallibilist, however, law-like statements and the theories in which they become embedded are not hard core of science; they are never anything more than heuristic speculations. The distinguishing feature of the scientific attitude is the willingness—indeed, the desire—to confront these speculations with pertinent empirical observations, thereby exposing them to the possibility of refutations. A refutation, when it occurs, is a logically conclusive result in which inductive reasoning plays no part. For the fallibilist, therefore, a scientific enquiry has two distinct phases—'an imaginative, hypothetical, theory-proposing phase' and 'a critical, objective,

[1] Fallibilism, also known as 'falsificationism' and 'critical rationalism', is largely the creation of a single individual, namely, Sir Karl Popper (1959) whose critique of positivism developed into the philosophy of fallibilism.

experimental phase'. Inductive inferences may be involved in the former, but they are strictly banned from the latter.

In the traditional positivist view of science, general laws established by inductive reasoning are not regarded as speculations but as 'verified scientific knowledge'. For the positivist, induction is the very foundation of science. For the fallibilist, induction leads only to conjectures, and science does not begin in earnest until these conjectures are confronted by the threat of falsification.

Although the distinction between deduction and induction serves the important purpose of identifying opposite ways to go about theory-building, most investigators find that their scientific work entails a certain amount of both. Induction probably permits more of an opportunity to see theory in a dynamic state of emergence rather than as already given, as is the case with deduction, but the issue is open to some debate. Because there is neither a rigorously developed comprehensive theory from which to deduce particular relationships for testing, nor a sufficient accumulation of data to allow for systematic theory development through induction, even personal preferences for one or the other approach must remain flexible and adaptable (Black and Champion, 1976, 67).

'The methodological rejection of induction can only apply to certain aspects of the formulation of scientific knowledge. Science attempts to organize the propositions within a deductive frame of inference.... The deductive form of scientific theories must be regarded as the end-product of scientific knowledge, rather than as the mould into which all scientific thought is cast from the very initiation of an investigation. But even assuming that a deductive theoretical structure has been successfully evolved, induction still plays an important function at certain stages in the articulation and verification of such a theoretical structure.... However, it is misleading to regard deduction and induction as mutually exclusive forms of inference. Although it is generally agreed that scientific knowledge should be organised as a hypothetico-deductive system and that the law contained in that system can best be applied by a deductive explanatory procedure, there are many occasions when inductive steps may be used within these deductive frameworks' (Harvey, 1969, 38–41).

However, the method of induction was criticised on two grounds. First, it was evident that in many cases the observations were themselves made with pre-conceptions (right or wrong) as to what constituted characteristics worthy of observation and recording. Second, the exact form of the law-like statement derived was seldom free from theoretical presuppositions and *a priori* definitions. It seemed that pure induction seldom occurred. It was recognised that the logical justification of induction itself relied upon the inductive method. These conclusions are important not only for an inductivist theory of scientific method, but for any other empiricist school of thought which believes that facts should and can be allowed to speak for themselves. As mentioned earlier, Popper (1959) introduced falsification as a concept to replace verification, and under this new concept all theories were deemed to be provisional—coherent systems of not yet falsified hypotheses about the nature of phenomena.

The falsificationist position had two other implications: (1) It required that all scientific statements should have the logical possibility of being proved false: in this case the claim that 'All swans are white' is incapable of being proved conclusively true (the next swan observed might be black), but could be proved false (by the observation of one truly black swan); (2) the form of empirical enquiries: no longer was it necessary to find evidence in support of a hypothesis, it became part of the scientific enterprise to find evidence which disapproved it. However, a complex and well-tested theory predicts an effect that fails to happen, but this failure may be due to a fundamental falsity of the theory or to some low level error of logic in deriving the research hypothesis. The falsification of the hypothesis requires a search for the false step, not the immediate abandonment of the theory as a whole (Chalmers, 1978; Marshal 1982, 1985).

The normal scientific methods have been subjected to critical analysis and criticism, especially on the issues of reductionism and objectivity. Reductionism has been defined earlier as the idea that the laws and theories of one discipline can, and should, be reformulated as special cases of the outworking of a more fundamental discipline, often linked to a change of scale of enquiry. But such a reduction appears to be fallacious (Hay and Johnston, 1983).

With regard to the fallacy in the objectivity of scientific knowledge, it is sometimes asserted and assumed that whereas the humanities have important areas which are matters of personal subjective value judgement, science is independent of such personal judgement, an independence assumed by its use of abstract logic and natural measurements. But this view is difficult to maintain when it is recognised that the type of questions investigated, the theory used and the observations conducted all depend on the research paradigm adopted, and the paradigm in turn reflects the value system of the society and its philosophical and scientific presuppositions (Hay, 1985, 136–37). If objectivity does exist, it is not the objectivity of the individual scientist, but a relative objectivity of the knowledge itself because it has been tested and corrected by many individuals working in different contexts and dimensions (Popper, 1972).

GEOGRAPHICAL APPLICATION OF SCIENTIFIC METHOD: SOME PROBLEMS

The post-War geography in the mid-twentieth century witnessed continuing debate with regard to the application of scientific method in geography. There were two mutually exclusive arguments on this issue. One side argued that scientific method should be introduced into both physical and human geography. On the other hand, some geographers had claimed that the discipline was in some sense an exceptional discipline which might be excused (if not completely excluded) from the constraints of scientific method. The debate, however, had its origin in the nineteenth century and over decades, it hardened to a great extent.

Despite the counter-arguments and dichotomies, the period from 1960 experienced a vigorous expansion of geographical research using quasi-scientific methods, with emphasis on the law-seeking approaches and model-based paradigms. The philosophical and methodological base for this was carried forward by many young geographers of the Anglo-American heritage and tradition. A number of textbooks in both human and physical geography emphasised the need for theory, laws, hypotheses, measurement and statistical testing. But the enthusiastic practitioners and protagonists of this approach were often unaware of the problems inherent in the scientific approach, and could not identify the additional problems posed by its geographical use. Most of these problems stemmed from the twin facts that '*geography as a whole deals with multi-variable open systems* and that human geography deals with knowing subjects'.

1. Geographers were, over many years, concerned with the term '*uniqueness*' because geographical phenomena on the surface of the Earth are unique and distinguishable, as well as complex in character and causation. The conclusion drawn is that geography deals with unique events, and generalisation in the form of laws and theories is doomed to failure or cannot be carried forth. It is the idiographic attitude that implies a concern with the uniqueness of individual phenomena or events whereas the nomothetic approach implies a desire to subsume individual cases under laws or law-like statements of very general, if not universal, applicability (Hartshorne, 1939/1961, 378–97).

This position, however, certainly provided a powerful argument against inductive methods in geography, and many geographers who stated for the uniqueness case, nevertheless argued for an inductive approach. It was then less clear that uniqueness was a valid objection to a theoretically based hypothesis-testing approach, because a collection of unique cases might nevertheless confirm or reject a hypothesised relationship: uniqueness was only an obstacle to that if it could be shown that causal relationships were themselves unique to each instance and changed inconsistently from place to place, and from time to time (spatial-temporal). It was, however, argued that uniqueness with respect to some trivial property (location of some geographical phenomenon may be a trivial property) or some peripheral relationship was insufficient reason to invalidate scientific method (Hay, 1985, 137–38).

2. A second consequence of geographical systems being large open systems is the *difficulty of carrying out experimental tests.* The sheer large size of a geographical system (the atmosphere, the river basin, a city) makes the laboratory experiment impossible. Scaling down the system may change and/or alter its properties in unknown ways. Even if the system is reproduced in the laboratory there is no assurance that all the variables relevant in reality have been included in the laboratory version. An alternative solution to it in scientific terms is the 'field experiment', but it is difficult to ensure that the only variables allowed to vary are those being investigated, and certain experiments in human ·geography would be politically or morally unacceptable. So in field experiments and certainly in field data collection much of the control of extraneous variables is achieved by purely statistical means which in theory allow the isolation of a two-variable relationship when 'all other variables are held constant'. Yet even such methods can only 'hold constant' recorded variables. There is no way of controlling for the possible effects of unrecognized and unrecorded additional variables (Hay and Johnston, 1983, 69–76).

3. A third consequence of the multi-variable nature of geographic system concerns the *use of theory from other discipline.* This may be applied and synthetic (the attempt to borrow theory from other disciplines to bear on a geographic problem) or reductionist (the attempt to interpret geographical relationships as special cases of more general theory in other disciplines). Such attempt to borrow is especially difficult if more than one other discipline is involved, each with a scale of analysis, a conceptual framework and definition which may not be compatible with each other or with the geographical terms of reference. A common geographical solution to this is to adopt one discipline as the source of a central theoretical framework (an 'economic approach' to urban geography, the 'physics' of slope development) and to use theory from other disciplines as modifying the central theory. A problem remains, however, that the best current theory in geography requires some knowledge of many natural and social sciences (Davidson, 1978). It is this, one suspects, that makes most geographers reluctant, if not unable, to pursue a wide slice of the discipline at the highest level.

4. Another problem that arises in applying scientific method in geography is *the interference by the observer with the phenomenon observed.* This problem is encountered in laboratory sciences, but it is usually possible to so design the experiment as to minimise the effect. It occurs in physical geography. In human geography, the same problem occurs in two more acute forms : (a) if the presence of an observer is known to the actors (either in a role of observer or as a simple stranger), it may lead to a short-run change in behaviour, conscious or unconscious; the results of observation will thus be untypical of normal behaviour, and (b) the interaction between observer and observed (at the time of observation or later by publication of research findings) may produce long-run changes which would not otherwise have occurred. If such changes are towards a research hypothesis, 'a false hypothesis

may be spuriously confirmed, if counter to the hypothesis a correct hypothesis may be mistakenly rejected'.

RELEVANCE OF SCIENTIFIC METHOD IN GEOGRAPHY

Despite criticisms with regard to the application of scientific method to geography as a way of obtaining useful and reliable knowledge; scientific methods hold relevance in geographical scholarship, research and training in both physical and human geography for three reasons.

1. Although scientism is a mistaken and dangerous ideology, the scientific method does have the ability to provide coherent and testable theories about the nature of geographical phenomena.
2. The scientific method remains appealing because it is in many respects a codified and logically corrected extension of thought structures developed in everyday life, including the willingness to correct theories or hypotheses in the light of experience.
3. Partly as a consequence of these two points, knowledge of a scientific type is required by society for its purpose of managing social and natural systems (and if geography fails to provide such knowledge; some other disciplines will develop to fill the gaps).

However, the scientific geography cannot remain untouched by critical evaluations and criticisms, and the elements which were required to be retained had to be modified. As a result, *many geographical theories were 'derivative'* in the sense that they attempted to specify the application of geographical theories which were established in cognate disciplines (in the physical and social sciences). The ability or inability of derivative theories to provide the basis for geographical explanations largely depended on the test of their overall value as research programmes. But, in addition to such derivative theories, there emerged a need for specifically geographical theories or laws which were essentially laws of composition, specifying the way in which these derivative laws appeared to have interacted to produce the multi-faceted phenomena that geographers sought to understand. The level at which these laws of composition operated, however, appeared to be much greater than the scale at which the derivative laws and theories operated.

It is possible to identify a number of examples of derivative theory already used in geography. Economic concepts have frequently been used as the foundation for geographic theory. Economics has, perhaps, been the most successful of the social sciences in developing formal theory (even if the empirical status of the theory is open to doubt). The central-place theory has frequently been described as the one relatively well-developed branch of theoretical economic geography. Central-place theory provides just one example out of many to demonstrate how geographical theory may be derived from the basic postulates of economics. The existence of such postulates was undoubtedly an important necessary condition for the emergence of a theoretical human geography (Burton, 1963, 159; Harvey, 1969, 119).

Many of the postulates and theorems of economics have been absorbed into geographical theory. In particular, the whole of location theory, which has been 'especially' concerned with the development of the theoretical-deductive method in geography. Human geographers have long recognised that geographic patterns are the end-product of a large number of individual decisions made at different times for often very different reasons, and that it was necessary to employ some psychological notions in explaining those patterns which revealed that psychological and sociological postulates were introduced in the construction of geographical theory (Harvey, 1966, 370). Similarly, the geographical studies of weathering had applied the chemistry of ions and cations to the specific chemical composition of parent rocks under stated conditions of temperature and humidity. The examples suggest that although some geographical problems required 'spatial laws', not all derived theory in geography would be concerned with spatial relationships (Sack, 1977).

It seemed rather less easy to identify) laws of composition in current geographical work, but some recent applications of choice theory to the choice of destination and mode of travel brought together economic and non-economic variables in a choice calculus that seemed more convincing than explanation in economic terms alone.

The use of derivative and geographical theories and laws necessarily implied an openness to reductionism—a willingness to accept reduction, but not an assumption that all geographical problems could and must be solved in reductionist terms. It also implied an open approach to the 'types of logic' adopted. However, many of the cognate disciplines (physics, chemistry, economics) themselves appeared to depend on 'abstract logic', and that abstract languages have proved extremely powerful and made it likely that such languages would be used in geography. Alan Wilson (1974) attempted to show how an abstract mathematical language was capable of integrating quite different parts of urban and regional systems. But the openness to abstract logic (like mathematics) must not be allowed to exclude from geographic theory those variables and concepts (e.g. the quality of landscape) which were not capable of representation. Geography, however, should remain open to the introduction of new languages that might prove more flexible.

The advocates of the application of scientific method to geography feel that at the level of practice, geography would need to retain most of its elements so as to prove the 'scientific status' of the discipline (e.g. geography). The research hypothesis, the hypothesis test and prediction as testing device would be required. But the idea that any one set of observations conclusively proved or definitively falsified a theory could not be retained. It was the accumulation of such results (positive or negative) that led to the advance or decline of rival research programmes. However, it seems inescapable that scientific geography, in spite of being subjected to criticism, will require the retention, development and refinement of measuring devices, guided by the emergent geographical theories and by the cognate disciplines from which they have been derived, in the future with openness as to what levels and forms of measurement are the most appropriate for a given study. There will be a continuing need for statistical analysis to carry on the application of scientific method to geographical study, research and training.

In spite of a strong assertion for the relevance of scientific method in geography, the admirers and/or adherents of the method also accept the possibility of the relevance of humanistic or phenomenological approaches in geographical explanation that may yield new insights into the nature of geographical phenomena. Many geographers strongly believe in the desirability of methodological *heterodoxy*, e.g. allowing the co-existence of radically different approaches within the discipline (Nagel, 1953; Johnston, 1980). Nevertheless, without scientific method in the subject, 'geography would cease to offer a convincing interpretation of the Earth's surface and the activities of individuals upon it' (Hay, 1985, 141). There is no doubt that application of scientific method to geography has given a nomothetic basis to the discipline with a scientific status and saved it from the crisis of its identity that it suffered during the transition period.

20

Post-modernism, Deconstruction and Spatiality

HIGHLIGHTS

- Modernity and Post-Modernity
- Modernlsation and Modernism
- Post-Modernism
- Deconstruction
- Spatiality

Post-modernism is a short-hand for a heterogeneous movement which had its origin in architecture and literary theory. According to the Oxford Dictionary of Geography (2nd edition, 1997, 338), the term 'post-modernism' refers to 'an architectural style which is a composite of past styles, characterized by a variety of colours, stylistic details from many periods and what is claimed to be a return to a vernacular type.... It is a philosophical stance which claims that it is impossible to make grand statements—meta narratives—about the structures of society or about historic causation because everything we perceive, express and interpret is influenced by our gender, class and culture .... No interpretation is superior to another. It has brought to geographers a recognition that space, place and scale are social constructs, not external givens.... Some geographers claim that post-modernism challenges the dominance of time and history in social theories and instead stresses the significance of geography and spatiality'. Post-modernism, therefore, raises urgent questions about place, space and landscape in the production of social life. 'Post-modernism', says Gregory (1989), 'is in its fundamentals a critique of what is usually called the "Enlightenment Project"'.

Post-modernism may be said to represent a radical attack upon the mimetic theory of representation and search for truth. It is anti-foundational in that it explicitly rejects totalising ambitions of modern social science. Such an epistemology, if taken seriously, is inescapable and radically relativist (Duncan and Ley, 1993).

Now, let us begin by looking briefly at the following 'family' of terms: Modernity and Post-modernity, Modernization, Modernism and Post-modernism. These terms/words are often used in confusing and interchangeable ways. To understand the philosophy of post-modernism in the context of geography, particularly, and social sciences, in general, it is essential and necessary also to know the exactness of the terms and words, and the distinction between them.

MODERNITY AND POST-MODERNITY

Modernity is generally held to have come into being with the Renaissance and was defined in relation to Antiquity. It implies the progressive economic and administrative rationalisation and differentiation of the social world. Modernity can be taken as a summary term, referring to that cluster of social, economic and political systems brought into being in the West from somewhere around eighteenth century onwards. Marshal Berman (1982) has attempted to explore the changing concepts of '*Modernity*' from the formative sixteenth-century clash between the 'Ancients' and the 'Moderns' to the contemporary debates that still herald conceptual and social reconfiguration, another reconsideration of what it means to be modern. In the concatenation of modernities is a history of historicism that can now begin to be written from a post-modern geographical perspective. He defines modernity as *'a mode of vital experience', a collective sharing of a particularised sense of the 'self and others', of 'life's possibilities and perils'.* In this definition, there is a special place given to the ways we think about and experience time and space, history and geography, sequence and simultaneity, event and locality, the immediate period and region in which we live. Modernity is thus comprised of both context and conjecture. It can be understood as the specificity of being alive, in the world, at a particular time and space; a vital individual and collective sense of contemporaneity (Soja, 1993, 24–25).

The experience of modernity, however, captures a broad mesh of sensibilities that reflects the specific and changing meanings of the three most basic and formative dimensions of human existence—space, time and being. Herein lies its particular usefulness as a means of resituating the debates on history and geography in critical social theory and for defining the context and conjuncture of post-modernity. 'From around 1800 to the outbreak of World War I a series of sweeping changes in technology and culture created distinctive new modes of thinking about and experiencing time and space.... cultural developments such as the stream of novel consciouness, psycho-analysis, cubism, and the theory of relatively shaped consciousness directly. The result was a transformation of the dimensions of life and thought' (Kern, 1983, 1–2).

Post-modernity, however, suggests what came after modernity ; if refers to the incipient or actual dissolution of those social norms associated with modernity. Some thinkers assume and/or believe that it is a movement towards a post-industrial age. 'It is a movement "beyond" the modern and simultaneously, an invitation to construct our "own" human geographies' (Gregory, 1989). However, there are many ambiguities in this assertion: should the post-modern be regarded a part of the modern? Is it a continuity or a radical break? Is it a material change or does it indicate a mood, a state of mind? Post-modernity, however, stresses and emphasises diverse forms of individual and social identity. Instead of a coercive totality and a totalising politics, post-modernity stresses a pluralistic and open democracy.

MODERNISATION AND MODERNISM

The term *'modernisation'* is often used to refer to the stages of social development which are based upon industrialisation. Modernisation is a diverse unity of socio-economic changes generated by scientific and technological discoveries and innovations, industrial upheaval, population movements, urbanisation, the formation of national states and mass political movements, all driven by the expanding; capitalist world market.

'Modernization ... is a continuous process of societal restructuring that is periodically accelerated to produce a significant recomposition of space-time-being in their concrete forms, a change in the nature and experience of modernity that arises primarily from the historical and geographical dynamics of production. For the past four hundred years, these dynamics have been predominantly capitalist, as has been the very nature and experience of modernity during that time. Modernization is like all social processes, unevenly developed across time and space and thus invites quite different historical geographies across regional social formations. But on occasion, in the ever-accumulating past, it has become systematically synchronic, affecting all predominantly capitalist societies simultaneously' (Soja, 1993, 27).

Modernisation can be directly linked to many different 'objective' processes of structural change which have been associated with the ability of capitalism to develop and survive, to produce successfully its fundamental social relations of production and distinctive divisions of labour despite endogenous tendencies towards debilitating crisis. There is currently a complex and conflictual dialectic developing between urgent socio-economic modernisation sparked by the system-wide crisis affecting contemporary capital societies; and a responsive cultural and political modernism aimed at making sense of material changes taking place in the world and gaining control over their future directions. Modernization and modernism interact under these conditions of intensified crisis and restructuring to create a shifting and conflictual social context in which everything appears to be 'pregnant with its contrary', in which all that was once assumed to be solid 'melts into air', a description that represents an essential feature of the vital experience of modernity-in-transition.

'In its broadest sense, modernism is the cultural, ideological, reflective and theory-forming response to modernization. It encompasses a heterogeneous array of subjective visions and strategic action programmes in art, literature, science, philosophy and political practice which are unleashed by the disintegration of an inherited, established order and the awareness of the projected possibilities and perils of a restructured contemporary moment or conjuncture. Each era of accelerated modernization has been a fertile spawning ground for powerful new modernisms emerging in almost every field of discourse and creativity. Of particular interest to the present narrative are two "modern movements" which emerged around the him of the nineteenth century to define separate and competitive realms of critical social theorization, one centred in the Marxist tradition, the other in more naturalist and positivist social science.' (Ibid, 28)

Thus, modernism concerns a particular set of cultural or aesthetic styles associated with the artistic movement which originated around the turn of the 20th century and have dominated the various arts until recently. Modernism developed in 'conscious opposition to classicism', and it emphasized experimentation and the aim of finding an inner truth behind surface appearance. However, one of the problems with trying to understand modernism is that many of these features appear in definition of post-modernism as well. Another problem with defining modernism is the question of how far back into the nineteenth century.

POST-MODERNISM

The term post-modernism originated among artists and critics in New York in the 1960s and was taken up by European theorists in the 1970s. One of them, Jean-Francois Lyotard (1984), who is considered to have pioneered the word *post-modernism* attacked the legitimating myths of the modern age (the grand narratives), the progressive liberation of humanity through science, and the idea that philosophy can restore unity to learning and develop universally valid knowledge for humanity. Post-modern theory became identified with the critique of universal knowledge and foundationalism. According to Lyotard (1984), 'We can no longer talk about a totalizing idea of *reason* for there is *no reason,* only *reasons.*'

Among the central features associated with post-modernism in the arts are:(1) the deletion of the boundary between the art and everyday life; (2) the collapse of the hierarchical distinction between elite and popular culture; (3) a stylistic eclecticism and mixing of codes. There is parody, pastiche, irony and playfulness. Many people, however, stress that post-modernists espouse a model which emphasizes not depth but surface. They are highly critical of structuralism and Marxism and are antagonistic to any theory that 'goes beyond' the manifest to the latent. The decline of the originality and genius view of the artistic producer has been replaced by the assumption that art can only be repetitious.

It is often said that in post-modernism there is: (1) a shift of emphasis from content to form or style; (2) a transformation of reality into images; (3) the fragmentation of time into a series of perpetual presents. There have been continual references to eclecticism, reflexivity, self-referentiality, quotation, artifice, randomness, anarchy, fragmentation, pastiche and allegory. With the development of post-modernism in recent years, there has been a move to 'textualize' everything: geography, history, philosophy, jurisprudence, sociology and other disciplines are treated as so many optional 'kinds of writing' or discourses.

Post-modernism characterises scepticism towards the grand claims and grand theory of modern era and their claims to intellectual superiority. It (as compared to modernism) stresses and emphasises openness to range of options in social enquiry, artistic experimentation and political empowerment. Post-modernism represents and manifests a response to modernism as a homogenising form and to its totalising theoreticism. Instead, post-modernism lays emphasis on discontinuities and disjunctures characteristic of everyday life in real world. The homogeneity in the built-up landscape of the era of 'organized capitalism' is contrasted with the heterogeneity of economic, social and political life in the current phase of 'disorganized capitalism'. To some people, post-modernism may be regarded as a culture of late capitalism that appeared to have initiated an integral part of the new phase of post-Fordist flexible accumulation (Jameson, 1984; Harvey, 1989).

The Post-modern Conditions

Many people are aware that western societies since the Second World War have radically changed their nature in some way. A fashionable description of such societies is that they are post-modern. Post-modernism is in part a description of a new type of society, but also, in part, a new term for post-structuralism in the arts. During the past 50 years, the leading sciences and technologies have become increasingly concerned with language: theories of linguistics, problems of communication and cybernetics, computers and their languages, problems of translation, information storage and data banks. The technological transformations are having a considerable impact on knowledge. The miniaturisation and commercialisation of machines are already changing the way in which learning is acquired, classified, made available and exploited.

According to Lyotard (1984), the nature of knowledge cannot survive unchanged within this context of general transformation. The status of knowledge is altered as societies enter what is known as the post-modern age. He, however, predicts that anything in the constituted body of knowledge that is not translatable into quantities of information will be abandoned and the direction of new research will be dictated by the possibility of the eventual results being translatable into computer language. The old principle that the acquisition of knowledge is indissociable from the training of minds, or even of individuals, is becoming obsolete: Knowledge is already ceasing to be an end in itself. It is and will be produced in order to be sold.

It is widely accepted that computerised knowledge has become the principal 'force of production' over the last few decades. This has already had a noticeable effect on the composition of the work-force of the most highly developed countries. Knowledge will be the major component in the world-wide competition for power, and it is conceivable that nation-states will one day fight for control of information, just as they battled for control over territories in the past. In the post-modern age, science will probably strengthen its pre-eminence in the arsenal of productive capacities of the nation-states and the gap between developed and developing countries will grow even wider. Lyotard (1984) suggests that power and knowledge are simply two aspects of the same question: who decides what knowledge is? Who knows what needs to be decided?'

To post-modernists, knowledge is a question of competence that goes beyond the simple determination and application of the criterion of truth, extending to the determination of criteria of efficiency (technical qualification), of justice and/or happiness (ethical wisdom), of beauty (auditory or visual sensibility), etc. Knowledge is what makes someone capable of forming not only 'good' denotative utterances, but also 'good' prescriptive and 'good' evaluate utterances. But how are they to be assessed? They are judged to be good if they conform to the relevant criteria (of justice, beauty, truth and efficiency) accepted in the social circle of knower's interlocutors.

Narrative Knowledge and Scientific Knowledge

Scientific knowledge does not represent the totality of knowledge; it has always existed in competition and conflict with another kind of knowledge, called 'Narrative'. Narratives (popular stories, myths, legends and tales) bestow legitimacy upon social institutions, or represent positive or negative models of integration into established institutions. Narratives determine criteria of competence and/or illustrate how they are to be applied. What is transmitted through these narratives is the set of rules that constitute the social bond.

The main difference between scientific knowledge and narrative knowledge is that scientific knowledge requires that one language game, denotation, be retained and all others be excluded. 'Both science and non-scientific (narrative) knowledge are equally necessary. Both are composed of sets of statements; the statements are "moves" made by the players within the framework of generally applicable rules' (Lyotard, 1984, 38). It is therefore impossible to judge the existence or validity of narrative knowledge on the basis of scientific knowledge or vice versa: the relevant criteria are different. Narrative knowledge certifies itself without having recourse to argumentation and proof. However, scientific knowledge cannot know and make known that it is the true knowledge without resorting to the other, narrative kind of knowledge, which from its point of view is no knowledge at all. In short, there is a recurrence of the narrative in the scientific.

However, (older) master narratives no longer function in contemporary society. The grand narrative has lost its credibility, regardless of whether it is a speculative narrative or a narrative of emancipation. The decline of the unifying and legitimating power of the grand narrative of speculation and emancipation can be seen as an effect of the blossoming of techniques and technologies since the Second World War, which has shifted emphasis from the 'end' of action to its 'means'.

The Mercantilisation of Knowledge

As a result of the Industrial Revolution, science became a force of production, a moment in the circulation of capital. An important aspect of research is the production of proof which requires to be proven. A scientific observation depends on facts being registered by sense organs. But the range and power of discrimination are limited. Technical devices follow the principle of optimal performance, maximising output and input. The game of science becomes the game of the rich, in which whoever is wealthiest has the best chance of being right. It is thus that an equation between wealth, efficiency and truth is established.

The goal of science in another way is no longer truth, but performativity, i.e. the best possible input/output equation. Since performativity increases the ability to produce proof, it also increases the ability to be right; the technical criterion cannot fail to influence the truth criterion. The utility of knowledge, therefore, appears to be more important than the search for truth. The shift of attention from 'ends of action to its means', from 'truth to performativity', is reflected in present-day educational policy and the emphasis is, therefore, on skills rather than on ideals.

It is only in the context of the grand narratives of legitimation—the life of the spirit and/or emancipation of humanity—that the partial replacement of teachers by machines seems inadequate or even intolerable. According to Lyotard (1984), 'it is probable that these narratives are already no longer the principal driving force behind interest in acquiring knowledge. In the context of mercantilization of knowledge, more often than not this question is equivalent to: is it saleable? And in the context of power-growth: is it efficient?' Education, as it seems clear, must provide not only for the reproduction of skills, but also for students' progress.

What is vitally important is the capacity to actualise the relevant data for solving a problem here and now, and to organise the data into an efficient strategy. Data banks are the encyclopaedia of tomorrow; they are 'nature for post-modern men and women'. What is, therefore, important is to arrange the data in a new way. It is imagination, i.e. capacity to articulate, which allows one either to make a new move (a new argument) with the established rules or to invent new rules, i.e. a new game.

It is a historical fact that countless scientists have seen their invention and/or discovery of new rules ignored or repressed, sometimes for decades, because that invention and/or discovery too abruptly destabilised the accepted position, not in the scientific hierarchy, but also in the discipline. The more striking the invention, the more likely it is to be denied, precisely because it changes the rules of the game upon which consensus was based.

Modernism and Post-modernism

It is commonly believed by many that with the cessation of the Second World War, a new type of society began to emerge. The society was labeled in various ways, depending on the way it was analysed: consumer society, post-industrial society, society of the spectacle, post-modern society, etc. Post-structuralists, on the whole, argue that the new society is now outdated, and as such, it does not and cannot apply to the new social developments. This argument often overlaps with one another concerning modernism and post-modernism. The crucial question in these debates is: Has the Enligtenment Project failed? Should we, like the post-structuralists and post-modernists, declare the entire project of modernity a lost cause? Or should we seek to hold on the intentions and aims of the Enlightenment and of cultural modernism?

The project of modernity formulated in the eighteenth century by the philosophers of the Enlightenment consisted in their efforts to develop objective science; universal morality and law and autonomous art. Contemporary philosophers wanted to use this accumulation of specialised culture for the enrichment of everyday life. They hoped that the arts and sciences would promote not only the control of natural forces, but also understanding of the world and of the self, moral progress, the justice of institutions and even the happiness of human beings.

But what has happened is in marked contrast to the hopes and ideals of the Enlightenment. Gradually each domain has been institutionalised; science, morality and art have become autonomous domains separated from the life-world. The structures of cognitive-instrumental, of moral-practical and aesthetic-expressive rationality have come under the control of special experts (Habermas, 1981). In America, France and elsewhere, cultural modernism is now under attack from many different quarters. More recently, the Enlightenment Project has been denounced by the French 'new philosophers', such as Foucault, Derrida and Lyotard and their contemporary English and American counterparts. It has also been attacked by the post-structuralists, whose works, however, require to be included among the manifestations of post-modernism.

'The concept of post-modernism is ambiguous and is not yet widely understood. It has probably emerged as a specific reaction against the established form of high modernism. For some thinkers, post-modernism is a periodizing concept whose function is to correlate the emergence of new features in culture. The concept seems to be connected with the appearance, between the 1950s and the 1960s, of a new social and economic order. Sometimes a useful distinction is made between pre-modernists, those who want to withdraw to a position anterior to modernity, anti-modernists and post modernists. There are so many similarities between post-structliralist theories and post modern practices that it is difficult to make a clear distinction between them' (Sarup, 1993, 67–68).

Main Features of Post-modernism

To understand post-modernism, it is necessary to know what Lyotard (1984, 45–46) means by the term 'Modern'. He uses the term modern 'to designate any science that legitimates itself with reference to a meta discourse—making an explicit appeal to some grand narrative, such as the dialecties of the spirit, the hermeneutics of meaning, the emancipation of rational or working subject or creation of wealth', or to put it in another way: 'Societies which anchor the discourses of truth and justice in the great historical and scientific narratives (recits) can be called modern.'

Post-modernists distrust meta-narratives; there is a deep suspicion of Hegel, Marx and any form of universal philosophy. The post-modern condition is one in which the *'grands recits'* of modernity—the dialectic of spirit, the emancipation of the worker, the accumulation of wealth, the classless society —have all lost credibility. A discourse is defined as modern when it appeals to one or another of these 'grands recits' for its legitimacy. The 'grands recits' are *master narratives*—narratives of mastery, of man seeking his *'telos'* in the conquest of nature. The Marxist master narrative is only one version among many of a modern narrative of mastery. The advent of post-modernity signals a crisis in a narrative's legitimising function, its ability to compel consumers.

Lyotard is critical of Marxism because it seeks to create a homogeneous society which can only be brought about through the use of cbercion. He believes that the *individualistic fragmented* society that we have today is to stay. Yet, he seems to be nostalgic for a pre-modern (traditional) society, for traditional societies stress narratives, that is to say, myth, magic, folk wisdom and other attempts at explanation. According to him, there is a conflict between narrative and science (theoretical knowledge). Narrative is disappearing and there is nothing to replace it with (Sarup, 1993).

Lyotard, as it is often said, argues that art, morality and science (the beautiful, the good and the true) have become separated and autonomous. A characteristic of the contemporary time is the fragmentation of language games. There is no *meta-language.* No one can grasp what is going on in society as a whole. Lyotard seems to be saying that there is no one system of domination. There are parallels between these ideas and some right-wing theorists (like Hegel) who argue that society works much better in terms of micro-events; a society that is left to market forces is better than a consciously planned society.

The argument of some *post-modernists* and *post-structuralists,* in short, is that big stories are bad, but little stories are good. Instead of a truth/falsity distinction they, including Lyotard, prefer a small/grand narrative criterion. Grand narratives have become associated with a political programme or party, while little narratives are associated with localised creativity. These ideas are similar to those of Foucault, who was also against grand narratives and supported the idea of local struggles.

Fredric Jameson (1984) has identified two significant features of post-modernism: *'pastiche'* and *'schizophrenia'.* He explains that the great modernisms were predicted on the invention of a personal, private style. The modernist aesthetic was organically linked to the conception of an authentic self and a private identity which can be expected to generate its own unique vision of the world and to forge its own unmistakable style. The post-structuralists argue against this; in their view the concept of the unique individual and the theoretical basis of individualism are ideological. Not only is the bourgeois individual subject a thing of the *past,* it is also a *myth,* it never really existed in the first place; it was just a mystification. And so, in a world in which stylistic innovation is no longer possible all that is left, is 'pastiche'. The practice of 'pastiche', the imitation of dead styles, can be seen in the 'nostalgia film'. We have lost our ability to locate ourselves historically. As a society we have become incapable of dealing with time.

Post-modernism has a peculiar notion of *time.* Jameson (1991) seeks to explain what he means in terms of Lacan's theory of *schizophrenia.* The originality of Lacan's thought in this area is to have considered schizophrenia as a language disorder. It emerges from the failure of the infant to enter fully into the realm of speech and language. The experience of temporality, human time, past, present, memory, the persistence of personal identity is an effect of language. It is because language has a past and a future, because the sentence moves in time, that we can have what seems to us a concrete or lived experience of time. The schizophrenic, in short, experiences a fragmentation of time, a series of perpetual presents. Jameson (1991) contends that experiences of temporal discontinuity, similar to those maintained above, are evoked in post-modernist works.

However, Derek Gregory (1989) has also identified three basic features of post-modernism in geographic perspective. First, 'post-modernism is, in very real sense, *"post-paradigm"*, that is to say, post-modern writers are immensely suspicious of any attempt to construct a system of thought which claims to be complete and comprehensive. In geography, of course, there has been no end of attempts of this kind, and many of those who have mistakenly made use of Kuhn's notion of a paradigm have done so prescriptively. They have claimed the authority of *"positivism"*, *"structuralism"* and *"humanism"* or whatever as a means of legislating for the proper conduct of geographical inquiry and of excluding work which lies beyond the competence of these various systems. Others have preferred to transcend these, to them partial, perspectives and to offer some more general (meta-theoretical) framework in which all these competing claims are supposed to be reconciled. For over a decade this was usually assumed to be of some kind of systems approach and now, apparently, it is the philosophy of realism (perhaps coupled with some version of Habermas's critical theory) which holds out a similar promise. But post-modernism rejects all of these manoeuvres.

'... postmodern writers are hostile to the *"totalizing"* ambition of the conventional social sciences. Their critique points in two directions. First, they reject the notion that social life displays what could be called a "global coherence": that our day-to-day social practices are moments in the reproduction of a self-sustaining social system whose fundamental, so to speak "structural" imperatives necessarily regulate our everyday lives in some automatic, pre-set fashion. The post-modern writers do not, of course, deny the importance of the independencies which have become such a common-place of the late twentieth-century world.... But they do object to the concept of totality which informs much of modern social theory because it tacitly assumes that social life somehow adds up to ... a coherent system with its own subordinate logic' (ibid.).

'Second,...these writers reject the notion that social life can be explained in terms of some *"deeper"* structure. This was out of the premises of structuralism ... it still surfaces in some of the cruder versions of realism. It is largely through this opposition that post-modernism is sometimes identified with post-structuralism.... The post-modern critique will seem to echo the complaints of those who saw in structuralism a displacement of the human subject. In human geography, as elsewhere, many commentators were dismayed by the way in which various versions of structuralism replaced the concrete complexities of human agency by the disembodied transformation of abstract structures. But post-modernism is not another humanism. It objects to structuralism because its sharpened concept of structure points towards a "centre" around which social life revolves ... but it objects to humanism for the very same reason' (ibid.).

'Thirdly, the accent on *"difference"* which dominates the preceding paragraphs is a *"leit motif"* of post-modernism. One of the distinguishing features of post-modern culture is its sensitivity to heterogeneity, particularity and uniqueness. To some readers, this insistence on "difference" will raise the spectre of the idiographic, which is supposed to have been laid once and for all (in geography at any rate) by Hartshorne-Schaefer in the 1950s and by the consolidation of a generalizing spatial science during the 1960s' (ibid.).

'Several commentators have emphasized deep-seated continuities between the Hartshonian orthodoxy and the prospectus of the so-called "New geography". But Schaefer's clarion call for geography as a nomothetic science, compelled to produce morphological laws and to disclose the fundamental geometries of spatial patterns, undo ubtedly sounded a retreat from aerial differentiation which was heard in many quarters. Specificity became eccentricity, and the new conceptual apparatus made no secret of its confinement: it was, variously, a "residual", background "noise" to be "filtered out", a "deviation" from the "normal" .... In the 1980s other writers in other fields gave specificity a wider resonance' (ibid.).

'In philosophy, Lyotard claims that 'post-modern knowledge ... refines our sensitivity to "differences" in social theory De cartean wants to fashion 'a science of singularity ...that 'links everyday pursuits to particular circumstances' and in anthropology Geertz parades "the diversity of things" and seeks illumination from the light of local knowledge.'

In geography there has been a remarkable return to aerial differentiation. But it is a return with a difference. When Harvey speaks of the 'uneven development' of capitalism ... or when Hagerstrand talks about 'pockets of local order', they and others like them are attempting much more than the recovery of geography's traditional project. For they herald not so much the reconstruction of modern geography as its deconstruction... what Soja calls the 'post-modernization of geography' (ibid.).

Totality or Fragmentation

Lyotard (1984), repudiates and contradicts the big stories, the meta-narratives of Hegel and Marx. He believes that no one can grasp what is going on in a society as a 'whole'. Rejecting totality, Lyotard and other post-modernists stress fragmentation—of language games, of time, of the human subject, of society itself. One of the fascinating things about the rejection of organic unity and the espousal of the fragmentary is that this belief was also held by historic avante-garde (artistic) movements. They also sought for the dissolution of unity. In their activities, the coherence and autonomy of the work was deliberately called into question or even methodically destroyed.

Walter Benjamin's concept *of 'allegory'* has been used as an aid to understanding avant-gardiste (non-organic) works of art. Allegory is essentially a fragment, the opposite of the organic symbol (Burger, 1984). Benjamin has described how the allegorist pulls an element out of the totality of the life context, isolating it, depriving it of its function. An allegorist is believed to join several isolated fragments, and thereby creates meaning. Montage presupposes fragmentation of reality; it breaks through the appearance of totality and calls attention to the fact that it is made up of reality fragments: in the organic work of art, the material is treated as a whole, while in the avant-gardiste work the material is torn out of the life totality and isolated. The aesthetic avant-gardist fragment challenges people to make it an integrated part of their reality and to relate it to their experience.

Sarup has identified three features of post-modernist thought. First, there is a tendency to reduce all truth-claims to the land of rhetorics, narrative strategies or Foucauldian discourses conceived as existing solely by virtue of the differences between them so that no single claimant can assert itself at the expense of any other.

Second, there is often an appeal to the Wittgenstenian notion of *'language games'* (sometimes called 'forms of life'). A belief in heterogeneous language games, each involving a different set of cognitive, historical or ethno-political criteria, implies that it is not possible to decide between rival interpretations.

Third, there is a turn towards the Kantian sublime as a means of devaluing cognitive truth-claims and elevating the notion of the unrepresented to absolute pride of place in the ethical realm. In other words, there has been *a move to aestheticise politics by removing ethical and political questions* as far as possible from the realm of truth and falsehood.

Lyotard, however, adopts a Wittgenstein (1958) language games approach to knowledge proposing that we conceive of various discourses as language games with their own rules, structure and moves. Different language games are thus governed by different criteria and rules, and none is privileged. Language games are indeed the social bond which bonds society together. Lyotard characterises social interaction primarily in terms of making a move in a game, playing a role and taking a part in various discrete language games. Lyotard's (1984) model of a post-modern society is one in which one struggles within various language games in an agonistic environment characterised by diversity and conflict.

According to Norris (1992), what results from Lyotard's post-modern interpretation of the Kantian sublime, is an outlook of extreme cognitive scepticism, along with a politics completely cut off from questions of real-world relevance and accountability.

Feminism and Post-modernism

Few women have participated and engaged in modernism and post-modernism debate and they have very little to say about post-modernism. However, Morris (1988) has argued that since feminism has acted as one of the enabling conditions of discourse about post-modernism, it is, therefore, appropriate to use feminist work to frame discussion of post-modernism.

Feminism and post-modernism have emerged as two most important politico-cultural currents of the last decade. Both, feminism and post-modernism have offered deep and far-reaching criticisms of philosophy, and of the relation of philosophy to the larger culture. Both have tried to develop new paradigms of social criticism which do not rely on traditional philosophical underpinnings.

There are also differences between feminism and post-modernism. Post-modernists offer sophisticated criticisms of foundationalism and essentialism, but their conceptions of social criticism tend to be anaemic. Feminists offer robust conceptions of social criticism, but they tend to lapse into foundationalism and essentialism.

Female post-modernists, like Linda Nicholson and Nancy Linda (1988) have suggested that each of these tendencies has much to learn from the other; each is in possession of valuable resources which can help remedy the deficiencies of the other. Lyotard (1984) offers a post-modern conception of what he calls the *'social bond'*. What holds the social bond is a weave of criss-crossing threads of discursive practices, no single one of which runs continuously throughout the whole. Individuals are the nodes where such practices intersect and, so, they participate in many simultaneously. It follows that social identities are complex and heterogeneous. They cannot be mapped on to one another nor on to the social totality. Indeed, strictly speaking, there is no social totality and there is no possibility of a totalising social theory.

Feminists, like post-modernists, have sought to develop new paradigms of social criticism which do not rely on traditional philosophical underpinnings. Political imperatives, however, have led some feminists to adopt modes of theorising which resemble the sorts of philosophical meta-narratives criticised by post-modernists.

In short, post-modern feminist theory would be pragmatic, and tailor its methods and categories to speak to the specific task at hand. Both, Fraser and Nicholson (1988), however, contend that there is decreasing interest in grand social theories on the one hand, and essentialist vestiges persist in the continued use of ahistorical categories without reflection as to how, when and why such categories originated and were modified over time, on the other hand. They suggest that this tension is symptomatically expressed in the work of French psychoanalytic feminists and believe that Kristena and others personally deny essentialism even as they performatively enact it.

Problems of Post-modernism

Post-modern is not without problems and criticisms. Lyotard, who pioneered post-modernism, is often confronted with some criticisms. His book *The Post-modern Condition* (1984) is on one level about the status of science and technology, about technocracy and the control of information. But on another level, it is a thinly veiled polemic against Habermas, who stands for a 'totalizing' and 'dialectic' tradition. Habermas thinks that the totality of life has become splintered and argues that the cognitive, ethical and political discourses should come close together. He wants, in fact, to defend modernity against the neo-conservative post-modernists. Post-modernist's main target is the Hegelian-Marxist concept of totality.

Post-modernists believe in the breaking up of the narratives without describing how and why this theoretical collapse has taken place and why they are themselves polemicising against these discourses. There are many kinds of grand narratives, but post-modernists tend to bring all of them together. Even if some of the narratives of legitimation are dubious, why reject all grand narratives? Many post-modernists fail to specify what causes the rupture in society and history that produces the post-modern conditions. Theorists who reject master narratives, or historical, periodising social theory, are naturally going to have difficulty producing such a narrative, and thus find themselves in an a poretic situation.

Post-modernists reject totalising social theories, the master narratives, because they believe that these theories and narratives are reductionist and simplistic, but, nevertheless, they offer a theory of post-modern condition that presupposes dramatic break from modernity. But surely the concept of post-modernism presupposes a master narrative, a totalising perspective. While post-modernists resist grand narratives, it is impossible to discern how one can have a theory of post-modernism without one.

Post-modernists insist that the field of the social is heterogeneous and non-totalisable. As a result, they rule out the sort of critical social theory which employs general categories like gender, race and class. From their perspective, such categories are too reductive of the complexity of social identities to be useful. In short, there is no place in post-modernists' universe for critique of relations of dominance and subordination along lines like gender, race and class. In their view, there is nothing to be gained in the critical analysis of large-scale institutions and social structures. They contend that sociological and geographical synthesis must be abandoned for playful deconstruction and the privileging of the aesthetic mode.

DECONSTRUCTION

Deconstruction, which has attained widespread recognition as one of the most important avant-garde intellectual movements in France and America, is essentially post-phenomenological and post-structuralist. In the history of contemporary deconstruction, the leading figure was Jacques Derrida, who published three books in 1973, 1975 and 1976.

Derrida was mainly concerned with the role and function of language, and was famous for having developed a process called 'Deconstruction'. This is a method of reading a text so closely that the author's conceptual distinctions on which the text relies are shown to fail on account of the inconsistent and paradoxical use made of these very concepts within the text as a whole. In other words, the text was seen to fail by its own criteria; the standards or definitions which the text set up were used reflexively to unsettle and shatter the original distinction.

The method of deconstruction is connected with what Derrida called the *'metaphysics of presence'*. It was Derrida's contention that Husserl, along with almost all other philosophers, relies on the assumption of an immediately available area of certainty. The origin and foundation of most philosophers' theories is 'presence'. In Husserl's case, the search for the form of pure expression is at the same time a search for that which is immediately present; thus implicitly, by being present in an unmediated way and present to itself, it is undeniably certain.

Derrida, however, denied the possibility of this 'presence', and in so doing removed the ground from which philosophers had in general proceeded. By denying 'presence' he was denying that there was a present in the sense of a single definable moment which, to him, was *'now'*. For most people, the present is the province of the known. By challenging access to the present, Derrida posed a threat to both positivism and phenomenology.

Having argued that a realm of the independent 'signified' does not exist, Derrida concluded, first, that no particular sign could be regarded as referring to any particular signified, and second, that we were unable to escape the system of signifiers. In combination, these combinations, however, imply that there can be no unqualified presence.

Phonocentrism-Logocentrism

It is because of the assumption of 'presence' that a priority has been given to speech over writing. Derrida called it 'phonocentrism'. Speech has been regarded as prior because it is closer to the possibility of 'presence'. It is closer because speech implies immediately. In speech, meaning is apparently immanent, above all when, using the inner voice of consciousness, we speak to ourselves. In the moment of speech, we appear to grasp its meaning and we are thereby able to capture 'presence' as if the meaning was decided once and for all. Thus, unlike writing, which is hopelessly mediated, speech is linked to the apparent moment and place of 'presence' and for this reason has had priority over writing. For Derrida, phonocentrism is one of the effects of presence. Derrida's attempt to 'deconstruct the opposition between speech and writing was linked to the uncovering of the metaphysics of presence as a whole'.

Derrida had also criticised Saussare for prescribing that linguistics should be a study of speech alone rather than of speech and writing. This was an emphasis shared by Jakabson and Levi-Strauss, indeed by all semiological structuralists. Derrida, however, suggested that this rejection of writing as an appendage, a mere technique and yet a measure built into speech—in effect, a scapegoat—is a symptom of a much broader tendency. He related this phonocentrism to logocentrism, the belief that the first and the last was the Logos, the World, the Divine Mind, the 'self-presence of full self-consciousness' (Sarup, 1993, 36).

Speech is, however, thought of as remaining closer to psychic interiority than writing that symbolises interiority only at one remove. Writing does not seem to be so direct, so natural or sincere. Compared with speech, writing seems mechanical, second-hand, a transcript of speech. Derrida argued that western philosophy has focused on speech, and emphasised the voice, while dismissing writing as a derivative form of expression. In this tradition, the phenomenological structure of the voice is regarded as the most immediate evidence of self-presence.

Besides being 'phonocentric', western philosophy is also 'logocentric'. Derrida used the term 'logocentric' as a substitute for metaphysics in order to foreground that which has determined metaphysical systems of thought, their dependence on a 'logos'. Western philosophy assumes that there is an essence, or truth which acts as the foundation of all our beliefs, hence there seems to be a disposition, a longing, for a 'transcendental signifier', which would directly relate, correspond, to a secure stable 'transcendental signified', i.e. 'logos'. Examples of such signs include: Idea, Matter, the World Spirit, God, etc. Each of these concepts acts as the foundation of a system of thought and forms an axis around which all other signs circulate. However, Derrida argued that any such transcendental meaning is a fiction. There are certain signifieds or meanings attached to such signifiers as Authority, Freedom, Order, which are highly valued in society. Sometimes, these meanings are thought of as if they were the origin of all the others. Nevertheless, these meanings are not always seen in terms of origin; they are often seen in terms of goal, towards which all other meanings are advancing. Conceiving of things in terms of their orientation towards a 'telos' or end point—teleology—is a way of organising meaning in a hierarchy of significance.

Derrida called 'metaphysical' any thought system that depends on a foundation, a ground or a principle. First principles are often defined by what they exclude, by a sort of 'binary opposition' to other concepts. These principles and their implied 'binary opposites' can always be *deconstructed.* All conceptual oppositions of metaphysics, according to Derrida, have for ultimate reference the 'presence of a present'. To him, the binary oppositions of metaphysics include signifier/signified, sensible/ intelligible, speech/writing, speech (parole)/ language (tongue), diachrony/synchrony, space/ time, passivity/active. One of Derrida's criticisms of the structuralists is that they have not put these concepts 'under erasure', that they have not put these binary oppositions into question.

It is known to all that ideologies draw up sharp distinctions between conceptual opposites such as truth and falsity, meaning and non-sense, centre and periphery. Derrida suggested that we should try to break down the oppositions by which we are accustomed to think and which ensure the survival of metaphysics in our thinking: matter/ spirit, subject/object, veil/truth, body/soul, text/meaning, interior/exterior, representation/ presence, appearance/essence, etc. Derrida's importance in this regard, however, lies in the fact that he suggested a method whereby one can subvert these oppositions and show that one term relies on and inheres within the other. He argued that 'phonocentrism and logocentrism' related to centrism itself the human desire to posit a 'central' presence at the beginning and end. The opposition between speech and writings, however, takes place within this pattern.

According to Sarup (1993), Derrida was attempting to subvert the logocentric theory of the sign. Traditionally, the signifier refers to the signified, that is, an acoustic image signifies an ideal concept, both of which are present to consciousness. In Derrida's view, 'the sign marks an absent present' rather than present the object we employ the sign: however, the meaning of the sign is always postponed or deferred.

Derrida developed a concept which he called 'difference' and that referred to 'to differ'—to be unlike or dissimilar in nature, quality or form—and to 'to defer'—to delay, to postpone. Spoken French makes no phonetic distinction between the endings '-ance' and '-ence'; the word registers as 'difference'. This undetected difference shows up only in writing. The concept of writing, then, is a challenge to the very idea of structure, for a structure always presumes a centre, a fixed principle, a hierarchy of meaning and a solid foundation, and it is just these notions which the endless differing and deferring of writing throws into question.

Language is the play of differences which are generated by signifiers which are themselves the product of those differences. Derrida attempted to incorporate into the meaning of 'Difference' the sense of deferring. Difference is itself endlessly deferred.

Metaphor

The study of metaphor is becoming important as it is being realised that language does not simply reflect reality, but helps to constitute it. Attention is now being increasingly given to how rhetorical devices shape our experience and our judgements, how language serves to promote the possibilities of certain kinds of action and exclude the practicability. In the past, metaphor was often studied as an aspect of the expressive function of language; but it is actually one of the essential conditions of speech. Language is essentially metaphorical, because metaphorical expressions are rooted in language itself. Derrida has shown that even philosophy is permeated with metaphor without knowing it.

An influential post-structuralist thinker Michel Foucault was particularly fond of using 'geographical' metaphors such as territory, domain, soil, horizon, archipelago, geopolitics, region, and landscape. He also made use of spatial metaphors—position, displacement, site, field. Similarly, Althusser also in 'Reading Capital' used many spatial metaphors such a terrain, site, etc. Foucault said and/or suggested earlier that there has been a devaluation of space. *'Space was treated as the dead, the fixed, the undialectical, the immobile; time: on the contrary, was richness, fecundity "life, dalectial"* (Foucault, 1980). But to talk in terms of space does not mean that one is hostile to time.

Althusser believed that the use of spatial metaphors in his work was necessary, but at the same time regressive, non-rigorous. Foucault, on the other hand, was more positive. He said that 'it is through these spatial obsessions that he came to what he was looking for': the relation's that are possible between power and knowledge. Once knowledge can be analysed in terms of region, domain, implantation, displacement, transportation, one is able to capture the process by which knowledge functions as a form of power and disseminates the effects of power (Foucault, 1980).

Metaphors determine to a large extent to what we can think in any field. Metaphors can help to make and defend a world view. It is important that the implications of the metaphor that we employ or accept are made explicit and that the ways in which they structure our thought and even our action are better understood. Metaphors can be productive of new insights and fresh illuminations. They can promote unexpected or subtle parallels or analogies. Metaphors can encapsulate and put forward proposals for another way of looking at things, for, we can have increased awareness of alternative possible world through metaphor. Thus the *metaphor is most often used by deconstructionists.*

Problems of Deconstruction

Deconstructive criticism takes the metaphoric structure of a text seriously. Since metaphors are not reducible to truth, their own structures 'as such' are part of the text. The deconstructive procedure is to spot the point where a text covers up its grammatical structure.

Derrida has provided a method of 'close-reading' a 'text' very similar to psychoanalytic approach to neurotic symptoms. Deconstructive 'close-reading', having 'interrogated' the text, breaks through its defences and shows that a set of binary oppositions can be found 'inscribed' within it in each of the pairs, private/public, masculine/feminine, same/other, rational/irrational, true/false, central/peripheral, etc.; the first term is privileged. Deconstructors show that the 'privileged' term depends for its identity on its excluding the other and demonstrate that primacy really belongs to the subordinate term instead.

Derrida attempted to stress the point that it is not enough simply to neutralise the binary opposition of metaphysics. Deconstruction involves reversal and displacement. Within the familiar philosophical opposition there is always a violent hierarchy. One of the two terms controls the other, holds the superior position. The first move in deconstructing the opposition is to overthrow the hierarchy. In the next phase this reversal must be displaced, the winning term put 'under erasure' (idea *of 'saus rature'*). Deconstruction, then, is an attempt '... to locate the promising marginal text, to disclose the undecidable moment, to pry it loose with the positive lever of the signifier, to reverse the resident hierarchy, only to displace it; to dismantle in order to reconstitute what is always already inscribed' (ibid.).

Derrida made closer and intensive study of many philosophers, and said that they had attempted to impose their systems of thought ignoring or suppressing the disruptive effects of language. His work draws attention to the ways in which language deflects the philosopher's project. Derrida did this by focusing on metaphor and other figurative devices at work in the texts of philosophy. In this way, he underlined the rhetorical nature of philosophical arguments.

Deconstruction stresses the irreducibility of metaphor, the difference at play within the very constitution of liberal meaning. It should be remembered that deconstruction is not simply a strategic reversal of categories which otherwise remain distinct and unaffected. It is an activity of reading in which texts must be read in a radically new way. Derrida discovered a set of paradoxical themes at odds with their manifest argument. His method attempted to reveal how the privileged term is held in place by the form of a dominant metaphor and not, as it might seem, by any conclusive logic. Metaphor often disrupts the logic of an argument (Ree, 1984).

However, Derrida's works on deconstruction reveal some problems, or it can be said that Derrida's works confront with some problems. Having argued that there cannot be a realm of the signified independent of the signifier, he opened up the vista of an endless play of signifiers that refer not to signified, but to other signifiers, as that meaning is always ultimately undecidable.

Deconstruction disarticulates traditional conception of the author and the work and undermines conventional notions of reading and history. Instead of mimetic, expressive and didactic theories of 'literature', it offers textuality. It kills the author, turns history and tradition into intertextuality and celebrates the reader.

One of the main features of post-structuralist theory is the deconstruction of the self. In place of a unified and stable being or consciousness, we get a multi-faceted and disintegrating play of selves. With deconstruction the categories 'criticism', 'philosophy', and 'literature' collapse, borders are overrun. The work, now called 'text', explodes beyond stable meaning and truth towards the radical and ceaseless play of infinite meanings. Critical writing, formerly analytical and coherent, becomes playfully fragmented.

The deconstructor's method often consists of deliberately inventing traditional oppositions and making the play of hitherto invisible concepts that reside unnamed in the gap between opposing terms. In the move from hermeneutics and semistics to deconstruction, there is a shift from identities to differences, unities to fragmentations, ontology to philosophy of language, espistemology to rhetoric, presence to absence. According to one recent commentator, deconstruction celebrates dissemination over truth, explosion and fragmentation over unity and coherence, undecidable spaces over prudent closures, playfulness and hysteria over care and rationality (Leitch, 1983).

It is said that every boundary, limit, division, frame or margin instals a line separating one entity or accept from another; that is to say, every border marks a difference. The question of border is a question of difference. According to Derrida, 'No border is guaranteed, inside or outside'. Applied to texts, this finding becomes 'no meaning can be fixed or decided upon'. According to deconstructionists, there is nothing other than interpretation. As there is neither an undifferentiated nor a literal bottom or ground, the activity of interpretation is endless. It is also a fact that every text tends itself to deconstruction and to further deconstruction, with nowhere any end in sight. However, the supreme irony of what Derrida has called logocentrism is that critique, deconstruction, is as insistent as monotonous and as inadvertently systematising as logocentrism (Said, 1985).

Deconstruction for Derrida is ultimately a political practice, an attempt to dismantle the logic by which a particular system of thought and, behind, that a whole system of political structures and social institutions maintains its force. But in practice it can be denied that his work has been grossly unhistorical and politically evasive (Eagleton, 1983).

Some people have said that deconstruction serves to indicate that our account of the world could be different, but that it cannot tell us how it would be different. Derrida seemed to believe that deconstruction is gradually able to shift the structure within which we operate little by little to modify the terrain of our work, and thereby produce new figurations.

According to Christopher Norris (1982, 96), deconstruction is inimical to Marxist thought. In his view, the insights of deconstruction are inevitably couched in a rhetoric which itself lies open to further deconstructive reading. 'Once criticism enters the labyrinth of deconstruction it is committed to a sceptical epistemology that leads back to Nietzsche, rather than Marx.'

Eagleton (1981), an English Marxist, for example, has argued that the main characteristics of deconstruction are that it rejects any notion of totality and that it is against privileging of the unitary subject. Deconstruction asserts that literary texts do not have relation to something other than themselves. Deconstruction, to Eagleton, is not only reformist but ultra-leftist too.

Norris (1992) argued that deconstruction has nothing in common with those forms of extreme anti-cognivist doctrine that would claim to have come out 'beyond' all distinction between truth and falsehood, reason and rhetoric, fact and fiction. Nevertheless, Derrida had a continuing critical engagement with the truth-claims and ethical values of Enlightenment thought. Derridean deconstruction supports the Enlightenment critique even while subjecting that tradition to a radical reassessment).

SPATIALITY

From the epistemological stand of geography viewpoint, the single most important contribution of post-modernism has been to connect the bias towards 'historicism' by putting 'spatial' at the centre of explanation, spatial dialectic alongside the historical dialectic. It is worth recalling that the last decades of the nineteenth century were an era of rising historicism and the parallel submergence of space in critical social thought (Singh, 2003). Foucault (1980, 70) once said, *'Space was treated as the dead, the fixed, the undialectical, the immobile. Time, on the contrary, was richness, fecundity, life, dialectic.'*

However, Focault (1986, 22) made the following observation, asserting the emergence of an 'epoch of space', contrary to what he said earlier: 'The great obsession of the nineteenth century was, ... history, with its themes of development and of suspension, of crisis and cycle, themes of the ever-accumulating past, with its great preponderance of dead men and the menacing glaciation of the world.... The present epoch will perhaps be, above all, the epoch of space.'

'As we move closer to the end of the twentieth century, ... Foucault's premonitory observations on the emergence of an 'epoch of space' assume a more reasonable cast. The material and intellectual context of modern critical social theory have begun to shift dramatically. In the 1980s, the hoary traditions of space-blinkered historicism are being challenged with unprecedented explicitness by convergent calls for a far-reaching spatialization of the critical imagination. A distinctively post-modern and critical human geography is taking shape, brashly reasserting the interpretive significance of space in the historically privileged confines of contemporary critical thought' (Soja, 1993).

'Geography may not have displaced history at the heart of contemporary theory and criticism, but there is a new animating polemic on the theoretical and political agenda, one which rings with significantly different ways of seeing time and space together, the interplay of history and geography, the "vertical" and "horizontal" dimensions of being in the world freed from the imposition of inherent categorical privilege.... The development of post-modern geographies has progressed far enough to have changed significantly both the material landscape of the contemporary world and the interpretive terrain of critical theory. The time has come ... for at least a first round of a responsive evaluation of these two changing contents of history and geography, modernity and post-modernity—one imprinted concretely on the empirical fabric of contemporary life (a post-modern geography of the material world), and the other treading through the ways we make practical and political sense of the present, the past and the potential future (a post-modern geography of critical social consciousness)' (ibid.).

The reassertion of 'space' in critical social theory is an excuse of both 'deconstruction' and 'reconstitution'. It cannot be simply accomplished by appending spatial highlights to inherited critical perspective and sitting back to watch them glow with logical conviction. The strangle-hold of a still addictive historicism must first be loosened,

Spatiality of Historicism

The post-modern critical human geography first appeared in the late 1960s, but it was barely heard against the then prevailing temporal din. For more than a decade, the spatialising project remained strangely muted by the untroubled reaffirmation of the primacy of history over geography that enveloped both Western Marxism and liberal social science in a virtually sanctified vision of the ever-accumulating past. C. Wright Mills (1959), in the late 1950s of the last century, had attempted to carry forward the historical contextualisation in his paradigmatic portrayal of the sociological imagination. His work provides a useful point of departure for specialising the historical narratives and reinterpreting the course of critical social theory. Mills maps out a sociological imagination that is deeply rooted in a historical rationality, that applies equally well to critical social science and to critical traditions of Marxism.

Mills (1959, 87) says: 'The sociological imagination enables us to grasp history and biography and the relation of the two within society. This is its task and its promise. To recognize this task and this promise is the mark of the classic social analyst.... No social study that does not come back to the problems of biography, of history, and of this intersection within society, has completed its intellectual journey'.

It is clear that Mills attempted to draw an essentially historical imagination to illustrate the alluring logic of historicism, the rational reduction of meaning and action to the temporal constitution and experience of social being. However, 'sociological' imagination has been changed to 'historical' that seemed to be an expression of Mills' disciplinary specialisation and socialisation. The nominal choice personally specifies what is a much more widely shared 'quality of mind' that he claims should pervade, indeed embody, all social theory and analysis, an emancipatory rationality grounded in the intersection of history, biography, and society. To be sure, these 'life-stories' have a geography too, they have milieus, immediate locales, provocative emplacements which affect thought and action.

The historical imagination is never completely spaceless and critical social historians have written, and continue to write, some of the best geographies of the past. But it is always time and history that provide the primary 'variable containers' in these geographies. This would be just clear whether the critical orientation is described as sociological or political or anthropological—for that matter phenomenological, existential, hermeneutic, or historical materialist. The particular emphasis may differ, but the encompassing perspective is shared. An already-made geography sets the stage, while the wilful making of history dictates the action and defines the story line. It is important to stress that this historical imagination has been particularly central to critical social theory, to the search for practical understanding of the world as a means of emancipation versus maintenance of the status quo.

Historicism has been defined in three different ways by Raymond Williams (1983): (1) Neutral—a method of study using facts from the past to trace the precedents of current events; (2) Deliberate—an emphasis on variable historical conditions and contents as a privileged framework for interpreting all specific events; (3) Hostile—an attack on all interpretations and prediction which is based on notion of historical necessity or general laws of historical development.

'Historicism is an overdeveloped historical contextualization of social life and social theory that actively submerges and peripheralizes the geographical or spatial imagination. This definition does not deny the extraordinary power and importance of historiography as a mode of emancipatory insight, but identifies historicism with creation of a critical silence, an implicit subordination of space to time that obscures geographical interpretation of the changeability of the social world and intrudes upon every level of theoretical discourse, from the most abstract autological concepts of being to the most detailed explanation of empirical events' (Soja, 1993, 15).

'This definition may appear rather odd when set against the long tradition of debate over historicism that has flourished for centuries. Even then, the main currents of critical social thought had become so spatially-blinkered that the most forceful reassertion of space versus time, geography versus history, had little effect. The academic discipline of modern geography had, by that time, been rendered theoretically inert and contributed little to these first reassertions' (ibid. 15–16).

The French Marxist philosopher, Henri Lefebvre, and critical theorists, Michel Foucault and John Berger have been identified to have pioneered the development of post-modern geography, and whose assertive post-modern geographies have been largely hidden from view by their more comforting and familiar identification as historians. The hidden history of spatialisation is actually illustrated and brought back into view by Henri Lefebvre. He was perhaps the most influential figure shaping the course and character of French Marxist theory from the early 1930s to at least the late 1950s. He became, after the 1950s, the leading spatial theoretician in Western Marxism and the most forceful advocate for the reassertion of space in critical social theory.

There seems to be a convincing basis for claiming that French Marxism, from the onset, was more inclined towards an explicit spatial perspective and theorisation than Marxism elsewhere. Throughout the twentieth century, French critical thought, whatever its primary source, preserved an ongoing spatial discourse in the contemporary post-structuralist and post-modernist debates. Lefebvre's spatial project, however, provides a new dimension about spatialised dialectic, an insistent demand for a fundamental change in the ways we think about space, time, and being; about geography, history and society; about the production of space, the making of history, and the constitution of social relations and practical consciousness.

Spatialised Dialectic

Lefebvre (1976, 14) says: 'The dialectic is back on the agenda. But it is no longer Marx's dialectic, just as Marx's was no longer Hegel's.... The dialectic today no longer clings to historicity and historical time, or to a temporal mechanism such as 'thesis-antithesis—synthesis' or 'affirmation—negation—negation of the negation.... To recognize space, to recognize what takes place there and what is used for, is to resume the dialectic : analysis will reveal the contradiction of space.'

He also attempted to recontextualise Marxism in theory and practice. And it is within this recontextualisation that one can discover many of the immediate sources of a materialist interpretation of spatiality and hence of the development of Marxist geography and historical-geographical materialism.

Lefebvre's theorisation of space cannot be summarised easily, for it is embedded in an extraordinary number of published works which touch upon every aspect of social theory and philosophy. In a response to the question as to how did he develop his interest in 'space', Lefebvre (1975, 28) says: 'The point of departure for me was the work of DATAR (Délégation a I' Amenagement du Territorize et a I' Action Regionale)—a new social and political practice emerged with the establishment of DATAR with the construction of the new town of Lacq-Mourenx in the Pyrenees Atlantiques).... Something new was happening, an idea of spatial planning and practice was born.... It was certainly an original French phenomena. I don't know many other countries that have gone beyond the stage of financial planning in their budgets to actually planning their space.'

Lefebvre seemed to have concentrated attention on the characteristic features of the modernised capitalism that consolidated around the turn of the century in what he called a 'bureaucratic society of controlled consumption', choreographed by the capitalist state—in essence, an instrumentalised spatial planning which increasingly penetrated into the recursive practice of daily life. Gradually, Lefebvre became concerned with 'urban', 'urbanism', and 'urbanization'. To him, urbanisation was a summative metaphor for the spatialisation of modernity, and the strategic 'planning' of everyday life that has allowed capitalism to survive, to reproduce successfully its essential relations of production.

The very survival of capitalism was built upon the creation of an increasingly embracing, instrumental, and socially mystified spatiality, hidden from critical view under thick veils of illusion and ideology. What distinguished capitalism's gratuitous spatial veil from the spatialities of other modes of production was its peculiar production and reproduction of geographically uneven development via simultaneous tendencies towards homogenisation, fragmentation, and hierarchisation—an argument that resembled in many ways Foucault's discourse on heterotopias, and the instrumental association of space, knowledge, and power (Soja, 1993, 49–50). 'This dialectical, conflictive space is where the reproduction of the relations of production is achieved. It is this space that produces reproduction, by introducing into it the multiple contradictions—contradictions that must be analytically and dialectically revealed to enable us to see what is hidden behind the spatial veil' (Lefebvre, 1976a).

Lefebvre's socially-produced space is a created structure comparable to other social constructions resulting from the transformation of given conditions inherent to being alive, in much the same way as human history represents a social transformation of time. Finally, Lefebvre (1976b, 35) says: 'Space is not a scientific object removed from ideology and politics; it has always been political and strategic. If space has an air of neutrality and indifference with regard to its contents and thus seems to be purely formal, the epitome of rational abstraction, it is precisely because it has been occupied and used, and has already been the focus of past processes whose traces are not always evident on the landscape. Space has been shaped and moulded from historical and natural elements, but this has been a political process. Space is always political and ideological. It is a product literally filled with ideologies.'

Heterotopia

Michel Foucault made significant contribution to the development of post-modern geography, particularly critical human geography. His most explicit and revealing observation on the relative significance of space and time, however, appears not in his major published works: *Madness and Civilization* (1961) and *The History of Sexuality* (1978), but almost innocuously in his lectures, and in two of his revealing interviews: 'Questions on Geography'(1980) and 'Space, Knowledge and Power' (Rabinow, 1984).

In these interviews and lectures, Foucault clarified and outlined his notion/concept of 'heterotopias' as the characteristic spaces of the modern world, superseding the hierarchic 'ensemble of places' of the Middle Ages and the enveloping 'space of emplacement' opened up by Galileo into an early-modern, infinitely unfolding, 'space of extension' and measurement. Moving away from both the 'internal space' of Bachelard's poetics (1969), and the intuitional regional description of the phenomenologists, Foucault focused attention on another spatiality of social life, an 'external space', the actually lived (and socially produced) space of sites and the relation between them.

Foucault says: 'The space in which we live, which draws us out of ourselves, in which the erosion of our lives, our time and history occurs, the space that claws and graws at us, is also, in itself, a heterogeneous space. In other words, we do not live in a kind of void, inside of which we could place individuals and things. We do not live inside a void that could be colored with diverse shades of light, we live inside a set of relations that delineates sites which are irreducible to one another and absolutely not superimposable on one another' (1986, 23).

These heterogeneous spaces of 'sites' and 'relations', called 'heterotopias' are constituted in every society, but take quite varied forms and change over time, as 'history unfolds' in its adherent spatiality. He identifies many such sites: the cemetery and the church, the theatre' and the garden, the museum and the library, the fairground and the 'vacation village', the barracks and the prison, the Moslem *hammam* and the Scandinavian sauna, the brothel and the colony. Foucault seems to contrast these 'real places' with the 'fundamentally unreal spaces' of utopias, which present society in either 'a perfected form' or else 'turned upside down'.

'The heterotopia is capable of juxtaposing in a single real place several spaces, several sites that are in themselves incompatible.... They have a function in relation to all the space that remains. This function unfolds between two extreme poles. Either their role is to create a space illusion that exposes every real space, all the sites inside of which human life is partitioned, as still more illusory.... Or else, on the contrary, their role is to create a space that is other, another real space, as perfect, as meticulous, as well arranged as ours is messy, ill constructed, and jumbled. The latter type would be the heterotopias, not of illusion, but of compensation, and I wonder if certain calories have not functioned somewhat in this manner' (ibid. 25–27).

Through these aforesaid remarks, Foucault attempted to raise a powerful argument against historicism and against the prevailing treatments of space in human sciences. His heterogeneous and relational space of heterotopias is neither a substanceless void to be filled by cognitive intuition nor a repository of physical forms to be phenomenologically described in all its resplendent variability. It is another space, actually lived and socially created spatiality, concrete and abstract at the same time, the habitus of social practices. It is a space rarely seen for it has been obscured by a bifocal vision that traditionally views space as either a mental construct or a physical form.

To account for his innovative interpretation of space and time, and to clarify some of the often confusing polemics which were growing around it, Foucault turned to the current debates on structuration, one of the twentieth century's most important avenues for the reassertion of space in critical social theory. He recognised the development of structuralism as a different and competing vision of history and geography, a critical reorientation that was connecting space and time in new and revealing ways.

Foucault, as it appears, refused to project his spatialisation as anti-history, but his history was provocatively spatialised from the very beginning. This was not just a shift in metaphorical preference, as it frequently seemed to be for Althusser and others more comfortable with the structuralist label than Foucault. He has strongly emphasised the centrality of space to the critical eye, regarding the contemporary moment. His spatialisation appeared to be more demonstrative rather than declarative stance.

When asked whether space was central to the analysis of power, Foucault (1986, 25–26) replied: 'Yes. Space is fundamental in any form of communal life; space is fundamental in any exercise of power. "It is said that Foucault persistently attempted to explore the 'total intersection of time with space" from the first to the last of his writings, and he provided and/or created an 'emerging perspective of a post-historicist and post-modern critical human geography.

However, Foucault admitted that space has been devalued by philosophers and social critics for generations, and reasserted the inherent spatiality of power/knowledge and ended with a *volte face.* He said to the interviewers: '... Now I can see that the problems you put to me about geography are crucial ones for me. Geography acted as a support, the condition of possibility for the passage between a series of factors I tried to relate. When geography itself was concerned, I either left the question hanging or established a series of arbitrary connections.... Geography must indeed lie at the heart of my concern' (1980, 77).

It is at this stage, Foucault makes the comment made earlier on the post-Bergsonian treatment of space as passive and lifeless, time as richness, fecundity, dialectic. He takes an integrative rather than deconstructive path, holding on to his history, but adding to it the crucial nexus that would flow through all his works: the linkage between space, knowledge, and power. He said: 'The spatializing description of discursive realities gives on to the analysis of related effects of power.... A whole history remains to be written of "spaces"—which would at the same time be the history of "powers" (both of these terms in the plural)—from the great strategies of geopolitics to the little tactics of the habitat' (Foucault, ed., 1980, Gordon, 77, 149).

Like Foucault, John Berger has also dealt with the intersection of time and space in virtually all his writings. Symbolising his insistent balancing of history and geography, lineage and landscape, period and region, Berger (1984) opens this slim volume: *A Question of Geography* by stating 'Part One is About Time, Part Two is About Space'. The embracing themes follow accordingly: the first part labelled 'Once', the second 'Here' both necessarily touched by other. Berger (1984) condenses the essence of 'post-modern geographies' in a spatially politicised aesthetic.

To make his point, Berger (1984) turns to an analogous change in modern novel, a shift in the context of meaning and interpretation which hinges around the simultaneity versus sequence, spatiality versus historicity, geography versus biography. In doing so, he begins to set into place a train of arguments that define the post-modern turn against historical determinations and vividly announce the need for an explicitly spatialised narrative (Soja, 1993, 24–25).

Berger (1984, 102) joins with Foucault in pushing towards a significant and necessary restructuring 'of critical social thought, a re-composition which enables us to see more clearly the long-hidden instrumentality of human geographies, in particular the encompassing and encaging spatialisation of social life that have been associated with historical development of capitalism. 'Foucault's path took him primarily into the micro spaces of power, discipline and surveillance, into the carceral city, the asylum, the human body, while Berger's path continues to open up new ways seeing art and aesthetics, portrait and landscapes, painters and peasants, in past (once) and in the present (here).' To crystallise and expand these spatial fields of insight and to attach post-modern critical human geography even more fearfully and explicitly to the instrumental spatiality of capitalism, the historical narrative must be re-entered at a different place and scale.

It is in the writings of Lefebvre, Foucault, and Berger that the geography of post-modernisation was most actually perceived although the links between them are not always direct and intentional, the intellectual trajectories of these three foundational post-modern geographers intersect in the contemporary deconstruction and reconstitution of modernity.

Convergence of Three Spatialisations

Soja (1993, 61) identifies three different paths of spatialisation that tend towards a creative convergence: *'post-historicism'*, *'post-fordism'*, and *'post-modernism'*. The first of these spatialisations—post-historicism is rooted in a fundamental reformulation of the nature and conceptualisation of social being, an essentially ontological struggle to rebalance the interplay between history, geography and society. Here the reassertion of space arises against the grain of an ontological historicism. That has privileged the separate constitution of being in time for at least the past century.

The second spatialisation—post-fordism—is directly linked to the political economy of the material world and, more specifically, to the 'fourth modernization' of capitalism, the most recent phase of far-reaching socio-spatial restructuring that has followed the end of the long post-War economic boom. The term 'post-fordism' is chosen to characterise the system of production marked by flexibility both of labour and machinery; by the vertical break-up of large corporations, by better use of links between firms so that subcontracting is increasingly used. Post-fordism is associated with agglomeration, which will simplify interaction between linked forms of economic activity. Here too it can be argued that space makes a critical difference, that revealing how spatial restructuring hides consequences from us is the key to making political and theoretical sense of the changing political economy of the contemporary world.

The third spatialisation—post-modernism—is linked to a cultural and ideological reconfiguration, a changing definition of the exponential meaning of modernity, the emergence of a new, post-modern culture of space and time. It is attuned to changes in the way we think about and respond to particularities—the perils and possibilities—of the contemporary moment via science, art, paleography, and programmes for political action. Post-modernism overlaps with post-historicism and post-fordism as a theoretical discourse and a periodising concept in which geography increasingly matters as a vantage point of critical insight.

The convergence of these three spatialisations is effectively exemplified in the work of Fredric Jameson. He captures the spatial specificity of the contemporary 'Zeitgeist': 'Post-modern (or multinational) space is not merely a cultural ideology or fantasy, but has genuine historical (and socio-economic) reality as a third great original expansion of capitalism around the globe (after the earlier expansion of the national market and the older imperialist system, which each had their own cultural specificity and generated new types of space appropriate to their dynamics).... We cannot (therefore) return to aesthetic practices elaborated on the basis of historical situations and dilemmas which no longer ours ... the conception of space that has been developed here suggests that a model of political culture appropriate to our own situation will necessarily have to raise spatial issues as its fundamental organizing concern' (Jameson, 1984, 88–89).

Jameson provisionally defines the spatialised model of radical political culture appropriate to the contemporary (post-modern) situation as an 'aesthetic of cognitive mapping', an ability to see in the cultural logic and form of post-modernism an instrumental cartography of power and social control; in other words a more accurate way of seeing how space hides consequence from us. This spatio-temporal patterning plays a significant role in Jameson's critique of reactionary post-modernism and a wide range of contemporary studies of post-fordist economic restructuring.

David Harvey (1969; 1973), the singular geographer who discarded his own defence of positivist geography in favour of Marxist interpretations (1973) mounted a post-modern critique of Marxist structuralism in his later works. His comments (1985) are self-revealing of his intention to the focus on spatiality as against the prevailing historicism—'there has been a strong and almost overwhelming predisposition to give time and history priority over space and geography'. David Harvey (1982), however, has given a post-modern outlook to Marxist geography. In *Limits to Capital* (1982), he reached out from the heart of Marxist geography to the wider realm of Western Marxism and modern critical social theory, presenting a demonstrative argument for a spatialised Marxism and a spatialised critique of capitalist development.

Looking back at the critical response to 'Limits', Harvey expressed his concern that most readers seemed to be somewhat confused over the message it was presenting. He said, 'Curiously, most reviewers passed by ... what I thought to be the most singular contribution of that work—the integration of the production of space and spatial configuration as an active element within the core of Marxist theorising. That was the key theoretical innovation that allowed me to shift from thinking about history to historical geography and so to open the way to theorizing about the urban process as an active moment in the historical geography of class struggle and capital accumulation' (1985).

Harvey (1985, 144) further says: 'Geographical space is always the realm of the concrete and the particular. Is it possible to construct a theory of the concrete and the particular in the context of the universal and abstract determination of Marx's theory of capitalist accumulation? This is the fundamental question to be resolved'.

Marvel Castells (1983, 3) also seems to have expressed his faith in Lefebvre's spatial project and says: 'Space is not a reflection of society, it is society.... Therefore, spatial forms at least on our planet, will be produced, as all other human objects are, by human object.... At the same time, spatial forms will be earmarked by the resistance from exploited classes, from oppressed subjects, from dominated women. And the work of such a contradictory historical process on the space will be accomplished on an already inherited spatial form, the product of former history and the support of new interests, projects, protests and dreams. Finally, from time to time, social movements will arise to challenge the meaning of spatial structure and therefore attempt new functions and new forms.' However, Castells falls short of the post-modern proclamation that it is now space more than time, geography more than history, that hides consequences from us.

Similarly, Dorren Massey (1984) has observed that: '... understanding geographical organization is fundamental to understanding economy and society The geography of society makes a difference in the way it works... If this is true analytically, it is also true politically. For there to be any hope of altering the fundamentally unequal geography of British economy and society (and that of other capitalist countries, too), a politics is necessary which links questions of geographical distribution to those of social and economic organization.'

Neil Smith (1984) notes: 'Geographical space is on the economic and political agenda as never before. The idea of the "geographical pivot of history" takes on a more modern and more profound meaning than Mackinder could have imagined'. Smith, as it seems, also strays consciously off the path of Marxist orthodoxy.

In spite of their admission of the significance of space over time, Harvey, Massey and Smith appear to be hesitant to engage deeply in the necessarily transformative deconstruction of historical materialism and its de-spatialising master narratives. But while historicism is shielded from a rigorous and systematic critique, there is a new confidence regarding the theoretical and political significance of space.

'The emerging post-modern critical human geography must continue to be built upon radical deconstruction, a deeper exploration of those critical sciences in the texts, narratives and intellectual landscapes of the past, an attempt to reinscribe and resituate the meaning and significance of space in history and in historical materialism. Spatial deconstruction aims to "reverse the imposing tapestry" of the past (Eagleton, 1986) ... exposing the dishevelled tangle of threads that constitutes the intellectual history of critical social thought.... Spatial deconstruction must therefore also be sufficiently flexible to parry the reactionary thrusts of historicism and avoid the simplistic defence of anti history or, even worse, a new and equally obfuscating capitalism.... Deconstruction alone is not enough, however.... It must be accompanied by ... tentative reconstruction grounded in the political and theoretical demands of the contemporary world and able to encompass all the scales of modern power, from the grand strategies of global geopolitics to the little tactics of the habitat' (Soja, 1993, 73–74).

References

Abler, P., J. S. Adams, and P. Gould. 1972. *Spatial Organization: The Geographer's View of the World,* London, Prentice-Hall.

Ackerman, E. A. 1945. 'Geographic Training, Wartime Research, and Immediate Professional Objectives', *Annals AAG. 35,* 121–43.

———. 1958. *Geography as a Fundamental Research Discipline,* University of Chicago, Department of Geography, Research Paper No. 53.

———. 1963: 'Where is a Research Frontier?' *Annals of the Association of American Geographers,* 53: 429–40.

———, et al. 1965. *The Science of Geography,* Washington, DC, National Academy of Science/National Research Council Pub. No. 1277.

Ahmad, N. 1947. *Muslim Contributions in Geography,* Lahore.

Alam, S. A. 1987. (December): 'Untitled Presidential Address at the 9th Indian Geography Congress of the National Association of Geographers', India (NAGI), Kharagpur.

Anuchin, V. A. 1957. 'The Nature of Geographical Environment and the Appearance of Indeterminism in Soviet Geography', *Voprosy geografie* 41, 52.

———. 1960. *Theoretical Problems of Geography.* Moscow.

———. 1964. 'The Problem of Synthesis in Geographic Science', *Soviet Geography,* 5, 39.

———. 1973. 'Theory of Geography'. In R. J. Chorley (ed.), *Directions in Geography,* London, Methuen.

Bagrow, L., and R. A. Skelton. 1964. *History of Geography,* Cambridge, Harvard University Press.

Baker, J. N. L. 1963. *The History of Geography,* New York, Barnes & Noble.

Baranskiy, N. N. 1926: *Economic Geography of Soviet Union.* Moscow

———. 1960. 'An Estimation of the Natural Environment in Economic Geography'. *Ekonomicheska a geographiya-Ekonomicheskaya Kartografia,* Moscow, *Gos Izd. Geog, literatury.*

Bennett, R. J. 1975. 'Dynamic Systems Modelling of the North-west Region', *Environment and Planning,* 7, 525–38, 539–66, 617–36, 887–98.

———. 1985. 'Quantification and Relevance'. In R. J. Johnston (ed.), *The Future of Geography,* London, Methuen.

Bennet, R. J., and R. J. Chorley. 1978. *Environmental Systems: Philosophy; Analysis and Control,* London, Methuen.

Berdoulay, V. 1976. 'French Possibilism as a Form of Neo-Kantian Philosophy'. *Procs. Ass. Am. Geog* 8, 176–79.

Berger, J. 1984. *And Our Faces, My Heart, Brief as Photos,* New York, Pantheon Books.

Berman, M. 1982. *All That is Solid Melts into Air: The Experience of Modernity.* New York, Simon and Schuster.

Berry, B. J. L. 1964. 'Cities as Systems within Systems of Cities Papers', *Regional Science Association, 13, 147–163.*

———. 1972. 'Revolutionary and Counter-revolutionary Theory in Geography—A Ghetto Commentary', *Antipode* 4 (2), 31–33.

———. 1973. *The Human Consequences of Urbanization,* London, Macmillan.

———. 1973. 'A Paradigm for Modern Geography'. In R. J. Chorley (ed.), *Directions in Geography,* London, Methuen, 3–22.

———. 1978. 'Geographical Theories of Social Change'. In B. J. L. Berry (ed.), *The Nature of Change in Geographical Ideas,* Northern Illinois University Press, 17–36.

Berry, B. J. L., W. L. Garrison. 1958. 'The Functional Bases of the Central Hierarchy'. In B. J. L. Berry and D. F. Marble (eds), *Special Analysis,* Englewood Cliffs, N. J. Prentice Hall, 91–100.

Bertalanfy, von L. 1968. *General Systems Theory—Foundations, Development, Application,* New York, George Brazilier.

Blowers, A. T. 1972. 'Bleeding Heart and Open Values'. *Area* 4, 290–92.

Bowman, I. 1934. *Geography in Relation to the Social Sciences,* New York, Charles & Scribner.

Briggs, D. J. 1981. 'The Principles and Practice of Applied Geography'. *Applied Geography* 1, 1–8.

Brigham, A. P. 1903. *Geographic Influences in American History,* Boston, Ginn & Co.

———. 1915. 'Problems of Geographic Influence', *Ann. Assoc. Amr. Geog.* 14, 109–16.

Broek, J. O. M. 1965. *Geography, Its Scope and Spirit,* Colombus Ohis, Merrill.

Brookfield, H. C. 1964. 'Questions on the Human Frontiers of Geography', *Economic Geography,* 40, 283–303.

Brunhes, J. 1910. *La Geographie humaine,* Paris, Armand Colin.

Bunbury, E. H., 1983. *A History of Ancient Geography among the Greeks and Romans from the Earliest Ages till the Fall of the Roman Empire,* 2 vols, London, John Murray.

Bunge, W. 1962. *Theoretical Geography,* Land Studies in Geography, Lund, C.W.K. Gleerup.

———. 1971. *Fitzgerald: Geography of a Revolution.* Cambridge, Mass, Schlenkrnan.

———. 1973. 'Ethics and Logic in Geography'. In R. J. Chorley (ed.), *Directions in Geography,* London, Methuen.

———. 1973. 'Spatial Prediction'. *Anns. Assoc. Amr. Geog,* 63, 227–95.

———. 1989. 'Our Planet is Big Enough for Peace but Too Small for War'. In R. J. Johnston and P. Taylor (eds), *A World in Crisis? Geographical Perspectives* (2nd edn), Oxford, Basil Blackwell.

Bunting, T. E., and L. Guelke. 1979. 'Behavioural and Perception Geography: A Critical Appraisal'. *Anns. Assoc. Amr. Geog,* 69, 448–62, 471–74.

Burton, I. 1963. 'The Quantitative Revolution and Theoretical Geography', *Canadian Geographer,* 7, 151–62.

Buttimer, A. 1971. *Society and Milieu in the French Geographical Tradition,* Chicago, Rand McNally.

———. 1976. 'Grasping the Dynamism of Lifeworld'. *Anns. Assoc. Amer. Geog.* 66, 277–92.

———. 1978. 'On People, Paradigms and Progress in Geography'. In D. R. Stoddart (ed.), *Geography, Ideology and Social Concern.* Oxford, Blackwe1l, 70–80.

———. 1978. 'Charism and Context: The Challenge of La Geographic Humaine'. In D. Lay and M. S. Samuels (eds), *Humanisitic Geography: Prospects and Problems.* Chicago, Maarofua Press, 58–76.

Capel, H. 1981. 'Institutionalization of Geography and Strategies of Change'. In D. R. Stoddart (ed.), *Geography, Ideology and Social Concern,* Oxford, Basil Blackwell.

Capelle, R. B. 1979. 'On the Periphery of Geography', *Journal of Geography,* 78, 64–68.

Castells, M. 1983. *The City and the Grass Roots.* Los Angeles, University of California Press.

Chisholm, M. 1962. *Rural Settlement and Land Use.* London, Hutchinson.

———. 1967. 'General Systems Theory and Geography'. *Trans., Inst. Br. Geog,* 42, 45–52.

———. 1971. 'Geography and Question of "Relevance"', *Area* 3, 65–68.

Chorley, R. J. 1962. *Geomorphology and General Systems Theory.* Professional paper 500-B. United States Geological Survey, Washington.

———. 1964. 'Geography and Analogue Theory', *Anns. Am. Geog.* 54, 127–37.

———. 1973. 'Geography as Human Ecology'. In R. J. Chorley (ed.), *Directions in Geography,* London, Methuen, 155–69.

Chorley R. J., and P. Haggett. 1965. *Frontiers in Geographical Teaching,* London, Methuen.

Chorley R. J., and P. Haggett (eds). 1967. *Models in Geography,* London, Methuen.

Chorley R. J., and B. A. Kennedy. 1971. *Physical Geography—A Systems Approach,* London, Prentice-Hall International.

Chorley R. J., and R. J. Bennett. 1981. 'Optimization: Control Models'. In N. Wrigley and R. J. Bennett (eds), *Quantitative Geography,* London, Routledge & Kegan Paul, 219–24.

Church, H. R. J. 1967. 'The French School of Geography'. In G. Taylor (ed.), *Geography in the Twentieth Century,* London, Methuen, 70–90.

Clark, G. L. 1982. 'Instrumental Reason and Policy Analysis'. In D. T. Herbert and R. J. Johnston (eds), *Geography and the Urban Environment,* vol. 5, Chichester, John Wiley, 41–42.

Coates, B. E. et al. 1977. *Geography and Inequality,* London, Oxford University Press.

Coppock, J. T. 1974. 'Geography and Public Policy: Challenges, Opportunities and Implications'. *Transactions,* Institute of British Geographers, 63, 1–16.

Cox, K. R. 1973. *Conflict, Power and Politics in the City: A Geographic View,* New York, McGraw Hill.

Crone, G. R. 1964. 'British Geography in the Twentieth Century', *Geog. Jour.* 130, 197–220.

Curry, L. 1967. 'Quantitative Geography'. *The Canadian Geographer,* 11, 265–74.

Dacey, M. F. 1973. 'Some Questions about Spatial Distributions'. In R. J. Chorley (ed.), *Directions in Geography,* London, Methuen, 127–52.

Darby, H. C. (ed.). 1936. *A Historical Geography of England before A. D. 1800.* Cambridge University Press.

Darby, H. C. 1951. 'The Changing English Landscape', *Geog, Journ,* 117, 337–398.

Darby, H. C. 1952. *The Domesday Geography of Eastern England,* Cambridge University Press.

Dayal, P. 1994 (December). 'The Agenda for Indian Geography: Presidential Address at the 17th Indian Geography Congress of the National Association of Geographers', India (NAGI), Hyderabad.

David, T. 1957. 'Against Geography'. *Universities Quarterly,* 13, 261–73.

Davies, W. M. 1899. 'The Geographical Cycle', *Geog. Journ.* 14, 481–504.

———. 1902. 'Systematic Geography'. *Proc. Amer. Phil. Soc.* 41, 235–59.

———. 1903. 'A Scheme of Geography'. *Geog. Journ.* 22, 413–23.

———. 1906. 'An Inductive Study of the Content of Geography'. *Bulletin of the American Geographical Society,* 38, 67–84.

———. 1919. 'Passarge's Principles of Landscape Description', *Anns. Assoc. Amer. Geog.* 5, 61–105.

Darwin, C. R. 1859. *On the Origin of Species by Means of Natural Selection or the Preservation of Favoured Races in the Struggle of Life,* London, John Murray.

de Martonne, E. 1902. *La Valachie, essai de monographie geographique,* Paris, Armand Colin.

———. 1909. *Traite de geographic physique.* Revised. Paris, Armand Colin.

Demolin, Edmond. 1901–03: *Essai de geographie sociale, Comment la route cree le type sociale.* Paris.

Derrida, J. 1973. *Speech and Phenomena, and Other Essays on Husserl's Difference,* Evanston, Northwestern University Press.

———. 1975. *Writing and Difference,* London, Routledge & Kegan Paul.

———. 1976. *Of Grammatology*, Baltimore/London, Johns Hopkins.

Deshpande, C. D. 1983: 'Discovering Indian Geography'. In R. P. Misra (ed.), *Contributions to Indian Geography: Concepts and Approaches*, New Delhi, Heritage Publishers.

Dickinson, R. E. 1969. *The Makers of Modern Geography*, Ludhiana, Lyall Book Depot.

Downs, R. M. 1970. 'Geographic Space Perception: Past Approaches and Future Prospects. In C. Board et al. (eds), *Progress in Geography*, 2, 65–108.

Dryer, C. R. 1915. 'Natural Economic Regions', *Anns. Assoc. Amer, Geog.* 5, 121–25.

Duncan, J., and D. Ley. 1993. *Representing Cultural Geography*, Anchlover, Chapmass and Hall.

———. 1993. *Place, Culture and Representation*, London, Routledge.

Eagleton, T. 1983. *Literary Theory: An Introduction*, Oxford, Basil Blackwell.

———. 1986. *Against the Grain: Essays: 1975-1985*, London, Verso.

———. 1935. *An Historical Geography of Europe*. London, Methuen.

———. 1951. 'Historical Geography'. In S. W. Woolridge and W. G. East (eds), *The Spirit and Purpose of Geography*, London, Hutchinson University Library, 80–102.

Eliot Hurst, M. E. 1980. 'Geography, Social Science and Society: Towards a Re-definition', *Australian Geographical Studies*, 8, 3–21.

Eyre, S. R., and G. R. J. Jones. 1966. *Geography as Human Ecology: Methodology by Example*, London, Arnold.

Fawcett, C. B. 1919. *The Provinces of England* (Rev. edn. 1960), London, Hutchinson.

Febvre, L. 1922. *A Geographical Introduction to History* (Eng. Trans. 1925), London, Knapf.

Fenneman, N. H. 1914. 'Physiographic Boundaries within United States', *Anns. Assoc. Amer. Geog*, 4, 84–134.

———. 1919. 'Circumference of Geography', *Anns. Assoc. Amer. Geog.* 9, 3–11.

Fischer, E., R. D. Campbell, and E. S. Miller. 1969. *A Question of Place, the Development of Geographic Thought*, Arlington, Virginia, Beatty.

Fleure, H. J. 1971. 'Regions humanies', *Annales de Geographie*, 26, 161–74.

Fotheringham, A. S. et al., (eds). 2000. *Quantitative Geography: Perspective on Spatial Data Analysis*, London, Sage Publication.

Foucault, M. 1980. 'Question on Geography'. In C. Gordon (ed.), *Power/Knowledge: Selected Interviews and Other Writings: 1972–1977*, 63–77.

———. 1986. 'Of Other Spaces', *Diatrics 16*, 22–27 (translated from the French by Jay Miskowiec).

Freeman, T. W. 1961. *A Hundred Years of Geography*, Duckworth, London.

———. 1967. *The Geographer's Craft*. New York, Barnes & Noble.

———. 1980. *A History of Modern British Geography*, New York, Longman.

Garrison, W. L. 1959. 'Spatial Structure of the Economy I, II, III', *Anns, Assoc. Amer, Geog*, 49, 50, 238–39, 471–82, 357–73.

Gerasimov, I. P. 1957. 'The Scientific Principles of Soviet Geography in the View of Canadian Geographer', *Izvestiya Vs oyuznogo geograpficheskogo obshchestva*, vol. LXXXIX, no. 4, 370–72.

———. 1962. *Soviet Geography* (trans.) L. Eckev (ed.) C. D. Haris (ed.), New York, American Geographical Society.

———. 1968. 'Constructive Geography: Aims, Methods and Results', *Soviet Geography*, 9, 739–55.

———. 1968. 'Fifty Years of Development of Soviet Geographic Thought', *Soviet Geography*, 9, 238–52.

Gilbert, E. W. 1967. 'Geography and Regionalism'. In G. Taylor (ed.), *Geography in the Twentieth Century*, London, Methuen.

Gilbert, G. K. 1878. *Report on the Geology of the Heavy Mountains*, Washington, DC, Department of Interior.

Glacken, C. J. 1956. 'The Changing Ideas of the Habitable World', In W. L. Thomas (ed.), *Man's Role in Changing the Face of the Earth*, Chicago, University Press, 70–92.

———. 1960. 'Count Buffon on Cultural Change of the Physical Environment', *Anns. Assoc. Am. Geog.* 50: 1–21.

———. 1967. *Traces on the Rhodian Shore: Nature and Culture in Western Thought from Ancient Times to the End of Eighteenth Century*, Berkeley, University of California.

Gold, J. R. 1980. *An Introduction to Behavioural Geography*, Oxford University Press.

Gold, R. J., and R. Chorley (eds). 1981. *Valued Environment*, London, Allen & Unwin.

Golledge, P. R., et al. 1972. 'Behavioural Approaches in Geography: An Overview'. *The Aust. Geog.* 12, 59–79.

Gordon, D. 1977. 'Class Struggle and the Stages of American Urban Development'. In D. Perry and A. Watkins (eds), *The Rise of the Sunbelt City*, Beverly Hills, Sage.

Gould, P. R. 1966. *On Mental Maps*, Michigan Inter-University of Mathematical Geographers, Discussion paper, 9, Michigan.

Grano, O. 1981. 'External Influence and Internal Changes in the Development of Geography'. In D. R. Stoddart (ed.), *Geography, Ideology and Social Concern*, Oxford, Blackwell, 17–36.

Gregory, D. 1978. *Ideology, Science and Human Geography*, London, Hutchinson.

———. 1981/1986 (2nd ed.). 'Environmental Determinism'. In R. J. Johnston et al. (ed.), *The Dictionary of Human Geography*, Oxford, Basel Blackwell.

———. 1989. 'Areal Differentiation and Post-modern Human Geography'. In D. Gregory and R. Walford (eds), *Horizans in Human Geography*, London, Macmillan.

Guelke, L. 1981. 'Idealism'. In M. E. Harvey and B. P. Holly (eds), *Themes in Geographic Thought*, London, Croom Helm, 133–47.

Guyot, A. 1849. *The Earth and Man: Lectures on Comparative Physical Geography in Its Relation to the History of Mankind.* Boston, Gould & Lincoln.

Habermas, J. 1981. 'Modernity versus Post-Modernity', *New German Critique*, 22.

Hagerstrand, T., 1953 (trans.). 1967. *Innovation Diffusion as a Spatial Process*, Chicago, University Press.

———. 1973. 'The Domain of Human Geography'. In R. J. Chorley (ed.), *Directions in Geography*, London, Methuen.

———. 1976. 'Geography and the Study of Interaction between Nature and Society', *Geoforum*, 7, 329–34.

———. 1977. 'The Geographers' Contribution to Regional Policy: The Case of Sweden'. In D. R. Deskins et al. (eds), *Geographic Humanism, Analysis and Social Action: A Half Century of Geographic at Michigan*, Ann Arbor: Michigan Geographical Publications No. 17, 326–46.

Haggett, P. 1965/1966. *Locational Analysis in Human Geography*, London, Arnold/New York, St. Martin's Press.

———. 1972 (new edns 1975, 1979). *Geography—A Modern Synthesis*, New York, Harper and Row.

———. 1981. *Geography*. In R. J. Johnston et al. (eds), *The Dictionary of Human Geography.* Oxford, Basil Blackwell, 133–36.

Haggett, P., and R. J. Chorley. 1965. 'Frontier Movements and the Geographical Tradition'. In R. J. Chorely and P. Haggett (eds), *Frontiers in Geographical Teaching*, London, Methuen.

———. 1967. 'Models, Paradigms and the New Geography'. In R. J. Chorley and P. Haggett (eds), *Models in Geography*, London, Methuen, 79–82.

Haggett, P., A. D. Cliff, and A. Frey. 1977. *Locational Analysis in Human Geography.* London, Arnold.

H., Jadu, J. G. 1968. 'Toward a Definition of Post-war German Social Geography'. *Anns. Assoc. Am. Geog.*, 58, 397–410.

Hall, P. 1974. 'The New Political Geography'. *Transactions,* Institute of British Geographers, 63, 48–52.

———. 1981. 'The Geographer and Society', *Geog. Journ* 147, 145–52.

Hare, F. K. 1974. 'Geography and Public Policy', *Transactions,* Institute of British Geographers, 63, 25–28.

Harris, C. D. 1970. *Cities of the Soviet Union: Studies in Their Function, Size, Density, and Growth,* Chicago, Rand McNally.

Harries, K. D. 1974. *The Geography of Crime and Justice,* New York, McGraw-Hill.

Hartshorne, R. 1939 (new print 1976). 'The Nature of Geography', Lancaster, Pennsylvania, *Association of American Geographers.*

———. 1954. 'Comment on Exceptionalism in Geography', *Anns, Assoc. Amer. Geog.,* 44, 108–09.

———. 1955. 'Exceptionalism in Geography Re-examined', *Anns. Assoc. Amer. Geog,* 45, 205–44.

———. 1958. 'The Concept of Geography as a Science of Space from Kant and Humboldt to Hettner', *Anns. Assoc. Amer. Geog.* 48, 97–108.

———. 1959. *Perspective on Nature of Geography,* Chicago, Rand McNally.

Hartshorne, R. 1972. 'Review of Kant's Concept of Geography', *Canadian Geographer,* 16: 77–79.

Harvey, D. 1969. *Explanation in Geography,* London, Edward Arnold.

———. 1969. 'Conceptual and Measurement Problems in the Cognitive Behavioural Approach to Location Theory'. In K. R. Cox and R. G. Golledge (eds), *Behavioural Problems in Geography: A Symposium,* Northwestern University Studies in Geography, 17, 35–68.

———. 1973. *Social Justice and the City.* London, Edward Arnold.

———.1974. 'What Kind of Geography for What Kind of Public Policy?' *Transactions,* Institute of British Geographers, 63, 18–24.

———. 1982. *The Limits to Capital.* Oxford, Basil Blackwell.

———. 1989. *The Condition of Post-Modernity: An Enquiry into the Origins of Cultural Change,* Oxford.

Hay, A. M. 1979. 'Positivism in Human Geography: Response to Critics'. In D. T. Herbert and R. J. Johnston (eds), *Regions Geograpique et Regions d'amenagement collection Les hommes et. Les lettres,* 7 Lyon, Editions Herma.

Herbertson, A. J. 1905. 'The Major Natural Regions: An Essay in Systematic Geography', *Geographical Journal,* 25, 300–12.

———. 1913–14. 'Natural Regions', *Geographical Teacher* 7, 158–63.

———. 1915–16. 'Regional Environment, Heredity and Consciousness', *The Geographical Teacher,* 8, 147–53.

Holt -Jenson, Arild. 1981. *Geography: Its History and Concepts-A Student's Guide.* London, Harper & Row.

Hooson, D. J. M. 1959. 'Some Recent Development in the Content and Theory of Soviet Geography', *Anns. Assoc. Amer. Geog.,* 49, 73–82.

———. 1968. 'The Development of Geography in Pre-Soviet Russia', *Anns. Assoc. Amer. Geog.,* 58, 250–72.

House, J. M. 1973. 'Geographers, Decision-takers and Policy-makers'. In M. Chisholm & B. Rodgers (eds.), *Studies in Human Geography,* London, Heinemann, 272–305.

Hsieh, Chiao-min. 1968. 'The Chinese Exploration of the Ocean—Study in Historical Geography', *Chinese Culture* (Taiwan) 9, 123–31.

Huggett, R. J. 1980. *Systems Analysis in Geography.* Oxford University Press.

Humboldt, A. von. 1811. *Essai politique sur le Royaume de la Nouvelle Espagne,* and in later edition in 1826 with a supplement, he wrote: *Essai politique sur life de Cuba,* vol: 25 and 26, Paris, F. Schoell.

———. 1847. *Cosmos,* vol. 2. Berlin.

———. 1898. *Kosmos*, vol. 4, English transalation by E. C. Otte, London: H. C. Bohn.

Huntington, E. 1907. *The Pulse of Asia,* Boston, Houghton Mifflin.

———. 1915. *Civilization and Climate,* New Haven Conn., Yale University Press.

———. 1945. *Mainsprings of Civilization.*

Husain, M. 1984 (rev. edn. 1988). 'Evolution of Geographical Thought', Jaipur, Rawat Publication.

lbn-Batuta. 1958 (trans. C. Deframery and B. R Sanguinetti). *The Travels of Ibn-Batuta, A. D. 1325–1354*, Cambridge University Press.

lbn-Khaldun. 1958 (trans. F. Rosenthal). *The Muqaddimah*, New York, Pantheon Books.

lsachenko, A. G. 1968. 'Fifty Years of Soviet Landscape Science', *Soviet Geography* 9, 402–07.

lsard, W. 1956. *Location and Space Economy*, New York, John Wiley.

———. 1960. *Method of Regional Analysis: An Introduction in Regional Science*, New York, John Wiley.

James, P. E., and G. J. Martin. 1972. *All Possible Worlds: A History of Geographical Ideas*, Second Edition, New York: John Wiley & Sons.

James, P. E. 1969. 'Significance of Geography in American Education', *Journal of Geography*, 68, 473–83.

———. 1972 (2nd edn. with co-author G. J. Martin, 1981). *All possible Worlds: A History of Geographical Ideas*, Indianapolis, The Odyssey Press/New York, John Wiley & Sons.

James, P. E., and C. F. Jones. 1954. *American Geography: Inventory and Prospect*, Syracuse, Syracuse University Press.

Jameson, F. 1984. 'Post-modernism, or the Cultural Logic of Late Capitalism'. *New Left Review*, 146, 53–92.

Jefferson, M. 1939. 'The Law of the Primate City, *Geog. Rev.*, 29, 226–32.

Joerg, W. L. G. 1922. 'Recent Geographical Works in Europe', *Geog. Rev.* 12, 432–84.

Johnston, R. J. 1983. *Geography and Geographers: Anglo-American Human Geography since 1945* (2nd edn), London, Arnold.

———. 1983. *Philosophy and Human Geography: An Introduction to Contemporary Approaches* (1st edn), London, Edward Arnold.

Johnston, R. J. 1996. 'Nature, State and Economy: The Political Economy of Environmental Problems', Chichester: John Wiley.

Johnston, R. J. et al. (ed.). 1986. *The Dictionary of Human Geography* (2nd edn), Oxford, Basil Blackwell.

Johnston, R. J., and P. J. Taylor (eds). 1989. *A World in Crisis?Geographical Perspectives* (2nd edn), Oxford, Basil Blackwell.

Jones, D. R. C. 1983. 'The Challenges of Change in Geography Teaching', *Geography* 61, 196–205.

Jones, E. 1984. 'On the Specific Nature of Space', *Geoforum* 15(1), 5–9.

Kapur, A. 2002. *Indian Geography—Voice of Concern*, National Association of Geographers, India, New Delhi, Concept Publishing Company.

Kasperson, R. E. 1971. 'The Post-behavioural Revolution in Geography', *British Columbia Geographical Series*, 12, 5–20.

Kates, R. W. 1962. 'Hazard and Choice Perception in Flood plain Management', University of Chicago, Department of Geographical Research Paper 78, Chicago.

———. 1987. 'The Human Environment: The Road not Takers, the Road Still Beckoning, *Annals of the Association of American Geographers* 85, 623–40.

Kavitha, N. et al. 2010. 'Ground Water Pollution in Dindigul Town: An Impact Analysis of Tanneries Using GIS Techniques', *Annals of National Association of Geographers, India* 30(2): 95–107.

Kellner, L. 1963: *Alexander von Humboldt*, London: Oxford University Press.

Kem, S. 1983. *The Culture of Time and Space, 1880–1918*, Cambridge, Harvard University Press.

Kimble, G. H. T. 1938. *Geography in the Middle Ages*, London, Methuen.

Kirk, W. 1963. 'Problems of Geography', *Geography* 48, 357–71.

Knox, P. L. 1975. *Social Well-being: A Spatial Perspective*, London, Oxford University Press.

Konstantinov, O. A. 1968. 'Economic Geography in the USSR during the Soviet Period', *Soviet Geography* 9, 393–402.

Koppen, W. 1936. *Das geographische system der klimate*. In W. Koppen and R. Geiger (eds), *Handbuch der kilmatologic*, vol. I, part C, Berlin.

Kriesel, K. M. 1968. 'Montesquieu: possibilist politicar geographer', *Anns. Ass. Amer. Geog.* 58, 557–74.

Kuhn, T. S. 1962. 1970. *The Structure of Scientific Revolutions*, Chicago, Maaroufa Press.

Lacan, J. 1979. *The Four Fundamental Concepts in Psycho-Analysis*, London, Penguin.

Langton, J. 1972. 'Potentialities and Problems of Adopting a Systems Approach to the Study of Change in Human Geography', *Progress in Geography* 4, 125–79.

Lee, D. H. K. 1954. 'Physiological Climatology'. In P. E. James and C. F. Jones (eds), *American Geography, Inventory and Prospect,* Syracuse, NY, Syracuse University Press.

Lefebvre, H. 1976 a. *The Survival of Capitalism*, London, Allison and Busby.

———. 1976b. 'Reflection on the Politics of Space', translated by M. Enders. *Antipode* 8: 30–37.

Leitch, W. 1983. *Deconstructive Criticism: An Advanced Introduction*, London Hutchinson, 246.

Lewthwaite, G. R. 1966. 'Environmentalism and Determinism: A Search for Clarification', *Anns. Ass. Amer. Geog.* 56: 1–23.

Ley, D., and M. S. Samuels. 1978. 'Introduction: Context of Modern Humanism in Geography'. In D. Ley and M. S. Samuels (eds), *Humanistic Geography: Prospects and Problems*, Chicago, Maaroufa Press.

Ley, D. 1980. 'Geography without Man: A Humanistic Critique', Research paper 24, School of Geography, University of Oxford.

Losch, A. 1954. *The Economics of Location*, New Haven, Conn., Yale University Press.

Lukermann, F. 1958. 'Towards a More Geographic Economic Geography', *The Professional Geographer*, 10, 2–10.

Lyotard, J. F. 1984. *The Post-modern Condition: A Report on Knowledge.* Manchester, Manchester University Press.

Mackinder, H. J. 1887. 'On the Scope and Methods of Geography', *Proceedings of the Royal Geographical Society*, 9, 141–74.

———. 1904. 'The Geographical Pivot of History', *Geog. Jn.* 23, 421–37.

———. 1919. *Democratic Ideals and Reality*, New York, Henry Hall.

———. 1943. 'The Round World and the Winning of Peace', *Foreign Affairs*, 21, 595–605.

Marsh, G. P. 1864. *Man and Nature, or Physical Geography as Modified by Human Action.* New York, Charles Scribner.

Martin, A. F. 1951. 'The Necessity for Determinism', *Trans Inst. Br. Geog.* 17, 1–12.

Martin, G. J. 1968. 'Mark Jefferson', *Geographer*, Ypsilanti, Eastern Michigan University State.

Marx, K. 1967 (new edn), *Capital* 3 vol, New York.

Massey, D. 1984. *Spatial Divisions of Labour: Social Structures and the Geography of Production*, London, Macmillan.

Matley, I. M. 1966. 'The Marxist Approach to the Geographical Environment', *Ann, Assoc. Amer, Georg*, 56, 97–111.

———. 1982. 'Nature and Society: The Continuing Soviet Debate', *Prog. Hum. Geog*, 6, 367–96.

Maury, M. F. 1855. *The Physical Geography of the Sea*, New York, Harper Brothers.

McCarty, H. H., and J. B. Lindberg. 1966. *A Preface to Economic Geography*, Englewood Cliffs, N. J., Prentice Hall.

Meadows, D. H. et al. 1972. *The Limits to Growth*, New York, Universal Books.

Mechinikov, V. 1924. *Civilization and the Great Historical Reviews*, Moscow, Gosudarstvennoye 13d-vo.

Menon, K. P. R. et al. 2008. 'Mumbai Monsoon Floods—2005: Impacts and Possible Mitigation', *Transaction of the Institute of Indian Geographers* 31 (2): 155–65.

Mikesell, M. W. 1974. 'Geography as the Study of Environment: An Assessment of Some Old and New Comments'. In I. R. Manners and M. W. Mikesell (eds), *Perspectives on Environment*, Washington DC Commission on Collage Geography, Association of American Geographers.

Mills, C. W. 1959. *The Sociological Imagination*, New York, Oxford University Press.

Mill, H. R. 1896. 'A Proposed Geographical Description of the British Isles', *Geog. In.* 7, 345–65.

Minshull, R. 1970. *The Changing Nature of Geography*, London, Hutchinson.

Misra, R. P. 1996 (October). 'Indian Geography: Problems, Challenges and Opportunities', Presidential Address at the 19th Indian Geography Congress of the National Association of Geographers, India (NAGI), Shillong.

Mitchell, B., and D. Draper. 1982. *Relevance and Ethics in Geography,*

London, Longman.

Montefiore, A., and W. Williams. 1955. 'Determinism and Possibilism', *Geogr. Stud.* 2: I–II.

Morrill, R. L., and E. W. Wholenberg. 1971. *The Geography of Poverty in the United States*, New York, McGraw Hill.

Moris, M.1988. *The Pirate's Fiancee: Feminism, Rending, Post-modernism*, London, Verso.

Nicholson, L., and N. Fraser. 1988. 'Social Criticism without Philosophy: An Encounter between Feminism and Post-Modernism'. In Andrew Ross (ed.), *Universal Abandon? The Politics of Post-modernism*, Edinburgh, University Press.

Nikitin, N. P. 1966. 'A History of Economic Geography in Pre-revolutionary Russia', *Soviet Geography* 7, 3–37.

Norris, C. 1982. *Deconstruction: Theory and Practice*, London, Methuen.

———. 1992. *Unicritical Theory: Post-Modernism, Intellectuals and Gulf War*, London, Lawrence & Wishart.

Odum, H. W., and H. F. Moore. 1938. *American Regionalism: A Cultural Historical Approach to National Integration*, New York, H. Holt & Company.

O' Riordian. 1981 *Environmentalism,* 2nd Edition. London.

———. 1989. 'The Challenge for Environmentalism'. In R. Peet and N. Thrift (eds), *'New Models in Geography* (vol. 1), London, Unwin Hymar.

Pacione, M. 1990. 'Conceptual Issues in Applied Urban Geography', *Tijdschrift Voor Economische in Sociale Geographie*, 81, 3–13.

Pannikar, M. 1955. *Geographical Factors in Indian History*, Bombay, Bhartiya Vidya Bhawan.

Pathak, C. R. 1985 (May). 'Geography, Regional Science and Regional Planning: Their Interpretation', Presidential Address at the 7th Indian Geography Congress of the National Association of Geographers, India (NAGI), Kharagpur.

Patil, B. S. 2010. 'Spatio-temporal Analysis of Cropping Intensity in Northern Part of Nandurbar District (Maharashtra): Using GIS Technique', *Bhootal Digdarshan (Earth Surface Review)* 01 (2): 57–64.

Peet, J. R. 1971. 'Poor, Hungry America', *The Professional Geographer,* 23, 99–104.

———. 1975. 'Inequality and Poverty: A Marxist Geographic Theory', *Annas. Assoc., Amr. Geog,* 65, 564–71.

———. 1977. 'The Development of Radical Geography in the United States', *Progress in Human Geography* 1/2, 240–63.

Peet, J. R. (ed.). 1978. *Radical Geography*, London, Methuen.

Peet, J. R., and D. Slater. 1980. 'Reply to the Soviet Review of Radical Geography', *Soviet Geography*, review and translation 21, 541–45.

Peschel O. 1865. *Geschichte der Erdkunde bis auf A. V. Humboldt und Carl Ritter*, Munich, J. G. Gotta.

———. 1879. *Physiche Erdukunde*, ed. Gustar Leipoldt, Leipzig, Dunker and Hurnbolt.

Phillips, D. R., and A. E. Joseph. 1984. *Accessibility and Utilization: Perspectives on Health Care Delivery*, London, Harper and Row.

Plekhanov, G. V. 1956. 'Essays on Historical Materialism', *Isbranneye filosofskiye protozvedeniya*, vol. 2, Moscow, Gos, Izd-vo. Polit 155.

———. 1957. *Basic Problems of Marxism, Isbranneye filosofskiye protozvedeniya*, vol. 3, Moscow.

Pokrovskiy, M. N. 1931. *A Brief Outline of Russian History*', Moscow, published in English as *Brief History of Russia*, New York, International Publisher (1933), 33–35.

Popper, K. R. 1945 and 1962. *The Open Society and Its Enemies*, New York, Harper & Row.

Powell, J. W. 1896. 'Physiographic Regions of the United States'. In *Physiography of the United States*, Washington, DC National Geog. Society, Monograph, 2.

Pred, A. 1967 and 1969. 'Behaviour and Location: Foundations for a Geographic and Dynamic Location Theory, 2 vols, Lund, C. W. K. Gleerup.

———. 1973. 'Urbanisation, Domestic Planning Problems and Swedish Geographic Research', *Progress in Geography*, 5, 1–76.

Rabinow, P. 1984. 'Space, Knowledge, and Power'. In P. Rabinow (ed.), *The Foucault Reader*, New York, Pantheon Books.

Rathore, N. P. 2009. 'Application of Multi-temporal Satellite Data for the Study of Water Resources Problems in Udaipur Basin, South Rajasthan', *Annals of National Association of Geographers, India* 29 (1): 72–88.

Ratzel, F., 1882–91. *Anthropogeographie*, vol. I (2nd ed. 1889, 3rd ed. 1909), vol. 2, 1891 (2n ed. 1912).

Ratzel, F. 1897. *Politische Geographie*, Munich and Berlin: R. Oldenbourg.

Ray, D. M. et al. 1974. 'Functional Prerequisites, Spatial Diffusion and Allometric Growth', *Economic Geography* 50, 341–51.

Raza, M. (ed.). 1979. 'Second Survey of Research in Geography', New Delhi: Indian Council of Social Science Research.

Reddy, N. B. K. 1982 (December). 'Perspective on Indian Geography', Presidential Address at the 5th Indian Geography Congress of the National Association of Geographers, India (NAGI), Tirupati.

Reddy, N. B. K., D. K. Singh, and J. K. Routary. 1999. 'Progress of Research Methods in Geography India'. In G. S. Gosal (ed.), *Fourth Survey Research in Geography*, New Delhi, Manak Publication (P.) Ltd., 8–29.

Ree, J. 1984. 'Metaphor and Metaphysics: The End of Philosophy and Derrida', *Radical Philosophy*, 38: 33.

Reynolds, R. B. 1956. 'Statistical Methods in Geographical Research', *Geographical Review*, 46, 129–32.

Richards, P. 1974. 'Kant's Geography and Mental Maps', *Transactions*, Institute of British Geographers, 61, 1–16.

Richthofen, F. Von. 1877, 1912. 'China: *Ergebnisse eigener und darauf gegrundte Studien*, 5 vols, Berlin Dietrich Reimer.

———. 1886. *Fuhrer fur Forschungsreisende*, Berlin, Robert Oppenheim.

Ritter, C. 1822–1859. *Die Erkunde*, 19 vols. Berlin: G. Reimer.

Roxby, P. M. 1925. 'The Distribution of Population in China', *Geog. Rev.* 15, 1–24.

———. 1938. 'The Terrain of Early Chinese Civilization', *Geography*, 23, 225–36.

Ryabchikov, A. M. 1968. 'Geography at Moscow University over the Last 50 Years (1917–1967), *Soviet Geography*, 9, 343–57.

Sack, R. D. 1972. 'Geography, Geometry and Explanation', *Annals,* Association of American Geographers, 64, 439–52.

Said, E. W. 1985. 'Opponents, Audiences, Constituencies and Community'. In H. Foster (ed.), *Post-modern Culture*, London, Pluto Press.

Salisbury, R. D. 1907. *Physiography*, New York, Henry Holt.

Santos, M. 1974. 'Geography, Marxism and Underdevelopment'. *Antipode,* 6 (3), 1–9.

Sarup, M. 1993. *An Introductory Guide to Post-structuralism and Post-modernism*, New York and Singaporei, Harvesta Wheatsted.

Sauer, C. O. 1924. 'The Survey Method in Geography and Its Objective', *Annals,* Association of American Geographers, 14, 17–33.

———. 1925. *The Morphology of Landscape*, University of California Publication in Geography, 2, 19–53.

———. 1927. 'Recent Developments in Cultural Geography'. In E. C. Hayes (ed.), *Recent Development in Social Sciences*, Philadelphia, J. B., Lippincott, 114–22.

Saushkin, Y. G. 1958. *Introduction to Economic Geography,* Moscow.

———. 1961. 'On Environmentalism and Possibilism', *Soviet Geography,* 2 (2) 74.

———. 1962. 'Economic Geography in the USSR', *Economic Geography,* 38, 28–37.

———. 1963. 'The Geographical Environment of Human Society', *Soviet Geography,* 4 (10) 14.

———. 1966. 'A History of Soviet Economic Geography', *Soviet Geography*, 7, 3–104.

Sayer, R. A. 1979. 'Epistemology and Conception of People and Nature in Geography', *Geoforum* 10: 19–43.

Schaeffer, F. K. 1953. 'Exceptionalism in Geography: A Methodological Examination, *Annals,* Association of American Geog. 43, 226–49.

Semenov Tyan-Shanski, V. P. 1928. 'Russia—Territory and Population, A Perpective on the 1926 Census', *Geographical Review,* 18, 616–40.

Semple, E. C. 1903. *American History and Its Geographic Conditions*. Boston: Houghton Mifflin.

———. 1911. *Influences of Geographic Environment*, New York: Henry Holt.

Shannon, G. W. and G. E. A. Dever. 1974. *Health Care Delivery: Spatial Perspectives,* New York, McGraw Hill.

Sharma, R. C. 1990. 'Models in Geography', *Employment News* 14 (50).

Singh, R. B. 2009. 'Remote Sensing and Geographical Information Systems'. In L. S. Bhat (ed.), *Geography in India-Selected Themes*, New Delhi: Pearson Publication, 203–57.

Singh, R. B., and S. Kumar. 2011. 'Mountain Risk in Downstream Water Resource Management in Upper Bhagirathi Basin, Indian Himalayas. Risk in Water Resources Management (*Proceedings of Symposium* H03, held during IUGG-IAHS publication: 49–54.

Singh, I. (ed.). 2000. 'Progress in Indian Geography: 1996 –2004', A Status Report Prepared for the 29th International Geographic Congress, Seoul Korea. New Delhi: Indian National Science Academy.

Singh, J. 2003. 'Indian Geography; Vision 2020', Presidential Address at 25th Indian Geography Congress, *National Association of Geographers,* Ghaziabad 1–12.

Smith, D. M. 1971. 'America! America? Views on a Pot Melting, 2, Radical Geography the Next Revolution?' *Area* 3, 153–57.

———. 1973. *The Geography of Social Well-being in the United States,* New York, McGraw Hill.

———. 1977. *Human Geography: A Welfare Approach*, London, Edward Arnold.

———. 1985. 'The New Blood Scheme and Its Application to Geography', *Area* 17, 237–43.

Smith, N. 1984. *Uneven Development*, Oxford, Basil Blackwell.

Soja, E. W. 1993. *Post-modern Geographers*, London and New York, Verso.

Spate, O. H. K. 1957. 'How Determined is Possibilism?' *Geographical Studies* 3–12.

Stalin, L. V. 1952. *Dialectical and Historical Materialism*, Voprosy leninizma, 11th ed. Moscow Gos. 12d-vi. polit. literatury 587–89.

Stamp, L. D. 1946. *The Land of Britain*, London, Longman.

———. 1947. *The Land of Britain: It Use and Misuse* (2nd ed. 1950), London, Methuen.

Stamp, L. D. 1960. *Applied Geography*, Harmondsworth: Penguin Books.

Stevenson, W. I. 1978. 'Patrick Geddes'. In T. W. Freeman and P. Pincheneal (eds), *Geographer, Bibilographical Studies*, vol. II, London, Mansell, 53–65.

Stewart, J. Q. 1956. 'The Development of Social Physics', *American Journal of Physics*, 18: 242–51.

Stoddart, D. R. 1965. 'Geography and the Ecological Approach: The Ecosystem as a Geographic Principle and Method', *Geography* 50, 242–51.

———. 1966. 'Darwin's Impact on Geography', *Annals*, Association of American Geographers, 56, 683–98.

———. 1975. 'Kropotkin, Reclus and Relevant Geography', *Area* 7, 188–90.

———. 1987. 'To Claim the High Grounds: Geography for the End of the Century', *Transactions*, Institute of British Geographers, NS 12, 327–36.

Strabo. 1917. *The Geography of Strabo* (trans. H. L. Jones), New York.

Stewart, J. Q. 1947. 'Empirical Mathematical Rules Concerning the Distribution and Equilibrium of Population', *Geographical Review* 37, 461–85.

———. 1956. 'The Development of Social Physics', *American Journal of Physics*, 18, 239–53.

Taaffe, E. J. (ed.). 1970. *Geography*, Englewood Cliffs, N. L., Prentice Hall.

Tatham, G. 1952 (reprint, 1967). 'Geography in the Nineteenth Century'. In G. Taylor (ed.), *Geography in Twentieth Century*, London, Methuen, 28–69.

———. 1952 (reprint, 1967). 'Environmentalism and Possibilism'. In G. Taylor (ed.), *Geography in Twentieth Century*. London, Methuen, 128–64.

Taylor, E. G. R. 1930. *Tudor Geography* (1485–1593), London, Methuen.

———. 1934. *Late Tudor and Early Stuart Geography*, London, Methuen.

Taylor, E. G. R., et al. 1942. 'Discussion on the Geographical Aspect of Planning', *Geographical Journal* XCIX, 61–80.

Taylor, G. 1941. *Australia: A Study of Warm Environments and Their Effects on British Settlements*, London: Methuen.

Taylor, G. 1952 reprint. 'Introduction: Scope of the Volume', *Geography in the Twentieth Century*, London, Methuen.

———. 1967. 'Urban Geography'. In G. Taylor (ed.), *Geography in the Twentieth Century*, London, Methuen, 500–27.

Taylor, P. J. 1985. 'The Value of a Geographical Perspective'. In R. J. Johnston (ed.) *The Future of Geography*, London, Methuen, 92–110.

Thakur, B. 1994. 'Indian Geography; Development, Trends and Prospects', *Transactions*, Institute of Indian Geographers 16 (1), 67–85.

Tillman, J. P. 1971. 'An Appraisal of the Geographical Works of Alertus Magnus and His Contributions to Geographical Thought', *Ann. Arbor*, University of Michigan.

Thomson, T. Q. 1965. *History of Ancient Geography*, New York, Biblo & Tennen.

Tozer, H. F. 1987. *A History of Ancient Geography*, Cambridge University Press (Reprint, New York, Biblo & Tannen, 1964).

Trewartha, G. T. 1953. 'A Case for Population Geography', *Annals of the Association Geographers*, 77–97.

———. 1969. *A Geography of Population: World Patterns*, New York, John Willey.

Troll, C. 1949. 'Geographical Science in Germany during the Period 1933–1945: A Critique and Justification, trans. and ed. by E. Fisher, *Annals,* Association of American Geographers, 39, 100–37.

Tuan, Yi-Fu. 1971. 'Geography, Phenomenology, and the Study of Human Nature', *The Canadian Geographer,* 15, 181–92.

———. 1976. 'Humanistic Geography', *Annals,* Association of American Geographers, 66, 226–76.

Ullman, E. J. 1953. 'Human Geography and Area Research', *Annals,* Association of American Geographers, 43, 54–66.

Unstead, J. F. 1916. 'A Synthetic Method of Determining Geographical Regions', *Geographical Journal,* 43, 230–39.

Vaidhyanadhan, R. 1987 (December). 'Whither Indian Geography: Presidential Address at the 10th Indian Geography Congress of the National Association of Geographers, India (NAGI), Pune.

Van Valkenberg. 1951 (reprint 1952, 1967). 'The German School of Geography'. In G. Taylor (ed.), *Geography in he Twentieth Century,* London, Methuen, 91–116.

Vidal de la Blache, P. 1899: 'Lecon d' ouverture du cours de geographie', *Annales de geographie* 8: 97–109.

———. 1913. 'Des caracteres distinctifs de la geographic', *Annals de Geographie*, 22, 289–99.

———. 1917. *La France De L'Est: Lorraine-Alsace,* Paris, Armand Colin.

———. 1921. *Principles de la Geographie Humaine*, trans. 1926 as *Principles of Human Geography*, Londori, Constable.

Visher, S. S. 1952 reprint. 'Climatic Influences'. In G. Taylor (ed.), *Geography in the Twentieth Century.*

Warntz, W. 1959. 'Geography of Mid-twentieth Century', *World Politics II*, 442–54.

Watson, J. W., 1953 (reprint, 1967). 'The Sociological Aspects of Geography'. In G. Taylor (ed.), *Geography in the Twentieth Century,* London, Methuen, 463–99.

Webber, M. J. 1972. *Impact of Uncertainty on Location*, Canberra, Australia, National University Press.

———. 1977. 'Pedagogy Again: What is Entropy?' *Annals,* Association of American Geographers 67, 254–66.

Westcoatt, J. L. 1992. 'Common Themes in the Work of Gilbert White and John Dewey: A Pragmatic Appraisal', *Annals* of the Association of American Geographers, 82, 587–607.

Whitback, R. H. 1926. 'Adjustments to Environment in South America: An Interplay of Influences', *Anns. Ass. Amer. Geog.* 44: 231–44.

White, C. I., and G. T. Renner, *Geography, an Introduction to Human Ecology.* New York, D. Appleton Century Company.

White, G. 1974. 'Natural Hazards Research'. In R. J. Chorley (ed.), *Directions in Geography*, London, Methuen, 193–216.

White, G. F. 1972. 'Geography and Public Policy', *The Professional Geographers* 24, 101–04.

White, R., and P. R. Gould. 1974. *Mental Maps*, Harmondsworth: Penguin.

Wilson, A. G. 1970. *Entropy in Urban and Regional Modelling*, London, Pion.

Williams, R. 1983. *Keywords: A Vocabulary of Culture and Society*, London, Fontanes.

Wittelessy, D. 1929. 'Sequent Occupation', *Annals,* Association of American Geographers, 19, 162–65.

Wittgenstein, L. 1958. *Philosophical Investigations*, Oxford, Basil Blackwell.

Wolpert, J. 1965. *Behavioural Aspects of the Decision to Migrate*, papers and proceedings, Regional Science Association, 15, 605–35.

———. 1970. 'Departures from the Usual Environment in Locational Analysis', *Annals*, Association of American Geographers, 60, 220–29.

Wooldenberg, M. J., and B. J. I., Berry. 1967. 'Rivers and Central Places: Analogous Systems?' *Journal of Regional Science*, 7, 129–40.

Woolridge, S. W., and W. G. East. 1951. *The Spirit and Purpose of Geography*, London, Hutchinson University Library.

———. 1958. *The Spirit and Purpose of Geography*, London, Hutchinson.

Wrigley, E. A. 1965. 'Changes in the Philosophy of Geography'. In R. J. Chorley and P. Haggett (eds), *Frontiers in Geographical Teaching*, London, Methuen, 3–20.

Yeates, M. 1968. *An Introduction to Quantitative Analysis in Economic Geography*, New York, McGraw-Hill.

Young, J., and B. L. Sukhwal. 1984. *Radical Geography: A New Philosophical Approach in the Development of Geographic Thought in the United States*, 51 (2), 183–95.

Zelinsky, W. 1970. 'Beyond the Exponentials: The Role of Geography in the Transition', *Economic Geography* 46, 499–535.

Zipf, G. K. 1949. *Human Behaviour and the Principle of Least Effort*, New York, Hafner.

Zvonkova, T. V., and Y. G. Saushkin. 1968. 'Problems of Long-term Geographic Prediction', *Soviet Geography*, 9, 755–65.